Hybrid Nanomaterials

Design, Synthesis, and Biomedical Applications

Hybrid Nanomaterials

Design, Synthesis, and Biomedical Applications

Editors

Feng Chen

Departments of Radiology
University of Wisconsin – Madison
Madison, WI
USA

Weibo Cai

Departments of Radiology and Medical Physics
University of Wisconsin – Madison
Madison, WI
USA

CRC Press
Taylor & Francis Group
Boca Raton London New York

CRC Press is an imprint of the
Taylor & Francis Group, an **informa** business

A SCIENCE PUBLISHERS BOOK

Cover illustration has been provided by the editors of the book and is reproduced with their kind permission.

CRC Press
Taylor & Francis Group
6000 Broken Sound Parkway NW, Suite 300
Boca Raton, FL 33487-2742

First issued in paperback 2021

© 2017 by Taylor & Francis Group, LLC
CRC Press is an imprint of Taylor & Francis Group, an Informa business

No claim to original U.S. Government works

ISBN-13: 978-0-367-78238-2 (pbk)
ISBN-13: 978-1-4987-2092-2 (hbk)

Visit the Taylor & Francis Web site at
http://www.taylorandfrancis.com

and the CRC Press Web site at
http://www.crcpress.com

Preface

This book provides the latest advances in the design, synthesis and biomedical applications of hybrid nanomaterials, which are emerging as powerful platforms for next generation cancer diagnosis and therapy. A total of 35 contributors who are experts in their respective fields have provided their great insights on the engineering and biomedical applications of various kinds of hybrid nanomaterials, including red blood cell-mimicking hybrid nanoparticles, hybrid nanovesicles formed by the self-assembly of porphyrin-phospholipids, core@shell structured hybrid nanoparticles (such as quantum dots, gold nanoparticles, etc.) synthesized by using epitaxial growth technique, nanoparticle-antibody hybrids, organic–inorganic hybrid nanoparticles, combinatorial nanoparticles, hybrid upconversion nanoparticles, radiolabeled hybrid nanomedicine, hybrid nanomaterials that are designed for cardiovascular disease imaging and photothermal therapy, nanoscale metal-organic frameworks based hybrids, silica-based hybrid nanomaterials, etc. It is our great hope that the book will inspire and help researchers who are interested in nanotechnology and its biomedical applications to foster more research in this exciting field.

Feng Chen and Weibo Cai
University of Wisconsin–Madison

Contents

1

Hybrid Nanomaterials:
An Introduction

Feng Chen[1,*,#] and *Weibo Cai*[1,2,3,4,*,#]

Hybrid nanomaterials with well-selected organic/inorganic components and integrated functionalities are believed to have a great impact on next generation cancer diagnosis and therapy. The last decade has witnessed spectacular advances in the engineering of various kinds of hybrid nanomaterials for per-clinical biomedical applications as well as first-in-human clinical trials. Aiming to provide scientists/students/clinicians who are interested in nanomedicine with a great resource, we worked closely with an international ensemble of leading experts in this field to organize this book entitled *"Hybrid Nanomaterials: Design, Synthesis, and Biomedical Applications"*.

The book starts with a comprehensive review of red blood cell (RBC)-mimicking hybrid nanoparticles by Dr. Cai and co-workers (*University of Wisconsin – Madison*). The rapid recognition and opsonization of intravenously administered nanoparticles (with a size larger than the renal clearance cut-off, i.e., 5.5 nm (Soo Choi et al. 2007) by components of the reticuloendo-thelial system (RES) has presented a major challenge to their systemic and targeted drug delivery (Alexis et al. 2008). RBCs are the most abundant, long circulating cells in the human body functioning as biological oxygen carri-ers. Harnessing and mimicking these remarkable properties, scientists have engineered new-generation hybrid nanosystems for future biomedical appli-cations (Hu et al. 2012). Very recent progress made in RBC-mimicking hybrid nanosystems is discussed with special focus on their fabrication methods,

---

[1] Department of Radiology, University of Wisconsin – Madison, WI, USA.
[2] Department of Medical Physics, University of Wisconsin – Madison, WI, USA.
[3] Materials Science Program, University of Wisconsin – Madison, WI, USA.
[4] University of Wisconsin Carbone Cancer Center, Madison, WI, USA.
[#] Address for Correspondence: Departments of Radiology and Medical Physics, University of Wisconsin – Madison, Room 7137, 1111 Highland Avenue, Madison, WI 53705-2275, USA.
[*] Corresponding authors: chenf@mskcc.org and wcai@uwhealth.org

unique attributes and applications as targeted and prolonged drug delivery systems. The authors conclude the chapter by addressing the great opportunities and current challenges in this field.

The third chapter is written by Dr. Wang and Dr. Lovell (*University at Buffalo*), who provide an overview of the current research status of using porphyrin-phospholipids (PoPs) for biomedical applications. Porphyrins are macrocyclic organic molecules composed of four linked pyrrole subunits and hold the ability to efficiently chelate metals (such as manganese ions, copper-64, etc.) for imaging purposes (Huynh and Zheng 2014). Phospholipids are amphiphilic organic molecules, which have hydrophilic phosphate based "head" groups and hydrophilic "tails" of fatty acid chains. The conjugation of porphyrins to phospholipid side chains could form PoPs, which can self-assemble into hybrid nanovesicles with intrinsic utility in both imaging and therapy. Examples of using PoPs for positron emission tomography (PET) imaging, magnetic resonance imaging (MRI), optical imaging, photoacoustic imaging, photodynamic therapy (PDT), photothermal therapy (PTT) and radiotherapy are presented.

Written by Dr. Cai and co-workers (*University of Wisconsin–Madison*), Chapter 4 is an in-depth overview of the engineering of heterostructured hybrid nanoparticles using epitaxial growth technique. Well-established synthetic strategies and biomedical applications of several heterostructures, including core@shell quantum dots (QDs), core@shell upconversion nanoparticles (UCNPs) and Au-based heterostructures are discussed in detail. Heterostructured hybrid nanoparticles that combine two or more components could offer enhanced chemical, optical, photothermal and magnetic properties as compared to the conventional single-component nanoparticles, making them ideal nanoplatforms for biomedical applications especially in cancer diagnosis and therapy.

The potential of using hybrid nanoparticles for either imaging or therapy will be greatly limited without suitable targeting strategy. Nanoparticle-antibody hybrids combine the properties of carrier materials capable of transporting high payloads of therapeutic and/or imaging agents with the specificity of targeting conferred by antibodies or antibody-like moieties. In Chapter 5, Dr. Vallis and her co-workers (*University of Oxford*) have conducted a detailed survey into the most promising nanomaterial-antibody hybrids in the field. The current status of the design, development and clinical applications of antibody-tagged liposomes, polymeric nanoparticles and gold nanoparticles are reviewed in depth. The authors also highlight the existing challenges in optimizing (parameters such as size, stability, biodegradability, blood half-life and the number of antibodies on the surface) these hybrid systems, developing strategies to overcome immunogenicity and opsonization and creating cost-effective techniques for large-scale production.

In Chapter 6, Dr. Liang and co-workers (*University of Science and Technology of China*) have given a thorough review of the applications of various types of organic–inorganic hybrid nanoparticles for cancer imaging. These hybrid nanoparticles have been designed based on both functional nanocrystals (such as magnetic nanoparticles, QDs, gold nanoparticles, etc.) and structural nanoparticles (such as polymer, silica, dendrimer, liposome, etc.). Significant advances have been made in small animal cancer diagnosis by using organic–inorganic hybrid nanomaterials. The authors have further commented on the existing challenges in translating these nanomaterials from bench to the bedside.

Dr. Cheng and co-workers (*Stanford University*) have summarized the topic of combinatorial (or hybrid) nanoparticles and their applications in biomedicine in Chapter 7. Colloid chemistry has become a reliable approach for obtaining high-quality hybrid nanomaterials. Successful efforts have led to plenty of novel synthesis approaches, capable of producing exquisite control over a wide range of sizes, shapes and compositions of nanomaterials (Mukerjee et al. 2012). The chapter starts with a description of different kinds of combinatorial nanoparticles (such as core@shell nanoparticles, polymer-coated nanoparticles, dumbbell structured nanoparticles, etc.), followed by an overview of the biomedical applications developed for targeted imaging, cell separation and therapy. The authors finally conclude the chapter by sharing their insights on the potential as well as the challenges of using combinatorial nanoparticles for future cancer imaging and therapy.

UCNP is an interesting type of nanocrystal which utilizes the sequential absorption and energy transfer steps in order to emit a higher energy photon after absorbing two or more low-energy excitation photons (Zhou et al. 2015). In Chapter 8, Dr. Bu and co-workers (*Shanghai Institute of Ceramics, Chinese Academy of Sciences*) have provided a detailed review of the research status of hybrid UCNPs for biodetection and cancer imaging. Different strategies that have been used for UCNP surface modifications and luminescence resonance energy transfer detections are discussed in detail. The authors have also provided an overview of using hybrid UCNPs for multi-modality cancer imaging applications. Addressing challenges such as potential long-term toxicity, low quantum yield, low *in vivo* tumor active targeting efficacy, etc. will continue to be the focus of future research.

Radio-nanomedicine merged with hybrid nanomaterials can eliminate the concerns about the possible hazards of the pharmacologic amount of nanomaterials administered and the trace technology behind radio-nanomedicine will help speed the progress of the clinical translation. Written by Dr. Lee and co-workers (*Seoul National University*), Chapter 9 presents an in-depth overview of the topic of engineering hybrid nanomaterial based radio-nanomedicine (including radiolabeled UCNPs, Au and QDs) for biomedical applications. Followed by this is a chapter presented by Dr. Chakravarty and his co-workers discussing the role of nanosorbents in the development of column chromatographic radionuclide generators. The present status of nanomaterial based radionuclide generators is highlighted and the great potential and intriguing opportunities for future development are discussed.

Cardiovascular disease (CVD) refers to a group of pathological conditions affecting the heart and its vessels. In Chapter 11, Dr. Cheng and co-workers (*Fudan University*) have covered the design and application of multifunctional nanoparticles in the diagnosis of CVD (such as atherosclerosis, vasculitis, myocardial infarction, aneurysm and valvular heart diseases). Multifunctional nanoparticles that are built by using both organic (such as liposome, micelles, dendritic polymers, etc.) and inorganic (such as magnetic nanoparticles, Au, QDs, etc.) nanocomponents are well discussed. The applications of photothermal therapy (PTT) have been widely explored in the last two decades as a means of treating malignancies, while effectively limiting possible damage to healthy tissues (Hirsch et al. 2003). Dr. England and Dr. Cai have given another comprehensive review on developing hybrid nanoparticles for photothermal therapy (PTT) of cancer in the following chapter. Metallic nanoparticles, core@shell structured nanoparticles and more complex nanoplatforms that are designed for PTT are reviewed in detail.

Metal-organic frameworks have been well-known for their outstanding performance in gas storage and separation, catalysis and their great potential in magnetism, nonlinear optics, light harvesting, etc. In Chapter 13, Dr. Ma and Dr. Hoang (*University of South Florida*) have given a comprehensive review of the biomedical applications of nanoscale metal-organic frameworks (NMOFs). The interaction between the guest molecules and the host framework can be enhanced through functionalization of the organic linkers, which make NMOF a very attractive and promising candidate in biomedical applications. Strategies used for the preparation of NMOFs, stability, toxicity, drug and gasotransmitter storage and delivery properties, the potential of using NMOFs as biosensors and imaging contrast agents are well discussed in this chapter.

The last three chapters are about silica-based hybrid nanomaterials. Written by Dr. Su and Dr. Jokerst (*University of California, San Diego*), Chapter 14 focuses on silica/gold hybrid nanoparticles for imaging and therapy. The chapter starts with a brief history of these materials and explains the benefits that they offer to both researchers and physicians, followed by a section discussing how shape would affect a nanoparticle's fate *in vivo*. The imaging (including photoacoustic imaging, surface enhanced Raman scattering imaging, X-ray computed tomography, MRI and fluorescence imaging) and therapeutic (including PDT, PTT, drug delivery and surgery) features of gold/silica hybrid nanomaterials are then discussed in detail in the following sections. The chapter closes with a discussion about the challenges to future clinical translation as well as potential solutions. Followed by this, Dr. Chen and Dr. Zhang (*Shanghai Institute of Ceramics, Chinese Academy of Sciences &*

Fudan University) have presented a thorough review of mesoporous silica-based hybrid nanosystems for MRI-based cancer theranostics in Chapter 15. Among various organic or inorganic nanosystems, inorganic mesoporous silica nanoparticles (MSNs) have attracted tremendous attention especially in various biomedical fields (Ambrogio et al. 2011). MSNs-based nanosystems can precisely provide the MRI signal of tumor tissues, which, at the same time, can also be used for targeted drug delivery, monitoring the therapeutic process and evaluating the therapeutic efficiency. The final chapter in this book contains a summary of the recent advances in the engineering of silica-based core@shell structured hybrid nanoparticles. A detailed survey into the latest development of four different types of silica-based hybrid nanomaterials was reviewed by Dr. Cai and I, with focus on the clinical translation history of ultrasmall renal clearable fluorescent dye-doped silica hybrid nanoparticles (also known as C dots (Benezra et al. 2011; Phillips et al. 2014). We also shared our insights on the current challenges and solutions during the translation of silica-based hybrid nanomaterials.

It is our great hope that this book will inspire and help researchers at all levels (from students to professors in academia and industry who are interested in nanotechnology and its biomedical applications) to foster more research in this exciting field. We are deeply grateful to all of the 35 contributors for their tremendous effort in writing these exceptional chapters. We also foresee that an increased understanding of the challenges in hybrid nanomaterial engineering, pre-clinical imaging/therapy applications, long-term toxicity concerns, clinical translation, etc., will lead to novel strategies for fighting cancers in a more effective way.

References

Alexis, F., Pridgen, E., Molnar, L.K. and Farokhzad, O.C. 2008. Factors affecting the clearance and biodistribution of polymeric nanoparticles. Mol Pharm 5: 505-515.

Ambrogio, M.W., Thomas, C.R., Zhao, Y.L., Zink, J.I. and Stoddart, J.F. 2011. Mechanized silica nanoparticles: a new frontier in theranostic nanomedicine. Acc Chem Res 44: 903-913.

Benezra, M., Penate-Medina, O., Zanzonico, P.B., Schaer, D., Ow, H., Burns, A., et al. 2011. Multimodal silica nanoparticles are effective cancer-targeted probes in a model of human melanoma. J Clin Invest 121: 2768-2780.

Hirsch, L.R., Stafford, R.J., Bankson, J.A., Sershen, S.R., Rivera, B., Price, R.E., et al. 2003. Nanoshell-mediated near-infrared thermal therapy of tumors under magnetic resonance guidance. Proc Natl Acad Sci USA 100: 13549-13554.

Hu, C.M., Fang, R.H. and Zhang, L. 2012. Erythrocyte-inspired delivery systems. Adv Healthc Mater 1: 537-547.

Huynh, E. and Zheng, G. 2014. Porphysome nanotechnology: a paradigm shift in lipid-based supramolecular structures. Nano Today 9: 212-222.

Mukerjee, A., Ranjan, A.P. and Vishwanatha, J.K. 2012. Combinatorial nanoparticles for cancer diagnosis and therapy. Curr Med Chem 19: 3714-3721.

Phillips, E., Penate-Medina, O., Zanzonico, P.B., Carvajal, R.D., Mohan, P., Ye, Y., et al. 2014. Clinical translation of an ultrasmall inorganic optical-PET imaging nanoparticle probe. Sci Transl Med 6: 260ra149.

Soo Choi, H., Liu, W., Misra, P., Tanaka, E., Zimmer, J.P., Itty Ipe, B., et al. 2007. Renal clearance of quantum dots. Nat Biotech 25: 1165-1170.

Zhou, J., Liu, Q., Feng, W., Sun, Y. and Li, F. 2015. Upconversion luminescent materials: advances and applications. Chem Rev 115: 395-465.

2

Red Blood Cell-Mimicking Hybrid Nanoparticles

Shreya Goel,[1,a] *Feng Chen*[2,]* and *Weibo Cai*[1,2,3,4,]*

INTRODUCTION

Through decades of research and development, nanomaterials like micelles, liposomes as well as polymeric and inorganic nanoparticles are now making revolutionary changes in the fields of diagnostics, imaging, therapy and drug delivery (Yoo et al. 2010). The rapid recognition and opsonization of intravenously administered nanoparticles by components of the reticuloendothelial system (RES) has presented a major challenge to their systemic and targeted drug delivery (Alexis et al. 2008). As a result, tremendous efforts have been put into the engineering of long circulating nanocarrier systems which can effectively evade Kupffer cell uptake and prolong the circulation of therapeutic agents with short elimination half-lives, allowing them to adequately reach their therapeutic targets (Moghimi et al. 2001; Yoo et al. 2010). Regardless of the type and function of the nanoparticles, their blood circulation time directly influences the *in vivo* outcomes. For example, Kupffer cell-evading drug carriers act as reservoirs from which drugs can be released controllably and continuously for a longer duration. This becomes more significant for small molecule therapeutic agents with short elimination half-lives. In addition, studies have shown that longer residence time aid in enhanced accumulation of nanoparticles in target sites in both active and passive targeting scenarios (Moghimi and Hunter 2001).

Erythrocytes are Nature's long circulating delivery vehicles. Harnessing and mimicking their remarkable properties, scientists have engineered

[1] Materials Science Program, University of Wisconsin – Madison, WI, USA.
[a] E-mail: goel6@wisc.edu
[2] Department of Radiology, University of Wisconsin – Madison, WI, USA.
[3] Department of Medical Physics, University of Wisconsin – Madison, WI, USA.
[4] University of Wisconsin Carbone Cancer Center, Madison, WI, USA.
[*] Corresponding authors: chenf@mskcc.org and wcai@uwhealth.org

new-generation nanosystems with extensive biomedical applications (Hu et al. 2012). Designs that mimic RBCs' functions, particularly those enabling their passage through narrow constrictions while maintaining a long *in vivo* survival, have been integrated into the engineering of drug carriers, resulting in novel delivery systems with improved drug tolerability, circulation lifetime and efficacy (Hu et al. 2012). Herein, we review the recent progress made in red blood cell-based hybrid nanosystems, focusing on their fabrication methods, unique attributes and applications as targeted and prolonged drug delivery systems.

Current Strategies in RES Evasion

Impairment of Phagocytic Function

Phagocytic activity of the Kupffer cells can be repressed either by receptor depletion/saturation (by overdose of phagocytosable material) or by inaccessibility to the ligands (Moghimi and Davis 1994). Intravenous injection of certain agents has been found to impair the endocytosis of the subsequently administered nanocarriers in a number of studies. For example, dextran sulfates, alkyl esters of fatty acids, poloxamines as well as salts of lanthanides such as gadolinium chloride have been demonstrated to effectively induce hepatic blockade (Diagaradjane et al. 2010; Diluzio and Wooles 1964; Lazar et al. 1989; Moghimi 1999; Sauer et al. 1997; Souhami et al. 1981; Weissleder et al. 1994). These agents act by inducing membrane changes which result in receptor loss or interference in the surface attachment and endocytotic functions of the hepatic Kupffer cells. For example, gadolinium chloride pretreatment was shown to inactivate RES macrophages, thereby increasing circulatory time and amplifying tumor-specific signal of epidermal growth factor (EGF)-conjugated quantum dot nanoprobes *in vivo* (Fig. 2.1) (Diagaradjane et al. 2010). However, concerns arising from $GdCl_3$ mediated toxicity as well as concomitantly increased phagocytic activity observed in smaller liver and splenic macrophages (Souhami et al. 1981), have rendered this technique therapeutically unacceptable (Moghimi and Davis 1994). To address the toxicity issues, researchers have proposed the saturation of the receptor sites by pre-injection of a large dose of the nanocarrier (Fernandez-Urrusuno et al. 1996; Moghimi et al. 2001; Moghimi and Szebeni 2003). However, reticuloendothelial blockade has shown only marginal improvement in the blood circulation and biodistribution profile of liposomes (Goto et al. 1991; Liu 1997). In addition, suppression of the essential defense mechanisms in the body raises concerns over the long-term effects and feasibility in clinical settings where repeated dosing and extended therapeutic regimens are common. Unfortunately neither strategy has produced effective or reproducible results, calling for more reliable techniques for RES evasion.

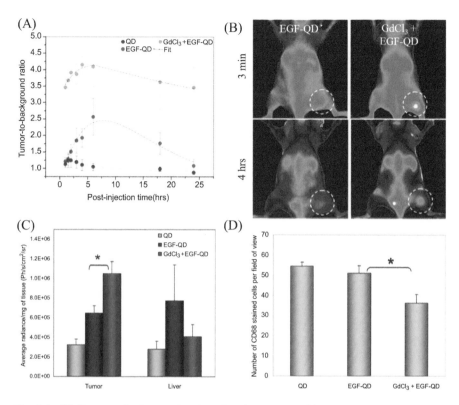

Fig. 2.1 (A) Tumor-to-background ratios after dynamic equilibrium in mice injected with QD, EGF-QD or GdCl$_3$ followed by EGF-QD. (B) *In vivo* fluorescence images of HCT116 tumor bearing animals at 3 min and 4 h after intravenous injection of EGF-QD and GdCl$_3$ pretreatment followed by EGF-QD. (C) Fluorescence levels (mean ± SE) from the extracts of tumor and liver tissues 4 h post-injection (*represents the statistical significance ($p < 0.001$) based on an unpaired Student's t test). (D) Quantification of anti-CD68 immunofluorescence staining activated Kupffer cells in liver tissues, 4 h after treatment with either QD, EGF + QD or GdCl$_3$ + EGF + QD. Reprinted with permission from (Diagaradjane et al. 2010).

Stealth Coatings and Surface Modifications

Modifying the surface of the nanoparticles with a range of nonionic surfactants or polymeric macromolecules to prevent opsonization of serum proteins and phagocytosis is currently the most widely used and successful strategy for prolonging the blood residence time of the particles (Moghimi and Szebeni 2003). Surface modifications can be performed by direct physical adsorption, incorporation during the nanocarrier synthesis or by covalent attachment via reactive surface groups. Stealth particles have been reported to circulate in the blood for 2 to 24 h in rodent models. In this regard, polysaccharides, polyethylene glycol (PEG) and PEG derived polymers have been most extensively studied (Alexis et al. 2008; Beletsi et al. 2005; Gao et al. 2013a; Harper et al. 1995; Yoo et al. 2010). Although polysaccharides such as dextran, heparin, hyaluronic acid and chitosan have been shown to alter the opsonization process due to their biomimetic natures, their use has been limited by the triggered immunogenic responses (Moghimi et al. 2001).

In contrast, polymeric coatings such as PEG create a steric barrier around the nanoparticles, thereby preventing their aggregation and subsequent RES clearance through complement activation. PEG is a biocompatible, flexible, hydrophilic polymer of ethylene glycol units and is the current gold standard for stealth coatings (Hu et al. 2011; Jokerst et al. 2011). Numerous studies ranging from biodegradable polymeric nanoparticles to liposomes and inorganic nanoparticles report surface modification with PEG chains of varied lengths, surface densities, thickness and configurations, all of which are critically important for the stealth property of PEGylated nanocarriers (Yoo et al. 2010). However, there is growing evidence that reduced protein opsonization or increased repulsion between particles and macrophage membranes may not be directly responsible for the increased circulation time of sterically protected nanoparticles (Knop et al. 2010). For example, PEGylated nanoparticles still clear within 24 h post injection, likely due to desorption/degradation of the PEG layer or activation of certain components of the complement system. Further research is warranted to fully elucidate the mechanisms behind PEGylated stealth nanoparticles, their extended blood residence time and eventual clearance by the RES organs (Zeng et al. 2015).

Hybrid Nanoparticles Mimicking Nature

Although the conventional strategies designed and implemented for prolonging *in vivo* residence time of the nanoparticles and enhancing drug delivery have advanced the science immensely, they fail to match the sophistication demonstrated by nature (Yoo et al. 2011). This has prompted scientists to either use biological entities such as macrophages (Dou et al. 2009), red blood cells (RBCs) (Doshi et al. 2009; Hu et al. 2012), lymphocytes (Stephan et al. 2010), stem cells (Roger et al. 2010; Stuckey and Shah 2014) or even tumor cells (Fang et al. 2014; Tang et al. 2012) etc. as drug carriers or model novel nanoparticles after them (Mitragotri and Lahann 2009; Yoo et al. 2011).

RBCs or erythrocytes are the most abundant (total number approaching 30 trillion in the human body), long circulating cells in the human body functioning as biological oxygen carriers. With a unique non-nucleated, biconcave discoidal structure, optimal size (7-8 µm diameter and 90 µm³ volume), mechanical flexibility and protein composition, RBCs have been studied for over 40 years for their role as drug delivery vehicles (Fang et al. 2012; Hamidi et al. 2007; Ihler et al. 1973; Muzykantov 2013). They are inherently biocompatible, biodegradable and non-immunogenic. The natural compartmentalization protects the encapsulated cargo from degradation and allows prolonged circulation lives (120 days) (Muzykantov 2010). In fact, RBC lifespan and size markedly exceed those of other drug delivery systems (e.g., < 10 h and < 100 nm for PEG modified 'stealth' liposomes) (Gao and Zhang 2015). Their highly flexible structure allows the RBCs to crossover narrow (2-3 μm) capillaries, undamaged, while maintaining a constant surface area (136 μm² for biconcave mature RBCs) (Pierige et al. 2008). Moreover, they possess a semi-permeable membrane which allows the slow and sustained release of their cargo over a long period, while preventing the encapsulated proteins from leaching out.

Carrier Erythrocytes: Traditional Roles

With a high plasma membrane surface area and extended lifespans, RBCs represent potentially attractive and in some aspects, unique cargo carriers for intravascular delivery, prolonging drug circulation and restricting unintended extravasation (Kolesnikova et al. 2013; Muzykantov 2010; Rossi et al. 2005). As discussed already, traditional roles for erythrocyte carriers involve the transportation and delivery of small molecule drugs, enzymes, peptides, nucleic acids etc. (Millan et al. 2004). Major functions of traditional carrier erythrocyte systems include:

(a) facilitating slow and sustained release of encapsulated small molecule drugs, peptides, proteins and nucleic acids

(b) site-specific delivery of the entrapped drugs, such as selective delivery to the RES of the liver, spleen and bone marrow

(c) servicing as cellular transplants in enzyme or hormone replacement therapy and lastly,

(d) acting as bioreactors to either house pro-drugs and metabolically convert them into active molecules or synthesize molecules capable of exerting a therapeutic effect.

Drug Encapsulation

For a compound to be successfully encapsulated into the erythrocytes, it should be: water soluble to a certain extent, inert to physical or chemical interactions with RBC components including the membrane and resistant to inactivation within the erythrocyte. There are several procedures that permit the encapsulation of drugs, enzymes, peptidic payloads and proteins up to 180 kDa molecular weights into the erythrocytes (Gao and Zhang 2015; Hamidi et al. 2007). Osmosis-based methods are the standard procedures which include hypotonic hemolysis, hypotonic dilution, hypotonic dialysis, hypotonic pre-swelling and osmotic pulse (Jain and Jain 2003; Magnani et al. 2012). These methods are physical in nature, with the majority being developed in the 1970s (Hamidi et al. 2007; Ihler et al. 1973; Ihler and

Tsang 1987). The general principle involves swelling and the resultant increase in the membrane permeability of erythrocytes when exposed to a hypotonic solution containing the drug to be encapsulated (Deloach and Ihler 1977; el-Kalay et al. 1989). The loaded erythrocytes are subsequently placed in iso- and hyper-osmotic solutions consecutively, in order to anneal and reseal the membrane (Millan et al. 2004).

Besides the mainstream process of hypotonic loading, some non-orthodox loading techniques include electroporation or inducing pores in the membrane using electrical pulse (Mangal and Kaur 1991), chemical perturbation (Deuticke et al. 1973), intrinsic endocytosis of substances by RBCs (Berman and Gallalee 1985), drug-induced endocytosis (Colin and Schrier 1991; Schrier et al. 1992), lipid fusion (Nicolau and Gersonde 1979) and laser loading (Mulholland et al. 1999). The performance of each of these methods varies, with the efficacy of encapsulation varying between 3.8 to 80 percent. However, the focus of all these methods is the same, i.e., to maximize the therapeutic performance of the entrapped molecules while maintaining the erythrocytes' biological configuration. Maintenance of normal structure and function is essential to the proper survival and circulation of the loaded RBCs (Hu et al. 2012).

Drug Delivery with Erythrocytes

Drug efflux kinetics plays an important role in designing therapeutic erythrocytes and planning the treatment regimens. According to Hamidi et al. (Hamidi and Tajerzadeh 2003), the size and polarity of the drug influences its release kinetics to a great extent. While certain drugs, such as lipophilic compounds; methotrexate (Flynn et al. 1994b), phenytoin (Hamidi et al. 2011) and vitamin B_{12} (Eichler et al. 1985) show a higher rate of release than the rate of hemolysis (or release of hemoglobin) owing to diffusion, certain others show release kinetics similar to that of hemoglobin. This indicates, that the drugs release only upon hemolysis which could provide a slower and more controlled drug release profile. Examples include polar drugs such as heparin (Eichler et al. 1986b), gentamycin (Eichler et al. 1986a) and peptides like L-lysine-L-phenylalanine and urogasterone (Hamidi and Tajerzadeh 2003). The potential therapeutic applications of drug loaded erythrocytes vary greatly with the encapsulated cargo and range from slow intravenous diffusion of proteins (Zolla et al. 1990, 1991), prodrugs such a phototherapeutics (Smith et al. 2014), anti-viral and anti-cancer drugs (Biagiotti et al. 2011); to targeted delivery to RES (Dale et al. 1979) and non-RES organs (Chiarantini et al. 1995) to hormone and enzyme replacement therapies (Berman and Gallalee 1985; Flynn et al. 1994a; Ihler et al. 1973; Ito et al. 1989), with some under clinical trials (Godfrin et al. 2012). Taking cue from the success of erythrocytes in the encapsulation and delivery of drugs, protein and nucleic acid payloads, scientists have focused their attention towards loading nanomaterials into the RBCs to create carriers that are useful in biomedical and diagnostic applications (Antonelli et al. 2013).

Erythrocyte Mimetic Hybrid Nanomaterials

Molecular imaging and therapy have benefitted immensely from the advent of nanotechnology. Researchers have been trying to design complementary imaging probes or contrast agents to increase the sensitivity and detectability of the current imaging modalities, using the vast repertoire of tools that nanotechnology has provided us with (Goel et al. 2014). However, the biological barriers set forth by the RES against foreign entities, including injected therapeutics and contrast agents, restrict nanoparticle function by blocking their destinations (Antonelli and Magnani 2014). Ideas and approaches developed for the encapsulation and release of drugs in RBC carriers have supported the design of sophisticated synthetic carriers of the new generation, combining potential advantages otherwise unattainable for the 'good old' RBCs or the conventional nanomaterials separately (Muzykantov 2013).

Whole Cell Hybrid Nanomaterials

Owing to their intrinsic properties such as biocompatibility, non-immunogenicity and prolonged circulation time, RBCs have been the most frequently investigated whole cells, studied as drug delivery vehicles for nanoparticles in order to avoid their rapid clearance by the RES (Yoo et al. 2011). The most common method for loading nanoparticles inside erythrocytes is through a reversible lysis pore that forms in the membrane when placed in hypotonic media. Hypotonic hemolysis relies on the remarkable capacity of erythrocytes for reversible shape changes with or without accompanying volume changes and for reversible deformation under stress (Ihler and Tsang 1987). The erythrocytes have little capacity to resist volume increases greater than 50%-75% of the initial volume and when placed in solutions less than about 150 mOsm/kg (corresponding to ~0.4% NaC1), the membranes rupture, permitting escape of the cellular contents (Baker 1967; Ihler and Tsang 1987). This technique has been harnessed by the researchers for the "Trojan horse" entry of nanomaterials into the RBCs. The membranes can then be easily resealed by raising the salt concentration to its original level and then upon incubation, the resealed erythrocytes resume their normal biconcave shape and impermeability. After resealing, the nanoparticles are trapped in the erythrocytes, which may then serve *in vivo* as carriers for the entrapped compounds. Remarkably, the resulting resealed erythrocytes can survive in the circulation with nearly normal life span.

RBC encapsulation of microparticles and colloids was envisaged two decades ago when Sprandel et al. reported the loading of ferromagnetic micromolecules, so-called ferrofluids in human erythrocyte ghosts for *in vitro* studies (Sprandel et al. 1987). Since then this strategy was adopted by researchers worldwide to load a variety of nanoparticles into the RBCs to develop novel contrast agents or drug delivery systems. Literature teems with reports on magnetic erythrocytes resulting from the encapsulation of some ferrofluids such as cobalt–ferrite and magnetite, sometimes along with drugs, which have then been reported to serve as magnetic resonance imaging (MRI) contrast agents or to direct the encapsulated drug predominantly to the desired sites of the body by means of an external magnetic field (Danilov Iu et al. 1985; Ferrauto et al. 2014; Jain and Vyas 1994; Markov et al. 2010; Orekhova et al. 1990; Sprandel et al. 1987; Sternberg et al. 2012; Vyas and Jain 1994). In addition, superparamagnetic nanoparticles such as iron oxide (SPIO) have been extensively encapsulated into the RBCs to extend the circulation time *in vivo* (Antonelli and Magnani 2014; Antonelli et al. 2013; Antonelli et al. 2011; Antonelli et al. 2008; Brahler et al. 2006; Cinti et al. 2011; Markov et al. 2010; Rahmer et al. 2013; Sternberg et al. 2012), which have then served variously as MRI contrast agents for cardiac, cancer and blood pool imaging, drug delivery and theranostics (Fig. 2.2) as well as functional micromotors, possessing highly efficient, ultrasound-powered, magnetically guided propulsion for a variety of *in vitro* and *in vivo* biomedical applications (Wu et al. 2014). For example, in an attempt to improve the targeting of such constructs and to allow more efficient, intracellular release of encapsulated compounds, a novel magnetic field-controlled erythrocyte-based drug delivery system was reported (Cinti et al. 2011). A viral spike fusion glycoprotein (hemagglutinin) was attached on the erythrocytes' membrane and superparamagnetic nanoparticles and the drug, 5-Aza-2-dC were encapsulated into the erythrocytes to yield erythro-magneto-HA virosomes, with the potential for magnetically driven, site-specific localization and highly efficient fusion capability with the targeted HeLa cells (Cinti et al. 2011). The therapeutic dose of the encapsulated drug was reduced by 10 times when compared to the standard therapy, indicating that erythro-magneto-HA virosomes prevent rapid and premature degradation, thereby increasing the bioavailability of the drug. In a separate study, gold nanoparticle incorporated RBCs were utilized as contrast-enhanced tracers for dynamic X-ray imaging of blood flows, which in turn could be used for clinical diagnosis and treatment of circulatory disorders (Ahn et al. 2011).

RBCs represent a unique and promising platform for improved nanoparticle pharmacokinetics. However, issues associated with formulation, storage, limitations on the kind and size of nanoparticles loaded, low encapsulation efficiency, destruction of loaded RBCs (especially those following electroporation and hemolytic dialysis procedures) and consequently reduced blood circulation time due to RES opsonization, have decelerated the research in this area since the initial burst of activity. RBC is a good carrier for drug delivery to intravascular and RES targets, but many other important therapeutic targets (e.g., solid tumors, extravascular tissue components, CNS) are normally inaccessible to RBCs (Muzykantov 2010). Hence, such limitation on the spectrum of suitable targets negatively impacted the development of this drug delivery platform.

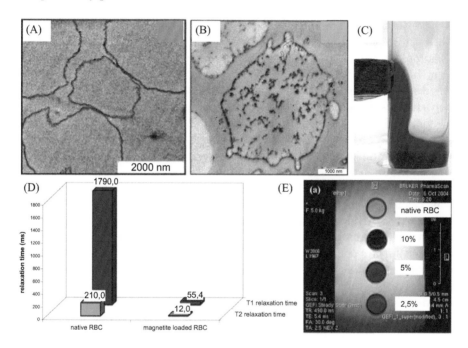

Fig. 2.2 (A) TEM pictures of non-loaded control RBCs and (B) SPION-loaded RBCs. (C) Magnetite loaded RBCs respond to an attached permanent magnet. (D) Relaxation time (T1 and T2) of native and SPION-loaded human RBCs. (E) MRI of the loaded RBCs. Top down: whole blood and loaded RBCs in whole blood, 10, 5 and 2.5%. Reproduced with permission (Brahler et al. 2006).

RBC Membrane Cloaked Hybrid Nanomaterials

The RBC membrane's relative simplicity, availability and ease of isolation have prompted researchers to use them as carriers for a wide variety of biologically active agents (Jain and Jain 2003; Luk and Zhang 2015). Red blood cell membranes were first isolated and developed as drug carriers by Desilets et al. (Desilets et al. 2001) and were patented under the name "nanoerythrosome" (nEryt). Small vesicles made with RBC membranes, nEryts were linked to anticancer drugs like daunorubicin and demonstrated higher anti-neoplastic activity than free drugs. In an effort to camouflage the nanoparticles suitably for longer circulation while retaining their functionalities, researchers have focused their efforts towards cloaking nanomaterials in extruded RBC membranes in order to create a new class of biomimetic nanoparticles. This method combines the biocompatibility of the RBCs with the easy tunability and multifunctionality of nanomaterials.

First developed by Zhang's group, in 2011, the preparation of RBC membrane-cloaked nanoparticles comprise of two processes: membrane vesicle derivation from RBCs and vesicle-particle fusion (Hu et al. 2011). The erythrocytes are isolated from the blood by centrifugation, followed by hypotonic rupture to release the cellular contents. The RBC ghosts (with little or no hemoglobin) are then extruded through 100 nm porous polycarbonate membranes to create RBC-membrane derived vesicles. Separately prepared poly (lactic-co-glycolic acid) or PLGA nanoparticles (~70 nm) are then mechanically co-extruded with the RBC ghosts through 100 nm pores repeatedly in order to yield fused nanoparticle-vesicle constructs (Fig. 2.3). RBC-membrane-coated PLGA nanoparticles in this study, showed uniform size distribution with ~70 nm core and 7-8 nm thick outer lipid shell (Hu et al. 2011). The elimination half-life of the membrane cloaked nanoparticles was calculated to be 39.6 h when compared to 15.8 h for PEG-coated nanoparticles. This top-down approach promises several advantages. Direct coating of the nanoparticles with the cellular membrane helps replicate the complex surface chemistry of the cells, bypassing labor-intensive processes of protein identification, purification and conjugation. The method also provides a bilayered medium for transmembrane protein anchorage and avoids chemical modifications that could compromise the integrity and functionalities of the target proteins (Copp et al. 2014; Hu et al. 2012; Hu et al. 2011). RBC-membrane-coated PLGA nanoparticles have been further harnessed by the same group for multifunctional roles: as doxorubicin loaded drug delivery vehicles (Aryal et al. 2013), as nanosponges for detoxification applications (Hu et al. 2013a; Pang et al. 2015), as nanosponge-hydrogel hybrids for local treatment of methicillin resistant bacteria (Fang et al. 2015; Wang et al. 2015) and as nanotoxoids for safe and effective vaccination (Hu et al. 2013c) against membrane damaging staphylococcal alpha-haemolysin (α-toxin). Furthermore Fang et al. (Fang et al. 2013) demonstrated the ability to specifically target such hybrid biomimetic nanoconstructs to cancer cells via ligand-linker-lipid tethered insertion into the RBC membrane followed by co-extrusion with PLGA nanoparticles. The robust and simple technique dispenses with complex, membrane damaging chemical reactions for functionalization of the RBC-nanoparticle construct and endows it with the capability of tailoring the density and kind of ligands for targeting specific diseases.

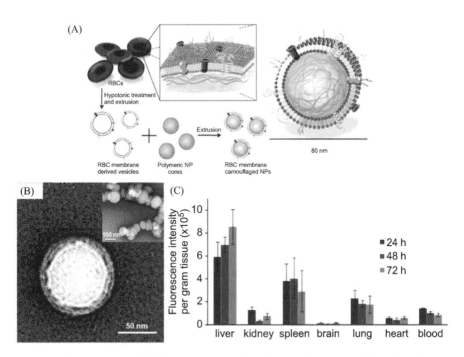

Fig. 2.3 Preparation and characterization of RBC membrane-camouflaged polymeric nanoparticles (NPs). (A) Schematics of the preparation process of the RBC membrane-coated polymeric nanoparticles. (B) Transmission electron microscopy (TEM) images of the resulting nanoparticles with negative staining using uranyl acetate. (C) Biodistribution of fluorescently labeled RBC-membrane-coated polymeric nanoparticles; Fluorescence intensity per gram of tissue (n = 6 per group). Reprinted with permission from (Hu et al. 2011).

This strategy was extended towards the encapsulation of other inorganic and organic nanoparticles for immune evasive camouflage and enhanced therapeutics. RBC membrane cloaked gold nanoparticles (~70 nm; RBC-AuNPs ~100 nm) (Gao et al. 2013b) and nanocages (~71.20 nm; RBC-AuNCs ~89.05 nm) (Piao et al. 2014) have been reported. RBC-AuNCs were found to retain the near infrared photothermal properties of the nanocages, with the cloaked dispersion exhibiting a temperature rise of ≥15°C within 2 min of 850 nm laser irradiation (Fig. 2.4). Repeated NIR irradiation partially ablated the RBC membrane covering which then rendered the RBC-AuNCs more effective in generating higher temperature profiles (Piao et al. 2014). The circulation half-life of RBC-AuNCs was calculated to be ~9.5 h when compared to ~1 h for control polyvinylpyrrolidone coated AuNCs (PVP-AuNCs). This prolonged blood circulation proved to be advantageous for photothermal ablation when tested in 4T1 murine breast cancer models. 250 μg RBC-AuNCs elevated the local tumor temperatures to 47.1°C, as compared to 41.2°C for PVP-AuNCs, when irradiated with 850 nm laser for 10 min. Moreover, based on the average tumor volume, body weight and survival, RBC-AuNCs were shown to consistently out-perform PVP-AuNCs, indicating their improved photothermal treatment efficacy. RBC membrane cloaking was also used to disguise the core-shell supramolecular gelatin nanoparticles (SGNPs) for adaptive and "on-demand" vancomycin delivery to infection sites (Li et al. 2014). Site-specific release of the antibiotic stemmed from the propensity of gelatin nanoparticles to disassemble in the presence of gelatinase known to be secreted by a wide spectrum of bacteria. Erythrocyte membrane coating served to reduce clearance by the immune system during the antibiotic delivery, as well as to absorb the bacterial exotoxin to relieve the symptoms caused by bacterial infection.

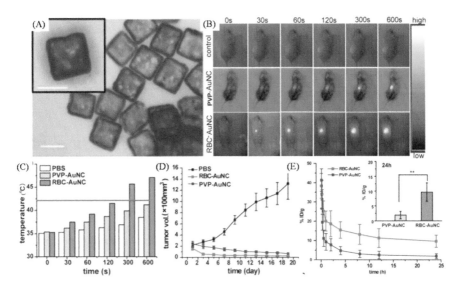

Fig. 2.4   (A) TEM image of vesicle fused Au nanocages (Scale bar = 50 nm). (B) Thermography images of mice injected with the RBC-AuNCs (100 μL RBC-AuNCs; Au content of 2.5 mg/mL) and the pristine PVP-AuNCs and at 48 h after injection, irradiated with an 850 nm laser at a power density of 1 W/cm². Control group comprises of mice injected with PBS only. (C) Plots of temperature at tumor site as a function of irradiation time based on the thermography images of mice in (B). Red line indicates 42°C, the lowest boundary of hyperthermia temperature for selectively ablating cancerous cells in minutes. (D) A plot of the average tumor volume as a function of time after initiation of NIR irradiation. (E) Blood retentions of intravenously injected RBC-AuNCs and the pristine PVP- AuNCs over a span of 24 h. Reprinted with permission from (Piao et al. 2014).

Deeper foray into the structural and interfacial aspects of these RBC membrane cloaked nanoparticles indicate that the membrane demonstrates a right side out orientation bias (Gao et al. 2013b; Hu et al. 2013b; Luk et al. 2014). Unilamellar coating of RBC-PLGA nanoparticles showed the presence of exposed extracellular domains of "marker-of-self" CD47 protein on the nanoparticle surface (Hu et al. 2013b) and glycan content which is similar to that on native RBC membranes (Luk et al. 2014). The membrane glycans and proteins not only served to prolong the *in vivo* circulation but also enhanced the steric stabilization of the nanoparticles under ionic conditions. Quantitative assessment of membrane orientation by Fan et al. indicated that at least 84% of the RBC-PLGA nanoconstructs maintained the correct outside-out membrane orientation, based on fluorescence quenching studies (Fan et al. 2014; Tan et al. 2015). The nanoparticles showed complete membrane coverage, which minimizes the risk of complement activation and immunological responses typically associated with foreign nanomaterials.

Cellular Hitchhiking and External Coupling of Cargo on RBC Membranes

Besides encapsulation, the large surface area of RBCs provides an opportunity to couple therapeutic agents and nanoparticles on the membrane, which can then evade RES uptake, owing to the extended blood circulation time of RBC-bound pathogens. External coupling strategies provide the advantage of avoiding damaging encapsulation procedures which in turn prevent compromising RBC integrity. Moreover, this method also resolves the diffusional limitations and problems associated with sub-optimal drug release seen in encapsulated payloads (Muzykantov 2010; Yoo et al. 2011). Importantly, coupling therapeutics to the RBC surface dispenses with the cumbersome logistics associated with the extraction of membranes for therapeutic encapsulation and reinfusion. However, the lack of isolation of the drug from the blood during transit to the targeted site presents an obvious disadvantage of this method when compared to the encapsulation of the therapeutics within the RBC membranes. This poses a significant hurdle for drugs that do not act in the blood stream and is therefore only limited to vascular drug or pro-drug formulations which are resistant to plasma inhibitors (Muzykantov 2010). Encapsulation of the therapeutics in nanoparticles provides enzymatic protection, thereby increasing the repertoire of drugs that can be effectively used and overcoming the disadvantages mentioned above. Moreover, circulating nanoparticles offer useful modality in blood pool and tissue imaging. However, the eventual recognition of nanoparticles by the RES has limited the applications. External coupling of polymeric nanoparticles to the erythrocyte membranes has been found to be effective in prolonging their circulation half-life (Anselmo et al. 2013; Chambers and Mitragotri 2004, 2007). In their seminal work, Chambers and Mitragotri (Chambers and Mitragotri 2004) demonstrated that erythrocyte attachment dramatically increased the circulation time of polystyrene nanoparticles (220 nm) as compared to that of unbound nanoparticles. Fluorescent dye-labeled polystyrene nanoparticles adhere onto the RBC membranes, likely as a result of van der Waals, electrostatic, hydrogen bonding and hydrophobic forces, without altering the morphology of the cells. The circulation time of bound nanoparticles was found to vary with the diameter and strength of binding. 220 and 450 nm nanoparticles exhibited the longest circulation time (~95% remained in circulation for T > 7 hours), while 830 nm and 1.1 µm particles were rapidly removed from circulation, though their circulation time was still higher than that of the unbound nanoparticles (~99.9% removed from circulation within 2 minutes) (Chambers and Mitragotri 2004). However, concerns over particle detachment and clearance within a few hours pose limitations on this strategy in drug delivery applications. In a follow-up study, it was demonstrated that surface modification of the RBC-nanoparticle hybrids using polyethylene glycol (PEG) increases the circulation half-life of bound particles to over 24 hours (Chambers and Mitragotri 2007).

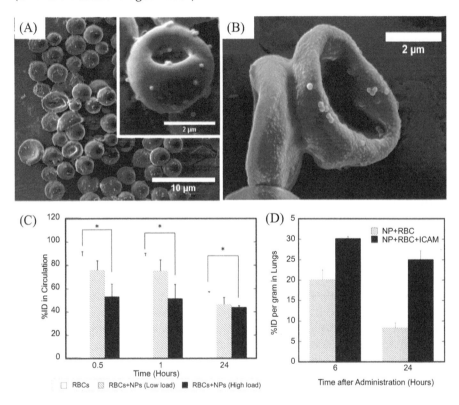

Fig. 2.5 Scanning electron micrographs depicting. (A) 200 nm polystyrene spheres attached to RBCs and (B) controlled detachment of nanoparticles from RBC surface on application of 5 Pa shear stress for 15 min at 37°C. (C) Time-dependent *in vivo* circulation of ^{51}Cr-RBCs; percentage of injected dose (%ID) for free ^{51}Cr-RBCs (white), ^{51}Cr-RBCs with nanoparticles attached at low loading of 1:10 (hatched) and 1:100 (black) incubation ratios (RBCs/NPs) over a 24 h period. (D) Lung accumulation of RBC/NP (hatched bars) and RBC/NP anti-ICAM-1 (black bars) complexes at 6 and 24 h. Both lung %ID/gram groups were statistically different (p < 0.05). Reprinted with permission from (Anselmo et al. 2013).

Cellular hitchhiking was also extended to deliver the nanoparticles exclusively to the lungs, while avoiding accumulation in the liver and spleen (Anselmo et al. 2013). RBC-nanoparticle complexes demonstrated ~3-fold increase in blood persistence and ~7-fold increased accumulation in the lungs when compared to free nanoparticles (Fig. 2.5). Accumulation in lungs was attributed to the mechanical transfer of particles from the RBC surface to lung endothelium. Furthermore, anti-ICAM-1 antibody was attached to the RBC coupled nanoparticles, which afforded increased lung and decreased liver accumulation as compared to RBC-NP with antibody.

Overall, cellular hitchhiking can potentially provide solutions to the two most sought after problems in nanotherapeutics, namely, the avoidance of MPS organs, liver and spleen and targeted delivery to the difficult-to-reach sites in the body such as brain and lungs. In another study, this hitchhiking of nanoparticles was applied in a nanoplasmonics-based opto-nanoporation technique of creating nanopores upon laser illumination for inducing diffusion and triggered release of small and large molecules from the RBCs (Delcea et al. 2012). Gold nanoparticle aggregates were adsorbed onto the RBCs loaded with the dye 5(6)-carboxyfluorescein (5(6)-CF) and a rhodamine labeled dextran (Rh-dextran). Near-infrared laser irradiation resulted in localized heating at the nanoparticles, which led to enhanced permeability of the membrane, thereby triggering the release of the drugs. This strategy potentially overcomes the issues related to sub-optimal diffusion kinetics of drug molecules encapsulated in the erythrocytes while preserving the shape of RBCs. Similarly, $Fe_2O_3@SiO_2$ core-shell magnetic nanoparticles (CSNMs) were also adsorbed onto human erythrocytes, for multimodal magnetic and fluorescence imaging (Laurencin et al. 2013). Detailed studies of this nanoparticle: RBC conjugates revealed that only aminated CSNMs could decorate the RBCs and that their adsorption interaction was mainly ruled by electrostatic attraction between the positively charged amino groups on CSMNs and the abundant sialic acid groups on the outer surface of RBCs (Mai et al. 2013). However, the weak nature of interaction and the potential detachment of the nanoparticles from the RBC surface in the studies thus far, present a major setback to the feasibility of their *in vivo* applications. Since, the coupling of nanoparticles in these reports was based on physical forces only, shear or direct RBC-endothelium contact may result in the nanoparticles falling off the membrane *in vivo* and subsequently accumulating in the RES organs leading to potential toxicity or misinterpretation of results. Increased adhesion strength may be achieved using nanoparticles modified with peptides, antibodies, etc. that can specifically interact with the RBC membrane proteins. Lectins such as wheat germ agglutinins are another alternative which can bind specifically to sialic acid residues on glycosylated RBC membrane proteins, for firm attachment of nanoparticles to RBCs (Chambers and Mitragotri 2004).

An elegant study by Wang et al. provides a simple solution to the above-mentioned hurdle. Multifunctional, theranostic iron oxide nanoparticle bearing erythrocytes were fabricated for magnetic field-enhanced drug delivery and image-guided combination therapy of cancer (Wang et al. 2014). The RBCs were loaded with doxorubicin drug via dilutional hypotonic haemolysis method, while iron oxide nanocrystals were pre-coated with a photodynamic agent, chlorine e6. The nanoparticles were then coupled with the RBCs using the mild and simple biotin-avidin reaction, ensuring minimal membrane disruption and cell damage. The final nanoconstructs after PEGylation (RBC-IONP-Ce6-PEG) exhibited a greatly prolonged blood-circulation half-life at 5.89 ± 1.75 h, when compared to IONP-Ce6-PEG only. Moreover, RBCs modified nanoparticles showed reduced retention in RES organs and greatly enhanced tumor targeting under magnetic field. *In vivo* combination therapy also displayed synergistically enhanced tumor growth inhibition at lower doses. This study sets a good precedence for rationally designed RBC-based theranostic agents, which in association with nanotechnology can offer novel, multifunctional platforms for future cancer treatment.

RBC-Mimicking Synthetic Nanoparticles

The remarkable properties of erythrocytes as long circulating, natural cargo carriers have motivated researchers to not only harness them as drug carriers, but also to design man-made vehicles with comparable features (Hu et al. 2012). Synthetic nanomaterials for drug delivery, therapy and diagnosis have advanced significantly in the past few decades, both in terms of functionality and diversity (Doshi et al. 2009). Combining the advantages of nanotechnology and molecular biology, researchers are now designing synthetic nanomaterials which mimic the mechanic or biological characteristics of RBCs in an effort to improve their drug delivery efficacy and effectiveness.

Mechano-Biological Mimicry of Erythrocytes

Special structural characteristics of RBCs, such as their distinct discoid shape and mechanical deformability play a major role in their unique ability to deliver cargo across narrow constrictions of sinusoids and capillaries, enhanced circulation time etc. (Bhateria et al. 2014; Hu et al. 2012). Such characteristics are hard to emulate for the current class of rigid, spherical nanostructures. The shape, size and rigidity of the nanoparticles have been found to bear unprecedented control over their longevity in circulation and targeting to selected cellular and subcellular locations (Simone et al. 2008). Lin et al. reported prolonged circulation time and improved mitigation of phagocytosis for non-spherical, mPEG-b-P (HEMA-co-histidine-PLA) polymeric nanoparticles (Lin et al. 2011). Elongated and flexible nanoparticles demonstrate flow alignment paradigm which results in reduced vascular collisions, capillary retention and RES clearance, which in the future may achieve superior drug delivery capabilities (Simone et al. 2008).

Various strategies have been reported for the synthesis of RBC-mimicking nanoparticles. Doshi et al. recreated the complex discoidal shape in hollow polystyrene nanosystems, taking cue from the biological genesis of bio-logical RBCs (Doshi et al. 2009). RBC precursors, reticulocytes are spherical structures with an elastic modulus of ~3 MPa, which undergo simultane-ous change in elastic modulus and shape in order to form mature discoidal RBCs. Mimicking this, biocompatible and biodegradable poly (lactic acid-co-glycolide) (PLGA) nanospherical templates were prepared using the electro-hydrodynamic jetting process. Treatment with isopropanol yielded RBC-like template used for layer-by-layer self-assembly of BSA and poly-allylamine hydrochloride. Tetrahydrofuran: Isopropanol solution collapsed the poly-mer core to yield soft, synthetic erythrocytic nanoparticles which mimicked many key attributes of RBCs including the size, shape, elastic modulus (92.8 ± 42 kPa; same order of magnitude as natural RBCs (15.2 ± 3.5 kPa), ability to deform under flow (~7 μm particles stretched themselves to pass through ~5 μm capillaries) and oxygen-carrying capacity (Doshi et al. 2009). In addition, these particles also retained the ability to encapsulate imaging agents such as iron oxide and therapeutic moieties (Fig. 2.6). Furthermore, in an innova-tive study, flexible RBC-like hybrids of Fe_3O_4 nanoparticle/fluorescence dye and cellulose derived polymer were synthesized for dual modality imag-ing (Hayashi et al. 2010). The flexible polymeric skeleton was fabricated by electrospraying based on electrospinning and magnetite nanoparticles and fluorescent dye were incorporated via hydrolysis-condensation of metal-organics during the electrospraying process. The RBC-like particles were able to pass through channels smaller than their size, indicating that the particles may have a shape suitable for smooth circulation in blood.

In addition, top-down approaches have also been employed in designing nanomaterials of different geometries for drug delivery applications (Hu et al. 2012). Of note is the stop-flow lithography technique demonstrated by Haghgooie et al. (Haghgooie et al. 2010) for synthesizing squishy PEG hydrogel colloids. Monomers mixed with photoinitiators were flowed through polydimethylsiloxane (PDMS) microfluidic channels and cross-linked under UV illumination. Different shapes ranging from monodisperse disks, rings, crosses to S-shapes were obtained by varying the photomasks used while modifying the deformability of the nanoparticles was tuned by altering the cross-linking density and monomer-to-initiator ratios. The PEG hydrogels could flow through 4 μm × 4 μm constrictions. In addition, the authors also demonstrated that several biologically important moieties could be incorporated into the nanostructures without affecting the other properties.

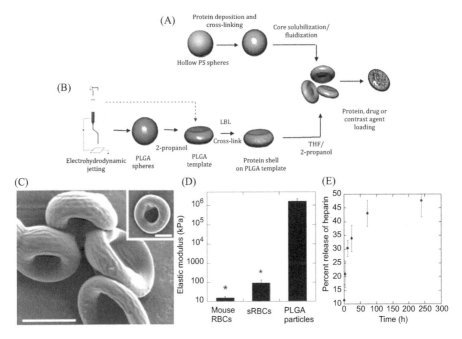

Fig. 2.6 Synthesis technique of RBC-mimicking particles. (A) RBC-shaped particles prepared from hollow PS template. (B) Biocompatible RBC-mimicking particles prepared from PLGA template particles. (C) SEM images of biocompatible sRBCs prepared from PLGA template particles by LbL deposition of PAH/BSA and subsequent dissolution of the polymer core. (D) Comparison of elastic modulus of sRBCs with mouse RBCs and PLGA particles (*, P < 0.001, n = 5). (E) Controlled release of radiolabeled heparin from sRBCs over a period of 10 days (n = 5). Reproduced with permission (Doshi et al. 2009).

Another top-down technique called particle replication in non-wetting templates (PRINT), has been employed by Merkel et al. (Merkel et al. 2011) to model hydrogel microparticles after 6 μm diameter mouse RBCs. 2-hydroxyethyl acrylate (HEA) hydrogel was lightly cross-linked with poly(ethylene glycol) diacrylate(PEGDA) with a photoinitiator (1-hydroxycyclohexyl phenyl ketone). Negatively charged RBC membrane was simulated by addition of 2-carboxyethyl acrylate (Fig. 2.7). PRINT is a high throughput technology that is able to generate uniform populations of organic micro- and nanoparticles with complete control of size, shape and surface chemistry (Gratton et al. 2008). Long-circulating behavior was observed for these RBC mimetics, showing the rapid distribution of the injected particles with an elimination half-life of 3.6 days, with 5% of the injected dose remaining in the blood after 5 days (Merkel et al. 2011). Easily tunable elastic module (from 63.9 kPa to 7.8 kPa or lower depending on the concentration of the cross-linker) and reversible deformability could be achieved, allowing the microparticles to circulate longer and bypass several organs including lungs, which are known to trap the rigid nanoparticles.

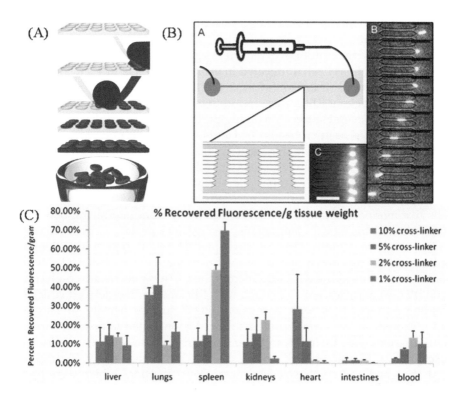

Fig. 2.7 (A) A graphical depiction of the PRINT process used to fabricate RBCMs. Elastomeric fluoropolymer mold with disc-shaped wells was covered by an aliquot of the prepolymer mixture. The mold was passed through a pressured nip covered by a high-energy sheet, wicking away excess liquid from the mold surface while filling the wells of the mold. The filled mold was cured photochemically, yielding cross-linked hydrogel particles, which were harvested from the mold by freezing onto a thin film of 1% poly (vinyl alcohol) in water (blue) and peeling away the mold. (B) Microfluidic evaluation of RBCM deformability; showing a schematic of the microfluidic device *A*, a 1% cross-linked RBCM (6.0-μm diameter) deforming to pass through a 3 × 3.5 μm channel *B* and Ten percent cross-linked RBCMs stuck in the entrance of the 3-μm pores in a microfluidic device *C*. (C) Biodistribution of RBCMs into various tissues 2 h post dosing by percent recovered fluorescence normalized for tissue weight. Reproduced with permission from (Merkel et al. 2011).

Chemo-Biological Mimicry of Erythrocytes

Not just the physical parameters such as shape, size and elasticity, but their complex membrane make up also contributes to the prolonged *in vivo* survival of the RBCs (Hu et al. 2013b). These self-marker membrane proteins inhibit the immune responses, namely complement activation and macrophage phagocytosis (Hu et al. 2012).

Since the discovery of ubiquitously expressed CD47 (integrin-associated transmembrane protein) that functions as a 'marker of self' on murine red blood cells (Oldenborg et al. 2000), several attempts have been made to develop immune-evasive, long circulating bioinspired stealth nanoparticles. CD47 binds to the inhibitory receptor signal regulatory protein alpha (SIRPalpha1) and has been implicated in *in vivo* survival of RBCs (Oldenborg et al. 2000), viruses (Cameron et al. 2005) and cancer cells (Jaiswal et al. 2009). CD47 knockout RBCs were found to be cleared rapidly by the splenic macrophages. Several studies involving perfluorocarbon emulsions (Hsu et al. 2003), polystyrene microbeads (Tsai and Discher 2008), polyvinyl chloride and polyurethane polymers (Finley et al. 2012; Stachelek et al. 2011) etc. have been carried out to minimize phagocytosis by appending CD47 marker or its derivatives to the nanoparticle surface. The studies also indicated that the presence of CD47 also prevented the attachment of inflammatory cells like monocytes. However, issues like suboptimal interfacing between synthetic and biological components, inconsistent protein surface density and orientations remain. This has prompted Hu et. al. (Hu et al. 2013b) to utilize a facile, top-down approach for functionalizing nanoparticles with native CD47 by cloaking with RBC membranes. 70 nm poly (lactic-co-glycolic acid) (PLGA) particles were extruded with RBC membrane-derived vesicles, yielding nanoparticles with equivalent CD47 surface density to natural RBCs. Moreover, the surface proteins were demonstrated to possess correct configuration, with their extracellular domains exposed, thereby reducing susceptibility to macrophage uptake *in vitro*. In another study, Rodriguez et al. designed and synthesized minimal "Self" peptides which were then bound to 160-nm polystyrene nanobeads (Rodriguez et al. 2013). Self peptides were shown to delay macrophage-mediated clearance of nanoparticles, which promoted persistent circulation and subsequently enhanced dye and drug delivery to tumors. In addition, Self peptides also potently inhibited nanoparticle uptake mediated by the contractile cytoskeleton, indicating their utility in nanotheranostics.

Conclusion

One of the major design considerations for nanoparticle-based drug delivery is prolonging the circulation half-life, which can lead to more effective targeting and improved therapeutic efficacies. While current stealth strategy employs synthetic PEG moieties, search for natural and more effective alternatives is imperative (Fang et al. 2012). Cell mediated drug delivery systems take advantage of cell properties, such as long circulation time, abundant surface ligands, flexible morphology, cellular signaling and metabolism, to offer a unique opportunity to maximize therapeutic outcomes as well as to minimize the side effects (Su et al. 2015). Besides erythrocytes, a wide plethora of cells and cellular membranes have been employed for camouflaging nanomaterials to prolong their *in vivo* circulation time and to improve their bioavailability and functionality. Whole cell carriers like stem cells (Pierige et al. 2008; Roger et al. 2010; Stuckey and Shah 2014), macrophages (Choi et al. 2007; Dou et al. 2009; Vinogradov et al. 2014), extracellular vesicles such as those derived from human umbilical vascular endothelial cells (Silva et al. 2013) as well as eukaryotic cell membranes such as cancer cell membranes (Fang et al. 2014) and leukocytes (Parodi et al. 2013) have all imparted excellent pharmacokinetic properties, biocompatibility and targeting capability to the encapsulated nanoparticles.

Erythrocyte inspired nanomedicine is a rapidly emerging field and has come a long way since its inception. As a step forward in personalized medicine, patients' own blood cells can be harnessed to camouflage foreign nanoparticulate entities, making them non-toxic and non-immunogenic while retaining their intrinsic functionalities. However, challenges still exist to this state-of-the-art approach. Procedures to incorporate nanomaterials into the intact cell or cell membranes often compromise the biological integrity of the hybrid system, resulting in rapid clearance. Hence, further optimization is required for non-disruptive and reliable RBC manipulations; to minimize structural alterations and boost delivery (Hu et al. 2012; Yoo et al. 2011). Effective isolation, functionalization and reinjection of cells are always prone to contamination. Detachment or leakage of nanoparticles from the hybrid systems is another concern. In addition, the introduction of cell carriers may disrupt the natural physiological balance, warranting careful optimization of the dose (Su et al. 2015). RBC-mimicking nanoparticles, which aim to replicate morphological and chemo-biological aspects of the blood cells will develop better with a clearer understanding of the exact structural characteristics and physiological delivery mechanisms of RBCs. Such hybrid nanoplatforms combine the advantages of synthetic systems, such as controllability and mass production with the extraordinary drug delivery capabilities of the biological RBCs. At the same time, they bypass the complicated and tiresome processes such as blood type matching and screening, cell/ cell membrane harvest etc. and have a broader applicability. With sustained efforts, RBC-based hybrid nanomaterials can be instrumental in advancing futuristic drug delivery technologies and next generation nanomedicine with broad medical applications.

Acknowledgements

This work is supported, in part, by the University of Wisconsin–Madison, the National Institutes of Health (NIBIB/NCI 1R01CA169365 & P30CA014520) and the American Cancer Society (125246-RSG-13-099-01-CCE).

References

Ahn, S., Jung, S.Y., Seo, E. and Lee, S.J. 2011. Gold nanoparticle-incorporated human red blood cells (RBCs) for X-ray dynamic imaging. Biomaterials 32: 7191-7199.

Alexis, F., Pridgen, E., Molnar, L.K. and Farokhzad, O.C. 2008. Factors affecting the clearance and biodistribution of polymeric nanoparticles. Mol Pharm 5: 505-515.

Anselmo, A.C., Gupta, V., Zern, B.J., Pan, D., Zakrewsky, M., Muzykantov, V., et al. 2013. Delivering nanoparticles to lungs while avoiding liver and spleen through adsorption on red blood cells. ACS Nano 7: 11129-11137.

Antonelli, A. and Magnani, M. 2014. Red blood cells as carriers of iron oxide-based contrast agents for diagnostic applications. J Biomed Nanotechnol 10: 1732-1750.

Antonelli, A., Sfara, C., Mosca, L., Manuali, E. and Magnani, M. 2008. New biomimetic constructs for improved in vivo circulation of superparamagnetic nanoparticles. J Nanosci Nanotechnol 8: 2270-2278.

Antonelli, A., Sfara, C., Manuali, E., Bruce, I.J. and Magnani, M. 2011. Encapsulation of superparamagnetic nanoparticles into red blood cells as new carriers of MRI contrast agents. Nanomedicine 6: 211-223.

Antonelli, A., Sfara, C., Battistelli, S., Canonico, B., Arcangeletti, M., Manuali, E., et al. 2013. New strategies to prolong the in vivo life span of iron-based contrast agents for MRI. PLoS One 8: e78542.

Aryal, S., Hu, C.M., Fang, R.H., Dehaini, D., Carpenter, C., Zhang, D.E., et al. 2013. Erythrocyte membrane-cloaked polymeric nanoparticles for controlled drug loading and release. Nanomedicine 8: 1271-1280.

Baker, R.F. 1967. Entry of ferritin into human red cells during hypotonic haemolysis. Nature 215: 424-425.

Beletsi, A., Panagi, Z. and Avgoustakis, K. 2005. Biodistribution properties of nanoparticles based on mixtures of PLGA with PLGA-PEG diblock copolymers. Int J Pharm 298: 233-241.

Berman, J.D. and Gallalee, J.V. 1985. Antileishmanial activity of human red blood cells containing formycin A. J Infect Dis 151: 698-703.

Bhateria, M., Rachumallu, R., Singh, R. and Bhatta, R.S. 2014. Erythrocytes-based synthetic delivery systems: transition from conventional to novel engineering strategies. Expert Opin Drug Deliv 11: 1219-1236.

Biagiotti, S., Paoletti, M.F., Fraternale, A., Rossi, L. and Magnani, M. 2011. Drug delivery by red blood cells. IUBMB Life 63: 621-631.

Brahler, M., Georgieva, R., Buske, N., Muller, A., Muller, S., Pinkernelle, J., et al. 2006. Magnetite-loaded carrier erythrocytes as contrast agents for magnetic resonance imaging. Nano Lett 6: 2505-2509.

Cameron, C.M., Barrett, J.W., Mann, M., Lucas, A. and McFadden, G. 2005. Myxoma virus M128L is expressed as a cell surface CD47-like virulence factor that contributes to the downregulation of macrophage activation in vivo. Virology 337: 55-67.

Chambers, E. and Mitragotri, S. 2004. Prolonged circulation of large polymeric nanoparticles by non-covalent adsorption on erythrocytes. J Control Release 100: 111-119.

Chambers, E. and Mitragotri, S. 2007. Long circulating nanoparticles via adhesion on red blood cells: mechanism and extended circulation. Exp Biol Med 232: 958-966.

Chiarantini, L., Rossi, L., Fraternale, A. and Magnani, M. 1995. Modulated red blood cell survival by membrane protein clustering. Mol Cell Biochem 144: 53-59.

Choi, M.R., Stanton-Maxey, K.J., Stanley, J.K., Levin, C.S., Bardhan, R., Akin, D., et al. 2007. A cellular Trojan Horse for delivery of therapeutic nanoparticles into tumors. Nano Lett 7: 3759-3765.

Cinti, C., Taranta, M., Naldi, I. and Grimaldi, S. 2011. Newly engineered magnetic erythrocytes for sustained and targeted delivery of anti-cancer therapeutic compounds. PLoS One 6: 0017132.

Colin, F.C. and Schrier, S.L. 1991. Spontaneous endocytosis in human neonatal and adult red blood cells: comparison to drug-induced endocytosis and to receptor-mediated endocytosis. Am J Hematol 37: 34-40.

Copp, J.A., Fang, R.H., Luk, B.T., Hu, C.M., Gao, W., Zhang, K., et al. 2014. Clearance of pathological antibodies using biomimetic nanoparticles. Proc Natl Acad Sci USA 111: 13481-13486.

Dale, G.L., Kuhl, W. and Beutler, E. 1979. Incorporation of glucocerebrosidase into Gaucher's disease monocytes in vitro. Proc Natl Acad Sci USA 76: 473-475.

Danilov Iu, N., Rudchenko, S.A., Samokhin, G.P., Orekhov, A.N. and Il'ina, M.B. 1985. Concentration of erythrocyte-based magnetic carriers in the vascular bed. Biull Eksp Biol Med 100: 701-702.

Delcea, M., Sternberg, N., Yashchenok, A.M., Georgieva, R., Baumler, H., Mohwald, H., et al. 2012. Nanoplasmonics for dual-molecule release through nanopores in the membrane of red blood cells. ACS Nano 6: 4169-4180.

Deloach, J. and Ihler, G. 1977. A dialysis procedure for loading erythrocytes with enzymes and lipids. Biochim Biophys Acta 496: 136-145.

Desilets, J., Lejeune, A., Mercer, J. and Gicquaud, C. 2001. Nanoerythrosomes, a new derivative of erythrocyte ghost: IV. Fate of reinjected nanoerythrosomes. Anticancer Res 21: 1741-1747.

Deuticke, B., Kim, M. and Zöllner, C. 1973. The influence of amphotericin B on the permeability of mammalian erythrocytes to nonelectrolytes, anions and cations. BBA – Biomembranes 318: 345-359.

Diagaradjane, P., Deorukhkar, A., Gelovani, J.G., Maru, D.M. and Krishnan, S. 2010. Gadolinium chloride augments tumor-specific imaging of targeted quantum dots in vivo. ACS Nano 4: 4131-4141.

Diluzio, N.R. and Wooles, W.R. 1964. Depression of phagocytic activity and immune response by methyl palmitate. Am J Physiol 206: 939-943.

Doshi, N., Zahr, A.S., Bhaskar, S., Lahann, J. and Mitragotri, S. 2009. Red blood cell-mimicking synthetic biomaterial particles. Proc Natl Acad Sci USA 106: 21495-21499.

Dou, H., Grotepas, C.B., McMillan, J.M., Destache, C.J., Chaubal, M., Werling, J., et al. 2009. Macrophage delivery of nanoformulated antiretroviral drug to the brain in a murine model of neuroAIDS. J Immunol 183: 661-669.

Eichler, H.G., Raffesberg, W., Gasic, S., Korn, A. and Bauer, K. 1985. Release of vitamin B12 from carrier erythrocytes in vitro. Res Exp Med 185: 341-344.

Eichler, H.G., Rameis, H., Bauer, K., Korn, A., Bacher, S. and Gasic, S. 1986a. Survival of gentamicin-loaded carrier erythrocytes in healthy human volunteers. Eur J Clin Invest 16: 39-42.

Eichler, H.G., Schneider, W., Raberger, G., Bacher, S. and Pabinger, I. 1986b. Erythrocytes as carriers for heparin. Preliminary in vitro and animal studies. Res Exp Med 186: 407-412.

el-Kalay, M.A., Koochaki, Z., Schutz, P.W. and Gaylor, J.D. 1989. Efficient continuous flow washing of red blood cells for exogenous agent loading using a hollow fiber plasma separator. Artif Organs 13: 515-524.

Fan, Z., Zhou, H., Li, P.Y., Speer, J.E. and Cheng, H. 2014. Structural elucidation of cell membrane-derived nanoparticles using molecular probes. J Mater Chem B 2: 8231-8238.

Fang, R.H., Hu, C.M. and Zhang, L. 2012. Nanoparticles disguised as red blood cells to evade the immune system. Expert Opin Biol Ther 12: 385-389.

Fang, R.H., Hu, C.M., Chen, K.N., Luk, B.T., Carpenter, C.W., Gao, W., et al. 2013. Lipid-insertion enables targeting functionalization of erythrocyte membrane-cloaked nanoparticles. Nanoscale 5: 8884-8888.

Fang, R.H., Hu, C.M., Luk, B.T., Gao, W., Copp, J.A., Tai, Y., et al. 2014. Cancer cell membrane-coated nanoparticles for anticancer vaccination and drug delivery. Nano Lett 14: 2181-2188.

Fang, R.H., Luk, B.T., Hu, C.M. and Zhang, L. 2015. Engineered nanoparticles mimicking cell membranes for toxin neutralization. Adv Drug Deliv 90: 69-80.

Fernandez-Urrusuno, R., Fattal, E., Rodrigues, J.M., Jr., Feger, J., Bedossa, P. and Couvreur, P. 1996. Effect of polymeric nanoparticle administration on the clearance activity of the mononuclear phagocyte system in mice. J Biomed Mater Res 31: 401-408.

Ferrauto, G., Delli Castelli, D., Di Gregorio, E., Langereis, S., Burdinski, D., Grull, H., et al. 2014. Lanthanide-loaded erythrocytes as highly sensitive chemical exchange saturation transfer MRI contrast agents. J Am Chem Soc 136: 638-641.

Finley, M.J., Rauova, L., Alferiev, I.S., Weisel, J.W., Levy, R.J. and Stachelek, S.J. 2012. Diminished adhesion and activation of platelets and neutrophils with CD47 functionalized blood contacting surfaces. Biomaterials 33: 5803-5811.

Flynn, G., Hackett, T.J., McHale, L. and McHale, A.P. 1994a. Encapsulation of the thrombolytic enzyme, brinase, in photosensitized erythrocytes: a novel thrombolytic system based on photodynamic activation. J Photochem Photobiol B 26: 193-196.

Flynn, G., McHale, L. and McHale, A.P. 1994b. Methotrexate-loaded, photosensitized erythrocytes: a photo-activatable carrier/delivery system for use in cancer therapy. Cancer Lett 82: 225-229.

Gao, H., Liu, J., Yang, C., Cheng, T., Chu, L., Xu, H., et al. 2013a. The impact of PEGylation patterns on the in vivo biodistribution of mixed shell micelles. Int J Nanomedicine 8: 4229-4246.

Gao, W. and Zhang, L. 2015. Engineering red-blood-cell-membrane–coated nanoparticles for broad biomedical applications. AIChE Journal 61: 738-746.

Gao, W., Hu, C.M., Fang, R.H., Luk, B.T., Su, J. and Zhang, L. 2013b. Surface functionalization of gold nanoparticles with red blood cell membranes. Adv Mater 25: 3549-3553.

Godfrin, Y., Horand, F., Franco, R., Dufour, E., Kosenko, E., Bax, B.E., et al. 2012. International seminar on the red blood cells as vehicles for drugs. Expert Opin Biol Ther 12: 127-133.

Goel, S., Chen, F., Ehlerding, E.B. and Cai, W. 2014. Intrinsically radiolabeled nanoparticles: an emerging paradigm. Small 10: 3825-3830.

Goto, R., Kubo, H. and Okada, S. 1991. Effect of reticuloendothelial blockade on tissue distribution of 99mTc-labeled synthetic liposomes in Ehrlich solid tumor-bearing mice. Chem Pharm Bull 39: 230-232.

Gratton, S.E., Ropp, P.A., Pohlhaus, P.D., Luft, J.C., Madden, V.J., Napier, M.E., et al. 2008. The effect of particle design on cellular internalization pathways. Proc Natl Acad Sci USA 105: 11613-11618.

Haghgooie, R., Toner, M. and Doyle, P.S. 2010. Squishy non-spherical hydrogel microparticles. Macromol Rapid Commun 31: 128-134.

Hamidi, M. and Tajerzadeh, H. 2003. Carrier erythrocytes: an overview. Drug Deliv 10: 9-20.

Hamidi, M., Zarrin, A., Foroozesh, M. and Mohammadi-Samani, S. 2007. Applications of carrier erythrocytes in delivery of biopharmaceuticals. J Control Release 118: 145-160.

Hamidi, M., Azimi, K. and Mohammadi-Samani, S. 2011. Co-encapsulation of a drug with a protein in erythrocytes for improved drug loading and release: phenytoin and bovine serum albumin (BSA). J Pharm Sci 14: 46-59.

Harper, G.R., Davis, S.S., Davies, M.C., Norman, M.E., Tadros, T.F., Taylor, D.C., et al. 1995. Influence of surface coverage with poly(ethylene oxide) on attachment of sterically stabilized microspheres to rat Kupffer cells in vitro. Biomaterials 16: 427-439.

Hayashi, K., Ono, K., Suzuki, H., Sawada, M., Moriya, M., Sakamoto, W., et al. 2010. Electrosprayed synthesis of red-blood-cell-like particles with dual modality for magnetic resonance and fluorescence imaging. Small 6: 2384-2391.

Hsu, Y.C., Acuna, M., Tahara, S.M. and Peng, C.A. 2003. Reduced phagocytosis of colloidal carriers using soluble CD47. Pharm Res 20: 1539-1542.

Hu, C.M., Zhang, L., Aryal, S., Cheung, C. and Fang, R.H. 2011. Erythrocyte membrane-camouflaged polymeric nanoparticles as a biomimetic delivery platform. Proc Natl Acad Sci USA 108: 10980-10985.

Hu, C.M., Fang, R.H. and Zhang, L. 2012. Erythrocyte-inspired delivery systems. Adv Healthc Mater 1: 537-547.

Hu, C.M., Fang, R.H., Copp, J., Luk, B.T. and Zhang, L. 2013a. A biomimetic nanosponge that absorbs pore-forming toxins. Nat Nanotechnol 8: 336-340.

Hu, C.M., Fang, R.H., Luk, B.T., Chen, K.N., Carpenter, C., Gao, W., et al. 2013b. 'Marker-of-self' functionalization of nanoscale particles through a top-down cellular membrane coating approach. Nanoscale 5: 2664-2668.

Hu, C.M., Fang, R.H., Luk, B.T. and Zhang, L. 2013c. Nanoparticle-detained toxins for safe and effective vaccination. Nat Nanotechnol 8: 933-938.

Ihler, G.M. and Tsang, H.C. 1987. Hypotonic hemolysis methods for entrapment of agents in resealed erythrocytes. Methods Enzymol 149: 221-229.

Ihler, G.M., Glew, R.H. and Schnure, F.W. 1973. Enzyme loading of erythrocytes. Proc Natl Acad Sci USA 70: 2663-2666.

Ito, Y., Ogiso, T., Iwaki, M. and Kitaike, M. 1989. Encapsulation of porcine insulin in rabbit erythrocytes and its disposition in the circulation system in normal and diabetic rabbits. J Pharmacobiodyn 12: 193-200.

Jain, S. and Jain N.K. 2003. Engineered nanoerythrocytes as a novel drug delivery system. pp. 77-90. *In*: M. Magnani [ed.]. Erythrocyte Engineering for Drug Delivery and Targeting. Plenum Publishers, New York, USA.

Jain, S.K. and Vyas, S.P. 1994. Magnetically responsive diclofenac sodium-loaded erythrocytes: preparation and in vitro characterization. J Microencapsul 11: 141-151.

Jaiswal, S., Jamieson, C.H., Pang, W.W., Park, C.Y., Chao, M.P., Majeti, R., et al. 2009. CD47 is upregulated on circulating hematopoietic stem cells and leukemia cells to avoid phagocytosis. Cell 138: 271-285.

Jokerst, J.V., Lobovkina, T., Zare, R.N. and Gambhir, S.S. 2011. Nanoparticle PEGylation for imaging and therapy. Nanomedicine 6: 715-728.

Knop, K., Hoogenboom, R., Fischer, D. and Schubert, U.S. 2010. Poly (ethylene glycol) in drug delivery: pros and cons as well as potential alternatives. Angew Chem Int Ed Engl 49: 6288-6308.

Kolesnikova, T.A., Skirtach, A.G. and Mohwald, H. 2013. Red blood cells and poly-electrolyte multilayer capsules: natural carriers versus polymer-based drug delivery vehicles. Expert Opin Drug Deliv 10: 47-58.

Laurencin, M., Cam, N., Georgelin, T., Clement, O., Autret, G., Siaugue, J.M., et al. 2013. Human erythrocytes covered with magnetic core-shell nanoparticles for multimodal imaging. Adv Healthc Mater 2: 1209-1212.

Lazar, G., van Galen, M. and Scherphof, G.L. 1989. Gadolinium chloride-induced shifts in intrahepatic distributions of liposomes. Biochim Biophys Acta 10: 2-3.

Li, L.L., Xu, J.H., Qi, G.B., Zhao, X., Yu, F. and Wang, H. 2014. Core-shell supramo-lecular gelatin nanoparticles for adaptive and "on-demand" antibiotic delivery. ACS Nano 8: 4975-4983.

Lin, S.Y., Hsu, W.H., Lo, J.M., Tsai, H.C. and Hsiue, G.H. 2011. Novel geometry type of nanocarriers mitigated the phagocytosis for drug delivery. J Control Release 154: 84-92.

Liu, D. 1997. Biological factors involved in blood clearance of liposomes by liver. Adv Drug Del Rev 24: 201-213.

Luk, B.T. and Zhang, L. 2015. Cell membrane-camouflaged nanoparticles for drug delivery. J Control Release (In Press).

Luk, B.T., Hu, C.M., Fang, R.H., Dehaini, D., Carpenter, C., Gao, W., et al. 2014. Interfacial interactions between natural RBC membranes and synthetic poly-meric nanoparticles. Nanoscale 6: 2730-2737.

Magnani, M., Pierige, F. and Rossi, L. 2012. Erythrocytes as a novel delivery vehicle for biologics: from enzymes to nucleic acid-based therapeutics. Ther Deliv 3: 405-414.

Mai, T.D., d'Orlye, F., Menager, C., Varenne, A. and Siaugue, J.M. 2013. Red blood cells decorated with functionalized core-shell magnetic nanoparticles: elucida-tion of the adsorption mechanism. Chem Commun 49: 5393-5395.

Mangal, P.C. and Kaur, A. 1991. Electroporation of red blood cell membrane and its use as a drug carrier system. Ind J Biochem Biophys 28: 219-221.

Markov, D.E., Boeve, H., Gleich, B., Borgert, J., Antonelli, A., Sfara, C., et al. 2010. Human erythrocytes as nanoparticle carriers for magnetic particle imaging. Phys Med Biol 55: 6461-6473.

Merkel, T.J., Jones, S.W., Herlihy, K.P., Kersey, F.R., Shields, A.R., Napier, M., et al. 2011. Using mechanobiological mimicry of red blood cells to extend circulation times of hydrogel microparticles. Proc Natl Acad Sci USA 108: 586-591.

Millan, C.G., Marinero, M.L., Castaneda, A.Z. and Lanao, J.M. 2004. Drug, enzyme and peptide delivery using erythrocytes as carriers. J Control Release 95: 27-49.

Mitragotri, S. and Lahann, J. 2009. Physical approaches to biomaterial design. Nat Mater 8: 15-23.

Moghimi, S.M. 1999. Re-establishing the long circulatory behaviour of poloxamine-coated particles after repeated intravenous administration: applications in can-cer drug delivery and imaging. Biochim Biophys Acta 18: 1-2.

Moghimi, S.M. and Davis, S.S. 1994. Innovations in avoiding particle clearance from blood by Kupffer cells: cause for reflection. Crit Rev Ther Drug Carrier Syst 11: 31-59.

Moghimi, S.M. and Hunter, A.C. 2001. Capture of stealth nanoparticles by the body's defences. Crit Rev Ther Drug Carrier Syst 18: 527-550.

Moghimi, S.M. and Szebeni, J. 2003. Stealth liposomes and long circulating nanoparticles: critical issues in pharmacokinetics, opsonization and protein-binding properties. Prog Lipid Res 42: 463-478.

Moghimi, S.M., Hunter, A.C. and Murray, J.C. 2001. Long-circulating and target-specific nanoparticles: theory to practice. Pharmacol Rev 53: 283-318.

Mulholland, S.E., Lee, S., McAuliffe, D.J. and Doukas, A.G. 1999. Cell loading with laser-generated stress waves: the role of the stress gradient. Pharm Res 16: 514-518.

Muzykantov, V.R. 2010. Drug delivery by red blood cells: vascular carriers designed by mother nature. Expert Opin Drug Deliv 7: 403-427.

Muzykantov, V.R. 2013. Drug delivery carriers on the fringes: natural red blood cells versus synthetic multilayered capsules. Expert Opin Drug Deliv 10: 1-4.

Nicolau, C. and Gersonde, K. 1979. Incorporation of inositol hexaphosphate into intact red blood cells. I. Fusion of effector-containing lipid vesicles with erythrocytes. Naturwissenschaften 66: 563-566.

Oldenborg, P.A., Zheleznyak, A., Fang, Y.F., Lagenaur, C.F., Gresham, H.D. and Lindberg, F.P. 2000. Role of CD47 as a marker of self on red blood cells. Science 288: 2051-2054.

Orekhova, N.M., Akchurin, R.S., Belyaev, A.A., Smirnov, M.D., Ragimov, S.E. and Orekhov, A.N. 1990. Local prevention of thrombosis in animal arteries by means of magnetic targeting of aspirin-loaded red cells. Thromb Res 57: 611-616.

Pang, Z., Hu, C.M., Fang, R.H., Luk, B.T., Gao, W., Wang, F., et al. 2015. Detoxification of organophosphate poisoning using nanoparticle bioscavengers. ACS Nano 9: 6450-6458.

Parodi, A., Quattrocchi, N., van de Ven, A.L., Chiappini, C., Evangelopoulos, M., Martinez, J.O., et al. 2013. Synthetic nanoparticles functionalized with biomimetic leukocyte membranes possess cell-like functions. Nat Nanotechnol 8: 61-68.

Piao, J.G., Wang, L., Gao, F., You, Y.Z., Xiong, Y. and Yang, L. 2014. Erythrocyte membrane is an alternative coating to polyethylene glycol for prolonging the circulation lifetime of gold nanocages for photothermal therapy. ACS Nano 8: 10414-10425.

Pierige, F., Serafini, S., Rossi, L. and Magnani, M. 2008. Cell-based drug delivery. Adv Drug Deliv Rev 60: 286-295.

Rahmer, J., Antonelli, A., Sfara, C., Tiemann, B., Gleich, B., Magnani, M., et al. 2013. Nanoparticle encapsulation in red blood cells enables blood-pool magnetic particle imaging hours after injection. Phys Med Biol 58: 3965-3977.

Rodriguez, P.L., Harada, T., Christian, D.A., Pantano, D.A., Tsai, R.K. and Discher, D.E. 2013. Minimal "Self" peptides that inhibit phagocytic clearance and enhance delivery of nanoparticles. Science 339: 971-975.

Roger, M., Clavreul, A., Venier-Julienne, M.C., Passirani, C., Sindji, L., Schiller, P., et al. 2010. Mesenchymal stem cells as cellular vehicles for delivery of nanoparticles to brain tumors. Biomaterials 31: 8393-8401.

Rossi, L., Serafini, S., Pierige, F., Antonelli, A., Cerasi, A., Fraternale, A., et al. 2005. Erythrocyte-based drug delivery. Expert Opin Drug Deliv 2: 311-322.

Sauer, J.M., Waalkes, M.P., Hooser, S.B., Kuester, R.K., McQueen, C.A. and Sipes, I.G. 1997. Suppression of Kupffer cell function prevents cadmium induced hepatocellular necrosis in the male sprague-dawley rat. Toxicology 121: 155-164.

Schrier, S.L., Zachowski, A. and Devaux, P.F. 1992. Mechanisms of amphipath-induced stomatocytosis in human erythrocytes. Blood 79: 782-786.

Silva, A.K., Di Corato, R., Pellegrino, T., Chat, S., Pugliese, G., Luciani, N., et al. 2013. Cell-derived vesicles as a bioplatform for the encapsulation of theranostic nanomaterials. Nanoscale 5: 11374-11384.

Simone, E.A., Dziubla, T.D. and Muzykantov, V.R. 2008. Polymeric carriers: role of geometry in drug delivery. Expert Opin Drug Deliv 5: 1283-1300.

Smith, W.J., Oien, N.P., Hughes, R.M., Marvin, C.M., Rodgers, Z.L., Lee, J., et al. 2014. Cell-mediated assembly of phototherapeutics. Angew Chem Int Ed Engl 53: 10945-10948.

Souhami, R.L., Patel, H.M. and Ryman, B.E. 1981. The effect of reticuloendothelial blockade on the blood clearance and tissue distribution of liposomes. Biochim Biophys Acta 674: 354-371.

Sprandel, U., Lanz, D.J. and von Horsten, W. 1987. Magnetically responsive erythrocyte ghosts. Methods Enzymol 149: 301-312.

Stachelek, S.J., Finley, M.J., Alferiev, I.S., Wang, F., Tsai, R.K., Eckells, E.C., et al. 2011. The effect of CD47 modified polymer surfaces on inflammatory cell attachment and activation. Biomaterials 32: 4317-4326.

Stephan, M.T., Moon, J.J., Um, S.H., Bershteyn, A. and Irvine, D.J. 2010. Therapeutic cell engineering with surface-conjugated synthetic nanoparticles. Nat Med 16: 1035-1041.

Sternberg, N., Georgieva, R., Duft, K. and Baumler, H. 2012. Surface-modified loaded human red blood cells for targeting and delivery of drugs. J Microencapsul 29: 9-20.

Stuckey, D.W. and Shah, K. 2014. Stem cell-based therapies for cancer treatment: separating hope from hype. Nat Rev Cancer 14: 683-691.

Su, Y., Xie, Z., Kim, G.B., Dong, C. and Yang, J. 2015. Design strategies and applications of circulating cell-mediated drug delivery systems. ACS Biomater Sci Eng 1: 201-217.

Tan, S., Wu, T., Zhang, D. and Zhang, Z. 2015. Cell or cell membrane-based drug delivery systems. Theranostics 5: 863-881.

Tang, K., Zhang, Y., Zhang, H., Xu, P., Liu, J., Ma, J., et al. 2012. Delivery of chemotherapeutic drugs in tumour cell-derived microparticles. Nat Commun 3: 1282.

Tsai, R.K. and Discher, D.E. 2008. Inhibition of "self" engulfment through deactivation of myosin-II at the phagocytic synapse between human cells. J Cell Biol 180: 989-1003.

Vinogradov, S., Warren, G. and Wei, X. 2014. Macrophages associated with tumors as potential targets and therapeutic intermediates. Nanomedicine 9: 695-707.

Vyas, S.P. and Jain, S.K. 1994. Preparation and in vitro characterization of a magnetically responsive ibuprofen-loaded erythrocytes carrier. J. Microencapsul 11: 19-29.

Wang, C., Sun, X., Cheng, L., Yin, S., Yang, G., Li, Y., et al. 2014. Multifunctional theranostic red blood cells for magnetic-field-enhanced in vivo combination therapy of cancer. Adv Mater 26: 4794-4802.

Wang, F., Gao, W., Thamphiwatana, S., Luk, B.T., Angsantikul, P., Zhang, Q., et al. 2015. Hydrogel retaining toxin-absorbing nanosponges for local treatment of methicillin-resistant staphylococcus aureus infection. Adv Mater 27: 3437-3443.

Weissleder, R., Heautot, J.F., Schaffer, B.K., Nossiff, N., Papisov, M.I., Bogdanov, A., Jr., et al. 1994. MR lymphography: study of a high-efficiency lymphotrophic agent. Radiology 191: 225-230.

Wu, Z., Li, T., Li, J., Gao, W., Xu, T., Christianson, C., et al. 2014. Turning erythrocytes into functional micromotors. ACS Nano 8: 12041-12048.

Yoo, J.W., Chambers, E. and Mitragotri, S. 2010. Factors that control the circulation time of nanoparticles in blood: challenges, solutions and future prospects. Curr Pharm Des 16: 2298-2307.

Yoo, J.W., Irvine, D.J., Discher, D.E. and Mitragotri, S. 2011. Bio-inspired, bioengineered and biomimetic drug delivery carriers. Nat Rev Drug Discov 10: 521-535.

Zeng, G., Ogaki, R. and Meyer, R.L. 2015. Non-proteinaceous bacterial adhesins challenge the antifouling properties of polymer brush coatings. Acta Biomater 17: 00285-00288.

Zolla, L., Lupidi, G., Marcheggiani, M., Falcioni, G. and Brunori, M. 1990. Encapsulation of proteins into human erythrocytes: a kinetic investigation. Biochim Biophys Acta 9: 5-9.

Zolla, L., Lupidi, G., Marcheggiani, M., Falcioni, G. and Brunori, M. 1991. Red blood cells as carriers for delivering of proteins. Ann Ist Super Sanita 27: 97-103.

3

CHAPTER

Engineering Porphyrin-Phospholipid Nanostructures for Medical Imaging and Therapy

Sophie L. Wang[a] and *Jonathan F. Lovell* *

INTRODUCTION

The recent development of porphyrins conjugated to phospholipid side chains has opened new possibilities for novel techniques in medical imaging and therapy applications. Porphyrin-phospholipids (PoPs) can self-assemble into nanovesicles such as liposomes or porphysomes which have intrinsic utility in theranostics (Carter et al. 2014; Huynh and Zheng 2014a; Komatsu et al. 2002; Lovell et al. 2011a). With the recent rise of nanomedical engineering (Luo et al. 2015), PoPs promise to be versatile tools for this purpose since they can form or coat other nanoscale structures in order to introduce additional capabilities (Ng et al. 2013; Rieffel et al. 2015a; Tam and McVeigh 2012). As such, PoPs can be a useful and straightforward tool to develop new types of advanced hybrid nanomaterials for medical applications.

Porphyrin-Phospholipids

PoPs are composed of a phospholipid backbone and single side-chain, along with a single esterified porphyrin group. Phospholipids are amphipathic organic molecules which have hydrophilic phosphate based "head" groups

Department of Biomedical Engineering, University at Buffalo, Buffalo, New York-14260, USA.
[a] E-mail: swang9@buffalo.edu
* Corresponding author: jflovell@buffalo.edu

and hydrophobic "tails" of fatty acid chains. When hydrated in an aqueous solution, they form bilayers wherein the tails are sequestered from the aqueous environment by the hydrophilic heads. As phospholipids are a major component of biological cell membranes, nanostructures comprised of phospholipids tend to show low toxicity and high biocompatibility (Akbarzadeh et al. 2013; Allen and Cullis 2013). Furthermore, alterations of phospholipid structure in chain lengths, molecular charge or conjugated subunits can change the membrane properties and pharmacokinetics of the resultant nanostructures (Chang et al. 1997; Ishida et al. 2002; Levchenko et al. 2002).

Porphyrins are macrocyclic organic molecules composed of four linked pyrrole subunits. The basic core of the porphyrin macrocycle has 22π electrons, which gives rise to strong photophysical properties and the central nitrogens of the pyrroles confer the ability to efficiently chelate metals (Chou et al. 2000; Drain et al. 2009; Huynh and Zheng 2014a). By modifying the pyrrole subunits or by chelating different metals, it is possible to tune the chemical and photophysical properties of the porphyrin molecule (Chandra et al. 2000). However, due to their aromatic ring structure, porphyrins can sometimes aggregate in aqueous solutions and thus may exhibit limited bioavailability or bioactivity as independent molecules (Huynh and Zheng 2014a; Komatsu et al. 2002). They are commonly found in nature incorporated within protein structures with notable examples being chlorophyll and heme, which hold vital roles in photosynthesis and oxygen transport (Zhang and Lovell 2012). Owing to their numerous inherent biological roles, porphyrins are usually biocompatible and highly biodegradable.

Efforts to improve porphyrin bioavailability have taken at least two pathways: incorporation into various delivery nanoparticles such as liposomes, or conjugation with more hydrophilic helper molecules (Bonnett 1995). As noted by Komatsu et al. the incorporation of porphyrins into nanoparticles such as liposomes faces the challenge of achieving a sufficiently high porphyrin load without destabilization of the particle (Huynh and Zheng 2014b; Komatsu et al. 2002). Many of these efforts have involved conjugating porphyrins to various molecules in order to improve particle incorporation and stability (Liang et al. 2011; Postigo et al. 2004; Tam and McVeigh 2012; Temizel et al. 2014).

Porphyrin-phospholipid (PoP) molecules were designed to improve porphyrin solubility while maintaining biocompatibility by conjugation to natural phospholipids (Lovell et al. 2012). This molecular addition provides an amphipathic characteristic to the molecule while maintaining the strong photophysical character of the porphyrin. Additionally, the phospholipid subunit can serve to improve the membrane uptake and loading capacity of the porphyrin in both liposomes and cells as the porphyrin does not prevent the phospholipid from self-assembling into lipid bilayers.

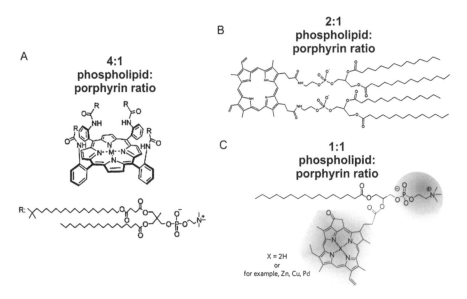

Fig. 3.1 Various porphyrin-lipid configurations. (A) Synthesized by Komatsu et al. 2002, the single porphyrin is conjugated to four phospholipids at the acyl chains. (B) Synthesized by Riske et al. 2009, the porphyrin is conjugated to two phospholipids at the head groups. (C) Synthesized by Lovell et al. 2011a, the porphyrin is conjugated to a single phospholipid at the glycerol backbone. Image used with permission from Huynh and Zheng 2014b.

As illustrated in Fig. 3.1A, Komatsu et al. conjugated the porphyrin to the phospholipid in a 4:1 phospholipid to porphyrin ratio with the porphyrin attached to the acyl chain of the phospholipid. They were able to demonstrate that the molecule could assemble into liposome-like nanovesicles with useful photophysical properties characteristic to porphyrins which included: strong fluorescence, extremely short triplet lifetime and a red-shifted Soret absorbance band (Komatsu et al. 2002).

Riske et al. synthesized porphyrin-phospholipids by conjugating one porphyrin to two phosphate head groups, resulting in a 2:1 phospholipid to porphyrin ratio (Fig. 3.1B). These molecules demonstrated the ability to form large stable micron-sized vesicles of bilayer membranes (Riske et al. 2009). Their aim was to investigate the effect of singlet oxygen generation from light irradiation upon bilayer shape.

Both of the molecules created by Komatsu et al. and Riske et al. required several phospholipid molecules per porphyrin – limiting the maximum levels and densities of the porphyrin incorporated into the lipid bilayers. In comparison, Lovell et al. developed a PoP where the porphyrin was conjugated to the phospholipid through an acylation reaction at the glycerol backbone (Fig. 3.1C). This meant that the porphyrin to phospholipid ratio was at 1:1, allowing an increase in both the maximum amount of porphyrin within a bilayer, as well as the packing density of the porphyrin within the bilayer. These molecules demonstrated the ability to self-assemble into liposome like nanovesicles called porphysomes and exhibited photophysical qualities characteristic to both liposomes as well as porphyrins. The porphysomes exhibited novel thermal and fluorescent qualities previously unknown to those structures from the high packing density of the porphyrin (Lovell et al. 2009, 2011a, 2011b, 2012) (Fig. 3.2).

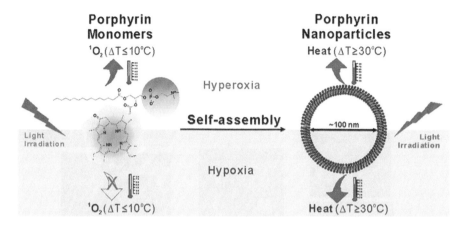

Fig. 3.2 Demonstration of structure dependent properties of PoPs. The monomers are fluorescent and exhibit photodynamic singlet oxygen generation with no photothermal qualities, whereas the PoPs self-assembled into porphysomes are FRET self-quenching and show photothermal heating. Image used with permission from Jin et al. 2013.

Porphyrin-Phospholipid Applications

So what are the desired applications of these PoP molecules? Coverage of the properties and characteristics may be useful, but the desirability of these qualities is difficult to understand without a concurrent understanding of the potential applications (Fig. 3.3).

Therapeutic Modalities

The first applications that may come to mind for molecules showing useful properties are therapeutic in nature. Given that PoPs show properties known to be useful in several therapeutic modalities, it would be worthwhile to cover how these therapeutic modalities work – especially in the context of PoPs.

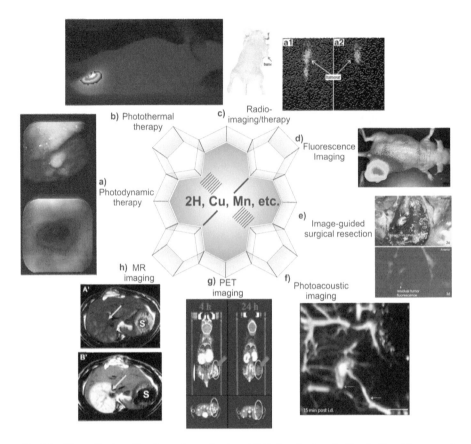

Fig. 3.3 Biomedical applications of porphyrins. (A) Before (top) and after (bottom) images of human esophageal cancer treated with photodynamic therapy. (B) Photothermal therapy of a xenograft tumor bearing mouse treated with porphysomes. (C) Radio images of melanoma tumors in mice at (a1) 8 hours and (a2) 24 hours. (D) Near-infrared fluorescent images utilizing laser activated porphysomes in KB xenograft mice. (E) Image-guided resection of rabbit brain tumor, surgical cavity (top) and a fluorescent image of surgical cavity showing tumor margins (bottom). (F) Photoacoustic images of lymph nodes in rats treated with intradermal injections of porphysomes. (G) Positron emission tomography images of tumors at 4 and 24 hours after copper-64 porphyrin injections. (H) Magnetic resonance images of infarcted liver lobe before (a) and after (b) Gadophrin-2 contrast agent injection. Image used with permission from Zhang and Lovell 2012.

Photodynamic Therapy (PDT)

The synergistic effect of photosensitizer administration concurrent with light exposure on cell death was observed by Raab and Von Tappeiner in the early 1900's. By testing the effects of the acridine red alone, light exposure alone, light exposed acridine and acridine with light exposure on paramecia, Raab was able to demonstrate that the highest levels of cell death occurred when the acridine was exposed to light and postulated that the effect was due to a product of fluorescence. Von Tappeiner later worked with Jodlbauer to elucidate the role of oxygen in the reaction and coined the term photodynamic actuation. The early therapeutic application to the treatment of skin cancer showed good results, though it was not until the late 1970's that PDT was systematically explored and categorized (Ackroyd et al. 2001; Berg et al. 2005; Roelandts 2002; Zhang and Lovell 2012). Porphyrins have since taken on a prominent role in photomedicine as PDT photosensitizers and metallic ion chelators (Huang et al. 2015).

The photodynamic reaction is initiated when a molecule of photosensitizer absorbs light. The photosensitizer's electrons are excited from the ground state to a long-lived triplet state. The triplet state can dissipate the energy through two types of reactions. In the type 1 reaction, the triplet state reacts to form radicals through hydrogen transfers – which in turn interact with oxygen to form reactive oxygen species (ROS). Alternatively, in the preferred type 2 reaction, the triplet state can directly interact with the molecular oxygen to form singlet oxygen radicals. Thus the components necessary for phototherapy are the photosensitizer, the light and the oxygen (Allison and Sibata 2010; Berg et al. 2005; Dolmans et al. 2003; Hopper 2000; Pandey et al. 2006).

In the 1970s and 1980s, heme-based porphyrin photosensitizers were developed, the best-known example being Photofrin, which remains a common clinically-used photosensitizer in cancer treatments. Next, other porphyrin derivatives and other chemicals were developed in the 1980s with an example being ALA, which induces endogenous expression of protoporphyrin IX and is commonly used in dermatological applications. Most recently, the complex targeting of next generation photosensitizers through conjugation with other molecules has been described. These conjugations allow for nanoparticle incorporation or other capabilities with the PoPs developed by Lovell et al. as an example (Huang 2005; Shao et al. 2015).

Another component in the application of PDT, the light source, is required to have wavelengths that match the excitation spectra of the photosensitizer and are able to deliver adequate power efficiently. The main classes are lasers, LEDs and filtered lamps. Diode lasers have excellent efficiencies and can easily be coupled with optical fibers for convenient internal light delivery and are single wavelength devices – so different photosensitizers will required different diode lasers. LEDs are monochromatic sources easily arranged into various irradiation geometries by changing the LED array configurations, they generate more heat than lasers do, but are also much less costly in comparison to diodes. Lamp systems can be filtered to match any desired wavelength and are flexible in terms of irradiation geometries. However, lamps are much less efficient than either LEDs or lasers and cannot be coupled with optical fibers for easy internal light delivery (Wilson and Patterson 2008).

While the role of oxygen in the photochemical reactions of PDT is known, the importance of tumor oxygenation to PDT effectiveness is still under investigation. Current research suggests that the type of photochemical reaction generated during PDT is dependent on oxygen and photosensitizer levels in the tumor. While well-oxygenated tumors generate more singlet oxygen, hypoxic tumors may be destroyed through nutrient and oxygen starvation from the microvascular damage (Ethirajan et al. 2011).

PDT is attractive as a clinical therapy as it is a minimally invasive repeatable treatment that preserves the biological function of the treated tissues (Jin et al. 2014). Its excellent cosmetic outcomes render it useful for the treatment of dermatological lesions, cancerous malignancies such as squamous cell carcinoma and melanoma, psoriasis and other cutaneous issues (Roelandts 2002). Other promising clinical applications of PDT have included treatments of: age-related macular degeneration; choroidal neovascular disease; oral, nasal, pharynx, larynx, intracranial, esophageal and gastroenterological tumors along with other mesothelial cancers; and cardiovascular disease (Brown et al. 2004; Ethirajan et al. 2011; Hopper 2000; Huang 2005).

Photothermal Therapy

Given that PDT, like radiotherapy, is more effective in highly oxygenated environments that can produce singlet oxygen free radicals, PDT is less effective in tumors with hypoxic and necrotic regions (Jin et al. 2013). With the development of inorganic nanoparticles came the development of photothermal therapy (PTT). The relationship between the particle size and optical qualities of metals was first studied by Faraday in 1857. Faraday demonstrated that a gold leaf on a flat surface that was heated until it decomposed into spherical particles changed in opacity and color, while the electrical conductivity of the metal remained the same (Faraday 1857).

Later research revealed that the colors exhibited by metallic nanoparticles were from the coherent optical resonances of their valence electrons (also known as surface plasmons). The resonance wavelength is dependent upon factors such as the size and shape of the particle, the type of metal and the local environment. As such, altering these factors allows the tuning of the optical characteristics for the nanoparticles: smaller particles are better optical absorbers and larger particles are better optical scattering agents (Lal et al. 2008). Nanoparticles with diameters of 100 nm are both strong absorbers and good scattering agents (Lal et al. 2008), which is useful in therapeutic applications as particles in the 100 nm range show the best retention characteristics. Due to the absorptive properties of these nanoparticles, laser illumination results in highly localized photothermal heating where the laser excited nanoparticles release energy in the form of heat. This effect has been utilized in cancer treatment as a noninvasive method to induce localized cell death through laser ablation (Gobin et al. 2007; Lal et al. 2008).

Despite the success of these inorganic nanoparticles in PTT, there are clinical concerns of long term toxicity due to particle bioaccumulation due to lack of biodegradability. Thus the development of organic nanoparticles with optical properties comparable to inorganic nanoparticles with high optical extinction coefficients and absorption is ideal (Huynh and Zheng 2014b). Of particular interest are porphysomes composed of PoPs self-assembled into liposome-like vesicles. The organic nature of porphysomes has been shown to be enzymatically degradable and it induced minimal acute toxicity at intravenous doses of 1000 milligrams per kilogram (Lovell et al. 2011a). The densely packed porphyrin groups in each particle are highly efficient at absorbing light and the density modulates the excited porphyrins to release the energy in the form of heat. The results from Jin et al. showed that the PTT treatment was effective regardless of oxygen presence within the tumors, while the PDT treatment was effective only in non-hypoxic tumors (Jin et al. 2013).

Radiotherapy

Radiotherapy uses ionizing radiation to cause damage to living tissues and is the most common treatment modality used to treat cancer. As human tissue contains 80% water, the radiation mostly generates aqueous free radicals such as OH, H, e_{aq}^-, HO_2, H_3O^+, which react with vital cellular macromolecules such as phospholipids in the membrane, DNA, RNA and proteins. The accumulation of damage from the reactions of the free radicles with the cells causes dysfunctions which trigger cell death. As the damage to tissue is non-selective, the best therapeutic results depend on balancing the dosage delivered to the treatment site and the dosage limit of the surrounding critical tissue (Nair et al. 2001).

Radiation sensitivity determines the dosage and is dependent on factors such as the cellular replication rate and the presence of oxygen. Cells that replicate quickly are more vulnerable to radiation, as the vital macromolecules involved in DNA replication are exposed to the destructive effects of radiation more often (Duggar 1936). The presence of oxygen in the cellular environment enhances the production of free radicals, increasing cellular sensitivity to radiation (Nair et al. 2001). Thus well-oxygenated tumors that are rapidly replicating as compared to the normal cells in their surroundings will be more vulnerable and responsive to radiation therapy (Gerweck et al. 2006). However, tumor cells often overgrow the abilities of their vascular supply, which produces regions of necrosis and hypoxia. These areas become less sensitive to radiation as compared to the surrounding normal tissue, which means that increasing the dose of radiation becomes counterproductive (Nair et al. 2001).

Exogenous agents may be used to increase the effectiveness of radiation – either by protecting the surrounding tissue (radio protectors) or by increasing the sensitivity of the tumor (radio sensitizers). Radio protectors work by scavenging the free radicles produced by radiation which protects the surrounding normal cells from free radicle damage. Radio sensitizers increase the free radicles produced by radiation and increase the free radicle cellular damage in tumors. As porphyrins were shown to accumulate in tumors during the development of PDT, porphyrins labeled with radionuclides and metalloporphyrin complexes have been used as radio sensitizers in targeted tumor radiotherapy to good effect (Ali and van Lier 1999; Huynh and Zheng 2014b; Jia et al. 2008; Zhang and Lovell 2012).

Imaging Modalities

In conjunction with therapeutic applications, imaging applications also serve important roles in the disease diagnosis and in the elucidation of disease mechanisms. Advancing and multimodal imaging modalities can help assess details of pathology, enabling more specific treatments with fewer unwanted side effects (Rieffel et al. 2015b).

Radiotracers

There is a long history of using radio-isotopes to trace the progress of molecules through biological processes in order to diagnose diseases and treat disorders. Radiotracers are radioactive species used to follow metabolic reactions or the systemic distribution and kinetics for molecules of interest and should contribute minimally to the background through non-target tissue interactions or uptake by excretion systems such as the renal or hepatic systems (Weissleder and Pittet 2008). Post tracer administration, auto-radiographic images can be recorded on film or phosphor-imaging devices for computer storage (Ruth 2008).

As compared to the traditional methods of radiotracers which are subject to instability and altered pharmacokinetics, PoPs show a marked improvement as they natively form stable high-affinity complexes with a variety of metallic ions – most notably with copper-64. Experiments in the 1980s demonstrated that the pharmacokinetics and biodistribution of porphyrins are not changed by the chelation of radioisotopes, as the chelated metals insert into the center of the porphyrin ring and thus do not interact with the side chains which determine the *in vivo* pharmacokinetics. Liu et al. developed a PoP chelated with copper-64 which produced highly stable radiolabeled nanoparticles with unprecedented specific activity (Liu et al. 2012). PoP can also be efficiently labeled with the SPECT tracer 99mTc while preserving its ability to form stable nanovesicles (Lee et al. 2014).

Positron emission tomography (PET) is a nuclear imaging technique that uses radiotracers to create three-dimensional distributions of radioactivity in the scanned object. The PET detectors take measurements of the photon pairs that are emitted, traveling in paths oriented at 180 degrees from each other, which are the result of the self-collimating nature of positron decay. The detected lines of radioactivity indicate where positrons decayed and mathematical algorithms use these lines of response to generate density maps that show the distribution of the positron emitters. However, the positron travels from its site of emission before it decays from a collision with an electron so the site of measured decay is not the exact site of positron emission, which limits the resolution when determining the origin of decay. Additionally, the positron is in motion when it decays through an electron collision so conservation of momentum dictates that the photons emitted are not perfectly collinear – causing measurement uncertainty over large imaging distances (Ruth 2008). The advantages of PET imaging are that it has excellent sensitivity and deep penetration with low background noise and is highly useful in analyzing biodistribution in living subjects. However, it tends to be low in resolution and is often partnered with other imaging modalities such as Computer Tomography (CT) and Magnetic Resonance (MR) for high resolution anatomical data that helps contextualize the molecular data from PET (Wang et al. 2014).

The copper-64 PoP radiotracer developed by Liu et al. was evaluated for PET/CT imaging with prostate tumors and showed good tumor specific uptake through clearly delineated images in both modalities with little background (Liu et al. 2012, 2013a). Another PET probe was developed by chelating copper-64 into a porphyrin-peptide-folate (PPF) probe, resulting in ^{64}Cu-PPF. The ^{64}Cu-PPF demonstrated high stability over 24 hours in saline and serum, good metabolic stability *in vivo* and folate-mediated tumor uptake based on folic acid blocking tests (Shi 2011). Further tests with ovarian cancer showed similarly good results of selective folate-mediated uptake by the tumors, with the benefit of highlighting the metastatic locations (Liu et al. 2013b).

Magnetic Resonance Imaging

Magnetic resonance (MR) is based on the principle that certain atomic nuclei placed in a strong magnetic field with pulsed radiofrequency energy will emit radio signals. In nuclear magnetic resonance (NMR) spectroscopy, the amplitude and frequency of the emitted signals can be used to determine the chemical composition of a given sample. When Dalmadian noticed that the spin-lattice relaxation time differed between normal and cancerous tissue, he suggested that the difference would be useful in identifying those tissues. Based on this, Lauterbur proposed a method that became the basis of creating images from NMR and the origin of MR imaging (Hendee and Morgan 1984).

The three factors involved in signal strength for NMR include the number of protons in the sample, the constant time it takes for the atom to realign with the field after being energized by the pulse or the spin-lattice relaxation time and the constant time it takes for the net magnetic moment to break down or the spin-spin relaxation time (Hendee and Morgan 1984) However, while the two time constants and the radio frequency emitted may be unique to a given atomic element, this is not the case with tissues which are composed of many different elements. Thus while MR imaging reveals soft tissue structures with clarity and sensitivity, it is relatively non-specific due to the majority of the signal being based on the protons of the water within the tissues.

Contrast agents that add specificity to the MR imaging can be useful in distinguishing the malignant from the benign soft tissue structures. Metalloporphyrins have been investigated as attractive contrast agents due to the selective uptake of porphyrins in malignant tumors and the porphyrins' ability to efficiently chelate metallic ions (paramagnetic ions in specific) which increase MR contrast (Furmanski and Longley 1988). The study by Furmanski et al. showed results useful for MR image enhancement with favorable qualities in terms of toxicity, solubility and stability (Furmanski and Longley 1988). Recently, porphysomes have been described with MR active properties (MacDonald et al. 2014).

Optical Imaging

Fluorophores are molecules that emit light of a longer wavelength, when excited by light themselves. This property was utilized in medicine since at least 1924 when tumors were observed to autofluoresce under UV light from the presence of endogenous porphyrins (Zhang and Lovell 2012). In 1942, tumors were observed to fluoresce red after the administration of porphyrins, which led to the use of fluorescein in brain tumor detection in 1948. By the 1980s, image reconstruction after passing light through tissue became possible after the observation that near-infrared photons between 650 and 900 nm suffered less scattering and traveled more efficiently than shorter wavelength photons. Concurrently, targeted fluorescent probes were developed for fluorescence imaging to improve specificity (Weissleder and Pittet 2008). Optical imaging is popular due to the simplicity of operation, ease of implementation and ability to provide real-time images during surgery (Liu et al. 2013a).

A few intrinsic limitations to fluorescence imaging include tissue background interference and depth penetration. As fluorescence imaging is based on light transmission, the ability to achieve the desired image is dependent on how much the tissue absorbs or scatters the excitation and emission, along with how much the tissue may autofluoresce and generate background signals (Leblond et al. 2010). Near-infrared fluorescence partially addresses these issues, as the tissue absorbance and scatter is reduced at near infrared wavelengths. Additionally, much tissue autofluorescence occurs when excited by visible light wavelengths and the use of near infrared reduces this (Frangioni 2003; Leblond et al. 2010).

Characteristics of successful fluorophores must include an excitation wavelength that can penetrate deeply enough to excite the fluorophore, brightness to allow for a strong signal generation, *in vivo* stability and suitable pharmacokinetics. If the tissue of interest is at the surface, it might not be essential to use near-infrared fluorophores, if visible light fluorophores will be better at eliciting fluorescence. Conversely, the use of visible light range fluorophores is not appropriate for imaging deeper tissues within the body, due to the limited penetration of light. Quantum dots are larger than antibodies and are bright, but have additional concerns of toxicity from bioaccumulation of their heavy metal components. Green fluorescent proteins are large and bright, but are immunogenic and difficult to conjugate due to their size. *In vivo* stability is necessary as fluorescent probes that are degraded too quickly will be difficult to work with, although stability can be unpredictable when fluorophores are conjugated to molecules for targeting or tracking. Lastly, conjugating fluorophores with other molecules may alter the pharmacokinetics of both molecules (Kobayashi et al. 2010).

Handily, porphyrins exhibit near-infrared fluorescence and have been used to differentiate normal tissues from tumors in imaging studies (Zhang and Lovell 2012). PoPs show similar near-infrared fluorescence, but also show structurally mediated fluorescence qualities. Free PoPs are fluorescent, while those that are incorporated into lipid bilayers in porphysomes are highly quenched. When porphysomes were injected, initial tumor imaging showed very low fluorescence. After 2 days, tumor fluorescence had become very high, indicating a restoration of PoP fluorescence from porphysomes degrading and releasing free porphyrins (Huynh and Zheng 2014a, 2014b; Lovell et al. 2011a).

Photoacoustic Imaging

The photoacoustic (PA) effect is the generation of acoustic waves from the absorption of light energy. This effect is attractive for imaging as it combines the advantages of optical imaging's contrast with ultrasound imaging's depth penetration and produces images of high contrast and resolution in large volumes of biological tissue. Using electromagnetic wavelengths that show good tissue penetrance, the optical scattering within the tissue is highly sensitive to irregularities and abnormalities. As PA imaging can detect photons absorbed by the tissue through the ultrasound signals generated from the absorption, it can take images at deeper tissue depths as compared to optical methods as which cannot detect deeply absorbed photons deep in the tissue (Xu and Wang 2006).

PA techniques have been used in numerous applications including to detect circulating cancer cells in blood vessels and in sentinel lymph nodes. There is potential for PA monitoring of drug delivery (Xia et al. 2015). Lovell et al. demonstrated that porphysomes can generate strong PA signals to visualize lymph nodes in mice over time (Lovell et al. 2011a), while Rieffel et al. demonstrated that the combination of PoPs with a nanoshell was sufficient to convey PA properties to the multimodal nanoparticle (Rieffel et al. 2015a). PoPs have also been used to create porphyrin shell microbubbles with intrinsic PA properties (Huynh and Zheng 2014b). Thus PoPs appear to be useful in adding modalities to nanoparticles without intrinsic PA properties.

Multimodal Theranostics

Theranostics is the combination of therapeutic and diagnostic functions within a single treatment meant to maximize treatment safety and efficiency and minimize adverse effects. The ability to monitor delivery, distribution, efficacy and kinetics increases the ability to make effective treatment choices. With the addition of targeting functionality, it is possible to achieve site-specific imaging and therapy which reduce unwanted side effects from non-specific drug interactions (Heidel and Davis 2011; Rai et al. 2010; Wang et al. 2014).

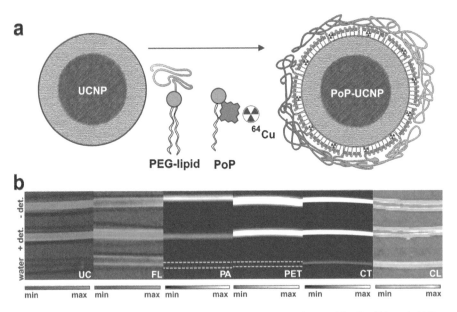

Fig. 3.4 PoP-coated up-conversion nanoparticle (UCNP) synthesized by Rieffel et al. 2015a. The UCNP has a core-shell structure made of NaYbF$_4$: Tm-NaYF$_4$ and was designed for near-infrared to near-infrared upconversion (UC) imaging wherein the particles are sequentially excited by multiple photons with wavelengths longer than the photon emitted. The UCNP is also suitable for imaging with computer tomography. The PoP coating confers near-infrared fluorescence and photoacoustic properties, as well as positron emission tomography and Cerenkov luminescence imaging abilities once incubated with copper-64 for post labeling. Image used with permission from Rieffel et al. 2015a.

Using nanoparticles as theranostic platforms can improve the circulation time of the conjugated therapeutic agent and selectively deliver the agent to targeted tissues through active or passive targeting. Once on site, the nanoparticles can control drug release rates through stimulus-triggered delivery systems and thus create site-specific drug concentrations (Heidel and Davis 2011; Jin et al. 2014; Lovell et al. 2010; Wang et al. 2014). Furthermore, it is possible to synthesize nanoparticles with multimodal imaging characteristics. The various imaging modalities have different advantages and disadvantages in terms of penetration, resolution, sensitivity and sensitivity. Combining imaging modalities compensates for the disadvantages in sensitivity or resolution and compounds the imaging information that results. A few common combinations that pair high sensitivity modalities with high-resolution modalities already in use include: PET with computed tomography, optical, or MRI; and optical imaging with CT or MR (Lee et al. 2012; Wang et al. 2014).

The original porphysomes demonstrated intrinsic multimodal qualities with structure-dependent fluorescence, photothermal and photoacoustic properties (Lovell et al. 2009, 2011a, 2011b). It has been demonstrated that PoP-doped liposomes were able to load therapeutic compounds such as doxorubicin and then were able to quickly release the cargo when triggered by near-infrared light to selectively accumulate in irradiated tumors (Carter et al. 2014; Kress et al. 2015; Luo et al. 2016). The usage of PoPs in multimodal nanoparticles by Rieffel et al. demonstrated the ability of PoPs to maintain the fluorescence and photoacoustic properties seen in the porphysomes, along with the simplicity of adding additional PET imaging capability through incubation with copper-64 due to the porphyrin intrinsic chelation capacity (Rieffel et al. 2015a) (Fig. 3.4). Jin et al. depended on the innate fluorescent qualities of the PoP to track the success of folate-targeting porphysomes concentrations (Jin et al. 2014).

Conclusion

PoPs are unique molecules that demonstrate many qualities useful for multimodal theranostics. In addition to the intrinsic characteristics of fluorescence and photosensitization (useful in imaging, PDT and PTT), the ability to chelate with metals opens the door to additional imaging modalities such as MR, PET and CT. The ability of PoPs to form unique structures and lipid bilayers allows for the development of new nanoscale materials that can include targeting moieties, while coating nanoparticles with bilayers can introduce additional theranostic qualities to the original nanoparticle.

Acknowledgements

This work was supported by the National Institutes of Health (R01EB017270 and DP5OD017898).

References

Ackroyd, R., Kelty, C., Brown, N. and Reed, M. 2001. The history of photodetection and photodynamic therapy. Photochem Photobiol 74: 656-669.

Akbarzadeh, A., Rezaei-Sadabady, R., Davaran, S., Joo, S.W., Zarghami, N., Hanifehpour, Y., et al. 2013. Liposome: classification, preparation and applications. Nanoscale Res Lett 8: 102.

Ali, H. and van Lier, J.E. 1999. Metal complexes as photo- and radiosensitizers. Chem Rev 99: 2379-2450.

Allen, T.M. and Cullis, P.R. 2013. Liposomal drug delivery systems: from concept to clinical applications. Adv Drug Deliv Rev 65: 36-48.

Allison, R.R. and Sibata, C.H. 2010. Oncologic photodynamic therapy photosensitizers: a clinical review. Photodiagnosis Photodyn Ther 7: 61-75.

Berg, K., Selbo, P.K., Weyergang, A., Dietze, A., Prasmickaite, L., Bonsted, A., et al. 2005. Porphyrin-related photosensitizers for cancer imaging and therapeutic applications. J Microsc 218: 133-147.

Bonnett, R. 1995. Photosensitizers of the porphyrin and phthalocyanine series for photodynamic therapy. Chem Soc Rev 24: 19.

Brown, S.B., Brown, E.A. and Walker, I. 2004. The present and future role of photodynamic therapy in cancer treatment. Lancet Oncol 5: 497-508.

Carter, K.A, Shao, S., Hoopes, M.I., Luo, D., Ahsan, B., Grigoryants, V.M., et al. 2014. Porphyrin-phospholipid liposomes permeabilized by near-infrared light. Nat Commun 5: 3546.

Chandra, R., Tiwari, M., Kaur, P., Sharma, M., Jain, R. and Dass, S. 2000. Metalloporphyrins—applications and clinical significance. Indian J Clin Biochem 15: 183-199.

Chang, C.W., Barber, L., Ouyang, C., Masin, D., Bally, M.B. and Madden, T.D. 1997. Plasma clearance, biodistribution and therapeutic properties of mitoxantrone encapsulated in conventional and sterically stabilized liposomes after intravenous administration in bdf1 mice. Br J Cancer 75: 169-177.

Chou, J.-H., Kosal, M.E., Nalwa, H.S., Rakow, N.A. and Suslick, K.S. 2000. The Porphyrin Handbook, Volume 6: Applications Past, Present and Future.

Dolmans, D.E.J.G.J., Fukumura, D. and Jain, R.K. 2003. Photodynamic therapy for cancer. Nat Rev Cancer 3: 380-387.

Drain, C.M., Varotto, A. and Radivojevic, I. 2009. Self-organized porphyrinic materials. Chem Rev 109: 1630-1658.

Duggar, B.M. 1936. Biological effects of radiation. Soil Sci 42: 399.

Ethirajan, M., Chen, Y., Joshi, P. and Pandey, R.K. 2011. The role of porphyrin chemistry in tumor imaging and photodynamic therapy. Chem Soc Rev 40: 340-362.

Faraday, M. 1857. The bakerian lecture: experimental relations of gold (and other metals) to light. Philos Trans R Soc London 147: 145-181.

Frangioni, J.V. 2003. In vivo near-infrared fluorescence imaging. Curr Opin Chem Biol 7: 626-634.

Furmanski, P. and Longley, C. 1988. Metalloporphyrin enhancement of magnetic resonance imaging of human tumor xenografts in nude mice. Cancer Res 48: 4604-4610.

Gerweck, L.E., Vijayappa, S., Kurimasa, A., Ogawa, K. and Chen, D.J. 2006. Tumor cell radiosensitivity is a major determinant of tumor response to radiation. Cancer Res 66: 8352-8355.

Gobin, A.M., Lee, M.H., Halas, N.J., James, W.D., Drezek, R.A. and West, J.L. 2007. Near-infrared resonant nanoshells for combined optical imaging and photothermal cancer therapy. Nano Lett 7: 1929-1934.

Heidel, J.D. and Davis, M.E. 2011. Clinical developments in nanotechnology for cancer therapy. Pharm Res 28: 187-199.

Hendee, W.R. and Morgan, C.J. 1984. Magnetic resonance imaging. part i–physical principles. West J Med 141: 491-500.

Hopper, C. 2000. Photodynamic therapy: a clinical reality in the treatment of cancer. Lancet Oncol 1: 212-219.

Huang, H., Song, W., Rieffel, J. and Lovell, J.F. 2015. Emerging applications of porphyrins in photomedicine. Front Phys 3: 1-15.

Huang, Z. 2005. A review of progress in clinical photodynamic therapy. Technol Cancer Res Treat 4: 283-293.

Huynh, E. and Zheng, G. 2014a. Porphysome nanotechnology: a paradigm shift in lipid-based supramolecular structures. Nano Today 9: 212-222.

Huynh, E. and Zheng, G. 2014b. Organic biophotonic nanoparticles: porphysomes and beyond. IEEE J Sel Top Quantum Electron 20: 27-34.

Ishida, T., Harashima, H. and Kiwada, H. 2002. Liposome clearance. Biosci Rep 22: 197-224.

Jia, Z.Y., Deng, H.F., Pu, M.F. and Luo, S.Z. 2008. Rhenium-188 labelled meso-tetrakis [3, 4-bis(carboxymethyleneoxy)phenyl] porphyrin for targeted radiotherapy: preliminary biological evaluation in mice. Eur J Nucl Med Mol Imaging 35: 734-742.

Jin, C.S., Lovell, J.F., Chen, J. and Zheng, G. 2013. Ablation of hypoxic tumors with dose-equivalent photothermal, but not photodynamic, therapy using a nanostructured porphyrin assembly. ACS Nano 7: 2541-2550.

Jin, C.S., Cui, L., Wang, F., Chen, J. and Zheng, G. 2014. Targeting-triggered porphysome nanostructure disruption for activatable photodynamic therapy. Adv Healthc Mater 3: 1240-1249.

Kobayashi, H., Ogawa, M., Alford, R., Choyke, P.L. and Urano, Y. 2010. New strategies for fluorescent probe design in medical diagnostic imaging. Chem Rev 110: 2620-2640.

Komatsu, T., Moritake, M., Nakagawa, A. and Tsuchida, E. 2002. Self-organized lipid-porphyrin bilayer membranes in vesicular form: nanostructure, photophysical properties and dioxygen coordination. Chem A Eur J 8: 5469-5480.

Kress, J., Rohrbach, D.J., Carter, K.A., Luo, D., Shao, S., Lele, S., et al. 2015. Quantitative imaging of light-triggered doxorubicin release. Biomed Opt Express 6: 3546-3555.

Lal, S., Clare, S.E. and Halas, N.J. 2008. Photothermal therapy: impending clinical impact. Acc Chem Res 41: 1842-1851.

Leblond, F., Davis, S.C., Valdés, P. A. and Pogue, B.W. 2010. Pre-clinical whole-body fluorescence imaging: review of instruments, methods and applications. J Photochem Photobiol B Biol 98: 77-94.

Lee, D.-E., Koo, H., Sun, I.-C., Ryu, J.H., Kim, K. and Kwon, I.C. 2012. Multifunctional nanoparticles for multimodal imaging and theragnosis. Chem Soc Rev 41: 2656.

Lee, J.-H., Shao, S., Cheng, K.T., Lovell, J.F. and Paik, C.H. 2014. ^{99m}Tc-labeled porphyrin–lipid nanovesicles. J Liposome Res 2104: 1-6.

Levchenko, T.S., Rammohan, R., Lukyanov, A.N., Whiteman, K.R. and Torchilin, V.P. 2002. Liposome clearance in mice: the effect of a separate and combined presence of surface charge and polymer coating. Int J Pharm 240: 95-102.

Liang, X., Li, X., Yue, X. and Dai, Z. 2011. Conjugation of porphyrin to nanohybrid cerasomes for photodynamic diagnosis and therapy of cancer. Angew Chemie Int Ed 50: 11622-11627.

Liu, T.W., MacDonald, T.D., Shi, J., Wilson, B.C. and Zheng, G. 2012. Intrinsically copper-64-labeled organic nanoparticles as radiotracers. Angew Chemie Int Ed 51: 13128-13131.

Liu, T.W., MacDonald, T.D., Jin, C.S., Gold, J.M., Bristow, R.G., Wilson, B.C., et al. 2013a. Inherently multimodal nanoparticle-driven tracking and real-time delineation of orthotopic prostate tumors and micrometastases. ACS Nano 7: 4221-4232.

Liu, T.W., Stewart, J.M., MacDonald, T.D., Chen, J., Clarke, B., Shi, J., et al. 2013b. Biologically-targeted detection of primary and micro-metastatic ovarian cancer. Theranostics 3: 420-427.

Lovell, J.F., Chen, J., Jarvi, M.T., Cao, W.G., Allen, A.D., Liu, Y., et al. 2009. Fret quenching of photosensitizer singlet oxygen generation. J Phys Chem B 113: 3203-3211.

Lovell, J.F., Liu, T.W.B., Chen, J. and Zheng, G. 2010. Activatable photosensitizers for imaging and therapy. Chem Rev 110: 2839-2857.

Lovell, J.F., Jin, C.S., Huynh, E., Jin, H., Kim, C., Rubinstein, J.L., et al. 2011a. Porphysome nanovesicles generated by porphyrin bilayers for use as multimodal biophotonic contrast agents. Nat Mater 10: 324-332.

Lovell, J.F., Chan, M.W., Qi, Q., Chen, J. and Zheng, G. 2011b. Porphyrin fret acceptors for apoptosis induction and monitoring. J Am Chem Soc 133: 18580-18582.

Lovell, J.F., Jin, C.S., Huynh, E., MacDonald, T.D., Cao, W. and Zheng, G. 2012. Enzymatic regioselection for the synthesis and biodegradation of porphysome nanovesicles. Angew Chemie Int Ed 51: 2429-2433.

Luo, D., Carter, K.A. and Lovell, J.F. 2015. Nanomedical engineering: shaping future nanomedicines. WIREs Nanomed Nanobiotechnol 7: 169-188.

Luo, D., Carter, K.A., Razi, A., Geng, J., Shao, S., Giraldo, D., et al. 2016. Doxorubicin encapsulated in stealth liposomes conferred with light-triggered drug release. Biomaterials 75: 193-202.

MacDonald, T.D., Liu, T.W. and Zheng, G. 2014. An mri-sensitive, non-photobleachable porphysome photothermal agent. Angew Chemie 126: 7076-7079.

Nair, C.K., Parida, D.K. and Nomura, T. 2001. Radioprotectors in radiotherapy. J Radiat Res 42: 21-37.

Ng, K.K., Lovell, J.F., Vedadi, A., Hajian, T. and Zheng, G. 2013. Self-assembled porphyrin nanodiscs with structure-dependent activation for phototherapy and photodiagnostic applications. ACS Nano 7: 3484-3490.

Pandey, R.K., Goswami, L.N., Chen, Y., Gryshuk, A., Missert, J.R., Oseroff, A., et al. 2006. Nature: a rich source for developing multifunctional agents. Tumor-imaging and photodynamic therapy. Lasers Surg Med 38: 445-467.

Postigo, F., Mora, M., De Madariaga, M.A., Nonell, S. and Sagristá, M.L. 2004. Incorporation of hydrophobic porphyrins into liposomes: characterization and structural requirements. Int J Pharm 278: 239-254.

Rai, P., Mallidi, S., Zheng, X., Rahmanzadeh, R., Mir, Y., Elrington, S., et al. 2010. Development and applications of photo-triggered theranostic agents. Adv Drug Deliv Rev 62: 1094-1124.

Rieffel, J., Chen, F., Kim, J., Chen, G., Shao, W., Shao, S., et al. 2015a. Hexamodal imaging with porphyrin-phospholipid-coated upconversion nanoparticles. Adv Mater 27: 1785-1790.

Rieffel, J., Chitgupi, U. and Lovell, J.F. 2015b. Recent advances in higher-order, multimodal, biomedical imaging agents. Small 11: 4445-4461.

Riske, K.A., Sudbrack, T.P., Archilha, N.L., Uchoa, A.F., Schroder, A.P., Marques, C.M., et al. 2009. Giant vesicles under oxidative stress induced by a membrane-anchored photosensitizer. Biophys J 97: 1362-1370.

Roelandts, R. 2002. The history of phototherapy: something new under the sun?. J Am Acad Dermatol 46: 926-930.

Ruth, T.J. 2008. The uses of radiotracers in the life sciences. Reports Prog Phys 72: 016701.

Shao, S., Geng, J., Ah Yi, H., Gogia, S., Neelamegham, S., Jacobs, A., et al. 2015. Functionalization of cobalt porphyrin–phospholipid bilayers with his-tagged ligands and antigens. Nat Chem 7: 438-446.

Shi, J. 2011. Transforming a targeted porphyrin theranostic agent into a pet imaging probe for cancer. Theranostics 363.

Tam, N. and McVeigh, P. 2012. Porphyrin–lipid stabilized gold nanoparticles for surface enhanced Raman scattering based imaging. Bioconjug Chem 23: 1726-1730.

Temizel, E., Sagir, T., Ayan, E., Isik, S. and Ozturk, R. 2014. Delivery of lipophilic porphyrin by liposome vehicles: preparation and photodynamic therapy activity against cancer cell lines. Photodiagnosis Photodyn Ther 11: 537-545.

Wang, D., Lin, B. and Ai, H. 2014. Theranostic nanoparticles for cancer and cardiovascular applications. Pharm Res 31: 1390-1406.

Weissleder, R. and Pittet, M.J. 2008. Imaging in the era of molecular oncology. Nature 452: 580-589.

Wilson, B.C. and Patterson, M.S. 2008. The physics, biophysics and technology of photodynamic therapy. Phys Med Biol 53: R61-R109.

Xia, J., Kim, C. and Lovell, J.F. 2015. Opportunities for photoacoustic-guided drug delivery. Curr Drug Targets 16: 571-581.

Xu, M. and Wang, L.V. 2006. Photoacoustic imaging in biomedicine. Rev Sci Instrum 77: 041101.

Zhang, Y. and Lovell, J.F. 2012. Porphyrins as theranostic agents from prehistoric to modern times. Theranostics 2: 905-915.

4

Epitaxial Growth of Heterostructured Nanoparticles for Biomedical Applications

Sixiang Shi,[1,a] *Shreya Goel,*[1,b] *Feng Chen*[2,]* and *Weibo Cai*[1,2,3,4,]*

INTRODUCTION

With the development of nanotechnology, inorganic nanoparticles have attracted tremendous interest as agents for biomedical diagnosis and therapy due to their unique properties. In the past decade, numerous inorganic nanoparticles, such as quantum dots (QDs) (Reiss et al. 2009; Zhu et al. 2013), upconversion nanoparticles (UCNPs) (Chen et al. 2013; Chen et al. 2015; Wang et al. 2013a; Zhou et al. 2012), superparamagnetic iron oxide nanoparticles (SPIONs) (Gobbo et al. 2015; Santhosh and Ulrih 2013; Singh and Sahoo 2014) and gold nanoparticles (Cai et al. 2008; Dreaden et al. 2012), have been employed for different biomedical applications. Among these inorganic nanomaterials, heterostructured nanoparticles which contain two or more components have been increasingly explored and are expected to play increasingly important roles in preclinical/clinical research in the future (He and Lin 2015; Sailor and Park 2012). Different hybrid nanostructures including core/shell, heterodimer and dumbbell structures have been designed and applied for biomedical purposes. As compared to conventional inorganic nanoparticles, the incorporation of second (or multiple) components offers heterostructured nanoparticles enhanced properties and enriched

---

[1] Materials Science Program, University of Wisconsin – Madison, WI, USA.
[a] E-mail: sshi9@wisc.edu
[b] E-mail: goel6@wisc.edu
[2] Department of Radiology, University of Wisconsin – Madison, WI, USA.
[3] Department of Medical Physics, University of Wisconsin – Madison, WI, USA.
[4] University of Wisconsin Carbone Cancer Center, Madison, WI, USA.
* Corresponding authors: chenf@mskcc.org and wcai@uwhealth.org

functionalities. For example, the growth of a shell or multiple shells onto QDs have successfully enhanced their fluorescence quantum yield and reduced their cytotoxicity (Ghosh Chaudhuri and Paria 2012). Similarly, the core/shell structure has also been widely studied to enhance emission efficiency and tune optical properties of UCNPs (Chen et al. 2015). In addition, dumbbell structured Au-Fe_3O_4 nanohybrids have been designed for simultaneous optical imaging and magnetic resonance imaging (MRI) (Yu et al. 2005).

The heterostructured nanoparticles can be generated either by non-epitaxial or epitaxial techniques. In the non-epitaxial approach, the heterostructures are prepared by polycrystalline or amorphous growth process without matching of the lattice, covalent conjugation of surface functional groups, or physical absorption. One example of non-epitaxial surface coating is the silica surface engineering, which has been widely used to modify QDs, Fe_3O_4, MnO_2 and other nanoparticles in order to improve their biocompatibility and solubility (Fang et al. 2013; Shi et al. 2013). Non-epitaxial growth is generally considered as a facile approach which can be conducted under mild conditions and low temperatures. On the other hand, epitaxial growth is a technique by which new crystalline grows layer by layer on a single crystal substrate where the lattice of the growing layer mimic the arrangement of the substrate (Stringfellow 1982). This technique impacts heterostructured nanoparticles with tunable properties, as well as with a more precise control of the size, shape, chemical composition and functionalities, thereby becoming one of the promising methods to synthesize noble inorganic materials. The common epitaxial growth methods for bulk materials and thin films include vapor-phase epitaxy (VPE), liquid-phase epitaxy (LPE), solid-phase epitaxy (SPE) and molecular beam-epitaxy (MBE) (Stringfellow 1982). However, due to the nanoscale size and solubility requirement for biomedical applications, heterostructured nanoparticles are usually produced by solution-phase chemical synthesis. In this chapter, we will introduce the synthesis and applications of several heterostructures grown through the epitaxial process, including core/shell QDs, core/shell UCNPs and Au-based heterostructures.

Core/Shell QDs

Quantum dots are nanocrystals made of semiconductor materials with a few hundred or a few thousand atoms (Reiss et al. 2009). Due to the small size of the QDs (typically less than 10 nm), quantum confinement effect can be observed as increasing band gaps with discrete energy levels, leading to unique electronic and optical properties. The optical property of QDs is size-dependent. In brief, the smaller size results in blue-shift emission spectrum, while the larger size results in red-shift emission spectrum (Reiss et al.

2009). As compared to organic fluorophores, QDs possess several advantages such as continuous absorption spectra spanning UV to near-infrared (NIR), narrow emission spectra (typically 20-30 nm full width at half maximum), large effective Stokes shifts (> 200 nm), high quantum yields and strong resistance to photobleaching and chemical degradation (Cai and Chen 2007). These ideal optical properties make QDs one of the best candidates for single molecule tracking. In addition, due to their ultrasmall size, QDs can be used as renally-cleared molecular imaging systems, which result in reduced side-effects and enhanced imaging contrast, making clinical translation possible. Therefore, biomedical imaging with QDs has become one of the major focuses of research in the biomedical realm.

The cores of QDs are generally prepared by two methods: the organometallic route and aqueous-phase route. In the organometallic route, QD cores were typically synthesized in a solvent with a high boiling point and high coordination capability to both metal and chalcogen elements, such as trioctylphosphine oxide (TOPO) (Qu et al. 2001; Smith et al. 2006; Talapin et al. 2001). The QDs prepared by this method have nearly perfect crystal structures, resulting in high quantum yields. In addition, narrow size distribution is another advantage. On the other hand, several aqueous-phase synthesis methods have also been developed, such as hydrothermal synthesis, ultrasonic methods and microwave-assisted synthesis (Law et al. 2009; Obonyo et al. 2010). The QDs prepared by the aqueous phase method are of improved simplicity and reduced toxicity. However, lower quantum yields and broader size distribution are the disadvantages of this method.

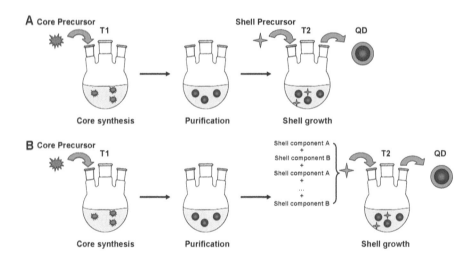

Fig. 4.1 Schematic illustration of the synthesis of core/shell QDs.

The synthesis of the QD cores is followed by the epitaxial growth of the shells, which can be accomplished by two methods: chemical bath deposition (CBD) and successive ionic layer adsorption and reaction (SILAR). In the CBD method, shell precursors are slowly injected into the reaction mixture via syringe pump and then epitaxially deposited on the core, followed by a purification step (Ghosh Chaudhuri and Paria 2012; Reiss et al. 2009) (Fig. 4.1A). To prevent self-nucleation of the shell precursors and uncontrolled ripening of the core, the reaction temperature for shell growth (T2) should be lower than that of core growth (T1). SILAR method, on the other hand, is based on the alternating injections of two shell components into the solution to grow a monolayer of the shell (Ghosh Chaudhuri and Paria 2012). The alternating injections can be repeated several times to form the heterostructured QDs with different thicknesses (Fig. 4.1B). As compared to the CBD method, SILAR offers better growth control, since it prevents independent nucleation of the shell material and favors isotropic and uniform growth, due to the low precursor concentration in the media throughout the reaction (Wang et al. 2014b). For both methods, the growth of shell is significantly important in order to increase the fluorescence efficiency and stability against photobleaching by improving the surface passivation. In addition, the core/shell structure can also assist in tuning the optical properties and reducing the toxicity of QDs. The synthesis and applications of several types of core/shell QDs are exemplified in this section.

Cd-Based Core/Shell QDs

Among core/shell QDs, CdSe/ZnS heterostructure is one of the most commonly studied nanosystems, which was first reported in 1996 (Hines and Guyot-Sionnest 1996). The CdSe core was synthesized in TOPO solvent with Cd/Se/TOP (tri-n-octylphosphine) stock solution as the precursor at 350°C. Subsequently, the ZnS shell was grown by injecting Zn/S/TOP solution into the same reaction pot at 300°C. As-prepared CdSe/ZnS QDs exhibited strong and stable luminescence with a 50% quantum yield at room temperature. In the following work, the shell was grown in TOPO solvent with Diethylzinc (ZnEt$_2$) and hexamethyldisilathiane ((TMS)$_2$S) as the precursor at different temperatures (ranging from 140°C to 220°C), which resulted in different shell thicknesses (Dabbousi et al. 1997). The influence of the thickness on the fluorescence spectrum was investigated, suggesting a red shift when the thickness increased. Enhanced quantum yield (50%) was also demonstrated after shell coating in comparison with that of the bare core (30%). Due to the excellent properties of CdSe/ZnS, a great number of studies have been conducted to examine their applications in optical imaging (Chen et al. 2008; Liu et al. 2007; SalmanOgli 2011; Zhang et al. 2012).

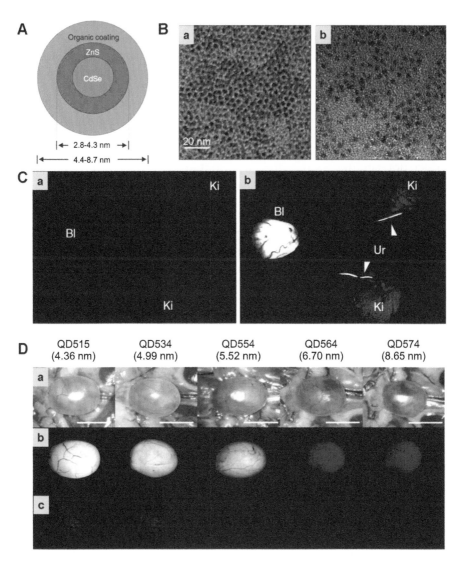

Fig. 4.2 Design, characterization and optical imaging of renal clearable CdSe/ZnS QDs. (A) Chemical compositions of CdSe/ZnS QDs with organic coatings. (B) TEM pictures of QD515 (a) with the size of 2.85 nm and QD574 (b) with the size of 4.31 nm. Scale bar, 20 nm. (C) *In vivo* fluorescence imaging of intravenously injected CdSe/ZnS QDs. Kidneys (Ki), ureters (Ur; arrowheads) and bladder (Bl) either T = 0 (a) or T = 2 h (b) after intravenous injection of QD515 into the rat. Scale bar, 1 cm. (D) Surgically exposed CD-1 mouse bladders after intravenous injection of QD515, QD534, QD554, QD564 or QD574 of defined hydrodynamic diameter. Shown are color video (a) and fluorescence images (c) for uninjected control bladder and 4 h after injection (b) for each quantum dot. Reproduced with the permission from (Choi et al. 2007).

In order to improve the *in vivo* properties, ultrasmall CdSe/ZnS QDs have been developed for renal-clearable optical imaging (Choi et al. 2007). The cores were grown in the mixture of hexadecylamine (HDA) and TOPO/TOP solvent at a lower temperature (280°C). To epitaxially grow the shell, precursor solutions of $ZnEt_2$ and $(TMS)_2S$ were added into a degassed solution of TOPO and n-hexylphosphonic acid (HPA) at 130°C and annealed overnight at 80°C. The resulting QDs had diameters ranging from 4.36 nm to 8.65 nm and emission spectra from 515 nm to 574 nm (Fig. 4.2A and 2B). After intravenous injection into healthy rats, QD515 exhibited kidney and bladder uptake after 2 h post-injection, indicating a rapid renal clearance (Fig. 4.2C). In addition, QDs of different sizes were also injected in the mice to investigate the influence of size on renal clearance (Fig. 4.2D). Stronger signal in bladder was observed after injection with smaller-size QDs, suggesting that a final hydrodynamic diameter < 5.5 nm results in rapid and efficient urinary excretion and elimination of quantum dots from the body. This result has not only provided a promising nanoplatform for *in vivo* diagnosis, but has also defined an important criterion for designing renal clearable nanoparticles for the future studies.

Although optical imaging possesses numerous advantages, such as high sensitivity, low cost and short acquisition time, QDs and other optical imaging probes still face a significant challenge for *in vivo* applications due to their limited tissue penetration. To overcome this drawback, novel nanosystems that do not require excitation light have attracted great attention in the research community. In a recent study, a self-illuminating QD system was constructed by doping a positron-emitting radionuclide ^{64}Cu into CdSe/ZnS core/shell QDs (QD580) via a cation-exchange reaction (Sun et al. 2014). The doping process can be accomplished simply by directly mixing ^{64}Cu and QDs in ammonium acetate buffer (NH_4Ac, pH 5.5) at 60°C. As-prepared ^{64}Cu-doped QDs can be used for both positron emission tomography (PET) and optical imaging via Cerenkov resonance energy transfer (CRET) without the need of external excitation. As compared to the physical mixture of ^{64}Cu and QDs, ^{64}Cu-doped QDs exhibited stronger CRET. Subsequently, ^{64}Cu-doped QD580, which exhibited the strongest luminescence intensity, was employed for *in vivo* dual-modality imaging. Decent tumor uptake in a U87MG glioblastoma xenograft model was observed by both PET and luminescence imaging, suggesting excellent *in vivo* properties of ^{64}Cu-doped QDs.

CdSe/CdS is another commonly used core/shell nanosystem, which was first prepared by the CBD method (Peng et al. 1997). In brief, CdSe stock solution [$Cd(CH_3)_2$ in Se powder dissolved tributylphosphine (TBP)] was injected into hot TOPO (360°C) and reacted at 300°C for CdSe core synthesis. Then the CdS stock solution [$(TMS)_2S$ in $Cd(CH_3)_2$ dissolved TBP/toluene solution] was injected into the same reaction pot for CdS shell growth. Three monolayers were successfully grown on cores ranging in diameter from 2.3

nm to 3.9 nm. A quantum yield of 50% was achieved in this nanosystem. Besides the CBD method, SILAR has also been used to epitaxially grow CdS shells (Li et al. 2003; van Embden et al. 2009). In one study, 1-octadecene (ODE) was selected as the noncoordinating solvent for both core nanocrystals and the two injection solutions of shell precursors (Li et al. 2003). To synthesize the CdSe core, Se dissolved TBP was injected into CdO dissolved ODE/TOPO/octadecylamine (ODA) solution at 280°C and reacted at 250°C. Then two injection solutions (cadmium solution: CdO dissolved in ODE/oleic acid (OA) solution; sulfur solution: sulfur dissolved in ODE solution) were successively injected into the CdSe solution. The reaction temperature was varied between 120°C and 260°C. The average thickness of each monolayer of CdS was determined to be about 0.35 nm, so the growth of an additional layer increased the diameter by 0.7 nm, endowing superior size control. Similar methods have also been used to epitaxially synthesize core/shell structures with other chemical compositions, such as PbSe/PbS QDs (Lifshitz et al. 2006).

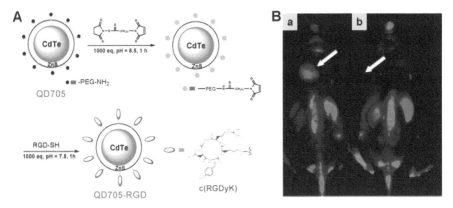

Fig. 4.3   RGD-labeled QD705 for imaging tumor vasculature in living mice. (A) Synthesis of QD705-RGD, PEG denotes poly(ethylene glycol) (MW 2000). (B) *In vivo* NIR fluorescence imaging of U87MG tumor-bearing mice (left shoulder, pointed by white arrows) injected with 200 pmol of QD705-RGD (a) and QD705 (b), respectively. The mice autofluorescence is color coded green while the unmixed QD signal is color coded red. Reproduced with the permission from (Cai et al. 2006).

Another core/shell QD system, red or NIR emitting CdTe nanocrystal, has been synthesized to overcome the limitation of tissue penetration, making it more suitable for *in vivo* biomedical imaging. The synthesis of CdTe nanocrystals was achieved by adding freshly prepared oxygen-free NaHTe solution to N_2-saturated $Cd(ClO_4)_2 \cdot 6H_2O$ solution at pH 11.2 in the presence of thiols as stabilizing agents (Rogach et al. 1997). However, the CdTe nanocrystals prepared in aqueous solution were cytotoxic, unstable and easily photobleachable without the protection of a shell. To grow the shells, the CdTe cores were phase-transferred to dodecanethiol/toluene and injected into TBP/TOPO solution. $ZnEt_2/(TMS)_2S/TOPO$ solution was used as the shell precursor (Tsay et al. 2004). By coating ZnS shells, highly luminescent and stable heterostructured CdTe/ZnS QDs were prepared with a high quantum yield of over 50% and far-red emission region (670–700 nm). Due to the long emission wavelengths, CdTe/ZnS QDs have been widely used for tumor imaging *in vivo* (Cai et al. 2006; Ding et al. 2011; Liu et al. 2012; Meng et al. 2011; Pang et al. 2014). In one study, QD705 was modified with polyethylene glycol (PEG) to improve its *in vivo* stability and conjugated with arginine-glycine-aspartic acid (RGD) peptide for tumor vasculature targeting (Fig. 4.3A) (Cai et al. 2006). After intravenous injection of QD705-RGD nanohybrids, enhanced tumor uptake was achieved based on active targeting, highlighting the potential of QD-based tumor diagnosis (Fig. 4.3B).

Cd-Free Core/Shell QDs

Although Cd-containing QDs possess excellent optical properties and have been widely used in optical imaging and other biomedical applications, the potential toxicity of Cd^{2+} is a common concern which limits their clinical translation. A number of studies have found Cd-containing QDs to be highly toxic, although the toxicity of the QD nanostructures may depend on the size, charge, surface modification, concentration and oxidative condition (Hardman 2006; Hoshino et al. 2011; Pelley et al. 2009; Soenen et al. 2013). To address the safety concerns, Cd-free QD systems have become an important research direction (Soenen et al. 2014). InP nanocrystals with zinc blende structure were first synthesized with indium oxalate and $P(SiMe_3)_3$ solutions as the precursors (Micic et al. 1994). The reaction was conducted in TOPO/TOP system by the CBD method, resulting in well-crystallized QDs (diameter ~2.5 nm). However, as-prepared InP nanocrystals suffered from low quantum yield (1%) and broad emission spectrum. To improve its optical properties, a ZnS shell was epitaxially grown onto the InP core. Similar to the Cd-based QDs, ZnS coating was realized in TOPO/TOP system with $ZnEt_2/(TMS)_2S$ as the precursor. As-prepared core/shell InP/ZnS QDs were found to have a significantly enhanced quantum yield (23%) (Haubold et al. 2001). Later on, the quantum yield was further enhanced in many studies (Li et al. 2008; Li and Reiss 2008; Xie et al. 2007; Xu et al. 2008b). Due to the relatively lower toxicity, InP/ZnS core/shell QDs have been applied in optical imaging for disease diagnosis (Chin et al. 2010; Gao et al. 2012; Yong et al. 2009). After introducing Ln(III) ions into the crystal lattice, InP/ZnS could also be used for multimodality imaging, which combines the advantages of different imaging techniques (Rosenberg et al. 2010; Stasiuk et al. 2011). Besides InP/ZnS, several other core/shell QDs have also been developed by similar synthesis techniques for biomedical applications, such as InAs/InP/ZnSe (Gao et al. 2010a; Gao et al. 2010b), ZnSe/ZnS (Martynenko et al. 2015;

Shu et al. 2013) and $CuInS_2/ZnS$ (Guo et al. 2015; Le Ngoc and Kim 2013) heterostructured nanosystems. These studies on Cd-free heterostructured QDs might potentially benefit their clinical translation, by reducing the toxicity and conducting appropriate surface coating.

Core/Shell UCNPs

UCNPs constitute a new generation of lanthanide-doped luminescent probes, which convert low energy continuous-wave NIR light into high-energy light through the sequential absorption of two or multiple photons or energy transfers (Chen et al. 2015). These novel nanoparticles possess outstanding properties, such as sharp emission bandwidth, tunable emission spectrum, large anti-Stokes shift, long excited-state lifetime and high resistance to photobleaching (Chen et al. 2013; Chen et al. 2015). More importantly, both excitation and emission spectra of UCNPs are in the range of NIR windows, which leads to absence of autofluorescence and reduced side-effects for living subjects, making them especially suitable for *in vivo* optical imaging (Zhou et al. 2012). Numerous studies have been conducted to investigate their applications in cell tracking, tumor imaging and lymphatic imaging (Zhou et al. 2012). Of note, unlike other luminescent probes, the upconversion properties are not associated with the nanoparticle size, since the spectral conversion of UCNPs is attributed to the electronic transitions within the 4f orbitals of special lanthanide ions rather than quantum confinement (Bunzli 2010). To the contrary, their upconversion properties are highly dependent on the chemical composition and surface states (Chen et al. 2015). Therefore, the concept of core/shell structure has been raised to improve the surface states and boost the upconversion efficiency. In addition, core/shell structure has also enriched the functionalities of UCNPs, inspiring a wave of excellent designs to apply them in multimodality imaging and cancer theranostics.

A typical core/shell UCNP comprises three components: host, activator and sensitizer. The host should typically have low energy photons (< 500 cm^{-1}) to minimize non-radiative losses and maximize radioactive emission and must be able to accommodate a high concentration of lanthanide dopant ions. Rare earth contained halides, oxides, oxysulfides, phosphates and vanadates can be the host candidates. Many host materials have been developed to build core/shell UCNPs, such as $NaYF_4$, $NaYbF_4$, $NaGdF_4$, $NaLaF_4$, LaF_3, GdF_3, GdOF, La_2O_3, Lu_2O_3, Y_2O_3 and Y_2O_2S (Zhou et al. 2012). Among them, fluorides are considered to be one of the best candidates due to the low photon energies and high chemical stability and have been widely used for biomedical applications. The activators are usually dopant ions such as Er^{3+}, Tm^{3+} and Ho^{3+}, which can generate upconversion luminescence (UCL) emission under NIR excitation. Due to a larger absorption cross-section at around 980 nm, Yb^{3+} is the most commonly used sensitizer to enhance the UCL efficiency.

Synthesis of Core/Shell UCNPs

To synthesis monodispersed UCNPs, hydrothermal reaction and thermal decomposition are two of the most popular methods. Hydrothermal reaction is performed by adding rare-earth precursors and fluoride precursors in an aqueous solution and then sealing and heating them in an autoclave. Liquid, solid and solution (LSS) strategy, which is based on phase transfer and separation occurring at the interfaces of the ethanol-linoleic acid mixture (liquid), metal linoleate (solid) and water-ethanol solution containing metal ions (solution) at different designated temperatures, was first reported in 2005 (Wang et al. 2005). Briefly, aqueous solutions of $Ln(NO_3)_3$ (Ln = La, Ce, Pr, Nd, Sm, Eu, Gd, Th, Dy, Ho, Er, Tm, Yb and Y) and NaF/or NH_4HF_2 were added into sodium linoleate, linoleate acid and ethanol mixture solution and then transferred to an autoclave for the hydrothermal reaction at 100-200°C. The final products were found to disperse well in a nonpolar solvent and then deposited in a polar solvent by adding ethanol in the reaction pot (Wang and Li 2006; Wang et al. 2006). Following this method, a great many UCNPs have been developed, such as $NaYF_4$ (Chen et al. 2011a; Zhang et al. 2009; Zhang et al. 2007a), LaF_3 (Hu et al. 2008), YF_3 (Yan and Li 2005), GdF_3 (Xianping et al. 2006), etc. LSS-based hydrothermal reaction is compatible with a wide variety of inorganic precursors, making it a versatile approach to synthesize monodisperse UCNPs. However, this method usually takes a prolonged reaction time, resulting in relatively large nanoparticles.

On the other hand, thermal decomposition is based on the thermolysis of metal trifluoroacetates (TFA) in OA/OM/ODE system at elevated temperatures. In one study, precursors $Na(CF_3COO)$ and $RE(CF_3COO)_3$ (RE = Pr to Lu, Y) were added into OA/OM/ODE solution (Mai et al. 2006). After removing water and oxygen, the reaction was carried out by heating to 250-330°C under an Ar atmosphere to generate $NaREF_4$ UCNPs. Following this method, size and shape control was further improved (Boyer et al. 2007; Boyer et al. 2006; Yi and Chow 2006) and many types of UCNPs have also been generated, such as $NaYF_4$ (Chen et al. 2010; Ye et al. 2010; Zhang et al. 2011), $NaGdF_4$ (Bogdan et al. 2010), $NaYbF_4$ (Zhan et al. 2011), $LiYF_4$ (Du et al. 2009; Mahalingam et al. 2009), etc.

The core/shell structure can be prepared by a similar chemical reaction based on two strategies: hot-injection method and heat-up method. Hot-injection method was carried out by injecting shell precursors into a proceeding reaction (Fig. 4.4A). One example is the epitaxial growth of $NaYF_4$: Yb, Er@$NaYF_4$ (or $NaYF_4$: Yb, Tm@$NaYF_4$) (Yi and Chow 2007). After the preparation of $NaYF_4$: Yb, Er (or $NaYF_4$: Yb, Tm) cores, shell precursor containing NaTFA and YTFA in OM solution was directly injected into the same reaction pot to grow a $NaYF_4$ shell layer. The hot-injection method allows one-pot synthesis of multiple shells through successive injection of shell precursors, which is its most important advantage (Chen et al. 2015). Unlike the hot-injection method, the heat-up method was carried out by synthesizing the cores and shells in separate reactions (Fig. 4.4B). In brief, after the synthesis of UCNP cores, they are transferred to a fresh reaction pot and mixed with shell precursors to epitaxially grow shell layers following an identical protocol for the core synthesis (Mai et al. 2007; Qian and Zhang 2008). As an advantage, the same bath of UCNP cores can be used for coating different shell structures via the heat-up method, which offers great flexibility for the further design of their properties and functionalities. Of note, the size and shape can be easily tuned during shell coating by changing the amount of precursors and the reaction conditions (Fig. 4.5) (Wang et al. 2014a). In addition, the epitaxial growth of core/shell heterostructures has been found to successfully enhance the emission efficiency, easily tune the optical properties and increase the functional multiplicity. Detailed information about the effect of core/shell structures on upconversion properties can be found in the excellent cited review (Chen et al. 2015).

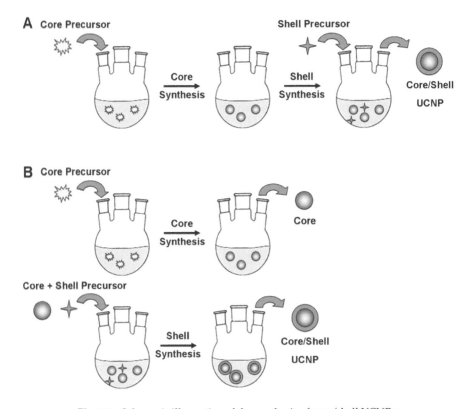

Fig. 4.4 Schematic illustration of the synthesis of core/shell UCNPs.

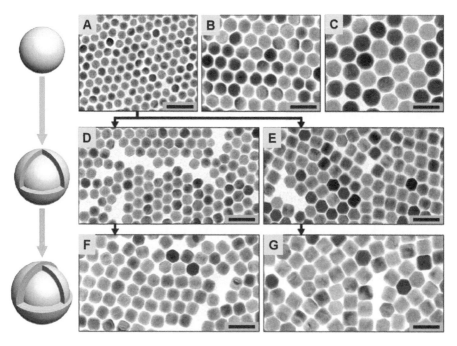

Fig. 4.5 Size-tuning of the NaGdF₄-based UCNPs. (A-C) TEM images of NaGdF₄: Yb, Tm nanocrystals. (D, E) TEM images of NaGdF₄: Yb, Tm@NaGdF₄: Tb core/shell UCNPs. (F, G) TEM images of NaGdF₄: Yb, Tm@NaGdF₄: Tb@NaGdF₄ core/shell/shell UCNPs. Scale bars, 50 nm. Reproduced with the permission from (Wang et al. 2014a).

In Vivo **Multimodality Imaging**

Imaging techniques, including optical imaging, MRI, X-ray computed tomography (CT) and radionuclide imaging, have become increasingly important in early and accurate diagnosis, therapeutic monitoring and post-therapy assessment. However, no single modality is perfect and sufficient enough to obtain all the necessary information for a particular question. For example, optical imaging has a high sensitivity and short acquisition time, whereas poor tissue penetration property confines its usage to animal research and only a few clinical applications. CT has excellent tissue penetration, but low detection sensitivity and soft tissue resolution. MRI has a high resolution for soft tissues, but it is limited by the high cost, low sensitivity and long acquisition time. Radionuclide imaging is sensitive and highly quantitative without tissue penetration drawbacks, but it fails to render anatomic information. Therefore, the combination of several imaging modalities into a single nanoparticle is warranted for more accurate imaging and diagnosis.

Among inorganic nanoparticles, core/shell UCNPs are one of the best candidates for multimodality imaging. To introduce MRI imaging modality into the construct, Gd^{3+} ions were doped into the core and/or shell of UCNPs. In one study, Gd^{3+} ions were doped by adding Gd_2O_3 into both core and shell precursors via hot-injection thermal decomposition reaction in OA/OM/ODE system (Park et al. 2009). As-prepared core/shell UCNPs, $NaYF_4$: Yb, Er@$NaYF_4$, exhibited strong T_1-weighted MRI signal. A similar approach was also reported to generate $NaYF_4$: Er, Yb@$NaGdF_4$ core/shell structures (Guo et al. 2010). To improve the r_1 relaxivity, a core/shell/shell structure $NaYF_4$: Er, Yb, Gd@ $NaYF_4$: Tm, Yb@ $NaGdF_4$ was prepared by seed-mediated re-growth process, resulting in significantly high r_1 value (6.18 mM^{-1s-1} per Gd^{3+} ion) among other reported counterparts by decreasing the $NaGdF_4$ shell thickness down to < 1 nm (Chen et al. 2011b). Besides MRI, due to heavy metal ions like Gd^{3+}, Yb^{3+} and Lu^{3+} in $NaYF_4$ matrix, core/shell UCNPs can be also employed for CT, which displays a high degree of spatial resolution of the hard tissues. In addition, by introducing radioisotopes, UCNPs can be also used for radionuclide imaging. For example, rapid ^{18}F-labeling was accomplished by incubating [18]F- with UCNPs via a specific inorganic reaction between [18]F- and rare-earth ions for positron emission tomography (PET) (Liu et al. 2011a; Liu et al. 2011b; Sun et al. 2011). In addition, the doping of [153]Sm into $NaLuF_4$: Yb, Tm@$NaGdF_4$ was reported in another study for single-photon emission computed tomography (Sun et al. 2013).

To investigate the *in vivo* properties of core/shell UCNPs, the above mentioned radioactive $NaLuF_4$: Yb, Tm@$NaGdF_4$([153]Sm) was used for UCL/MRI/CT/SPECT four-modality imaging (Fig. 4.6A) (Sun et al. 2013). Based on the images, detailed information on biodistribution was achieved at the whole-body level. However, without sufficient surface coating, the majority of the UCNPs were retained by the liver and spleen, suggesting the significantly important role of surface engineering for *in vivo* applications. Passive tumor targeting was also achieved in tumor-bearing mice (Fig. 4.6B). In order to improve the solubility and biocompatibility of core/shell UCNPs, surface modification is essential. In one study, PEGylated $NaYF_4$: Yb, Er/Tm@ $NaGdF_4$ was proven to be water-soluble with low toxicity (Xiao et al. 2012). As-prepared UCNPs were then conjugated with radiopaque tantalum oxide (TaO_x, x ≈ 1) for enhanced CT contrast. Importantly, the UCL signal was not affected, since TaO_x is fluorescence-transparent. High intensity/contrast CT/MRI/UCL trimodal imaging *in vitro* and *in vivo* was achieved simultaneously without detectable imaging interference among the three modalities. In another study, $NaGbF_4$: Yb, Er@$NaGdF_4$ was covalently conjugated to methylphosphonate functionalized silica nanospheres (pSi) and modified with Pluronic F127 copolymer for UCL/MRI/CT imaging, showing excellent stability and relatively low toxicity (Liu et al. 2013a).

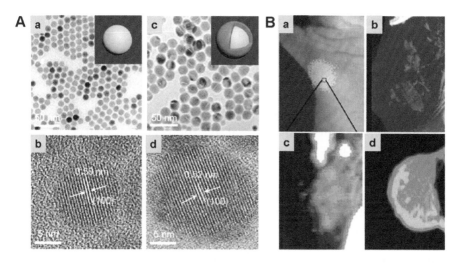

Fig. 4.6 Four-modality *in vivo* imaging using core/shell lanthanide UCNPs. (A) TEM image (a) and enlarged TEM image (b) of the NaLuF$_4$: Yb, Tm. TEM image (c) and enlarged TEM image (d) of the NaLuF$_4$: Yb, Tm@NaGdF$_4$(^{153}Sm). (B) UCL (a), CT (b), SPECT (c) and MRI (d) four-modality imaging of the focused tumor from the tumor-bearing nude mouse 1 h after intravenous injection of NaLuF$_4$: Yb, Tm@NaGdF$_4$(^{153}Sm). Reproduced with the permission from (Sun et al. 2013).

In order to further enhance the *in vivo* imaging properties, multi-shell heterostructured UCNPs have also been designed and tested *in vivo*. One example is NaYF$_4$: Yb, Tm@NaLuF$_4$@NaYF$_4$@NaGdF$_4$ (Shen et al. 2013). The core NaYF$_4$: Yb, Tm provides strong UCL, while the shell of NaLuF$_4$ is epitaxially grown on the core not only to provide an optically inert layer for enhancing the UCL but also to serve as a contrast agent for CT. The outermost NaGdF$_4$ shell is grown as a thin layer to render high r_1 desired for MR imaging. The transition shell layer of NaYF$_4$ functions as an interface to facilitate the formation of NaGdF$_4$ shell and also inhibits the energy transfer from inner upconversion activator to the surface paramagnetic Gd^{3+} ions. As prepared multi-shell UCNPs were then modified with Poly-(acrylic acid) (PAA) as the stabilizing agents and folic acid (FA) as the targeting ligand for *in vivo* tumor targeting and UCL/MRI/CT imaging, demonstrating great potential to serve as a multimodality imaging probe that combines the advantages of different imaging techniques.

Photodynamic and Photothermal Therapy

Photodynamic therapy (PDT) has emerged as an alternative approach to chemotherapy and radiotherapy to treat various diseases including cancer. PDT utilizes tissue oxygen and photosensitizers (PS) that are excited by light energy to generate highly cytotoxic singlet oxygen ($^1O_2$) and other reactive oxygen species (ROS) for killing cancer cells. However, most of the conventional PS agents are excited by visible light, which limits their applications only to a few cancer types, such as skin or breast cancers. In this regard, core/shell UCNPs are especially advantageous for PDT due to their NIR excitation/emission spectra. Appropriate loading of PS molecules onto UCNPs is important for effective resonance energy transfer from donors to acceptors in NIR-induced PDT. The loading methods include silica encapsulation, non-covalently physical adsorption and covalent conjugation via chemical linkages. The details of loading PS onto UCNPs can be found in the cited excellent review (Wang et al. 2013a). Furthermore, the selection of suitable PS agents is also important. The first study on UCNP-based PDT was accomplished by encapsulating Merocyanine-540 (M540) into a silica matrix (Chen et al. 2013; Zhang et al. 2007b). However, the loading capacity of M540 was limited due to the electrostatic repulsion between negatively charged M540 and the tetraethyl orthosilicate (TEOS) precursor. In the following studies, many other PS agents which can be excited under 980 nm NIR light were utilized for UCNP-based PDT, such as Zinc phthalocyanine (ZnPc) (Xia et al. 2014), tetrasubstituted carboxy aluminum phthalocyanine (AlC$_4$Pc) (Zhao et al. 2012) and 5-aminolevulinic acid (ALA) (Punjabi et al. 2014). Noteworthy among them is ALA, which is FDA-approved and already in clinical use. In this study, ALA was covalently conjugated to the surface of NaYF$_4$: Yb, Er@NaYF$_4$ core/shell UCNPs. The as-prepared PDT system exhibited strong singlet oxygen generation and ~70% cell death within 20 min of NIR irradiation. Reduced growth of tumor was also observed from *in vivo* studies. Although 980-excited NIR PS molecules exhibited excellent PDT efficiency, a huge concern has been raised since 980 nm NIR has significant water absorption that might overheat the surrounding tissue and kill non-cancer cells. To address this problem, 808 nm NIR-excited Chlorin e6 (Ce6) has been tested for UCNP-based PDT (Ai et al. 2015). At around 800 nm, water absorption is minimal and limited overheating effects are observed. The Nd^{3+} → Yb^{3+} → activator energy transfer in NaYbF$_4$: Nd@NaGdF$_4$: Yb, Er@NaGdF$_4$ core/shell/shell structure enables excitation at a shorter wavelength of 808 nm (Wang et al. 2013b). Besides organic PS molecules, TiO$_2$ nanoparticles have also been decorated on the surface of core/shell UCNPs for *in vivo* PDT study (Hou et al. 2015).

Different from PDT, photothermal therapy (PTT) is another type of phototherapy that utilizes the heat generated from NIR absorbing agents to kill cancer cells. Very few studies have investigated the PTT properties of core/shell UCNPs, partially due to their suboptimal photothermal conversion efficiency. To enhance the photothermal effect, polydopamine (PDA) was coated onto NaGdF$_4$: Yb, Er@NaGdF$_4$ UCNPs (Liu et al. 2015a). The PDA shell can be also used to improve the *in vivo* stability and serves as a carrier to load anticancer drugs, such as doxorubicin. Based on the results of *in vivo* studies, significantly reduced tumor growth was achieved by both PTT and PTT combined chemotherapy.

Au-Based Heterostructures

Au nanoparticles (AuNPs) have been applied to various biomedical applications due to their excellent properties. Similar to other inorganic nanoparticles, the nanoscale size allows AuNPs to preferentially accumulate at the tumor sites via enhanced permeability and retention (EPR) effects, making them an inherent passive targeting agent. Furthermore, synthesis and surface modification of AuNPs involves facile chemistry, which can be employed to conjugate various biomolecules for the targeted cancer diagnosis and therapy (Giljohann et al. 2009; Javier et al. 2008; Kim et al. 2010; Lee et al. 2009; Rink et al. 2010). The most important feature of AuNPs is the localized surface plasmon resonance (LSPR) (Xia et al. 2011). At a certain excitation frequency, the incident light couples onto the plasma resonance of the conduction electron resulting in a strong collective oscillation on the surface of AuNPs. Through LSPR, AuNPs can strongly absorb the irradiation light energy and convert it into heat, endowing these nanomaterials with outstanding photothermal properties (Nie and Chen 2012). By appropriate control of size and shape, the AuNPs can be easily tuned to absorb wavelengths in the NIR window. Therefore, PTT has become one of the most promising and most widely studied applications of AuNPs. Besides PTT, AuNPs can be also used for photoacoustic imaging and CT, owing to the efficient light-heat conversion capability and absorption of copious amounts of X-ray, respectively (Dreaden et al. 2012).

AuNPs are generally synthesized by the reduction of an Au-containing solution (e.g., $HAuCl_4$) using different reducing agents in the presence of surface-capping stabilizers (Cai et al. 2008). The most commonly used reducing agent is citrate, which was first reported in 1951 to form monodispersed Au nanospheres with the size ranging from 5 nm to 200 nm in aqueous phase (Turkevich et al. 1951). Besides citrate, other reducing agents such as sodium borohydride ($NaBH_4$), or diborane (B_2H_6) can be also used depending on the polarity of the solvent (Leifert et al. 2013). The stabilizers are very important in preventing aggregations of AuNPs during synthesis. The common stabilizers comprise carboxylates, amines, phosphines, thiols and polymers (Leifert et al. 2013). Among these stabilizers, bromide or chloride hexadecyl ammonium (CTAB or CTAC) are the most common ones generally used in the phase-transfer synthesis and shape-controlled seed-mediated growth. By changing the ratio of stabilizers to Au precursors/reducing agents, the size of AuNPs can be easily tuned. A larger amount of stabilizers results in smaller particle size and vice versa (Cai et al. 2008).

Au-Ag Core/Shell Structure

In addition to single component, binary noble metallic hybrid nanoparticles have also attracted great attention due to their enhanced properties and enriched functionalities. Many biometallic hybrid nanoparticles, including Au-Ag, Au-Al, Al-Ga, Au-Pd and Au-Pt, have been developed for various applications in biomedicine, catalysts, electronics, energy and optics (Fan et al. 2008). Among them, Au-Ag core/shell nanostructure has been applied for near-infrared photothermal therapy due to its enhanced LSPR (Bai et al. 2014; Gong et al. 2012). In one study, core/shell Au@Ag nanoparticles were prepared by the sequential reduction of $AuCl_4^-$ and Ag^+ ions with the use of citrate ions (Wu et al. 2013). Under this strategy, the Au core was synthesized by adding trisodium citrate solution into $HAuCl_4$ solution in water phase at boiling point. After 20 min, $AgNO_3$ solution and trisodium citrate solution were added into the same reaction pot for Ag shell growth. As-prepared nanoparticles had a size of about 30 nm (20 nm core and 5 nm-thick shell). Au@Ag nanoparticles were further conjugated with aptamers to target A549 lung adenocarcinoma cells with high affinity and specificity. Under irradiation of 808 nm NIR laser, Au@Ag nanoparticles exhibited excellent photothermal obliteration of A549 cells at a very low irradiation power density (0.20 W cm^{-2}) without any adverse effects on the healthy cells and the surrounding normal tissues. In a following study, another Au@Ag/Au structure was also conjugated with aptamers for PTT of A549 lung adenocarcinoma cells (Shi et al. 2014). This Au@Ag/Au structure was prepared by coating Au nanorod seeds with an Ag layer via reaction between $AgNO_3$ and ascorbic acid and then reacting with $HAuCl_4$ to yield an Au layer outside (Fig. 4.7A). Of note, the Au@Ag/Au nanostructures not only function as an excellent PTT agent, but also serve as a quenching agent for activatable cancer imaging. Upon intravenous injection of Au@Ag/Au nanostructures, a strong fluorescent signal and an obvious lesion were observed in A549 tumors, which could not be detected in the control groups (Fig. 4.7B).

In a recent study, a novel hollow nanocage Au-Ag@Au has been reported for enhanced PTT (Jiang et al. 2015). Au-Ag nanocages were synthesized by using Ag nanocubes as the sacrificial template through the galvanic replacement reaction. The Ag nanocube cores were prepared by sulfide-mediated polyol method via a reaction between CF_3COOAg and NaHS in ethylene glycol solution with poly (vinyl pyrrolidone) (PVP) as the stabilizer. To increase the chemical stability of nanocages against etching, a layer of Au was deposited on the surface by reaction with $HAuCl_4$ in PVP solution. The Au-Ag@Au nanocages were then formed by reacting with $HAuCl_4$ in an aqueous solution containing NaOH, PVP and L-ascorbic acid. By changing the amount of the precursors, as-prepared Au-Ag@Au nanocages exhibited excellent shape control (Fig. 4.7C). Irradiated by 808-nm laser, high photothermal transduction efficiency was achieved. Furthermore, the hollow structure can serve as a carrier for drug loading and delivery.

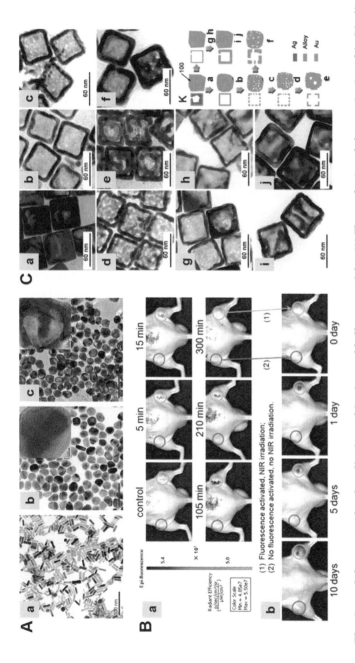

Fig. 4.7   The size and shape control and applications of Au–Ag core/shell nanoparticles. (A) The synthesis and characterization of Au@Ag/Au NPs with TEM images of the Au nanorods (a), Au@Ag NPs (b) and Au@Ag/Au NPs (c). Scale bar: 100 nm (insets: TEM images with higher magnification. Scale bar: 5 nm). (B) *In vitro* activatable fluorescence imaging (a) and the guided site-specific NIR PTT (b) of the tumors through an intravenous injection of aptamer-conjugated Au@Ag NPs (Left circle: SPCAi tumor as the control group; right circle: A549 tumor as the positive group). (C) TEM images of Ag nanocubes after reaction with different volumes of 0.24 mM HAuCl₄: 3 mL (a), 6 mL (b), 8 mL (c), 10 mL (d) and 12 mL (e). TEM image of Au–Ag nanocages (f) corresponding to (c) deposited by Au atoms (3.2 × 10⁻⁴ mmol). TEM images of Au–Ag nanobox by incubating (a) with 260 μL H₂O₂ (g), 380 μL H₂O₂ (h), 260 μL H₂O₂ and the deposition of Au atoms (1.6 × 10⁻⁴ mmol) (i), 380 μL H₂O₂ and the deposition of Au atoms (3.2 × 10⁻⁴ mmol) (j). Schematic illustrations (k) show the reaction processes corresponding to each TEM image. Reproduced with the permission from (Jiang et al. 2015; Shi et al. 2014).

Dumbbell-Like Au-Fe$_3$O$_4$

To enrich the functionality of AuNPs, Fe$_3$O$_4$ nanoparticles which can introduce MRI properties are perhaps the best candidates. Many studies have been conducted to integrate AuNPs and Fe$_3$O$_4$ nanomaterials via non-epitaxial approach with small molecules, polymers and silica coating as the connection (Dong et al. 2011; Hu et al. 2013; Jin 2014; Li et al. 2015; Li et al. 2014; Smolensky et al. 2011). Via the epitaxial approach, a novel design of dumbbell-like Au-Fe$_3$O$_4$ nanoparticles was first reported in 2005, by the decomposition of iron pentacarbonyl, Fe(CO)$_5$ precursor, over the surface of the Au nanoparticles (Yu et al. 2005). The Au seeds were generated by a reducing reaction with HAuCl$_4$ similar to the methods mentioned above and then were reacted with Fe(CO)$_5$ in ODE/OA/oleylamine solution at 300°C followed by room-temperature oxidation under air (Fig. 4.8A). The size of Au nanoparticles was tuned by controlling the temperature at which the HAuCl$_4$ was injected, or by controlling the HAuCl$_4$/oleylamine ratio. The size of the Fe$_3$O$_4$ nanoparticles was controlled by adjusting the ratio between Fe(CO)$_5$ and Au (Fig. 4.8B). By controlling the size of Au and Fe$_3$O$_4$ components, the magnetic and optical properties of dumbbell-like Au-Fe$_3$O$_4$ could be readily optimized. In addition, the presence of Au and Fe$_3$O$_4$ surfaces allows the attachment of different chemical functionalities for target-specific imaging and delivery applications (Xu et al. 2008a). Due to these distinct advantages of dumbbell-like Au-Fe$_3$O$_4$ over single component and conventional core/shell structures, dumbbell-like Au-Fe$_3$O$_4$ provides a promising integrated theranostic nanoplatform.

Follow-up studies have variously investigated the *in vitro* and *in vivo* applications of dumbbell-like Au-Fe$_3$O$_4$ nanoparticles. In one study, the Au particle was protected with HS-PEG-NH$_2$ (M$_r$ = 2204) with the thiol moiety attaching to Au to improve its stability and biocompatibility. In the same nanosystem, the Fe$_3$O$_4$ surfaces were functionalized with epidermal growth factor receptor antibody (EGFRA) through polyethylene glycol (PEG, Mr = 3000) linker and dopamine for active tumor cell targeting (Xu et al. 2008a). As a result, as-prepared nanoconjugates exhibited excellent stability in phosphate buffered saline (PBS) and higher specificity in their attachment to EGFR-positive A431 cells than control cells. *In vitro* MRI property was also tested in this study. The same PEGylated dumbbell-like Au-Fe$_3$O$_4$ nanosystem was modified with Her2-specific monoclonal antibody Herceptin and PDT agent cisplatin for target-specific platin delivery to Her2-positive breast cancer cells in a separate study (Xu et al. 2009). In a recent study, *in vivo* dual-modality CT and MRI were accomplished with PEGylated dumbbell-like Au-Fe$_3$O$_4$ in a rabbit model, showing its feasibility for real *in vivo* applications (Fig. 4.8C and 8D) (Zhu et al. 2014). Furthermore, radioactive isotope [99mTc] and several fluorescent species have also been used to functionalize dumbbell-like Au-Fe$_3$O$_4$ for radionuclide imaging and optical imaging in different studies (Felber and Alberto 2015; Liu et al. 2015b; Liu et al. 2013b).

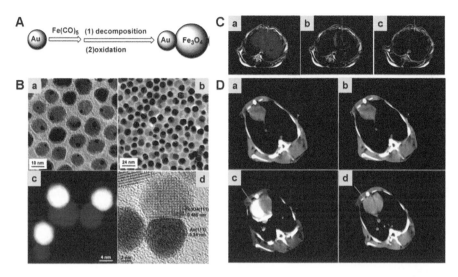

Fig. 4.8 The synthesis and applications of dumbbell-like Au-Fe₃O₄ nanoparticles. (A) Schematic illustration of the synthesis of dumbbell-like Au-Fe₃O₄ nanoparticles. (B) TEM image of the 3-14 nm (a) and 9-14 nm (b) Au-Fe₃O₄, HAADF-STEM image of the 8-9 nm Au-Fe₃O₄ (c) and HRTEM image of one 8-12 nm Au-Fe₃O₄ (d). (C) T_2-weighted MR images of the rabbit liver following ear vein injection of Au-Fe₃O₄ obtained at 0 min (a), 5 min (b) and 10 min (c). (D) *In vivo* CT images of a rabbit heart before injection (a), 5 s (b), 25 s (c) and 45 s (d) after intravenous injection of Au-Fe₃O₄. Reproduced with the permission from (Yu et al. 2005; Zhu et al. 2014).

Conclusion

Heterostructured nanoparticles that combine two or more components offer enhanced chemical, optical, photothermal and magnetic properties as compared to conventional single-component nanoparticles, making them ideal nanoplatforms for biomedical applications especially in cancer diagnosis and therapy. To establish controllable and straightforward approaches towards the synthesis of these highly desirable heterostructures, great efforts have been made to epitaxially grow the surface coatings or attach hetero-nanoparticles. Furthermore, numerous studies have been conducted to explore the *in vivo* applications of such multicomponent systems. With the protection of coating shells, the surface states of the nanoparticles can be significantly improved, thereby leading to higher fluorescent efficiency. In addition, the shells also endow them with tunable morphology and reduced toxicity. By incorporating magnetic species, fluorescent dyes, radionuclides and photosensitizers in the core or shell components, we can further enrich the functionalities of heterostructured nanoparticles, resulting in multifunctional theranostic agents, which combine the advantages of different diagnosis/therapy techniques into a single platform and can potentially reform the regimes of cancer management in the near future.

Acknowledgements

This work is supported, in part, by the University of Wisconsin – Madison, the National Institutes of Health (NIBIB/NCI 1R01CA169365 & P30CA014520) and the American Cancer Society (125246-RSG-13-099-01-CCE).

References

Ai, F., Ju, Q., Zhang, X., Chen, X., Wang, F. and Zhu, G. 2015. A core-shell-shell nano-platform upconverting near-infrared light at 808 nm for luminescence imaging and photodynamic therapy of cancer. Sci Rep 5: 10785.

Bai, T., Sun, J., Che, R., Xu, L., Yin, C., Guo, Z., et al. 2014. Controllable preparation of core-shell Au-Ag nanoshuttles with improved refractive index sensitivity and SERS activity. ACS Appl Mater Interfaces 6: 3331-3340.

Bogdan, N., Vetrone, F., Roy, R. and Capobianco, J.A. 2010. Carbohydrate-coated lan-thanide-doped upconverting nanoparticles for lectin recognition. J Mater Chem 20: 7543-7550.

Boyer, J.C., Vetrone, F., Cuccia, L.A. and Capobianco, J.A. 2006. Synthesis of colloidal upconverting NaYF4 nanocrystals doped with Er3+, Yb3+ and Tm3+, Yb3+ via thermal decomposition of lanthanide trifluoroacetate precursors. J Am Chem Soc 128: 7444-7445.

Boyer, J.C., Cuccia, L.A. and Capobianco, J.A. 2007. Synthesis of colloidal upconvert-ing NaYF4: Er3+/Yb3+ and Tm3+/Yb3+ monodisperse nanocrystals. Nano Lett 7: 847-852.

Bunzli, J.C. 2010. Lanthanide luminescence for biomedical analyses and imaging. Chem Rev 110: 2729-2755.

Cai, W. and Chen, X. 2007. Nanoplatforms for targeted molecular imaging in living subjects. Small 3: 1840-1854.

Cai, W., Shin, D.W., Chen, K., Gheysens, O., Cao, Q., Wang, S.X., et al. 2006. Peptide-labeled near-infrared quantum dots for imaging tumor vasculature in living subjects. Nano Lett 6: 669-676.

Cai, W., Gao, T., Hong, H. and Sun, J. 2008. Applications of gold nanoparticles in can-cer nanotechnology. Nanotechnol Sci Appl 1: 17-32.

Chen, D., Huang, P., Yu, Y., Huang, F., Yang, A. and Wang, Y. 2011a. Dopant-induced phase transition: a new strategy of synthesizing hexagonal upconversion NaYF4 at low temperature. Chem Commun (Camb) 47: 5801-5803.

Chen, F., Bu, W., Zhang, S., Liu, X., Liu, J., Xing, H., et al. 2011b. Positive and negative lattice shielding effects co-existing in Gd (III) ion doped bifunctional upconver-sion nanoprobes. Adv Funct Mater 21: 4285-4294.

Chen, F., Bu, W., Cai, W. and Shi, J. 2013. Functionalized upconversion nanoparticles: versatile nanoplatforms for translational research. Curr Mol Med 13: 1613-1632.

Chen, G., Ohulchanskyy, T.Y., Kumar, R., Agren, H. and Prasad, P.N. 2010. Ultrasmall monodisperse NaYF(4): Yb(3+)/Tm(3+) nanocrystals with enhanced near-infra-red to near-infrared upconversion photoluminescence. ACS Nano 4: 3163-3168.

Chen, L.D., Liu, J., Yu, X.F., He, M., Pei, X.F., Tang, Z.Y., et al. 2008. The biocompat-ibility of quantum dot probes used for the targeted imaging of hepatocellular carcinoma metastasis. Biomaterials 29: 4170-4176.

Chen, X., Peng, D., Ju, Q. and Wang, F. 2015. Photon upconversion in core-shell nanoparticles. Chem Soc Rev 44: 1318-1330.

Chin, P.T., Buckle, T., Aguirre de Miguel, A., Meskers, S.C., Janssen, R.A. and van Leeuwen, F.W. 2010. Dual-emissive quantum dots for multispectral intraoperative fluorescence imaging. Biomaterials 31: 6823-6832.

Choi, H.S., Liu, W., Misra, P., Tanaka, E., Zimmer, J.P., Itty Ipe, B., et al. 2007. Renal clearance of quantum dots. Nat Biotechnol 25: 1165-1170.

Dabbousi, B.O., Rodriguez-Viejo, J., Mikulec, F.V., Heine, J.R., Mattoussi, H., Ober, R., et al. 1997. (CdSe)ZnS core–shell quantum dots: synthesis and characterization of a size series of highly luminescent nanocrystallites. J Phys Chem B 101: 9463-9475.

Ding, H., Yong, K.T., Law, W.C., Roy, I., Hu, R., Wu, F., et al. 2011. Non-invasive tumor detection in small animals using novel functional pluronic nanomicelles conjugated with anti-mesothelin antibody. Nanoscale 3: 1813-1822.

Dong, W., Li, Y., Niu, D., Ma, Z., Gu, J., Chen, Y., et al. 2011. Facile synthesis of monodisperse superparamagnetic Fe3O4 Core@hybrid@Au shell nanocomposite for bimodal imaging and photothermal therapy. Adv Mater 23: 5392-5397.

Dreaden, E.C., Alkilany, A.M., Huang, X., Murphy, C.J. and El-Sayed, M.A. 2012. The golden age: gold nanoparticles for biomedicine. Chem Soc Rev 41: 2740-2779.

Du, Y.P., Zhang, Y.W., Sun, L.D. and Yan, C.H. 2009. Optically active uniform potassium and lithium rare earth fluoride nanocrystals derived from metal trifluroacetate precursors. Dalton Trans 8574-8581.

Fan, F.R., Liu, D.Y., Wu, Y.F., Duan, S., Xie, Z.X., Jiang, Z.Y., et al. 2008. Epitaxial growth of heterogeneous metal nanocrystals: from gold nano-octahedra to palladium and silver nanocubes. J Am Chem Soc 130: 6949-6951.

Fan, X., Pi, D., Wang, F., Qiu, J. and Wang, M. 2006. Hydrothermal synthesis and luminescence behavior of lanthanide-doped GdF/sub 3/nanoparticles. IEEE Trans Nanotechnol 5: 123-128.

Fang, X., Zhao, X., Fang, W., Chen, C. and Zheng, N. 2013. Self-templating synthesis of hollow mesoporous silica and their applications in catalysis and drug delivery. Nanoscale 5: 2205-2218.

Felber, M. and Alberto, R. 2015. (99m)Tc radiolabelling of Fe3O4-Au core-shell and Au-Fe3O4 dumbbell-like nanoparticles. Nanoscale 7: 6653-6660.

Gao, J., Chen, K., Xie, R., Xie, J., Lee, S., Cheng, Z., et al. 2010a. Ultrasmall near-infrared non-cadmium quantum dots for in vivo tumor imaging. Small 6: 256-261.

Gao, J., Chen, K., Xie, R., Xie, J., Yan, Y., Cheng, Z., et al. 2010b. In vivo tumor-targeted fluorescence imaging using near-infrared non-cadmium quantum dots. Bioconjug Chem 21: 604-609.

Gao, J., Chen, K., Luong, R., Bouley, D.M., Mao, H., Qiao, T., et al. 2012. A novel clinically translatable fluorescent nanoparticle for targeted molecular imaging of tumors in living subjects. Nano Lett 12: 281-286.

Ghosh Chaudhuri, R. and Paria, S. 2012. Core/shell nanoparticles: classes, properties, synthesis mechanisms, characterization and applications. Chem Rev 112: 2373-2433.

Giljohann, D.A., Seferos, D.S., Prigodich, A.E., Patel, P.C. and Mirkin, C.A. 2009. Gene regulation with polyvalent siRNA-nanoparticle conjugates. J Am Chem Soc 131: 2072-2073.

Gobbo, O.L., Sjaastad, K., Radomski, M.W., Volkov, Y. and Prina-Mello, A. 2015. Magnetic nanoparticles in cancer theranostics. Theranostics 5: 1249-1263.

Gong, J., Zhou, F., Li, Z. and Tang, Z. 2012. Synthesis of Au@Ag core-shell nanocubes containing varying shaped cores and their localized surface plasmon resonances. Langmuir 28: 8959-8964.

Guo, H., Li, Z., Qian, H., Hu, Y. and Muhammad, I.N. 2010. Seed-mediated synthesis of NaYF4: Yb, Er/NaGdF4 nanocrystals with improved upconversion fluorescence and MR relaxivity. Nanotechnology 21: 125602.

Guo, W., Sun, X., Jacobson, O., Yan, X., Min, K., Srivatsan, A., et al. 2015. Intrinsically radioactive [64Cu]CuInS/ZnS quantum dots for PET and optical imaging: improved radiochemical stability and controllable cerenkov luminescence. ACS Nano 9: 488-495.

Hardman, R. 2006. A toxicologic review of quantum dots: toxicity depends on physicochemical and environmental factors. Environ Health Perspect 114: 165-172.

Haubold, S., Haase, M., Kornowski, A. and Weller, H. 2001. Strongly luminescent InP/ZnS core-shell nanoparticles. Chemphyschem 2: 331-334.

He, C. and Lin, W. 2015. Hybrid nanoparticles for cancer imaging and therapy. Cancer Treat Res 166: 173-192.

Hines, M.A. and Guyot-Sionnest, P. 1996. Synthesis and characterization of strongly luminescing ZnS-capped CdSe nanocrystals. J Phys Chem 100: 468-471.

Hoshino, A., Hanada, S. and Yamamoto, K. 2011. Toxicity of nanocrystal quantum dots: the relevance of surface modifications. Arch Toxicol 85: 707-720.

Hou, Z., Zhang, Y., Deng, K., Chen, Y., Li, X., Deng, X., et al. 2015. UV-emitting upconversion-based TiO2 photosensitizing nanoplatform: near-infrared light mediated in vivo photodynamic therapy via mitochondria-involved apoptosis pathway. ACS Nano 9: 2584-2599.

Hu, H., Chen, Z., Cao, T., Zhang, Q., Yu, M., Li, F., et al. 2008. Hydrothermal synthesis of hexagonal lanthanide-doped LaF(3) nanoplates with bright upconversion luminescence. Nanotechnology 19: 375702.

Hu, Y., Meng, L., Niu, L. and Lu, Q. 2013. Facile synthesis of superparamagnetic Fe3O4@polyphosphazene@Au shells for magnetic resonance imaging and photothermal therapy. ACS Appl Mater Interfaces 5: 4586-4591.

Javier, D.J., Nitin, N., Levy, M., Ellington, A. and Richards-Kortum, R. 2008. Aptamer-targeted gold nanoparticles as molecular-specific contrast agents for reflectance imaging. Bioconjug Chem 19: 1309-1312.

Jiang, T., Song, J., Zhang, W., Wang, H., Li, X., Xia, R., et al. 2015. Au-Ag@Au hollow nanostructure with enhanced chemical stability and improved photothermal transduction efficiency for cancer treatment. ACS Appl Mater Interfaces 7: 21985-21994.

Jin, Y. 2014. Multifunctional compact hybrid Au nanoshells: a new generation of nanoplasmonic probes for biosensing, imaging and controlled release. Acc Chem Res 47: 138-148.

Kim, J.H., Jang, H.H., Ryou, S.M., Kim, S., Bae, J., Lee, K., et al. 2010. A functionalized gold nanoparticles-assisted universal carrier for antisense DNA. Chem Commun (Camb) 46: 4151-4153.

Law, W.C., Yong, K.T., Roy, I., Ding, H., Hu, R., Zhao, W., et al. 2009. Aqueous-phase synthesis of highly luminescent CdTe/ZnTe core/shell quantum dots optimized for targeted bioimaging. Small 5: 1302-1310.

Le Ngoc, T. and Kim, J.S. 2013. Temperature dependent fluorescence of CuInS/ZnS quantum dots in near infrared region. J Nanosci Nanotechnol 13: 6115-6119.

Lee, J.S., Green, J.J., Love, K.T., Sunshine, J., Langer, R. and Anderson, D.G. 2009. Gold, poly(beta-amino ester) nanoparticles for small interfering RNA delivery. Nano Lett 9: 2402-2406.

Leifert, A., Pan-Bartnek, Y., Simon, U. and Jahnen-Dechent, W. 2013. Molecularly stabilised ultrasmall gold nanoparticles: synthesis, characterization and bioactivity. Nanoscale 5: 6224-6242.

Li, J., Hu, Y., Yang, J., Wei, P., Sun, W., Shen, M., et al. 2015. Hyaluronic acid-modified Fe3O4@Au core/shell nanostars for multimodal imaging and photothermal therapy of tumors. Biomaterials 38: 10-21.

Li, J.J., Wang, Y.A., Guo, W., Keay, J.C., Mishima, T.D., Johnson, M.B., et al. 2003. Large-scale synthesis of nearly monodisperse CdSe/CdS core/shell nanocrystals using air-stable reagents via successive ion layer adsorption and reaction. J Am Chem Soc 125: 12567-12575.

Li, L. and Reiss, P. 2008. One-pot synthesis of highly luminescent InP/ZnS nanocrystals without precursor injection. J Am Chem Soc 130: 11588-11589.

Li, L., Protière, M. and Reiss, P. 2008. Economic synthesis of high quality InP nanocrystals using calcium phosphide as the phosphorus precursor. Chem Mater 20: 2621-2623.

Li, W.P., Liao, P.Y., Su, C.H. and Yeh, C.S. 2014. Formation of oligonucleotide-gated silica shell-coated Fe(3)O(4)-Au core-shell nanotrisoctahedra for magnetically targeted and near-infrared light-responsive theranostic platform. J Am Chem Soc 136: 10062-10075.

Lifshitz, E., Brumer, M., Kigel, A., Sashchiuk, A., Bashouti, M., Sirota, M., et al. 2006. Air-stable PbSe/PbS and PbSe/PbSexS1-x core-shell nanocrystal quantum dots and their applications. J Phys Chem B 110: 25356-25365.

Liu, F., He, X., Liu, L., You, H., Zhang, H. and Wang, Z. 2013a. Conjugation of NaGdF4 upconverting nanoparticles on silica nanospheres as contrast agents for multimodality imaging. Biomaterials 34: 5218-5225.

Liu, F., He, X., Lei, Z., Liu, L., Zhang, J., You, H., et al. 2015a. Facile preparation of doxorubicin-loaded upconversion@polydopamine nanoplatforms for simultaneous in vivo multimodality imaging and chemophotothermal synergistic therapy. Adv Healthc Mater 4: 559-568.

Liu, J., Zhang, W., Zhang, H., Yang, Z., Li, T., Wang, B., et al. 2013b. A multifunctional nanoprobe based on Au-Fe3O4 nanoparticles for multimodal and ultrasensitive detection of cancer cells. Chem Commun (Camb) 49: 4938-4940.

Liu, J., Liu, W., Zhang, H., Yang, Z., Wang, B., Chen, F., et al. 2015b. Triple-emitting dumbbell fluorescent nanoprobe for multicolor detection and imaging applications. Inorg Chem 54: 7725-7734.

Liu, L., Yong, K.T., Roy, I., Law, W.C., Ye, L., Liu, J., et al. 2012. Bioconjugated pluronic triblock-copolymer micelle-encapsulated quantum dots for targeted imaging of cancer: in vitro and in vivo studies. Theranostics 2: 705-713.

Liu, Q., Chen, M., Sun, Y., Chen, G., Yang, T., Gao, Y., et al. 2011a. Multifunctional rare-earth self-assembled nanosystem for tri-modal upconversion luminescence/fluorescence/positron emission tomography imaging. Biomaterials 32: 8243-8253.

Liu, Q., Sun, Y., Li, C., Zhou, J., Yang, T., Zhang, X., et al. 2011b. 18F-Labeled magnetic-upconversion nanophosphors via rare-earth cation-assisted ligand assembly. ACS Nano 5: 3146-3157.

Liu, T.C., Wang, J.H., Wang, H.Q., Zhang, H.L., Zhang, Z.H., Hua, X.F., et al. 2007. Bioconjugate recognition molecules to quantum dots as tumor probes. J Biomed Mater Res A 83: 1209-1216.

Mahalingam, V., Vetrone, F., Naccache, R., Speghini, A. and Capobianco, J.A. 2009. Colloidal Tm3+/Yb3+ -doped LiYF4 nanocrystals: multiple luminescence spanning the UV to NIR regions via low-energy excitation. Adv Mater 21: 4025-4028.

Mai, H.X., Zhang, Y.W., Si, R., Yan, Z.G., Sun, L.D., You, L.P., et al. 2006. High-quality sodium rare-earth fluoride nanocrystals: controlled synthesis and optical properties. J Am Chem Soc 128: 6426-6436.

Mai, H.X., Zhang, Y.W., Sun, L.D. and Yan, C.H. 2007. Highly efficient multicolor upconversion emissions and their mechanisms of monodisperse NaYF4: Yb, Er core and core/shell-structured nanocrystals. J Phys Chem C 111: 13721-13729.

Martynenko, I.V., Kuznetsova, V.A., Orlova capital A, C., Kanaev, P.A., Maslov, V.G., Loudon, A., et al. 2015. Chlorin e6-ZnSe/ZnS quantum dots based system as reagent for photodynamic therapy. Nanotechnology 26: 055102.

Meng, H., Chen, J.Y., Mi, L., Wang, P.N., Ge, M.Y., Yue, Y., et al. 2011. Conjugates of folic acids with BSA-coated quantum dots for cancer cell targeting and imaging by single-photon and two-photon excitation. J Biol Inorg Chem 16: 117-123.

Micic, O.I., Curtis, C.J., Jones, K.M., Sprague, J.R. and Nozik, A.J. 1994. Synthesis and characterization of InP quantum dots. J Phys Chem 98: 4966-4969.

Nie, X. and Chen, C. 2012. Au nanostructures: an emerging prospect in cancer theranostics. Sci China Life Sci 55: 872-883.

Obonyo, O., Fisher, E., Edwards, M. and Douroumis, D. 2010. Quantum dots synthesis and biological applications as imaging and drug delivery systems. Crit Rev Biotechnol 30: 283-301.

Pang, L., Xu, J., Shu, C., Guo, J., Ma, X., Liu, Y., et al. 2014. Characterization and cancer cell targeted imaging properties of human antivascular endothelial growth factor monoclonal antibody conjugated CdTe/ZnS quantum dots. Luminescence 29: 1177-1182.

Park, Y.I., Kim, J.H., Lee, K.T., Jeon, K.S., Bin Na, H., Yu, J.H., et al. 2009. Nonblinking and nonbleaching upconverting nanoparticles as an optical imaging nanoprobe and T1 magnetic resonance imaging contrast agent. Adv Mater 21: 4467-4471.

Pelley, J.L., Daar, A.S. and Saner, M.A. 2009. State of academic knowledge on toxicity and biological fate of quantum dots. Toxicol Sci 112: 276-296.

Peng, X.G., Schlamp, M.C., Kadavanich, A.V. and Alivisatos, A.P. 1997. Epitaxial growth of highly luminescent CdSe/CdS core/shell nanocrystals with photostability and electronic accessibility. J Am Chem Soc 119: 7019-7029.

Punjabi, A., Wu, X., Tokatli-Apollon, A., El-Rifai, M., Lee, H., Zhang, Y., et al. 2014. Amplifying the red-emission of upconverting nanoparticles for biocompatible clinically used prodrug-induced photodynamic therapy. ACS Nano 8: 10621-10630.

Qian, H.S. and Zhang, Y. 2008. Synthesis of hexagonal-phase core–shell NaYF4 nanocrystals with tunable upconversion fluorescence. Langmuir 24: 12123-12125.

Qu, L., Peng, Z.A. and Peng, X. 2001. Alternative routes toward high quality CdSe nanocrystals. Nano Lett. 1: 333-337.

Reiss, P., Protiere, M. and Li, L. 2009. Core/shell semiconductor nanocrystals. Small 5: 154-168.

Rink, J.S., McMahon, K.M., Chen, X., Mirkin, C.A., Thaxton, C.S. and Kaufman, D.B. 2010. Transfection of pancreatic islets using polyvalent DNA-functionalized gold nanoparticles. Surgery 148: 335-345.

Rogach, A.L., Katsikas, L., Kornowski, A., Su, D., Eychmüller, A. and Weller, H. 1997. Synthesis, morphology and optical properties of thiol-stabilized CdTe nanoclusters in aqueous solution. Ber Bunsen phys Chem 101: 1668-1670.

Rosenberg, J.T., Kogot, J.M., Lovingood, D.D., Strouse, G.F. and Grant, S.C. 2010. Intracellular bimodal nanoparticles based on quantum dots for high-field MRI at 21.1 T. Magn Reson Med 64: 871-882.

Sailor, M.J. and Park, J.H. 2012. Hybrid nanoparticles for detection and treatment of cancer. Adv Mater 24: 3779-3802.

SalmanOgli, A. 2011. Nanobio applications of quantum dots in cancer: imaging, sensing and targeting. Cancer Nanotechnol 2: 1-19.

Santhosh, P.B. and Ulrih, N.P. 2013. Multifunctional superparamagnetic iron oxide nanoparticles: promising tools in cancer theranostics. Cancer Lett 336: 8-17.

Shen, J.W., Yang, C.X., Dong, L.X., Sun, H.R., Gao, K. and Yan, X.P. 2013. Incorporation of computed tomography and magnetic resonance imaging function into NaYF4: Yb/Tm upconversion nanoparticles for in vivo trimodal bioimaging. Anal Chem 85: 12166-12172.

Shi, H., Ye, X., He, X., Wang, K., Cui, W., He, D., et al. 2014. Au@Ag/Au nanoparticles assembled with activatable aptamer probes as smart "nano-doctors" for image-guided cancer thermotherapy. Nanoscale 6: 8754-8761.

Shi, S., Chen, F. and Cai, W. 2013. Biomedical applications of functionalized hollow mesoporous silica nanoparticles: focusing on molecular imaging. Nanomedicine (Lond) 8: 2027-2039.

Shu, C., Huang, B., Chen, X., Wang, Y., Li, X., Ding, L., et al. 2013. Facile synthesis and characterization of water soluble ZnSe/ZnS quantum dots for cellar imaging. Spectrochim Acta A Mol Biomol Spectrosc 104: 143-149.

Singh, A. and Sahoo, S.K. 2014. Magnetic nanoparticles: a novel platform for cancer theranostics. Drug Discov Today 19: 474-481.

Smith, A.M., Ruan, G., Rhyner, M.N. and Nie, S. 2006. Engineering luminescent quantum dots for in vivo molecular and cellular imaging. Ann Biomed Eng 34: 3-14.

Smolensky, E.D., Neary, M.C., Zhou, Y., Berquo, T.S. and Pierre, V.C. 2011. Fe3O4@ organic@Au: core-shell nanocomposites with high saturation magnetisation as magnetoplasmonic MRI contrast agents. Chem Commun (Camb) 47: 2149-2151.

Soenen, S.J., Demeester, J., De Smedt, S.C. and Braeckmans, K. 2013. Turning a frown upside down: exploiting nanoparticle toxicity for anticancer therapy. Nano Today 8: 121-125.

Soenen, S.J., Manshian, B.B., Aubert, T., Himmelreich, U., Demeester, J., De Smedt, S.C., et al. 2014. Cytotoxicity of cadmium-free quantum dots and their use in cell bioimaging. Chem Res Toxicol 27: 1050-1059.

Stasiuk, G.J., Tamang, S., Imbert, D., Poillot, C., Giardiello, M., Tisseyre, C., et al. 2011. Cell-permeable Ln (III) chelate-functionalized InP quantum dots as multimodal imaging agents. ACS Nano 5: 8193-8201.

Stringfellow, G.B. 1982. Epitaxy. Rep Prog Phys. 45: 469.

Sun, X., Huang, X., Guo, J., Zhu, W., Ding, Y., Niu, G., et al. 2014. Self-illuminating 64Cu-doped CdSe/ZnS nanocrystals for in vivo tumor imaging. J Am Chem Soc 136: 1706-1709.

Sun, Y., Yu, M., Liang, S., Zhang, Y., Li, C., Mou, T., et al. 2011. Fluorine-18 labeled rare-earth nanoparticles for positron emission tomography (PET) imaging of sentinel lymph node. Biomaterials 32: 2999-3007.

Sun, Y., Zhu, X., Peng, J. and Li, F. 2013. Core-shell lanthanide upconversion nanophosphors as four-modal probes for tumor angiogenesis imaging. ACS Nano 7: 11290-11300.

Talapin, D.V., Haubold, S., Rogach, A.L., Kornowski, A., Haase, M. and Weller, H. 2001. A novel organometallic synthesis of highly luminescent CdTe nanocrystals. J Phys Chem B 105: 2260-2263.

Tsay, J.M., Pflughoefft, M., Bentolila, L.A. and Weiss, S. 2004. Hybrid approach to the synthesis of highly luminescent CdTe/ZnS and CdHgTe/ZnS nanocrystals. J Am Chem Soc 126: 1926-1927.

Turkevich, J., Stevenson, P.C. and Hillier, J. 1951. A study of the nucleation and growth processes in the synthesis of colloidal gold. Discuss Faraday Soc 11: 55-75.

van Embden, J., Jasieniak, J. and Mulvaney, P. 2009. Mapping the optical properties of CdSe/CdS heterostructure nanocrystals: the effects of core size and shell thickness. J Am Chem Soc 131: 14299-14309.

Wang, C., Cheng, L. and Liu, Z. 2013a. Upconversion nanoparticles for photodynamic therapy and other cancer therapeutics. Theranostics 3: 317-330.

Wang, F., Deng, R. and Liu, X. 2014a. Preparation of core-shell NaGdF4 nanoparticles doped with luminescent lanthanide ions to be used as upconversion-based probes. Nat Protoc 9: 1634-1644.

Wang, H., Barcelo, I., Lana-Villarreal, T., Gomez, R., Bonn, M. and Canovas, E. 2014b. Interplay between structure, stoichiometry and electron transfer dynamics in SILAR-based quantum dot-sensitized oxides. Nano Lett. 14: 5780-5786.

Wang, L. and Li, Y. 2006. Na(Y1.5 Na0.5)F6 single-crystal nanorods as multicolor luminescent materials. Nano Lett 6: 1645-1649.

Wang, X., Zhuang, J., Peng, Q. and Li, Y. 2005. A general strategy for nanocrystal synthesis. Nature 437: 121-124.

Wang, X., Zhuang, J., Peng, Q. and Li, Y. 2006. Hydrothermal synthesis of rare-earth fluoride nanocrystals. Inorg Chem 45: 6661-6665.

Wang, Y.F., Liu, G.Y., Sun, L.D., Xiao, J.W., Zhou, J.C. and Yan, C.H. 2013b. Nd(3+)-sensitized upconversion nanophosphors: efficient in vivo bioimaging probes with minimized heating effect. ACS Nano 7: 7200-7206.

Wu, P., Gao, Y., Lu, Y., Zhang, H. and Cai, C. 2013. High specific detection and near-infrared photothermal therapy of lung cancer cells with high SERS active aptamer-silver-gold shell-core nanostructures. Analyst 138: 6501-6510.

Xia, L., Kong, X., Liu, X., Tu, L., Zhang, Y., Chang, Y., et al. 2014. An upconversion nanoparticle–zinc phthalocyanine based nanophotosensitizer for photodynamic therapy. Biomaterials 35: 4146-4156.

Xia, Y., Li, W., Cobley, C.M., Chen, J., Xia, X., Zhang, Q., et al. 2011. Gold nanocages: from synthesis to theranostic applications. Acc Chem Res 44: 914-924.

Xiao, Q., Bu, W., Ren, Q., Zhang, S., Xing, H., Chen, F., et al. 2012. Radiopaque fluorescence-transparent TaOx decorated upconversion nanophosphors for in vivo CT/MR/UCL trimodal imaging. Biomaterials 33: 7530-7539.

Xie, R., Battaglia, D. and Peng, X. 2007. Colloidal InP nanocrystals as efficient emitters covering blue to near-infrared. J Am Chem Soc 129: 15432-15433.

Xu, C., Xie, J., Ho, D., Wang, C., Kohler, N., Walsh, E.G., et al. 2008a. Au-Fe3O4 dumbbell nanoparticles as dual-functional probes. Angew Chem Int Ed Engl 47: 173-176.

Xu, C., Wang, B. and Sun, S. 2009. Dumbbell-like Au-Fe3O4 nanoparticles for target-specific platin delivery. J Am Chem Soc 131: 4216-4217.

Xu, S., Ziegler, J. and Nann, T. 2008b. Rapid synthesis of highly luminescent InP and InP/ZnS nanocrystals. J Mater Chem 18: 2653-2656.

Yan, R.X. and Li, Y.D. 2005. Down/up conversion in Ln3+-doped YF3 nanocrystals. Adv Funct Mater 15: 763-770.

Ye, X., Collins, J.E., Kang, Y., Chen, J., Chen, D.T., Yodh, A.G., et al. 2010. Morphologically controlled synthesis of colloidal upconversion nanophosphors and their shape-directed self-assembly. Proc Natl Acad Sci USA 107: 22430-22435.

Yi, G.S. and Chow, G.M. 2006. Synthesis of hexagonal-phase NaYF4: Yb, Er and NaYF4: Yb, Tm nanocrystals with efficient up-conversion fluorescence. Adv Funct Mater 16: 2324-2329.

Yi, G.S. and Chow, G.M. 2007. Water-soluble NaYF4: Yb, Er (Tm)/NaYF4/polymer core/shell/shell nanoparticles with significant enhancement of upconversion fluorescence. Chem Mater 19: 341-343.

Yong, K.T., Ding, H., Roy, I., Law, W.C., Bergey, E.J., Maitra, A., et al. 2009. Imaging pancreatic cancer using bioconjugated InP quantum dots. ACS Nano 3: 502-510.

Yu, H., Chen, M., Rice, P.M., Wang, S.X., White, R.L. and Sun, S. 2005. Dumbbell-like bifunctional Au-Fe3O4 nanoparticles. Nano Lett 5: 379-382.

Zhan, Q., Qian, J., Liang, H., Somesfalean, G., Wang, D., He, S., et al. 2011. Using 915 nm laser excited Tm(3)+/Er(3)+/Ho(3)+ -doped NaYbF4 upconversion nanoparticles for in vitro and deeper in vivo bioimaging without overheating irradiation. ACS Nano 5: 3744-3757.

Zhang, F., Wan, Y., Yu, T., Shi, Y., Xie, S., Li, Y., et al. 2007a. Uniform nanostructured arrays of sodium rare-earth fluorides for highly efficient multicolor upconversion luminescence. Angew Chem Int Ed Engl 46: 7976-7979.

Zhang, F., Li, J., Shan, J., Xu, L. and Zhao, D. 2009. Shape, size and phase-controlled rare-earth fluoride nanocrystals with optical up-conversion properties. Chemistry 15: 11010-11019.

Zhang, H., Li, Y., Lin, Y., Huang, Y. and Duan, X. 2011. Composition tuning the upconversion emission in NaYF4: Yb/Tm hexaplate nanocrystals. Nanoscale 3: 963-966.

Zhang, M.Z., Yu, R.N., Chen, J., Ma, Z.Y. and Zhao, Y.D. 2012. Targeted quantum dots fluorescence probes functionalized with aptamer and peptide for transferrin receptor on tumor cells. Nanotechnology 23: 485104.

Zhang, P., Steelant, W., Kumar, M. and Scholfield, M. 2007b. Versatile photosensitizers for photodynamic therapy at infrared excitation. J Am Chem Soc 129: 4526-4527.

Zhao, Z., Han, Y., Lin, C., Hu, D., Wang, F., Chen, X., et al. 2012. Multifunctional core-shell upconverting nanoparticles for imaging and photodynamic therapy of liver cancer cells. Chem Asian J 7: 830-837.

Zhou, J., Liu, Z. and Li, F. 2012. Upconversion nanophosphors for small-animal imaging. Chem Soc Rev 41: 1323-1349.

Zhu, J., Lu, Y., Li, Y., Jiang, J., Cheng, L., Liu, Z., et al. 2014. Synthesis of Au-Fe3O4 heterostructured nanoparticles for in vivo computed tomography and magnetic resonance dual model imaging. Nanoscale 6: 199-202.
Zhu, Y., Hong, H., Xu, Z.P., Li, Z. and Cai, W. 2013. Quantum dot-based nanoprobes for in vivo targeted imaging. Curr Mol Med 13: 1549-1567.

5

Nanomaterial-Antibody Hybrids

Jyothi U. Menon,[1,a] *Lei Song,*[1,b]
Nadia Falzone[1,2,c] and *Katherine A. Vallis*[1,*]

INTRODUCTION

Research into the use of nanotechnology for drug delivery and imaging has witnessed remarkable growth in recent years. A major challenge is the design of nanomedicines that are specifically taken up by the cells being targeted while minimizing the toxicity that may result from the release of encapsulated therapeutics in the proximity of healthy cells. Targeting of diseased cells can be achieved by surface modification of nanomaterials using molecules such as antibodies, antibody fragments, aptamers and peptides, that specifically bind disease-associated receptors or antigens (Gu et al. 2007). In the case of cancer drug delivery, nanoparticle targeting occurs through passive accumulation via the enhanced permeability and retention (EPR) effect in tumor tissue. Accumulation in the tumor is also achieved by active targeting when small molecule ligands such as folate (Lee and Low 1995) or transferrin (Suzuki et al. 1997), peptides such as Asn-Gly-Arg (NGR) (Pastorino et al. 2003) or antibodies (Torchilin 2008; Vingerhoeds et al. 1994) are attached to the surface of nanoparticles (Fig. 5.1).

Many antibody-based medicines are now in clinical practice. Whole monoclonal antibodies (mAb) have the advantages of high avidity and stability during long-term storage. They also have the ability to initiate a signalling cascade resulting in cancer cell death when immune cells bind to the Fc domain of the antibody (Peer et al. 2007). The use of antibody fragments, although not as stable as whole mAbs, is advocated as a safer alternative for

[1] CRUK/MRC Oxford Institute for Radiation Oncology, Department of Oncology, University of Oxford, Oxford, OX3 7DQ, UK.
[a] E-mail: jyothi.menon@oncology.ox.ac.uk
[b] E-mail: lei.song@oncology.ox.ac.uk
[2] Department of Biomedical Science, Tshwane University of Technology, Pretoria, South Africa.
[c] E-mail: nadia.falzone@oncology.ox.ac.uk
[*] Corresponding author: katherine.vallis@oncology.ox.ac.uk

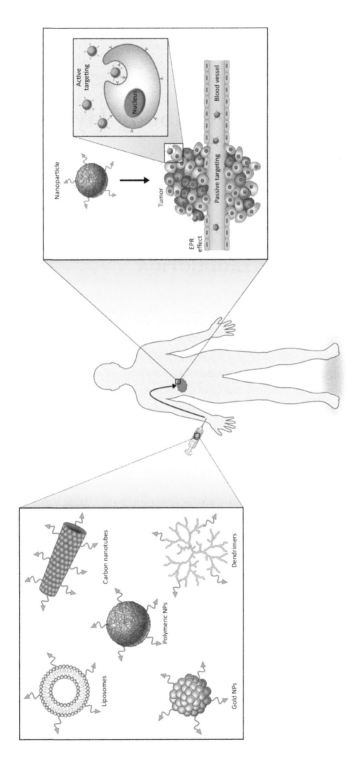

Fig. 5.1   An illustration of the different types of nanoparticle-antibody hybrids and their cancer-targeting strategies following *in vivo* administration.

systemic delivery due to reduced non-specific binding (Cole and Holland 2015). Chimeric and humanized antibodies have also been developed to overcome the immunogenicity that has hindered the clinical studies of mAbs (Adams and Weiner 2005). Studies of antibody-drug conjugates have demonstrated high intra-tumoral accumulation of free drug following binding to targeted cells while healthy cells are spared from exposure to the drug and therefore, toxicity (Alley et al. 2010). Several antibody-drug conjugates have now been approved by the Food and Drug Administration (FDA) and several more are currently in clinical trials particularly for cancer therapy (Adams and Weiner 2005). Antibodies are now also being studied extensively as components of nanomedicines because of their excellent specificity and the availability of a large number of functional groups for attachment to nanomaterials (Goodall et al. 2014). To improve drug bioavailability, drug-loaded nanoparticles surface-modified with antibodies have been developed and investigated as treatments for a range of diseases. The aim of this Chapter is to present a summary of the current status of research into the design, development and clinical applications of nanomaterial-antibody hybrid agents. Antibody-tagged liposomes, polymeric nanoparticles and gold nanoparticles (GNPs) are considered in depth, as these nanomaterials have been investigated most extensively and have demonstrated the greatest potential for clinical use in the near future.

Liposome-Antibody Hybrids

Liposomes, self-assembled colloidal vesicles with a characteristic lipid bilayered membrane are the most frequently studied nanocarriers for targeted drug delivery (Sawant and Torchilin 2012). Their popularity stems from their controllable size, biocompatibility, biodegradability, low toxicity and lack of immune system activation. Liposomes can modify the pharmacokinetic behavior of their drug cargo and so enhance specific drug targeting (Torchilin 2005). Being amphiphilic in nature, liposomes can also stably encapsulate and protect both water soluble or insoluble therapeutics in the core (Cole and Holland 2015). However, despite more than 30 years of use as nanocarriers, unmodified liposomes have some limitations including poor stability, low loading efficiency and poor release profiles, that have hampered their widespread use in selective molecular targeting of cells. To improve the targeting capabilities of liposomes, immunoliposomes have been developed through the conjugation of antibodies or antibody-based constructs (e.g., antigen-binding fragments (Fab), dimers of antigen-binding fragments (F(ab')$_2$), single-chain variable fragments (scFv) or other engineered fragments) to the liposomal surface (Hansen et al. 1995; Huang et al. 1980; Huang et al. 1981). The following section describes the techniques for synthesis of immunoliposomes and their biomedical applications.

Design and Synthesis of Liposome-Antibody Hybrids

Liposomes are prepared from lipid dispersions in water, comprising the addition of an aqueous phase to dry lipid layers followed by vortexing, reverse phase evaporation, detergent dialysis or ethanol injection and high-pressure homogenization. Size and lamellarity (the number of concentric bilayers) are controlled by sonication or extrusion through filters. Liposomal size varies from 10 nm to > 10 μm and depends on composition (up to 50 mol% cholesterol can be incorporated to increase stability) and method of preparation. Unilamellar liposomes are preferred for maximal encapsulated drug volume and controlled drug release (Sofou and Sgouros 2008). Liposomes prepared using the reverse phase evaporation method allow more than 50 percent encapsulation of the cargo substance from the water phase. Drug release occurs over a few minutes to hours depending on the membrane composition, cholesterol content and physiological environment.

A number of strategies have been developed to confer different properties to liposomes including stabilization, prolonged systemic circulation, increased accumulation at the target tissue, increased cellular internalization and organelle-specific drug delivery (Allen et al. 1991; Allen 1994; Lasic et al. 1991). Most notably, coating the carrier with hydrophilic polymers such as poly (ethylene glycol) (PEG) or amphiphilic polymers such as synthetic copolymers of polyethylene oxide (hydrophilic block) and propylene oxide (hydrophobic block), avoids protein adsorption on the surface and hence inhibits immune clearance by the reticulo-endothelial system (RES), thereby prolonging blood circulation and half-life (Torchilin 2011). This advance gave rise to the use of "smart multi-functional" nanocarriers. Multifunctional immunoliposomes consist of a phospholipid membrane that accommodates lipophilic compounds, an encapsulated aqueous phase containing the therapeutic agent (such as chemotherapeutic drug, magnetic particles or plasmids for gene therapy), surface-grafted polymer chains and surface-grafted targeting ligands such as antibodies. The optimal configuration for targeting *in vivo* is provided by placing the antibody ligands on the free termini of PEGylated lipids as this avoids steric hindrance of binding of the ligand to the target (Torchilin 2008).

Biomedical Applications of Liposome-Antibody Hybrids

The recent strategy of using multifunctional liposomes for the sustained release of drugs with increased plasma residence time and antibody-based targeting of tumors has attracted enormous scientific attention. Clinically available antibodies that are widely used as immunoliposome ligands include rituximab for lymphoma, trastuzumab for breast cancer, bevacizumab for inhibition of angiogenesis and cetuximab for the treatment of colorectal cancer (Cole and Holland 2015).

Imaging

Many cases of the adaption of liposome antibody hybrids for imaging applications exist. For example, gadolinium and quantum dots can be incorporated into immunoliposome constructs for MRI or optical imaging respectively, facilitating simultaneous targeting and imaging (Freedman et al. 2009; Weng et al. 2008). One area of research that has received considerable interest is the intra-operative administration of liposomes. In this regard a short circulation time due to the quick biological clearance of the nanoparticles could be beneficial, allowing prompt commencement of surgery and treatment soon after administration. In a recent study, Rüger and co-workers developed a fluorescence-activatable liposome (fluorescence-quenched during circulation and fluorescence activation upon cellular uptake), bearing scFv directed against fibroblast activation protein (FAP), a type-II transmembrane sialoglycoprotein for early detection of molecular changes preceding invasive growth of solid tumors (Rüger et al. 2014). The anti-FAP-scFv-modified liposomes accumulated selectively in FAP-expressing xenograft models. Kim and co-workers investigated the use of nitric oxide-loaded echogenic immuno-liposomes (NO-ELIP) as a pre-treatment strategy to optimize contrast agent delivery to the vascular wall for both diagnostic and therapeutic applications (Kim et al. 2013b, 2015). When atherosclerotic animals were treated with anti-VCAM-1-ELIP, it was found that pre-treatment with NO-ELIP plus ultrasound resulted in a significant increase in acoustic enhancement by anti-VCAM-1-ELIP. Immunoliposomes have also been used for the targeted delivery of paramagnetic molecular probes for electron paramagnetic resonance imaging (Burks et al. 2010, 2011). Hc7 cells, that selectively overexpress HER2, were targeted by anti-HER2 immunoliposomes encapsulating quenched nitroxides. HER2-dependent liposomal delivery enabled Hc7 cells to accumulate sufficient concentrations of nitroxide intra-cellularly for EPR imaging.

Therapy: Cancer Chemotherapeutics and Gene Therapy

Liposomes offer enhanced transport as potential carriers of bio-macromolecules such as anti-cancer proteins, drugs and siRNA (Gurudevan et al. 2013). Immunoliposome carriers are designed to combine the properties of long-circulating PEGylated liposomes with targeted drug delivery. Doxorubicin-loaded long-circulating liposomes (Doxil™) were modified with the nucleosome-specific mAb, 2C5, which recognizes cancer cells via cell surface-bound nucleosomes (Lukyanov et al. 2004). Elbayoumi demonstrated a 3-8 fold increase in binding and internalization of the mAb 2C5-modified PEGylated liposomes in several cancer cell lines, resulting in significant cytotoxicity, even in cells previously noted to be resistant to doxorubicin (Elbayoumi and Torchilin 2007). The recepteur d'orgine nantais (RON) plays a role in tumor progression. Targeting of this surface protein by anti-RON doxorubicin loaded immunoliposomes (Zt/g4 and Zt/c) has shown increased cytotoxic efficacy attributed to antibody-mediated RON activation and endocytosis of the immunoliposome (Guin et al. 2011).

Doxorubicin has also been used in the treatment of mesangial cell-mediated nephropathy, by targeting the Thy1.1 antigen (OX7) in a rat model (Tuffin et al. 2005). F(ab')$_2$-fragments of the OX7 mAb were coupled to doxorubicin-encapsulated immunoliposomes. Marked uptake in mesangial cell cytoplasm was noted after a single intravenous (i.v.) injection that lead to extensive glomerular damage with little effect on the normal kidney. The same target-antigen strategy was employed to deliver mycophenolate mofetil (MMF)-containing immunoliposomes in a rat model (Suana et al. 2011). Administration of MMF-loaded OX7-IL resulted in a reduction of mesangial cells and glomerular extracellular matrix expansion (a feature of nephropathy). These immunoliposomes therefore offer a promising delivery system in the treatment of mesangial cell mediated glomerulonephritis. Another biomarker, CD74 (a type II transmembrane protein expressed on B cells and upregulated in lymphocytic leukaemia), has also been investigated for active immunoliposome drug delivery. An anti-CD74 antibody (milatuzumab) was incorporated into liposomes designed as targeted dexamethasone carriers and was found to increase drug bioavailability and cytotoxicity in comparison to free drug in CD74$^+$ lymphoma cells (Mao et al. 2013).

Several studies have also reported antibody-selective targeting by immunoliposomes for the delivery of gene therapy (Zhou et al. 2012). The intercellular adhesion molecule (ICAM-1), a glycoprotein of the immunoglobulin superfamily, plays an important role in tumor extravasation and in the inflammatory process by mediating leukocyte binding to the endothelium. The delivery of the ICAM-1 gene through immunoliposomes for targeting endothelial cells in tumors, mediated anti-tumor immunity by overcoming tumor-mediated immunosuppression (Kanwar et al. 2003). In another approach, a cationic immunoliposome system directed by a lipid-tagged scFv against the human transferrin receptor (TfR) showed promising efficacy for systemic p53 gene therapy in a human breast cancer metastasis model (Xu et al. 2002). The scFv-cys targets the cationic liposome-DNA complex (lipoplex) to tumor cells and enhances transfection efficiency both *in vitro* and *in vivo* in a variety of human tumor models.

Therapy: Radiolabeled Immunoliposomes

Antibody-targeted liposomes have also been used for the delivery of radionuclides for molecularly targeted radiotherapy (MRT). MRT refers to the selective delivery of radionuclides that emit charged particles such as α-particles, β-particles or Auger-electrons to cancer cells via a targeting moiety that seeks functional and molecular targets within cancer cells whilst sparing normal tissues. The efficacy of MRT has been shown to critically depend on antibody-specific activity and antigen density in or on the target cell. By exploiting the large surface area or encapsulation techniques of nanoscale carrier systems, the number of radioactive atoms delivered to the tumor can be increased. Recently, liposome-based delivery of α-particles was achieved either through conjugation (Lingappa et al. 2010) or by encapsulation (Bandekar et al. 2014). Lingappa et al. showed that [213]Bi-radiolabeled immunoliposomes are effective in treating early micrometastasis in the rat/*neu* transgenic mouse model of breast cancer (Lingappa et al. 2010). Bandekar et al. used PEGylated liposomes loaded with an α-emitter, [225]Ac, to selectively kill prostate-specific membrane antigen (PSMA)-expressing cells in an assessment of their potential for targeted anticancer radiotherapy.

Theranostics

Immunoliposomes can be engineered to be multi-functional and several recent advances in their use in theranostic applications have been reported. Wang et al. used fluorescent probes (luciferase and red fluorescent protein) for optical detection of an anti-CD44 antibody-coupled, doxorubicin-containing liposome designed for the treatment of hepatocellular carcinoma (HCC) in a NOD/SCID mice model (Wang et al. 2012b). It was possible to visualize specific targeting of CD44[+] cells, which lead to apoptosis and delayed growth of HCC tumors. It was concluded that concurrent imaging provided a means to assess the targeting efficiency of the immunoliposomes. Li et al. investigated the potential of radioactive immunoliposomes to treat both the surgical cavity and draining lymph nodes in a rat breast cancer model that simulates positive surgical margins (Li et al. 2012). Immunoliposomes modified with either panitumumab (anti-EGFR) or bevacizumab (anti-VEGF) were loaded with ^{99m}Tc and injected into the surgical cavity with positive margins following lumpectomy in female nude rats. Both antibody immunoliposomes had a higher intra-cavity retention as compared to control liposomes. However accumulation on the residual tumor surface and in the lymph nodes was greater for the anti-EGFR construct. These immunoliposomes offer the potential of antibody targeted post-lumpectomy focal therapy with low systemic toxicity.

Clinical Experience with Liposome-Antibody Hybrids

To date, no active targeted immunoliposome construct has achieved FDA approval. Despite the advantages of active targeting, only a few immuno-liposome formulations have advanced to clinical trials (Perez-Herrero and Fernandez-Medarde 2015). Paclitaxel- or doxorubicin-encapsulated immuno-liposomes have been administered to patients with metastatic cancer (Kato et al. 2006; Matsumura et al. 2004) and have demonstrated an enhancement in activity compared to non-targeted versions with some reported clinical responses. Mamot and co-investigators studied the efficacy of anti-EGFR immunoliposomes loaded with doxorubicin (anti-EGFR IL-dox) in patients with solid tumors (Mamot et al. 2012). Anti-EGFR IL-dox nanoparticles were constructed by covalently linking PEGylated liposomes containing doxorubicin to antigen-binding fragments (Fab') of cetuximab. Anti-EGFR IL-dox was well tolerated (up to 50 mg doxorubicin per m^2) and clinical responses were observed warranting a phase 2 trial. In another recent report, Senzer et al. provided evidence of tumor specific targeting as well as clinical efficacy in a phase I trial of a liposomal nanoparticle vector encoding wild-type p53 (SGT-53), which was engineered with the aim of restoring the expression and pro-apoptotic function of p53 in tumor tissue (Senzer et al. 2013). The liposomal nano-delivery complex employed an anti-transferrin receptor (TfR) scFv as the targeting molecule for the delivery of a non-viral vector for gene therapy. The majority of patients treated with this agent demonstrated stable disease, with minimal side effects.

There is enormous potential for the use of multifunctional immunoliposomes in dual drug delivery and imaging theranostic strategies as well as in dual intraoperative (optical imaging) and conventional imaging (PET/CT, SPECT/CT, MRI, targeted ultrasound) modalities. The expectation is high that the next generation of immunoliposomes will fulfill the promise of delivering personalized and patient specific cancer treatments and diagnostics.

Polymer-Antibody Hybrids

Polymeric nanomaterials have been extensively investigated for the past four decades as drug carriers for various medical applications (Mohanty et al. 2015). Drug delivery via polymeric carriers potentially circumvents some disadvantages of conventional therapeutics such as poor solubility, premature drug degradation, short blood half-life and systemic toxicity due to non-specific treatment (Mohanty et al. 2015; Wang et al. 2012a). The most attractive properties of a polymeric drug delivery vehicle include biocompatibility, biodegradability, ability to incorporate hydrophobic and hydrophilic payloads, prolonged circulation half-life and sustained drug release kinetics (Kumari et al. 2010).

Recent advances in drug delivery have led to the development of hybrid polymeric nanomaterials for therapeutic applications, tissue regeneration and cell isolation as briefly described below. These hybrid materials incorporating antibodies and nano-sized components with varying physical and chemical properties would ideally provide synergistic and targeted diagnosis and treatment. Poly lactic-co-glycolic acid (PLGA) is an FDA-approved polymer extensively used in the development of polymeric hybrid nanomaterials (Haider et al. 2014; Pandita et al. 2014). For instance, Kocbek et al. developed PLGA-based nanoparticles tagged with a mAb that recognizes cytokeratins expressed in breast epithelial cells, that specifically targeted MCF-10A neoT cells *in vitro* both in mono- and co-culture (Kocbek et al. 2007). Similarly, camptothecin-loaded PLGA nanoparticles tagged with peripheral antibodies targeting Fas receptors were synthesized by McCarron et al. for cancer targeting and treatment (McCarron et al. 2008). Other commonly used polymers in nanomaterial-antibody hybrid formulations include polyethylene glycol (PEG), chitosan and albumin (Aktaş et al. 2005; Elzoghby et al. 2012). Sometimes it is desirable to use a combination of different materials to develop hybrid nanoparticles conjugated with antibodies for targeted delivery. A PLGA-lipid combination can result in better control of drug release while PLGA can overcome issues with lipid structural integrity. For example, a nanoparticle hybrid system consisting of a PLGA core, a lipid monolayer and PEG corona, tagged with anti-carcinoembryonic antigen (CEA) monoclonal half-antibody and loaded with paclitaxel was developed by Hu and colleagues for pancreatic cancer treatment (Hu et al. 2010). The reduced-size antibody moiety helped limit the nanoparticle size while retaining the targeting capabilities of the antibody. Inorganic components can also be incorporated in polymeric carriers to impart imaging capabilities to the final product.

Design and Synthesis of Polymer-Antibody Hybrids

Based on the polymer being used, different synthesis techniques can be employed to formulate nanoparticles. For example, PLGA-based nanoparticles can be prepared using a nanoprecipitation or emulsion solvent-evaporation technique while albumin- and chitosan-based particles can be prepared using desolvation (Kouchakzadeh et al. 2013) and ionic gelation techniques (Aktaş et al. 2005) respectively. Stimuli-sensitive or "smart" nanoparticles that can swell or shrink to release encapsulated agents in response to external stimuli, can be generated using polymers such as the thermosensitive poly N-isopropylacrylamide (PNIPA) using free radical polymerization (Kim et al. 2013a).

Following nanoparticle formation, the conjugation of antibodies to the surface can be carried out by various reactions. The most commonly used techniques include EDC-NHS carbodiimide chemistry, biotin-streptavidin chemistry or sulfhydryl-reactive crosslinker chemistry. These techniques rely on the availability of amine, carboxyl or thiol/sulfhydryl groups on the antibody and polymer of interest. Sulfhydryl-reactive human serum albumin (HSA) nanoparticles having a maleimide-PEG-NHS crosslinker were conjugated with thiolated cetuximab by sulfhydryl-based chemistry and the final product was used to target EGFR, overexpressed in colorectal cancer (Löw et al. 2011). EDC-NHS chemistry involving COOH end groups on PLGA and amine groups on anti-OX40 antibodies was employed in the development of PLGA nanoparticles designed to elicit a cytotoxic T lymphocyte response for cancer immunotherapy (Chen et al. 2014). Aktaş and colleagues on the other hand used biotin-streptavidin chemistry to conjugate OX26 mAb onto chitosan-PEG nanospheres for receptor-mediated transport of Z-DEVD-FMK, a caspase inhibitor peptide, across the blood brain barrier (Aktaş et al. 2005).

During synthesis untagged antibodies are usually separated from the antibody-conjugated nanoparticles by centrifugation (Chen et al. 2014), FPLC (Owen et al. 2013) or dialysis (Kim et al. 2013a). Successful conjugation of antibodies to the particle surface can be determined using protein assays such as Bradford assay with Coomassie dye (Moura et al. 2014), by HPLC (Aktaş et al. 2005; Löw et al. 2011) or gel electrophoresis (Xu et al. 2011). In some cases, the antibody is tagged to a fluorophore so that fluorescence can be measured and quantified against a standard curve (Hu et al. 2010).

Biomedical Applications of Polymer-Antibody Hybrids

Polymeric nanoparticle-antibody hybrids are being used for a wide range of biomedical applications. These hybrids can be used to target diseased cells at the site of interest so as to provide site-specific treatment thereby reducing systemic toxicity of the encapsulated payload. The nanoparticles can encapsulate therapeutic agents such as drugs, radionuclides, proteins and cDNA, as well as contrast agents and fluorophores to allow imaging following administration.

Targeted Drug Delivery and Treatment

The conjugation of antibodies to polymeric nanoparticles is advantageous as it can modulate the half-life of the antibody while improving the specificity of the polymeric nanoparticles. Lipid-PLGA hybrid nanoparticles conjugated with CEA half-antibody demonstrated enhanced specificity and cytotoxicity towards CEA-positive BxPC-3 pancreatic cancer cells as compared to CEA-negative XPA-3 pancreatic cancer cells *in vitro* (Hu et al. 2010). Greater *in vitro* specificity of antibody-conjugated nanoparticles was also observed by Song et al. who developed paclitaxel-loaded micelles using pluronic triblock copolymers conjugated with anti-HIF-1α antibody for cancer targeting (Song et al. 2010). These nanoparticles could specifically target and kill HIF-1α expressing MGC-803 stomach cancer cells *in vitro* but did not target normal fibroblasts. The targeting capabilities of nanoparticle-antibody hybrids have also been studied *in vivo* by several research groups. Iron oxide nanoparticles coated with an amphiphilic triblock polymer were conjugated with a single-chain anti-EGFR antibody (ScFvEGFR) and were administered intravenously to mice bearing pancreatic cancer xenografts. Accumulation of these stable, targeted particles in the tumor was established by a marked MR contrast decrease in the tumor up to 30 h following injection, as compared to animals given non-targeted particles. Prussian blue staining also confirmed the presence of iron oxide in the tumor sections of animals that received ScFv EGFR nanoparticles, indicating their potential as targeted nanocarriers for drug delivery to EGFR overexpressing tumor cells (Yang et al. 2009). Another example of *in vivo* targeting efficacy of antibody-nanoparticle hybrids was demonstrated by Kolhar and colleagues who developed polystyrene nanospheres and nanorods surface-modified with ICAM-1 antibody to target the endothelium in diseased tissues (Kolhar et al. 2013). Following intravascular injection, the ICAM-1 antibody-conjugated rods and spheres showed higher accumulation in the lung endothelium known to express basal levels of ICAM-1, than their counterparts coated only with IgG. Also relatively greater lung accumulation of the nanorods was observed as compared to the nanospheres, indicating that the effect of nanocarrier shape on their specificity requires further investigation.

Polymeric nanocarriers are versatile in that they can be used to encapsulate a wide range of hydrophobic or hydrophilic therapeutic agents according to the disease to be treated. Paclitaxel-loaded PLGA-montmorillonite nanoparticles surface-modified with a monoclonal HER2-specific antibody, Trastuzumab, showed a significantly higher chemotherapeutic effect *in vitro* on Caco-2 colon adenocarcinoma and SK-BR-3 breast cancer cells than unmodified particles and free Taxol (Sun et al. 2008). Anti-CD64 antibody conjugated PLGA nanoparticles containing methotrexate and superparamagnetic iron oxide (SPIO) were developed by Moura et al. to target the macrophage-specific receptor CD64, known to be overexpressed in rheumatoid arthritis (Moura et al. 2014). These particles showed greater toxicity towards RAW 264.7 murine macrophage cells than free drug. Paclitaxel-containing PLGA nanoparticles tagged with anti-CD133 antibody were used by Jin et al. to target liver cancer stem cells (Jin et al. 2013). *In vitro* and *in vivo* studies demonstrated that the particles were selectively taken up by liver cancer cells thus aiding in selective elimination of cancer cells overexpressing CD133. Antibody-nanomaterial hybrids have also been used in the isolation of circulating cells from blood (Xu et al. 2011). Iron oxide nanoparticles modified with amphiphilic polymers and conjugated with anti-HER2 antibody were developed by this group to target circulating cancer cells overexpressing HER2/*neu* receptor in the blood. *In vitro* studies confirmed that the nanoparticles could bind to HER2-positive SK-BR3 cells, which were spiked into fresh blood. These cells could then be separated using a low magnetic field gradient (100 T/m). Hybrids of polymeric nanoparticles and antibodies have also been developed for photodynamic therapy. An example is the chitosan/alginate nanoparticle formulation developed by Abdelghany et al. which was surface-modified with antibodies targeting the cell surface receptor death receptor 5 (DR5) and which encapsulated meso-Tetra (N-methyl-4-pyridyl) porphine tetra tosylate (TMP), a photosensitizer (Abdelghany et al. 2013). These particles showed enhanced cellular uptake over non-targeted particles when incubated with DR5-positive HCT116 cells *in vitro*. These particles also demonstrated greater cytotoxicity towards HCT116 cells on treatment with red light, than non-conjugated particles thus demonstrating their potential as a vehicle for carrying photosensitizers for photodynamic therapy. Another interesting application of antibody-polymeric nanoparticle hybrids is vaccine delivery. PLGA nanoparticles linked with PEG and antibodies specific to the carbohydrate-recognition domain (CRD) of the dendritic cell-specific receptor DC-SIGN could specifically bind to and be internalized by dendritic cells *in vitro*. These particles thus have potential as dendritic cell-targeting vaccine carriers (Cruz et al. 2011).

Imaging

Incorporation of contrast agents and dyes within polymer nanoparticle-antibody hybrids can aid in tracking them by various imaging modalities following administration, thus adding a theranostic aspect to the system. Paclitaxel-containing PLGA nanoparticles incorporating either Fe_3O_4 or quantum dots and conjugated with anti-Her2 antibody were developed for combined imaging, near-infrared photothermal therapy and chemotherapy. This system was further tagged with GNPs that can absorb NIR light and produce heat, so that the particles can be destroyed causing rapid paclitaxel release at the delivery site. On laser irradiation, these particles demonstrated significantly higher HeLa cell toxicity than cells treated with the drug-loaded particles without irradiation and cells treated with empty particles followed by irradiation. The quantum dots or iron oxide within the nanoparticle system can be used for imaging the particles after administration (Cheng et al. 2010). Alexa Fluor-488 labeled anti-CEA antibody was conjugated to lipid-PLGA nanoparticles (Hu et al. 2010) in order to visualize their uptake by CEA-positive BxPC-3 cells *in vitro*, indicating that these carriers could potentially be used for the delivery of fluorescent dyes for imaging applications. PLGA-based hybrid nanoparticles containing SPIO and docetaxel and conjugated with prostate stem cell antigen antibodies were synthesized for magnetic resonance imaging and cancer treatment (Ling et al. 2011). Feng and colleagues developed a multifunctional polymeric nanoparticle system prepared using four π-conjugated polymers that have distinct tunable emissions, which were then tagged with multiple antibodies for specific targeting of individual tumor cells (Feng et al. 2014). These particles showed good photostability and cytocompatibility with MCF-7 cells. Moreover, tagging the particles with both anti-EpCAM and anti-ErbB2 antibodies helped target the EpCAM-positive SKBR-3 and MCF-7 cells *in vitro*. Whereas when the particles were tagged only with anti-ErbB2 antibody, they could target only SKBR-3 cells and not MCF-7 cells. HeLa cells were negative for particles tagged with both antibodies. These novel particles could potentially be used for targeted imaging and for detecting different types of cells following administration.

Progress Towards Clinical Evaluation of Polymer-Antibody Hybrids

Although polymeric nanoparticle-antibody hybrids are being investigated extensively, this is still a relatively new area of study and hence has not currently entered clinical trials. However there are several preclinical studies that have shown encouraging results for these formulations. Preclinical studies on dextran-coated SPIO nanoparticles tagged with anti-VEGF antibody showed *in vivo* accumulation of the tagged particles in colon tumor, which could be visualized by MRI and Prussian blue staining. These particles could thus be potentially used for molecular-targeted imaging of tumors (Hsieh et al. 2012). SPIO nanoparticles in a dextran matrix with PEG surface groups were used by DeNardo for radiolabeling with ^{111}In and conjugation with ChL6 mAb for magnetic field-based cancer therapy (DeNardo et al. 2005). *In vivo* studies on the human breast cancer HBT 3477 tumor-bearing mouse model indicated that the particles had a good blood half-life and accumulated in the tumor following intravenous administration. These particles were visualized by radioimmuno-imaging due to the availability of ^{111}In radioactive tracer and could potentially be used for alternating magnetic field (AMF)-based thermal ablative therapy. The availability of several biocompatible polymers for the synthesis of nanoparticles containing contrast and therapeutic agents and surface modified with cell-specific antibodies demonstrates the multifunctionality of these systems and their potential for providing targeted combinational therapy.

Gold Nanoparticle-Antibody Hybrids

Gold has been used for medical purposes for centuries and has been exploited for various applications such as the treatment of epilepsy, arthritis and tumors (Dykman and Khlebtsov 2012). With the emergence of nanotechnology, gold nanoparticles (GNPs) have been intensively studied and a few products have been developed as cancer treatments and have progressed to clinical trials. GNPs of various sizes and shapes such as nanospheres, nanorods and nanocages, are of biomedical research interest because of their facile synthesis and functionalization, monodispersity, controllable size and lack of toxicity. GNPs are promising nanocarriers for the delivery of both small molecular drugs and biomolecules such as nucleic acids, proteins, peptides and carbohydrates (Cao-Milán and Liz-Marzán 2014; Ghosh et al. 2008; Rana et al. 2012; Song et al. 2013). In addition, due to their unique physicochemical and optical properties, GNPs have been exploited for imaging and phototherapy, as contrast agents and radiosensitizers (Cao-Milán and Liz-Marzán 2014; Hainfeld et al. 2013). To improve the specificity of GNP-based systems and therefore, to minimize toxicity to healthy tissues and to avoid serious side effects *in vivo,* the attachment of targeting ligands is critical. Among these, antibodies including antibody fragments, have been widely studied and are incorporated in several clinical products. GNP-antibody hybrids, known as 'immunogold', were first developed in 1971 (Faulk and Taylor 1971) and have since been extensively studied.

Design and Synthesis of GNP-Antibody Hybrids

Antibodies can be attached to GNPs non-covalently or covalently. For non-covalent attachment, antibodies are simply mixed with GNPs at the optimal pH for direct physisorption. Examples of antibodies bound to GNPs at their isoelectric points via hydrophobic interaction, have been proposed as cancer diagnostics and therapeutics (Sokolov et al. 2003; El-Sayed et al. 2005, 2006). It is a quick and simple process to produce GNP-antibody hybrids, but under non-covalent binding equilibrium both GNP-bound and free antibodies exist within the pool of GNP-antibody hybrids. Also the random orientation of the antibodies anchored to the GNP surface is a disadvantage. Poor stability and biocompatibility of such hybrids are due to nonspecific protein adsorption resulting in aggregation or antibody replacement by other biomolecules in biological samples (Eck et al. 2008; Kumar et al. 2008).

There are two methods to covalently bind antibodies to GNPs: direct binding or indirect binding via a linker. If the antibodies have accessible sulphur atoms (i.e., Cys residues), GNP-antibody hybrids are directly formed via Au-S bonds. For example, Ackerson et al. reported a GNP-antibody hybrid using a scFv with an accessible cysteine residue for direct GNP-scFv coupling (Ackerson et al. 2006). It has been suggested that with this method, the Au-S bond is not the only force to form GNP-antibody hybrids, as non-covalent interactions such as ionic attraction and/or hydrophobic attraction are also involved.

A common approach for covalent attachment of antibodies to GNPs is via indirect binding through a linker, such as heterobifunctional PEG. These linkers generally have a sulphur-containing functional group at one end for attachment to the GNP surface and another functional group at the other end for coupling to antibodies. For example, thiol/dithiol-PEG-COOH linkers and ortho pyridyl disulfide-polyethylene glycol-N-hydroxysuccinimide (OPSS-PEG-NHS) have been used to prepare Au-antibody/scFv hybrids (Eck et al. 2008; Loo et al. 2005a; Qian et al. 2008). Here the linkers were bound to GNPs via Au-S bonds and the antibodies were attached using NHS/EDC coupling chemistry. Joshi et al. reported another type of heterofunctional PEG linker with hydrazide and dithiol groups to attach antibodies to GNPs with controllable antibody orientation on the GNP surface (Joshi et al. 2013). In this case antibodies were conjugated specifically through the carbohydrate moiety on one of the heavy chains of the Fc region (Kumar et al. 2008). The carbohydrate was oxidized to an aldehyde group and then reacted with the hydrazide group of the PEG linker. As a result, the antigen-binding sites on the Fab region were directed outward, away from the GNP surface, facilitating binding to target. Due to the hydrophilic nature of PEG, gold-antibody hybrids using PEG linkers can suppress nonspecific binding. They are more stable and biocompatible both *in vitro* and *in vivo* and are highly selective for specific targets (Eck et al. 2008). Carbon linkers (Liao and Hafner 2005), N-succinimidyl-S-acetylthiopropionate (SATP) (Das et al. 2011), plasma-polymerized-allylamine shell (PPAA) (Marega et al. 2012), poly (acrylic acid) (PAA) (Popovtzer et al. 2008), oligonucleotide (Nitin et al. 2007) and avidin-biotin (Niemeyer and Ceyhan 2001) have also been reported for use with GNPs.

Biomedical Applications of GNP-Antibody Hybrids

GNP-targeting ligand combinations have numerous physicochemical properties that have been explored in biomedical applications such as imaging, drug delivery and photothermal therapy.

Electron Microscopic (EM) Imaging

GNP-antibody hybrids have been exploited as an immune marker for EM since the 1970s (Faulk and Taylor 1971). In this context GNPs are conjugated to secondary antibodies and because of the high electron density of GNPs, provide high contrast. With silver enhancement, GNP-antibody hybrids can also be applied to light microscopy.

Optical Imaging

GNPs can strongly absorb and scatter light through surface plasmon resonance (SPR), the collective oscillation of the conduction electrons on the surface of gold under excitation by the light of a specific wavelength. The scattering intensity of GNPs is much higher than that of non-plasmonic nanoparticles and has been applied to imaging in dark field microscopy. In addition, these light scattering properties can be tuned by changing the size or shape of GNPs. For example, anti-EGFR mAb conjugated gold nanospheres (El-Sayed et al. 2005) or nanorods (Huang et al. 2006) were used to visualize malignant oral epithelial cell lines (HOC 313 clone 8 and HSC 3) and to distinguish them from non-malignant epithelial cells (HaCat) under dark-field microscopy as a result of specific binding of the gold-antibody hybrid to the malignant cells. Using the same imaging technique, optical identification of resected human cancer tissue has been described (Eck et al. 2008). FAP-binding F19 mAb was covalently bound to GNPs to form GNP-F19 hybrids and employed to label human pancreatic cancer tissue. The GNP-F19 labeled cancer tissue was clearly imaged under dark-field microscopy and showed the same stromal structure seen with immunohistochemical secondary antibody staining of FAP-α. Given the high scattering intensity, lack of photobleaching (which allows imaging over long periods of time) and the option to use near-infrared (NIR) light to increase optical tissue penetration, imaging using GNP-antibody hybrids holds promise for clinical diagnosis and surgical applications.

X-Ray Imaging

Due to the high atomic number and X-ray absorption coefficient of gold, GNPs have been investigated as contrast agents for X-ray computed tomography (CT) (Lee et al. 2013; Liu et al. 2012; Xi et al. 2012). Following i.v. administration to mice, GNPs can be successfully imaged (Hainfeld et al. 2006; Kim et al. 2007) and it has been proposed that, by incorporating targeting ligands such as antibodies, GNP-antibody hybrids could potentially transform CT into a targeted molecular imaging modality. Eck et al. used anti-CD4 mAb-GNP hybrids as an X-ray contrast agent to target peripheral lymph nodes in mice (Eck et al. 2010). After i.v. administration, the anti-CD4 antibody-GNP hybrids produced distinctly enhanced X-ray contrast of the lymph nodes, which were clearly visualized on CT. In another example of the use of GNP-antibody hybrids for imaging, human squamous cell carcinoma head and neck tumor implanted into nude mice was clearly imaged by CT after the administration of anti-EGFR antibody-GNP hybrids (Reuveni et al. 2011).

Other Imaging Techniques

GNP-antibody hybrids have been applied to other detection techniques such as photoacoustic imaging (Agarwal et al. 2007; Joshi et al. 2013) and surface-enhanced Raman spectroscopy (SERS) and imaging (Lee et al. 2009; Qian et al. 2008). Photoacoustic imaging is a non-invasive technique based on the photoacoustic effect, a process of converting absorbed light into acoustic waves that can be recorded by ultrasonic transducers to form images with high resolution. GNPs can absorb light very efficiently due to SPR and the resonance wavelength can also be tuned to the range of 700-900 nm (tissue transparent window) to maximize light penetration depth in tissues. Thus, GNP-antibody hybrids are good candidates for targeting cancer cells for high contrast photoacoustic imaging. GNPs can significantly amplify the SERS efficiency of Raman reporters by 14-15 orders of magnitude. Qian and others have exploited this property, demonstrating that small Raman reporter molecules (e.g., malachite green) can be incorporated into and stabilized by GNP-ScFv B10 antibody hybrids for *in vivo* tumor detection by SERS (Qian et al. 2008). Furthermore, GNP-antibody hybrids have been applied to PET and SPECT by incorporating radioactive isotopes such as Zr^{89} and In^{111} (Kao et al. 2013; Karmani et al. 2014), to fluorescence imaging by exploiting the native fluorescence from GNPs (He et al. 2008) and to two-photon luminescence imaging by taking advantage of the highly efficient two-photon-induced luminescence by GNPs (Durr et al. 2007; Zhang et al. 2013).

Drug Delivery

GNP-antibody hybrids are characterized by a large surface area to volume ratio and ease of surface functionalization and so have been developed as platforms for both chemotherapy and radiotherapy delivery. Acetuximab (C225)-GNP hybrid for targeted delivery of gemcitabine caused significant growth inhibition of EGFR-positive pancreatic cancer cells *in vitro* and tumor growth *in vivo* (Patra et al. 2008). A GNP-C225 hybrid was also exploited to carry a radioactive isotope, ^{131}I, as a theranostic agent (Kao et al. 2013). ^{131}I-labeled C225-GNP, which was also surface-modified by PEG, exhibited an enhanced anti-proliferative effect in EGFR-positive A549 lung carcinoma cells as compared to ^{131}I and other unlabeled hybrids. After i.v. injection of ^{131}I-C225-GNP to an A549 xenograft-bearing mouse model, tumor uptake was clearly imaged by SPECT.

Photothermal Therapy

As noted above GNPs absorb light strongly (5-6 orders of magnitude higher than most organic dyes) and almost 100% of the absorbed light can be converted into heat over a picosecond time scale (El-Sayed et al. 2006; Huang and El-Sayed 2010). Thus GNP-antibody hybrids hold promise for targeted photothermal therapy. Of the different GNP shapes and structures, gold nanorods, nanoshells and nanocages are of particular interest for this application since their SPR bands are shifted to the near-infrared (NIR) region (650-900 nm), where light extinction by blood and tissue is minimized, making deeper tissue penetration of light possible. For example, two EGFR-overexpressing oral epithelial malignant cell lines (HOC 313 clone 8 and HSC 3) were exposed to a gold nanorod-based anti-EGFR antibody hybrid and red laser (800 nm) (Huang et al. 2006). These cells were photothermally destroyed by using only half of the laser energy required for the non-malignant EGFR-negative control cell line, HaCat. Loo et al. developed a GNP-antibody hybrid for photothermal therapy by using gold nanoshells combined with anti-HER2 antibody (Loo et al. 2005b). HER2-positive SKBr3 breast carcinoma cells were exposed to the GNP-anti-HER2 hybrid or to a control GNP-anti-IgG hybrid and then exposed to NIR light (820 nm; 0.008 W/m^2). GNP-anti-HER2 hybrid caused cancer cell death whereas cells exposed to the control GNP-IgG hybrid or NIR alone were unaffected. Both GNP-antibody hybrids described above acted as a multifunctional theranostics because along with their therapeutic effect, they also enabled simultaneous optical imaging.

Radiosensitization

GNPs absorb X-rays strongly, mainly via the photoelectric effect, resulting in the emission of electrons and characteristic X-rays. These emissions cause an enhancement of the applied radiation dose and therefore may lead to amplified damage to nearby biomolecules (Hainfeld et al. 2008). This phenomenon has led to the development of targeted GNP-based radiosensitizers. For example, a GNP-trastuzumab hybrid was administered to athymic mice by direct injection into MDA-MB-361 xenografts 24 h prior to X-ray exposure (11 Gy; 100 kV) (Chattopadhyay et al. 2013). In the presence of the GNP-trastuzumab hybrid, X-radiation reduced the tumor size by 46% during a 4-month study, while the tumor size increased by 16% following radiation alone.

Other Nanomaterial-Antibody Hybrids

In addition to liposomes, polymers and GNPs, many other nanoparticle formulations have been investigated in combination with antibodies for a range of imaging, therapeutic and sensing applications. While it is not possible to provide a complete compendium of these applications, some important examples are described in this section. Functionalized mesoporous silica nanoparticles have been PEGylated and conjugated with a CD105-specific antibody, TRC105, for breast cancer targeting and labeled with ^{64}Cu for PET imaging (Chen et al. 2013). *In vivo* studies showed good tumor uptake of these highly stable particles within 0.5 hr of administration. Cerium oxide nanoparticles which are potent antioxidants have also been extensively studied and in one report, were tagged to anti-Aβ antibodies for targeting of Aβ plaques and simultaneous protection against oxidative stress as a possible treatment for neurodegenerative disorders (Cimini et al. 2012). Dendrimers have also been used in nanohybrids with antibodies. Polyamidoamine dendrimers conjugated with fluorescein isothiocyanate and two antibodies, 60 bca to target CD14 and J591 to target PSMA, were developed (Thomas et al. 2004). *In vitro* studies demonstrated good time- and dose-dependent internalization of these constructs by cells expressing the antigens (PSMA-expressing LNCaP and CD14-positive HL60 cells). In contrast, the antibody-dendrimer conjugates did not bind the CD14-negative Jurkat cell line or PSMA-negative PC3 cell line, thus demonstrating the specificity of the formulation. Nanographene oxide (NGO) conjugated with TRC105, a CD105 specific mAb, which was developed by Hong et al. to target the tumor vasculature demonstrated good tumor specificity and uptake *in vitro*, *in vivo* and *ex vivo* (Hong et al. 2012). In another study, NGO tagged with trastuzumab and radiolabeled with ^{111}In was taken up by HER2-positive tumors *in vivo* and used successfully for SPECT imaging of breast cancer xenografts (Cornelissen et al. 2013). Carbon nanotube (CNT)-antibody hybrids are being tested as innovative cell-specific sensors. A recent example describes how single-walled CNTs decorated with antibodies directed against the prostate cancer marker, osteopontin (OPN) and then placed between electrodes might be used for immunosensing since the binding of OPN to the CNT resulted in a measurable change in resistance (Sharma et al. 2015). These examples represent only a small selection of the burgeoning literature on nanoparticle-antibody hybrids but serve to demonstrate the versatility of these constructs and their potential as precision medicines and imaging agents.

Outlook and Challenges

Nanoparticle-antibody hybrids combine the properties of carrier materials capable of transporting high payloads of therapeutic and/or imaging agents with the specificity of targeting conferred by antibodies or antibody-like moieties. Many parameters must be considered during optimization of these hybrid systems including size, stability, biodegradability, blood half-life and the number of antibodies on the surface for targeting. Several investigators have reported an increase in nanoparticle size following antibody tagging. Aktaş et al. reported an increase in the size of chitosan-PEG nanoparticles from 149 nm to 590 nm on surface modification with biotin and to 657 nm with addition of OX26 antibody (Aktaş et al. 2005). Similarly Moura et al. observed a 30-50 nm increase in particle size on surface modification of SPIOs-loaded PLGA nanoparticles with anti-CD64 antibody (Moura et al. 2014). The variation in size range obtained for the two formulations above following antibody conjugation could be due to the difference in molecular weight and length of the antibody being used. For most applications it is important that nanoparticle diameter is maintained at less than 200 nm to avoid renal filtration, to prolong circulation time and to promote tumor- or tissue-specific uptake (Park et al. 2009). The addition of antibody makes this goal more difficult to achieve in some cases. Another problem that has been identified is that steric interference, especially of multifunctional immunoliposomes, could hinder the binding to target receptors on the tumor cell surface (Sawant and Torchilin 2012). While nanomaterial-antibody hybrid technology requires further research, progress will be made possible by careful target selection, meticulous engineering of the targeted nanoparticulate construct and advancement of strategies to overcome immunogenicity and opsonization. A significant hurdle to the development of antibody-nanomaterial hybrids is affordable large-scale production (Cheng et al. 2012). The development of facile, inexpensive synthetic methods is crucial for nanomaterial-antibody hybrids to be taken into clinical trials and ultimately, for their adoption into practice.

Acknowledgments

The authors wish to acknowledge grant support from Cancer Research UK, the CR-UK/EPSRC Oxford Cancer Imaging Centre and the EPSRC Oxford Centre for Drug Delivery Devices.

References

Abdelghany, S.M., Schmid, D., Deacon, J., Jaworski, J., Fay, F., McLaughlin, K.M., et al. 2013. Enhanced antitumor activity of the photosensitizer meso-tetra (N-methyl-4-pyridyl) porphine tetra tosylate through encapsulation in antibody-targeted chitosan/alginate nanoparticles. Biomacromolecules 14: 302-310.

Ackerson, C.J., Jadzinsky, P.D., Jensen, G.J. and Kornberg, R.D. 2006. Rigid, specific and discrete gold nanoparticle/antibody conjugates. J Am Chem Soc 128: 2635-2640.

Adams, G.P. and Weiner, L.M. 2005. Monoclonal antibody therapy of cancer. Nat Biotechnol 23: 1147-1157.

Agarwal, A., Huang, S.W., O'Donnell, M., Day, K.C., Day, M., Kotov, N., et al. 2007. Targeted gold nanorod contrast agent for prostate cancer detection by photoacoustic imaging. J Appl Phys 102: 064701.

Aktaş, Y., Yemisci, M. Andrieux, K., Gürsoy, R.N., Alonso, M.J., Fernandez-Megia, E., et al. 2005. Development and brain delivery of chitosan-PEG nanoparticles functionalized with the monoclonal antibody OX26. Bioconjug Chem 16: 1503-1511.

Allen, T. 1994. The use of glycolipids and hydrophilic polymers in avoiding rapid uptake of liposomes by the mononuclear phagocyte system. Adv Drug Deliv Rev 13: 285-309.

Allen, T., Hansen, C., Martin, F., Redemann, C. and Yau-Young, A. 1991. Liposomes containing synthetic lipid derivatives of poly (ethylene glycol) show prolonged circulation half-lives *in vivo*. Biochim Biophys Acta 1066: 29-36.

Alley, S.C., Okeley, N.M. and Senter, P.D. 2010. Antibody-drug conjugates: targeted drug delivery for cancer. Curr Opin Chem Biol 14: 529-537.

Bandekar, A., Zhu, C., Jindal, R., Bruchertseifer, F., Morgenstern, A. and Sofou, S. 2014. Anti-prostate-specific membrane antigen liposomes loaded with [225]Ac for potential targeted antivascular alpha-particle therapy of cancer. J Nucl Med 55: 107-114.

Burks, S.R., Macedo, L.F., Barth, E.D., Tkaczuk, K.H., Martin, S.S., Rosen, G.M., et al. 2010. Anti-HER2 immunoliposomes for selective delivery of electron paramagnetic resonance imaging probes to HER2-overexpressing breast tumor cells. Breast Cancer Res Treat 124: 121-131.

Burks, S.R., Legenzov, E.A., Rosen, G.M. and Kao, J.P. 2011. Clearance and biodistribution of liposomally encapsulated nitroxides: a model for targeted delivery of electron paramagnetic resonance imaging probes to tumors. Drug Metab Dispos 39: 1961-1966.

Cao-Milán, R. and Liz-Marzán, L.M. 2014. Gold nanoparticle conjugates: recent advances toward clinical applications. Expert Opin Drug Deliv 11: 741-752.

Chattopadhyay, N., Cai, Z.L., Kwon, Y.L., Lechtman, E., Pignol, J.P. and Reilly, R.M. 2013. Molecularly targeted gold nanoparticles enhance the radiation response of breast cancer cells and tumor xenografts to X-radiation. Breast Cancer Res Treat 137: 81-91.

Chen, F., Hong, H., Zhang, Y., Valdovinos, H.F., Shi, S., Kwon, G.S., et al. 2013. *In vivo* tumor targeting and image-guided drug delivery with antibody-conjugated, radiolabeled mesoporous silica nanoparticles. ACS Nano 7: 9027-9039.

Chen, M., Ouyang, H., Zhou, S., Li, J. and Ye, Y. 2014. PLGA-nanoparticle mediated delivery of anti-OX40 monoclonal antibody enhances anti-tumor cytotoxic T cell responses. Cell Immunol 287: 91-99.

Cheng, F.-Y., Su, C.-H., Wu, P.-C. and Yeh, C.-S. 2010. Multifunctional polymeric nanoparticles for combined chemotherapeutic and near-infrared photothermal cancer therapy *in vitro* and *in vivo*. Chem Commun 46: 3167-3169.

Cheng, Z., Al Zaki, A., Hui, J.Z., Muzykantov, V.R. and Tsourkas, A. 2012. Multifunctional nanoparticles: cost versus benefit of adding targeting and imaging capabilities. Science 338: 903-910.

Cimini, A., D'Angelo, B., Das, S., Gentile, R., Benedetti, E., Singh, V., et al. 2012. Antibody-conjugated PEGylated cerium oxide nanoparticles for specific targeting of Aβ aggregates modulate neuronal survival pathways. Acta Biomater 8: 2056-2067.

Cole, J.T. and Holland, N.B. 2015. Multifunctional nanoparticles for use in theranostic applications. Drug Deliv Transl Res 5: 295-309.

Cornelissen, B., Able, S., Kersemans, V., Waghorn, P.A., Myhra, S., Jurkshat, K., et al. 2013. Nanographene oxide-based radioimmunoconstructs for *in vivo* targeting and SPECT imaging of HER2-positive tumors. Biomaterials 34: 1146-1154.

Cruz, L.J., Tacken, P.J., Fokkink, R. and Figdor, C.G. 2011. The influence of PEG chain length and targeting moiety on antibody-mediated delivery of nanoparticle vaccines to human dendritic cells. Biomaterials 32: 6791-6803.

Das, A., Soehnlen, E., Woods, S., Hegde, R., Henry, A., Gericke, A., et al. 2011. VEGFR-2 targeted cellular delivery of doxorubicin by gold nanoparticles for potential antiangiogenic therapy. J Nanopart Res 13: 6283-6290.

DeNardo, S.J., DeNardo, G.L., Miers, L.A., Natarajan, A., Foreman, A.R., Gruettner, C., et al. 2005. Development of tumor targeting bioprobes (^{111}In-chimeric L6 monoclonal antibody nanoparticles) for alternating magnetic field cancer therapy. Clin Cancer Res 11: 7087s-7092s.

Durr, N.J., Larson, T., Smith, D.K., Korgel, B.A., Sokolov, K. and Ben-Yakar, A. 2007. Two-photon luminescence imaging of cancer cells using molecularly targeted gold nanorods. Nano Lett 7: 941-945.

Dykman, L. and Khlebtsov, N. 2012. Gold nanoparticles in biomedical applications: recent advances and perspectives. Chem Soc Rev 41: 2256-2282.

Eck, W., Craig, G., Sigdel, A., Ritter, G., Old, L.J., Tang, L., et al. 2008. PEGylated gold nanoparticles conjugated to monoclonal F19 antibodies as targeted labeling agents for human pancreatic carcinoma tissue. ACS Nano 2: 2263-2272.

Eck, W., Nicholson, A.I., Zentgraf, H., Semmler, W. and Bartling, S. 2010. Anti-CD4-targeted gold nanoparticles induce specific contrast enhancement of peripheral lymph nodes in X-ray computed tomography of live mice. Nano Lett 10: 2318-2322.

El-Sayed, I.H., Huang, X. and El-Sayed, M.A. 2005. Surface plasmon resonance scattering and absorption of anti-EGFR antibody conjugated gold nanoparticles in cancer diagnostics: applications in oral cancer. Nano Lett 5: 829-834.

El-Sayed, I.H., Huang, X.H. and El-Sayed, M.A. 2006. Selective laser photo-thermal therapy of epithelial carcinoma using anti-EGFR antibody conjugated gold nanoparticles. Cancer Lett 239: 129-135.

Elbayoumi, T.A. and Torchilin, V.P. 2007. Enhanced cytotoxicity of monoclonal anti-cancer antibody 2C5-modified doxorubicin-loaded pegylated liposomes against various tumor cell lines. Eur J Pharm Sci 32: 159-168.

Elzoghby, A.O., Samy, W.M. and Elgindy, N.A. 2012. Albumin-based nanoparticles as potential controlled release drug delivery systems. J Control Release 157: 168-182.

Faulk, W.P. and Taylor, G.M. 1971. An immunocolloid method for the electron microscope. Immunochemistry 8: 1081-1083.

Feng, L., Liu, L., Lv, F., Bazan, G.C. and Wang, S. 2014. Preparation and biofunctionalization of multicolor conjugated polymer nanoparticles for imaging and detection of tumor cells. Adv Mater 26: 3926-3930.

Freedman, M., Chang, E.H., Zhou, Q. and Pirollo, K.F. 2009. Nanodelivery of MRI contrast agent enhances sensitivity of detection of lung cancer metastases. Acad Radiol 16: 627-637.

Ghosh, P., Han, G., De, M., Kim, C.K. and Rotello, V.M. 2008. Gold nanoparticles in delivery applications. Adv Drug Deliv Rev 60: 1307-1315.

Goodall, S., Jones, M.L. and Mahler, S. 2014. Monoclonal antibody-targeted polymeric nanoparticles for cancer therapy – future prospects. J Chem Tech Biotechnol 90: 1169-1176.

Gu, F.X., Karnik, R., Wang, A.Z., Alexis, F., Levy-Nissenbaum, E., Hong, S., et al. 2007. Targeted nanoparticles for cancer therapy. Nano Today 2: 14-21.

Guin, S., Ma Q, Padhye, S., Zhou, Y.Q., Yao, H.P. and Wang, M.H. 2011. Targeting acute hypoxic cancer cells by doxorubicin-immunoliposomes directed by monoclonal antibodies specific to RON receptor tyrosine kinase. Cancer Chemother Pharmacol 67: 1073-1083.

Gurudevan, S., Kanwar, R.K., Veedu, R.N., Sasidharan, S., Kennedy, R.L., Walder, K., et al. 2013. Targeted multimodal liposomes for nano-delivery and imaging: an avenger for drug resistance and cancer. Curr Gene Ther 13: 322-334.

Haider, A., Gupta, K.C. and Kang, I.-K. 2014. PLGA/nHA hybrid nanofiber scaffold as a nanocargo carrier of insulin for accelerating bone tissue regeneration. Nanoscale Res Lett 9: 1-12.

Hainfeld, J.F., Slatkin, D.N., Focella, T.M. and Smilowitz, H.M. 2006. Gold nanoparticles: a new X-ray contrast agent. Br J Radiol 79: 248-253.

Hainfeld, J.F., Dilmanian, F.A., Slatkin, D.N. and Smilowitz, H.M. 2008. Radiotherapy enhancement with gold nanoparticles. J Pharm Pharmacol 60: 977-985.

Hainfeld, J.F., Smilowitz, H.M., O'Connor, M.J., Dilmanian, F.A. and Slatkin, D.N. 2013. Gold nanoparticle imaging and radiotherapy of brain tumors in mice. Nanomedicine (Lond) 8: 1601-1609.

Hansen, C., Kao, G., Moase, E., Zalipsky, S. and Allen, T. 1995. Attachment of antibodies to sterically stabilized liposomes – evaluation, comparison and optimization of coupling procedures. Biochim Biophys Acta 1239: 133-144.

He, H., Xie, C. and Ren, J. 2008. Nonbleaching fluorescence of gold nanoparticles and its applications in cancer cell imaging. Anal Chem 80: 5951-5957.

Hong, H., Yang, K., Zhang, Y., Engle, J.W., Feng, L., Yang, Y., et al. 2012. *In vivo* targeting and imaging of tumor vasculature with radiolabeled, antibody-conjugated nanographene. ACS Nano 6: 2361-2370.

Hsieh, W.J., Liang, C.J., Chieh, J.J., Wang, S.H., Lai, I.R., Chen, J.H., et al. 2012. *In vivo* tumor targeting and imaging with anti-vascular endothelial growth factor antibody-conjugated dextran-coated iron oxide nanoparticles. Int J Nanomedicine 7: 2833-2842.

Hu, C.M., Kaushal, S., Tran Cao, H.S., Aryal, S., Sartor, M., Esener, S., et al. 2010. Half-antibody functionalized lipid-polymer hybrid nanoparticles for targeted drug delivery to carcinoembryonic antigen presenting pancreatic cancer cells. Mol Pharm 7: 914-920.

Huang, A., Huang, L. and Kennel, S.J. 1980. Monoclonal antibody covalently coupled with fatty-acid. A reagent for *in vitro* liposome targeting. J Biol Chem 255: 8015-8018.

Huang, A., Kennel, S.J. and Huang, L. 1981. Immunoliposome labeling: a sensitive and specific method for cell surface labeling. J Immunol Methods 46: 141-151.

Huang, X.H. and El-Sayed, M.A. 2010. Gold nanoparticles: optical properties and implementations in cancer diagnosis and photothermal therapy. J Adv Res 1: 13-28.

Huang, X.H., El-Sayed, I.H., Qian, W. and El-Sayed, M.A. 2006. Cancer cell imaging and photothermal therapy in the near-infrared region by using gold nanorods. J Am Chem Soc 128: 2115-2120.

Jin, C., Yang, Z., Yang, J., Li, H., He, Y., An, J., et al. 2013. Paclitaxel-loaded nanoparticles decorated with anti-CD133 antibody: a targeted therapy for liver cancer stem cells. J Nanopart Res 16: 1-15.

Joshi, P.P., Yoon, S.J., Hardin, W.G., Emelianov, S. and Sokolov, K.V. 2013. Conjugation of antibodies to gold nanorods through Fc portion: synthesis and molecular specific imaging. Bioconjug Chem 24: 878-888.

Kanwar, J.R., Berg, R.W., Yang, Y., Kanwar, R.K., Ching, L.M., Sun, X., et al. 2003. Requirements for ICAM-1 immunogene therapy of lymphoma. Cancer Gene Ther 10: 468-476.

Kao, H.W., Lin, Y.Y., Chen, C.C., Chi, K.H., Tien, D.C., Hsia, C.C., et al. 2013. Evaluation of EGFR-targeted radioimmuno-gold-nanoparticles as a theranostic agent in a tumor animal model. Bioorg Med Chem Lett 23: 3180-3185.

Karmani, L., Bouchat, V., Bouzin, C., Leveque, P., Labar, D., Bol, A., et al. 2014. (89) Zr-labeled anti-endoglin antibody-targeted gold nanoparticles for imaging cancer: implications for future cancer therapy. Nanomedicine (Lond) 9: 1923-1937.

Kato, K., Hamaguchi, T., Yasui, H., Okusaka, T., Ueno, H., Ikeda, M., et al. 2006. Phase I study of NK105, a paclitaxel-incorporating micellar nanoparticle, in patients with advanced cancer. ASCO Annual Meeting Proceedings. J Clin Oncol 24: 2018.

Kim, D., Park, S., Lee, J.H., Jeong, Y.Y. and Jon, S. 2007. Antibiofouling polymer-coated gold nanoparticles as a contrast agent for *in vivo* X-ray computed tomography imaging. J Am Chem Soc 129: 7661-7665.

Kim, D.H., Vitol, E.A., Liu, J., Balasubramanian, S., Gosztola, D.J., Cohen, E.E., et al. 2013a. Stimuli-responsive magnetic nanomicelles as multifunctional heat and cargo delivery vehicles. Langmuir 29: 7425-7432.

Kim, H., Kee, P.H., Rim, Y., Moody, M.R., Klegerman, M.E., Vela, D., et al. 2013b. Nitric oxide improves molecular imaging of inflammatory atheroma using targeted echogenic immunoliposomes. Atherosclerosis 231: 252-260.

Kim, H., Kee, P.H., Rim, Y., Moody, M.R., Klegerman, M.E., Vela, D., et al. 2015. Nitric oxide-enhanced molecular imaging of atheroma using vascular cellular adhesion molecule 1-targeted echogenic immunoliposomes. Ultrasound Med Biol 41: 1701-1710.

Kocbek, P., Obermajer, N., Cegnar, M., Kos, J. and Kristl, J. 2007. Targeting cancer cells using PLGA nanoparticles surface modified with monoclonal antibody. J Control Release 120: 18-26.

Kolhar, P., Anselmo, A.C., Gupta, V., Pant, K., Prabhakarpandian, B., Ruoslahti, E., et al. 2013. Using shape effects to target antibody-coated nanoparticles to lung and brain endothelium. Proc Natl Acad Sci USA 110: 10753-10758.

Kouchakzadeh, H., Shojaosadati, S.A., Tahmasebi, F. and Shokri, F. 2013. Optimization of an anti-HER2 monoclonal antibody targeted delivery system using PEGylated human serum albumin nanoparticles. Int J Pharm 447: 62-69.

Kumar, S., Aaron, J. and Sokolov, K. 2008. Directional conjugation of antibodies to nanoparticles for synthesis of multiplexed optical contrast agents with both delivery and targeting moieties. Nat Protoc 3: 314-320.

Kumari, A., Yadav, S.K. and Yadav, S.C. 2010. Biodegradable polymeric nanoparticles based drug delivery systems. Colloids Surf. B Biointerfaces 75: 1-18.

Lasic, D., Martin, F., Gabizon, A., Huang, S. and Papahadjopoulos, D. 1991. Sterically stabilized liposomes: a hypothesis on the molecular origin of the extended circulation times. Biochim Biophys Acta 1070: 187-192.

Lee, N., Choi, S.H. and Hyeon, T. 2013. Nano-sized CT contrast agents. Adv Mater 25: 2641-2660.

Lee, R. and Low, P. 1995. Folate-mediated tumor cell targeting of liposome-entrapped doxorubicin *in vitro*. Biochim Biophys Acta 1233: 134-144.

Lee, S., Chon, H., Lee, M., Choo, J., Shin, S.Y., Lee, Y.H., et al. 2009. Surface-enhanced Raman scattering imaging of HER2 cancer markers overexpressed in single MCF7 cells using antibody conjugated hollow gold nanospheres. Biosens Bioelectron 24: 2260-2263.

Li, S., Goins, B., Hrycushko, B.A., Phillips, W.T. and Bao, A. 2012. Feasibility of eradication of breast cancer cells remaining in postlumpectomy cavity and draining lymph nodes following intracavitary injection of radioactive immunoliposomes. Mol Pharm 9: 2513-2522.

Liao, H.W. and Hafner, J.H. 2005. Gold nanorod bioconjugates. Chem Mater 17: 4636-4641.

Ling, Y., Wei, K., Luo, Y., Gao, X. and Zhong, S. 2011. Dual docetaxel/superparamagnetic iron oxide loaded nanoparticles for both targeting magnetic resonance imaging and cancer therapy. Biomaterials 32: 7139-7150.

Lingappa, M., Song, H., Thompson, S., Bruchertseifer, F., Morgenstern, A. and Sgouros, G. 2010. Immunoliposomal delivery of ^{213}Bi for alpha-emitter targeting of metastatic breast cancer. Cancer Res 70: 6815-6823.

Liu, Y.L., Ai, K.L. and Lu, L.H. 2012. Nanoparticulate X-ray computed tomography contrast agents: from design validation to *in vivo* applications. Acc Chem Res 45: 1817-1827.

Loo, C., Hirsch, L., Lee, M.H., Chang, E., West, J., Halas, N.J., et al. 2005a. Gold nanoshell bioconjugates for molecular imaging in living cells. Opt Lett 30: 1012-1014.

Loo, C., Lowery, A., Halas, N., West, J. and Drezek, R. 2005b. Immunotargeted nanoshells for integrated cancer imaging and therapy. Nano Lett 5: 709-711.

Löw, K., Wacker, M., Wagner, S., Langer, K. and von Briesen, H. 2011. Targeted human serum albumin nanoparticles for specific uptake in EGFR-expressing colon carcinoma cells. Nanomedicine 7: 454-463.

Lukyanov, A., Elbayoumi, T.A., Chakilam, A.R. and Torchilin, V.P. 2004. Tumor-targeted liposomes: doxorubicin-loaded long-circulating liposomes modified with anti-cancer antibody. J Control Release 100: 135-144.

Mamot, C., Ritschard, R., Wicki, A., Stehle, G., Dieterle, T., Bubendorf, L., et al. 2012. Tolerability, safety, pharmacokinetics and efficacy of doxorubicin-loaded anti-EGFR immunoliposomes in advanced solid tumours: a phase 1 dose-escalation study. Lancet Oncol 13: 1234-1241.

Mao, Y., Triantafillou, G., Hertlein, E., Towns, W., Stefanovski, M., Mo, X., et al. 2013. Milatuzumab-conjugated liposomes as targeted dexamethasone carriers for therapeutic delivery in CD74(+)B-cell malignancies. Clin. Cancer Res 19: 347-356.

Marega, R., Karmani, L., Flamant, L., Nageswaran, P.G., Valembois, V., Masereel, B., et al. 2012. Antibody-functionalized polymer-coated gold nanoparticles targeting cancer cells: an *in vitro* and *in vivo* study. J Mater Chem 22: 21305-21312.

Matsumura, Y., Gotoh, M., Muro, K., Yamada, Y., Shirao, K., Shimada, Y., et al. 2004. Phase I and pharmacokinetic study of MCC-465, a doxorubicin (DXR) encapsulated in PEG immunoliposome, in patients with metastatic stomach cancer. Ann Oncol 15: 517-525.

McCarron, P.A., Marouf, W.M., Quinn, D.J., Fay, F., Burden, R.E., Olwill, S.A., et al. 2008. Antibody targeting of camptothecin-loaded PLGA nanoparticles to tumor cells. Bioconjug Chem 19: 1561-1569.

Mohanty, A., Dilnawaz, F., Mohanta, G. and Sahoo, S. 2015. Polymer-drug conjugates for targeted drug delivery. pp. 389-407. *In:* P.V. Devarajan and S. Jain [eds.]. Targeted Drug Delivery: Concepts and Design. Springer International Publishing, Switzerland.

Moura, C.C., Segundo, M.A., das Neves, J., Reis, S. and Sarmento, B. 2014. Co-association of methotrexate and SPIONs into anti-CD64 antibody-conjugated PLGA nanoparticles for theranostic application. Int J Nanomedicine 9: 4911-4922.

Niemeyer, C.M. and Ceyhan, B. 2001. DNA-directed functionalization of colloidal gold with proteins. Angew Chem Int Ed Engl 40: 3685-3688.

Nitin, N., Javier, D.J. and Richards-Kortum, R. 2007. Oligonucleotide-coated metallic nanoparticles as a flexible platform for molecular imaging agents. Bioconjug Chem 18: 2090-2096.

Owen, S.C., Patel, N., Logie, J., Pan, G., Persson, H., Moffat, J., et al. 2013. Targeting HER2+ breast cancer cells: lysosomal accumulation of anti-HER2 antibodies is influenced by antibody binding site and conjugation to polymeric nanoparticles. J Control Release 172: 395-404.

Pandita, D., Kumar, S. and Lather, V. 2014. Hybrid poly(lactic-co-glycolic acid) nanoparticles: design and delivery prospectives. Drug Discov Today 20: 95-104.

Park, J.H., Gu, L., von Maltzahn, G., Ruoslahti, E., Bhatia, S.N. and Sailor, M.J. 2009. Biodegradable luminescent porous silicon nanoparticles for *in vivo* applications. Nat Mater 8: 331-336.

Pastorino, F., Brignole, C., Marimpietri, D., Cilli, M., Gambini, C., Ribatti, D., et al. 2003. Vascular damage and anti-angiogenic effects of tumor vessel-targeted liposomal chemotherapy. Cancer Res 63: 7400-7409.

Patra, C.R., Bhattacharya, R., Wang, E.F., Katarya, A., Lau, J.S., Dutta, S., et al. 2008. Targeted delivery of gemcitabine to pancreatic adenocarcinoma using cetuximab as a targeting agent. Cancer Res 68: 1970-1978.

Peer, D., Karp, J.M., Hong, S., Farokhzad, O.C., Margalit, R. and Langer, R. 2007. Nanocarriers as an emerging platform for cancer therapy. Nat Nano 2: 751-760.

Perez-Herrero, E. and Fernandez-Medarde, A. 2015. Advanced targeted therapies in cancer: drug nanocarriers, the future of chemotherapy. Eur J Pharm Biopharm 93: 52-79.

Popovtzer, R., Agrawal, A., Kotov, N.A., Popovtzer, A., Balter, J., Carey, T.E., et al. 2008. Targeted gold nanoparticles enable molecular CT imaging of cancer. Nano Lett 8: 4593-4596.

Qian, X.M., Peng, X.H., Ansari, D.O., Yin-Goen, Q., Chen, G.Z., Shin, D.M., et al. 2008. *In vivo* tumor targeting and spectroscopic detection with surface-enhanced Raman nanoparticle tags. Nat Biotechnol 26: 83-90.

Rana, S., Bajaj, A., Mout, R. and Rotello, V.M. 2012. Monolayer coated gold nanoparticles for delivery applications. Adv Drug Deliv Rev 64: 200-216.

Reuveni, T., Motiei, M., Romman, Z., Popovtzer, A. and Popovtzer, R. 2011. Targeted gold nanoparticles enable molecular CT imaging of cancer: an *in vivo* study. Int J Nanomedicine 6: 2859-2864.

Rüger, R., Tansi, F.L., Rabenhold, M., Steiniger, F., Kontermann, R.E., Fahr, A., et al. 2014. *In vivo* near-infrared fluorescence imaging of FAP-expressing tumors with activatable FAP-targeted, single-chain Fv-immunoliposomes. J Control Release 186: 1-10.

Sawant, R.R. and Torchilin, V.P. 2012. Challenges in development of targeted liposomal therapeutics. AAPS J 14: 303-315.

Senzer, N., Nemunaitis, J., Nemunaitis, D., Bedell, C., Edelman, G., Barve, M., et al. 2013. Phase I study of a systemically delivered p53 nanoparticle in advanced solid tumors. Mol Ther 21: 1096-1103.

Sharma, A., Hong, S., Singh, R. and Jang. J. 2015. Single-walled carbon nanotube based transparent immunosensor for detection of a prostate cancer biomarker osteopontin. Anal Chim Acta 869: 68-73.

Sofou, S. and Sgouros, G. 2008. Antibody-targeted liposomes in cancer therapy and imaging. Expert Opin Drug Deliv 5: 189-204.

Sokolov, K., Follen, M., Aaron, J., Pavlova, I., Malpica, A., Lotan, R., et al. 2003. Real-time vital optical imaging of precancer using anti-epidermal growth factor receptor antibodies conjugated to gold nanoparticles. Cancer Res 63: 1999-2004.

Song, H., He, R., Wang, K., Ruan, J., Bao, C., Li, N., et al. 2010. Anti-HIF-1α antibody-conjugated pluronic triblock copolymers encapsulated with Paclitaxel for tumor targeting therapy. Biomaterials 31: 2302-2312.

Song, L., Ho, V.H.B., Chen, C., Yang, Z.Q., Liu, D.S., Chen, R.J., et al. 2013. Efficient, pH-triggered drug delivery using a pH-responsive DNA-conjugated gold nanoparticle. Adv Healthc Mater 2: 275-280.

Suana, A.J., Tuffin, G., Frey, B.M., Knudsen, L., Muhlfeld, C., Rodder, S., et al. 2011. Single application of low-dose mycophenolate mofetil-OX7-immunoliposomes ameliorates experimental mesangial proliferative glomerulonephritis. J Pharmacol Exp Ther 337: 411-422.

Sun, B., Ranganathan, B. and Feng, S.S. 2008. Multifunctional poly(d, l-lactide-co-glycolide)/montmorillonite(PLGA/MMT) nanoparticles decorated by Trastuzumab for targeted chemotherapy of breast cancer. Biomaterials 29: 475-486.

Suzuki, S., Inoue, K., Hongoh, A., Hashimoto, Y. and Yamazoe, Y. 1997. Modulation of doxorubicin resistance in a doxorubicin-resistant human leukaemia cell by an immunoliposome targeting transferrin receptor. Br J Cancer 76: 83-89.

Thomas, T.P., Patri, A.K., Myc, A., Myaing, M.T., Ye, J.Y., Norris, T.B., et al. 2004. *In vitro* targeting of synthesized antibody-conjugated dendrimer nanoparticles. Biomacromolecules 5: 2269-2274.

Torchilin, V. 2005. Recent advances with liposomes as pharmaceutical carriers. Nat Rev Drug Discov 4: 145-160.

Torchilin, V. 2008. Antibody-modified liposomes for cancer chemotherapy. Expert Opin Drug Deliv 5: 1003-1025.

Torchilin, V. 2011. Tumor delivery of macromolecular drugs based on the EPR effect. Adv Drug Deliv Rev 63: 131-135.

Tuffin, G., Waelti, E., Huwyler, J., Hammer, C. and Marti, H.P. 2005. Immunoliposome targeting to mesangial cells: a promising strategy for specific drug delivery to the kidney. J Am Soc Nephrol 16: 3295-3305.

Vingerhoeds, M.H., Storm, G. and Crommelin, D.J. 1994. Immunoliposomes *in vivo*. Immunomethods 4: 259-272.

Wang, A.Z., Langer, R. and Farokhzad, O.C. 2012a. Nanoparticle delivery of cancer drugs. Annu Rev Med 63: 185-198.

Wang, L., Su, W., Liu, Z., Zhou, M., Chen, S., Chen, Y., et al. 2012b. CD44 antibody-targeted liposomal nanoparticles for molecular imaging and therapy of hepatocellular carcinoma. Biomaterials 33: 5107-5114.

Weng, K.C., Noble, C.O., Papahadjopoulos-Sternberg, B., Chen, F.F., Drummond, D.C., Kirpotin, D.B., et al. 2008. Targeted tumor cell internalization and imaging of multifunctional quantum dot-conjugated immunoliposomes *in vitro* and *in vivo*. Nano Lett 8: 2851-2857.

Xi, D., Dong, S., Meng, X.X., Lu, Q.H., Meng, L.J. and Ye, J. 2012. Gold nanoparticles as computerized tomography (CT) contrast agents. RSC Adv 2: 12515-12524.

Xu, H., Aguilar, Z.P., Yang, L., Kuang, M., Duan, H., Xiong, Y., et al. 2011. Antibody conjugated magnetic iron oxide nanoparticles for cancer cell separation in fresh whole blood. Biomaterials 32: 9758-9765.

Xu, L., Huang, C.C., Huang, W., Tang, W.H., Rait, A., Yin, Y.Z., et al. 2002. Systemic tumor-targeted gene delivery by anti-transferrin receptor scFv-immunoliposomes. Mol Cancer Ther 1: 337-346.

Yang, L., Mao, H., Wang, Y.A., Cao, Z., Peng, X., Wang, X., et al. 2009. Single chain epidermal growth factor receptor antibody conjugated nanoparticles for *in vivo* tumor targeting and imaging. Small 5: 235-243.

Zhang, Z.J., Wang, J. and Chen, C.Y. 2013. Gold nanorods based platforms for light-mediated theranostics. Theranostics 3: 223-238.

Zhou, M., Wang, L., Su, W., Tong, L., Liu, Y., Fan, Y., et al. 2012. Assessment of therapeutic efficacy of liposomal nanoparticles mediated gene delivery by molecular imaging for cancer therapy. J Biomed Nanotechnol 8: 742-750.

6

Applications of Organic-Inorganic Hybrid Nanoparticles for Cancer Imaging

Wei Du[a]*, Yusi Cui*[b] *and Gaolin Liang**

INTRODUCTION

Cancer is a generic group of diseases that can affect different parts of the body. It is a leading cause of death worldwide. According to the statistics of the World Health Organization (WHO), cancer leads 8.2 million deaths in 2012. Since most solid cancers are not applicable to palpation and inspection, precise imaging of cancers has been one of the cornerstones of cancer theranostics. Indeed, imaging provides correct anatomical delineation and staging of the malignant process of cancers, which directly impact the therapeutic strategy as well as the patient's prognosis (Debergh et al. 2012). Medical imaging aims to noninvasively obtain anatomic and physiologic information of human body not only for the detection and diagnosis of disease or injury, but also for the design, delivery, or monitoring of treatment. Medical imaging was truly born in 1895 when Roentgen took a medical imaging radiograph of his wife's hand by using X-rays. Since then, people realized that X-rays could be used to look inside the human body in order to detect and characterize the disease. Only two months later, X-rays were used clinically. Over the past 120 years, there has been a huge increase in the number of imaging technologies and their applications (Fig. 6.1).

CAS Key Laboratory of Soft Matter Chemistry, Department of Chemistry, University of Science and Technology of China, Hefei, Anhui 230026, China.

[a] E-mail: duweift@mail.ustc.edu.cn
[b] E-mail: cuiyusi@mail.ustc.edu.cn
* Corresponding author: gliang@ustc.edu.cn

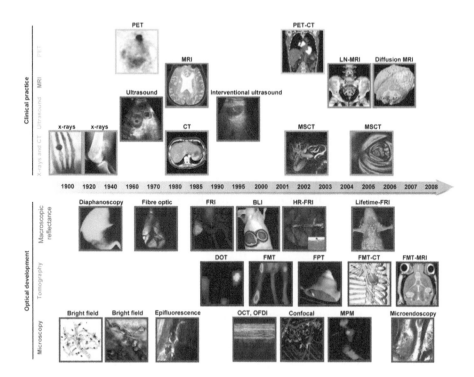

Fig. 6.1 Imaging technologies used in oncology. Many macroscopic imaging technologies (shown above the timeline) are in routine clinical use and there have been huge advances in their capabilities to obtain anatomical and physiological information since the beginning of the twentieth century. Shown here are some examples of bones (X-rays), soft tissue (ultrasound, MRI and CT rows), three-dimensional organs (CT and MRI rows) and physiological imaging (MRI and PET rows). Microscopic and other intravital optical techniques (shown below the timeline) have developed over the past decade and now allow studies of genetic, molecular and cellular events *in vivo*. Shown here are surface-weighted, whole-mouse, two-dimensional techniques (macroscopic reflectance row); tomographic three-dimensional techniques, often in combination with other anatomical modalities (tomography row); and intravital microscopy techniques (microscopy row). The timeline is approximate and is not to scale. BLI, bioluminescence imaging; CT, computed tomography; DOT, diffuse optical tomography; FMT, fluorescence-mediated tomography; FPT, fluorescence protein tomography; FRI, fluorescence reflectance imaging; HR-FRI, high-resolution FRI; LN-MRI, lymphotropic nanoparticle-enhanced MRI; MPM, multiphoton microscopy; MRI, magnetic resonance imaging; MSCT, multislice CT; OCT, optical coherence tomography; OFDI, optical frequency-domain imaging; PET, positron emission tomography. (Reprinted with permission from Weissleder, R. and Pittet, M. J., *Nature*, 452. 580–589. Copyright 2008, Nature Publishing Group.)

Broadly speaking, biomedical imaging can be categorized according to the electromagnetic spectrum as shown in Fig. 6.2: magnetic resonance, optical/near-infrared (NIR) and ionizing radiation (X-rays, γ-rays and particulate radiation) (Pansare et al. 2012). Imaging based on ionizing radiation generally refers to the detection of high-frequency emissions from radioactive isotopes (e.g., gamma rays from emitter ^{111}In or ^{99m}Tc) or the passage of X-rays through the body. The main techniques involved are positron emission tomography (PET), single-photon emission computed tomography (SPECT) and X-ray computed tomography (X-CT). Ultrasound imaging is the second-most widely used imaging modality in clinical practice after conventional X-ray radiography, in which sound waves are emitted from ultrasound devices and the echoes from diagnostic sites return to the sound transducer for disease screening or diagnosis. Comparison of the difference of the various existing imaging technologies in five main aspects (i.e., spatial and temporal resolution, depth penetration, energy expended for image generation (ionizing or nonionizing, depending on which component of the electromagnetic radiation spectrum is exploited for image generation), availability of injectable/biocompatible molecular probes and the respective detection threshold of probes for a given technology) is summarized and shown in Table 6.1.

Most of the clinically used imaging agents are small molecules (e.g., gadolinium complexes as T1 contrast agents for magnetic resonance imaging, MRI). However, short blood circulation time or non-specific biodistribution of these small molecules will affect the imaging results or cause unwanted side effects during their clinical applications. Recently, the tremendous development of nanomaterials with fine tuning of the composition, shape, size and chemical functionalities has brought hopes to overcome the limitations of small molecules. Nanoparticles (NPs) are among these new nanomaterials which possess large payloads, high signal intensity/stability, avidity and the capacity for multiple and simultaneous applications due to their high surface area to volume ratio (Jain et al. 2008). Compared with small molecules, the blood circulation times of nanoparticles can be significantly increased by size control and their imaging efficacy can be greatly enhanced by surface modification with targeting molecules such as antibodies and peptides (Kim et al. 2009). A wide range of substances, including organic materials (e.g., liposomes, natural and synthetic polymers including dendrimers and carbon nanotubes), inorganic materials (e.g., quantum dots (QDs, metallic nanostructures and metal oxides) and biomolecules (e.g., viruses, proteins, peptides, nucleotides), can be used for the preparation of nanoparticles for medical imaging (Fig. 6.3) (Kim et al. 2004; Lee et al. 2012b; Taylor-Pashow et al. 2010). Recently, due to the increased matching between organic and inorganic matters in structure, size and especially in material controllability at molecular and atomic level, organic-inorganic hybrid nanomaterials have gradually become one of the most promising platforms for medical imaging.

Table 6.1 **Comparison of Imaging Modalities** (Source: Massoud, T.F. and Gambhir, S.S. 2003. Molecular imaging in living subjects: seeing fundamental biological processes in a new light. *Genes. Dev.* 17. 545-580.)

Imaging technique	Source of imaging	Spatial resolution[a]	Temporal resolution[b]	Tissue penetrating depth	Sensitivity[c]	Types of probe	Advantages	Disadvantages
Magnetic resonance imaging (MRI)	Radiowave	25-100 um	minutes to hours	no limit	10^{-3} to 10^{-5} mol.L^{-1}	paramagnetic chelates (chelates of Mn (II), Mn (III) and Gd (III) ions), superparamagnetic iron oxide particles	highest spatial resolution, combines morphological and functional imaging	relatively low sensitivity, long scan and post-processing time, mass quantity of probe may be needed
X-ray computed tomography (CT)	X-rays	50-200 um	minutes	no limit	10^{-3} mol.L^{-1}	high atomic number atoms (iodine, barium sulfate, bismuth sulfide)	bone and tumor imaging, anatomical imaging	limited "molecular" applications, limited soft tissue resolution radiation
Positron emission tomography (PET)	Higher-energy γ-rays	1-2 mm	10 seconds to minutes	no limit	10^{-11}-10^{-12} mol.L^{-1}	short-lived positron emitting isotopes (^{18}F, ^{11}C, ^{13}N, ^{15}O, ^{124}I, ^{64}Cu)	high sensitivity, isotopes can substitute naturally occurring atoms, quantitative translational research	PET cyclotron or generator needed, relatively low spatial resolution, radiation to subject

Table 6.1 Contd.

Imaging technique	Source of imaging	Spatial resolution[a]	Temporal resolution[b]	Tissue penetrating depth	Sensitivity[c]	Types of probe	Advantages	Disadvantages
Single-photon emission computed tomography (SPECT)	lower-energy γ-rays	1-2 mm	minutes	no limit	10^{-10}-10^{-11} mol.L^{-1}	lower energy γ-emitting isotopes (^{123}I, ^{99m}Tc or ^{111}In-labeled compounds)	many molecular probes available, can image multiple probes simultaneously, may be adapted to clinical imaging systems	relatively low spatial resolution because of sensitivity, collimation, radiation
Optical florescence imaging	visible or near-infrared light	in vivo, 2-3 mm; in vitro, sub-um	seconds to minutes	<1 cm	10^{-9}-10^{-12} mol.L^{-1}	fluorescent dyes, quantum dots	high sensitivity, detects fluorochrome in live and dead cells	relatively low spatial resolution, surface-weighted
Ultrasound	high-frequency sound	50-500 um	seconds to minutes	mm-cm	—	microbubbles	real-time, low cost	limited spatial resolution, mostly morphological

[a] Spatial resolution is a measure of the accuracy or detail of graphic display in the images expressed in millimeters. It is the minimum distance between two independently measured objects that can be distinguished separately. It is a measure of how fine the image is.

[b] Temporal resolution is the frequency at which the final interpretable version of images can be recorded/captured from the subject once the imaging process is initiated. This relates to the time required to collect enough events to form an image and to the responsiveness of the imaging system to rates of any change induced by the operator or in the biological system at hand.

[c] Sensitivity, the ability to detect a molecular probe when it is present, relative to the background, measured in moles per liter.

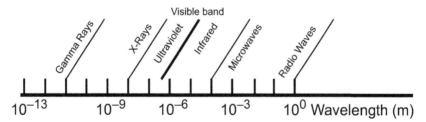

Fig. 6.2 Electromagnetic spectrum. (Reprinted with permission from Pansare, V., Hejazi, S., Faenza, W. and Prud'homme, R.K. Review of long-wavelength optical and NIR imaging materials: contrast agents, fluorophores and multifunctional nano carriers. *Chem. Mater.* 24. 812-827. Copyright 2012 American Chemical Society.)

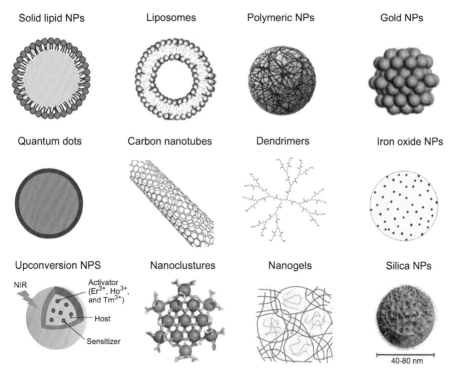

Fig. 6.3 Common nanoparticle platforms for medical imaging. Contrast agents may be developed as discrete crystalline geometries (spheres, rods, cubes, etc.) or incorporated into a variety of nanoparticle platforms, such as being entrapped in dendrimers, liposomes or micelles, or loaded into capsules, such as porous silica nanoparticles. (Adapted from *Nanomedicine: NBM.*, 8, Re, F., Gregori, M. and Masserini, M., Nanotechnology for neurodegenerative disorders, S51-S58, Copyright 2012, with permission from Elsevier.)

Hybrids are either homogeneous systems containing monomers and miscible organic and inorganic components or heterogeneous systems (nanocomposites) in which one or more of the components have a dimension ranging from A° to several nanometers. The properties of hybrid materials are influenced by both the chemical properties of the components and the interactions among these components. According to the interaction force of the inner interface, hybrids are generally categorized into two classes (Fig. 6.4). In class I, organic and inorganic components are embedded and cohere to each other via weak interactions (hydrogen bonds, ionic bonds or van der Waals force) to form the whole hybrid structure. In class II, at least a fraction of these two phases are linked together through strong chemical interactions (covalent bonds, iono-covalent bonds or Lewis acid-base bonds) (Sanchez et al. 2005). Hybrid organic–inorganic materials can not only retain the endogenous advantages of their inorganic and organic components, but most likely also introduce some special and innovative properties from the interactions between the organic and inorganic components for imaging applications. For example, the ability of hybrid organic–inorganic materials to combine a multitude of organic and inorganic components in a modular fashion allows them the control of the systematic tuning of the properties of the resultant hybrid nanomaterial. When organic/inorganic nanoparticles are functionalized with a variety of moieties, we get multifunctional hybrid nanoparticles.

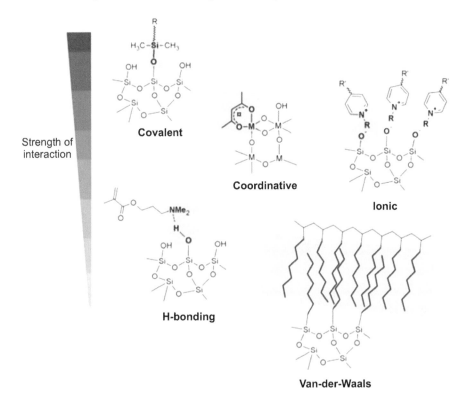

Fig. 6.4 Selected interactions typically applied in hybrid materials and their relative Strength. (Kickelbick, Guido. Hybrid materials: synthesis, characterization and applications. 4. 2007. Copyright Wiley-VCH Verlag GmbH & Co. KGaA. Reproduced with permission.)

Important Characteristics of Hybrid Nanomaterials for Cancer Imaging

Generally, for rapid and effective clinical translation, nanocarriers should have some basic properties as listed below. (Peer et al. 2007).

- They are made of materials which are biocompatible, well characterized and easily functionalized;
- They should exhibit high differential uptake efficiency in the target cells over normal cells (or tissues);
- They should be either soluble or colloidal in aqueous conditions for increased effectiveness;
- They should have an extended circulating half-life, a low rate of aggregation and a long shelf life.

In order to maximize the accumulation and detection accuracy of nanoparticles as imaging agents in the tumor, it is essential to take particle size, shape and surface chemistry as design parameters into consideration. First, there is more than one component in hybrid nanoparticles. As a result, the size will naturally be larger than an individual one-component nanoparticle. The size of a nanoparticle closely affects *in vivo* behaviors of the nanoparticle such as circulation time, circulation rate, extravasation, immunogenicity, degradation, clearance and cellular internalization. In biology, biomolecules such as amino acids, proteins, DNA and viruses are measured in nanometers and nanoscale devices and components are of the same basic size as biological entities (Fig. 6.5). Nanoscale devices smaller than 50 nm can easily enter most cells, whereas those smaller than 20 nm can transit out of blood vessels (Sekhon and Kamboj 2010). Based on the biokinetics of particles, the sizes of 10-100 nm are optimal for *in vivo* delivery because they escape the rapid renal clearance (< 10 nm) and sequester the reticuloendothelial system (RES) of spleen and liver (> 200 nm) (Fig. 6.6) (Sunderland et al. 2006).

Second, the mechanical flexibility of a hybrid nanoparticle influences its engagement with cellular surfaces. Hybrids based on flexible nanostructures such as liposomes, micelles and carbon nanotubes have enhanced diffusional processes and permeability as compared to the rigid nanostructures such as magnetic nanoparticles, gold nanoparticles and porous silica nanoparticles (Sailor and Park 2012). Therefore, surface chemistry seems to be crucially important to the rigid nanostructures. Modification of the nanoparticle's outer layer allows a large variety of chemical, molecular and biological entities to be covalently or non-covalently bound to it. For example, addition of polyethylene glycol (PEG) or other hydrophilic polymer to the surfaces of the nanoparticles will increase their *in vivo* compatibility and solubility. On the other hand, insoluble compounds can be attached, adsorbed or otherwise encapsulated in the hydrated nanoparticles (McNeil 2005).

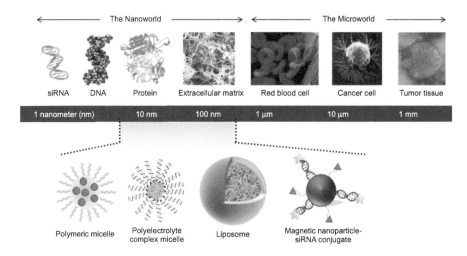

Fig. 6.5 Relative size of nanoparticles compared with familiar items. (Reprinted from Bae, K.H. et al., *Mol. Cells.*, 31, 295-302, 2011. With permission of The Korean Society for Molecular and Cellular Biology and Springer Netherlands.)

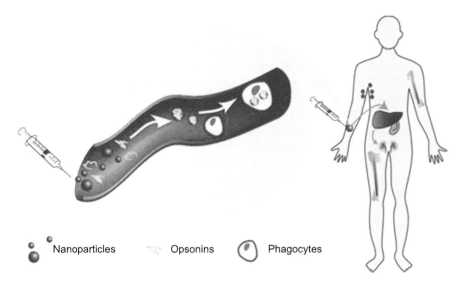

Fig. 6.6 Clearance of NPs by the reticuloendothelial system (RES). When larger NPs are injected into the blood stream, a group of proteins called opsonins cover their surface (opsonization), enabling phagocytic cells in the blood stream to recognize the NPs and remove them by degradation, renal excretion or accumulation in one of the RES organs (liver, spleen, etc.). (From Barreto, J.A. et al., Nanomaterials: applications in cancer imaging and therapy. *Adv. Mater.* 2011. 23. H18-40. Copyright Wiley-VCH Verlag GmbH & Co. KGaA. Reprinted with permission.)

Third, in order to achieve the desired multifunctionality, hybrid nanoparticles must contain multiple functional components that should all work in concert. However, the more complicated the modification of a hybrid nanoparticle, the tougher the fabrication process. The hybrid nanoparticle may dissociate into its sub-constituents before accomplishing its designed functions. For example, if an imaging component is stripped from the hybrid nanoparticle during circulation, the biodistribution of the hybrid will not be properly reported and this might lead to a false positive (or more likely negative) diagnosis. In addition, if the hybrid goes awry, leak of chemical and nanocrystal components could lead to unintended acute or long-term toxicity. Therefore, it is essentially necessary to carefully select nanocomponents and chemical agents to fabricate reliable and biocompatible hybrid nanoparticles and considerable attention should be paid to the design and *in vivo* behavior of the hybrid nanostructures.

Lastly, different functional nanocomponents incorporated into a hybrid nanodevice will interfere with each other. Thus, the imaging components co-loaded should share the limited space or volume available with other components in the hybrid. Therefore, careful selection of the incorporated functional components and loading strategy are required for the preparation of an efficient hybrid nanodevice. For example, a hollow hybrid, where the drugs are encapsulated in the interior and the reporting components are decorated on the exterior, is one possibility to maximize the therapeutic and diagnostic functions, respectively.

Successful delivery of nanoparticles to the tumor requires that the particle enters the tumor microcirculation, which can be achieved by two approaches, namely passive and active targeting methods. In the case of passive targeting, most nanoparticles are expected to accumulate in tumors due to the enhanced permeation and retention (EPR) effect. Thus, by controlling the size of nanoparticles, the homing vectors are designed to navigate through the tumor leaky vasculature (cutoff: 200-800 nm), enter into the tumor interstitium (extravasation step) and finally be delivered to the cytoplasmic targets in cancer cells (Ghosh et al. 2008; Toy et al. 2014). While in active targeting, the surface of nanoparticles is modified with functional groups with high affinity for the desired target. Targeting agents can be broadly classified into proteins (mainly antibodies and their fragments), nucleic acids (aptamers) or other receptor ligands (peptides, vitamins and carbohydrates). This strategy can enhance the specific recognition of hybrid nanoparticles with cell surface receptors. Thus, it is generally assumed that surface features on the nanoparticle are more important than those in the core, because the surface is in direct contact with the blood and organs (Choi et al. 2003). If blood is not the desired target, a strategy of active targeting and coating has to be developed for nanoparticles in order to minimize or delay the nanoparticle uptake by the mononuclear phagocyte system (MPS).

Types of Hybrid Organic–Inorganic Nanomaterials Currently Applied in Cancer Imaging

According to their structures, interior properties and roles in hybrid organic–inorganic imaging nanomaterials, nanocarriers can be categorized into 2 types:

Type I, functional nanoparticles: nanocarrier with endogenous contrast for imaging (e.g., iron oxide nanomaterials, quantum dots, gold nanoparticles (AuNPs), Fluorescent proteins) modified with organic functionalities.

Type II, structural nanoparticles: nanocarrier does not have imaging properties but can carry the imaging agent (e.g., radio-labeled silica, fluorescent silica, liposome, dendrimer, multimodal NPs). Functional nanocomponents can be either incorporated into the inner space of a structural nanocomponent ("tanker") or equipped on the surface of a structural nanocomponent ("barge").

Functional Nanoparticle-Based Hybrids

This type of nanoparticles are typically inorganic, including typical lanthanide upconversion nanoparticles (UCNPs), quantum dots (QDs), magnetic nanoparticles (MNPs), gold nanoparticles (AuNPs), etc. To prepare hybrid nanoparticles based upon these nanoparticles, three routine steps are necessary: synthesis, surface modification and bioconjugation, among which surface modification plays an important role for the whole procedure. Generally, these nanoparticles are hydrophobic, charged and possess large surface area to volume ratios. Moreover, naked metallic nanoparticles are chemically highly active and are easily oxidized in air, which generally results in loss of magnetism or dispersibility. In order to reduce the electrostatic and hydrophobic interactions among the NPs (agglomeration), to avoid their clearance by opsonization and to enhance their long circulation time and biocompatibility, it is necessary to introduce shielding groups onto their surfaces:

i. organic polymers, such as dextran, chitosan, polyethylene glycol, polysorbate, polyaniline and sugar.

ii. organic surfactants, such as sodium oleate and dodecylamine.

iii. bioactive molecules and structures, such as liposomes, peptides, viruses and ligands/receptors.

A suitable coating of a nanocarrier (e.g., with NH_2 or COOH groups available after coating) could be in favor of attaching functional ligands to the NPs using bioconjugation chemistries and "click" chemistries (Sun et al. 2006). Targeting agents, permeation enhancers, optical dyes and therapeutic agents can all be conjugated on the surfaces or incorporated within the interiors of the nanostructures (Fig. 6.7). The theranostic nanoparticulate systems are usually formed by using a contrast agent material (e.g., iron oxide nanoparticles) as the carrier for a therapeutic agent (e.g., small molecules, siRNA drugs, etc.).

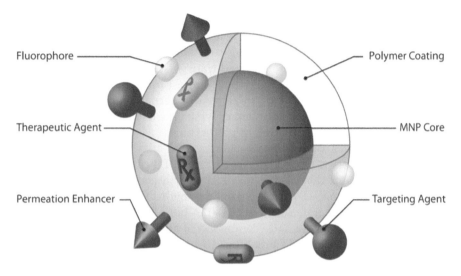

Fig. 6.7 MNP possessing various ligands to enable multifunctionality from a single nanoparticle platform. (Reprinted from *Ad. Drug. Deliv. Rev.*, 60, Sun, C. et al., Magnetic nanoparticles in MR imaging and drug delivery, 1252-265, Copyright 2008, with permission from Elsevier.)

Magnetic Nanoparticles (MNPs)-Based Hybrids

Magnetic nanoparticles, with a size generally between 3 and 10 nm, have been developed as contrast agents for both standard and functional MR imaging (Bonnemain 1998). According to the composition of the MNPs, they are normally classified into three major types: single metal oxide (iron oxide, manganese oxide, cobalt oxide etc.), dopant MNPs [MFe_2O_4, M = Mn, Ni, Co], metal-Alloy (FeCo, FePt, etc.) (Zhang 2015). Iron oxide nanoparticles (IONPs) are the most widely used magnetic NPs for *in vivo* bio-imaging applications. They can be subdivided into 3 categories: iron oxide NPs with polymeric coating-monocrystalline iron oxide nanocolloid (MION, 5-30 nm), cross-linked iron oxide (CLIO, 10-50 nm) and superparamagnetic iron oxide (SPIO, > 50 nm). Among them, superparamagnetic IONPs (SPIO) and ultrasmall SPIO (USPIO) developed in 1980s were approved by the U.S. Food and Drug Administration (FDA) for clinical applications (Huang et al. 2012). Currently, RES-mediated uptake of SPIOs has been applied for clinical imaging of liver tumors and metastases as small as 2-3 mm (Corot et al. 2006). In addition, the clinical application of USPIO MNPs to improve the delineation of brain tumor boundaries and quantify tumor volumes has been under evaluation (Enochs et al. 1999). The main advantages of MNPs can be summarized as follows (Nguyen et al. 2012). Firstly, MNPs have a unique ability to be guided by an external magnetic field and this has been utilized for targeted drug and gene delivery, tissue engineering, cell tracking and bioseparation. Secondly, with the ability of perturbing local magnetic field, they can serve as effective contrast enhancers in magnetic resonance imaging (MRI). It is well known that magnetic nanoparticles create extremely large microscopic field gradients in a magnetic field, causing substantial diphase and shortening of longitudinal relaxation time (T1) and transverse relaxation time (T2 and T2*) of nearby nuclei (e.g., ^{1}H) in the case of most MRI applications. Finally, MNPs can effectively adsorb energy from an external alternating magnetic field to create nanosized heating sources which are used as thermo-seeds in magnetic inductive heating (MIH) hyperthermia.

Surface properties of magnetic nanoparticles play important roles in their magnetic properties and efficiency for MRI, because the interactions between water and the MNPs occur primarily on the surface of the nanoparticles (Huang et al. 2012) but the hydrophobic surfaces of the MNPs tend to agglomerate (Lu et al. 2007). A biomedical organic–inorganic hybrid MNP platform is usually comprised of an inorganic MNP core and a biocompatible organic surface coating. The organic surface coating is used to passivate the surface of the nanoparticles during or after their synthesis in order to avoid agglomeration and to enhance the biocompatibility of the MNPs (Fig. 6.8). Polymers containing functional groups (e.g., carboxylic acids, phosphates and sulfates) can be used to coat the surface of the magnetite. Suitable polymers for coating include poly(pyrrole), poly(aniline), poly(alkylcyanoacrylates), poly(methylidene malonate) and polyesters such as poly(lactic acid), poly(glycolic acid), poly(e-caprolactone) and their copolymers (Lu et al. 2007). For examples, Weissleder and co-workers (Shen et al. 1993; Josephson et al. 1999) developed various formulations of dextran-coated iron oxide nanoparticles as monocrystalline iron oxide nanoparticles (MION) or cross-linked iron oxide nanoparticles (CLIO). Application of suitable surface chemistry allows for the additional integration of functional ligands. For example, chemical functionality was established by treating CLIO with ammonia to provide primary amino groups for the attachment of biomolecules such as proteins or peptides (Schellenberger et al. 2002; Wunderbaldinger et al. 2002).

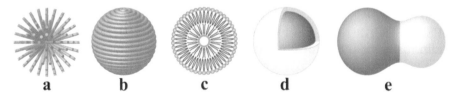

Fig. 6.8   MNP structures and coating schemes. (a) End-grafted polymer coated MNP. (b) MNP fully encapsulated in the polymer coating. (c) Liposome-encapsulated MNP. (d) Core–shell MNP. (e) Heterodimer MNP. (Reprinted from *Ad. Drug. Deliv. Rev.*, 60, Sun, C. et al., Magnetic nanoparticles in MR imaging and drug delivery, 1252-265, Copyright 2008, with permission from Elsevier.)

Although passive targeting of MNPs to tumors can be achieved through EPR effect, surface modification of the nanoparticles with targeting warheads to improve their tumor-targeting efficiency for MRI (or other) applications seems more attractive. When MNPs are functionalized with warheads targeting the biomarkers overexpressed in lesions (usually refer to cancers), effective ligand/receptor interactions (active targeting) will enhance the accumulation of the MNPs in malignant tissues. Using superparamagnetic iron oxide (SPIO) nanoparticles to target vascular integrins or VEGFR has been extensively exploited for MR imaging of tumor vasculature. SPIOs functionalized with deep tissue-targeting warheads such as urokinase plasminogen activator (uPA), transferrin receptors, HER2 receptors, chemokine receptor 4 have also been explored to MR image primary tumors (Toy et al. 2014). For example, Sun et al. demonstrated that the specific accumulation of chlorotoxin-targeted iron oxide nanoparticles (NP-PEG-CTX) in 9L glioma flank xenografts resulted in obviously enhanced MR contrast of the tumors, as compared to tumors in control groups (Fig. 6.9) (Sun et al. 2008).

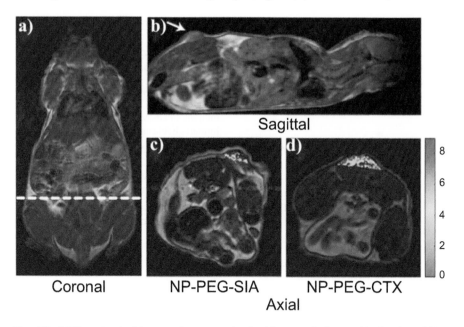

Fig. 6.9 MRI anatomical image of a mouse in the (a) coronal plane with the dotted line displaying the approximate location of the axial cross sections displayed in (c) and (d). Anatomical image in the (b) sagittal plane displaying the location of the 9L xenograft tumor. Change in R2 relaxation values for the tumor regions (superimposed over anatomical MR images) for mouse receiving (c) non-targeting PEG-coated iron oxide nanoparticles and (d) CTX-targeted PEG-coated iron oxide nanoparticles 3 h post nanoparticle injection. (From Sun, C. et al., *In vivo* MRI detection of gliomas by chlorotoxin-conjugated superparamagnetic nanoprobes. *Small.* 2008. 4. 372–379. Copyright Wiley-VCH Verlag GmbH & Co. KGaA. Reprinted with permission.)

MRI provides both anatomical and physiological images inside the human body with high spatial resolution, but current MRI agents are not sensitive enough. Optical or radionuclide imaging provides physiological and molecular information with high sensitivity, but their spatial resolutions are far below the requirement of good anatomical images (Key and Leary 2014). When MNPs are integrated with other imaging contrasts, the hybrid nanostructures exhibit paramagnetism alongside other features such as fluorescence. This type of dual-modality MNPs combine the advantages of two imaging modalities or compensate for the limitations of a single imaging modality. There are many possible combinations of dual-modality imaging modalities such as MR–Optical, MR–PET, MR–SPECT and MR-CT. These imaging contrast agents for hybridizing with MNPs can be either small molecules (e.g., ^{18}F, ^{64}Cu, CY5.5) or nanocomponents (e.g., quantum dots (QDs) or metallic nanoparticles) (Fig. 6.10). Some examples of dual-modality imaging contrast agents based on magnetic nanoparticles (MNPS) are summarized in Table 6.2.

Table 6.2 Representative MNPs-based Hybrid Nanoparticles for Dual-modality Cancer Imaging.

Imaging modality	Contrast	Coating	Application	Ref.
MR–optical	IONPs, IRDye800	PEG	targeted MR/FL imaging of U87MG human glioblastoma	(Chen et al. 2009)
	$Tm^{3+}/Er^{3+}/Yb^{3+}$ codoped $NaGdF_4$ upconversion nanophosphors	Azelaic acid	MR and NIR-to-NIR UCL dual-modality imaging of living whole-body animals	(Zhou et al. 2010)
	EGFRA-conjugated $Au–Fe_3O_4$ dumbbell nanoparticles	PEG	simultaneous magnetic and optical detection of A431 cells	(Xu et al. 2008)
	γ-Fe_2O_3–CdSe magnetic quantum dots	Silica layer, BMA	labeling of live cell membranes of 4T1 mouse breast cancer cells and HepG2 human liver cancer cells	(Lin et al. 2011)
MR–PET	^{124}I-SA-Mn-doped Fe_3O_4 NPs	SA		(Choi et al. 2008)
	^{69}Ge-SPION	PEG	lymph-node mapping	(Chakravarty et al. 2014)
	[^{64}Cu(dtcbp)$_2$]–Endorem	dextran		(de Rosales et al. 2011)

Table 6.2 Contd.

Imaging modality	Contrast	Coating	Application	Ref.
MR–SPECT	^{111}In-mAbMB-SPION conjugates	Carboxy methyl dextran	*in vivo* detection of mesothelin-expressing tumors	(Misri et al. 2012)
	[99m]TcSPIONs	PEG	sentinel lymph node (SLN) mapping *in vivo*	(Madru et al. 2012)
MR-CT	hybrid gold–iron oxide (Au–Fe_3O_4) nanoparticles	Poly(DMA-r-mPEGMA-r-MA)	*in vivo* hepatoma imaging	(Kim et al. 2011)
	Fe_3O_4@Au-mPEG-PEI.NHAc composite nanoparticles	mPEG-PEINHAc	MR imaging of mouse liver and CT imaging of rat liver and aorta	(Li et al. 2013)
	Strawberry-like Fe_3O_4-Au nanoparticles	DMSA	Accurately distinguish the grade of liver disease	(Zhao et al. 2015)

SA: serum albumin;
SPION: superparamagnetic iron oxide nanoparticle;
dtcbp: dithiocarbamatebisphosphonate;
Endorem: a superparamagnetic contrast agent for magnetic resonance imaging of the liver and spleen. The contrast agent consists of dextran-coated iron oxide particles with a size distribution between 120 and 180 nm and an iron concentration of 0.2 mol/L;
TMAH: Tetramethylammonium hydroxide;
GSH: Glutathione;
BMA: Oleyl-O-poly (ethyleneglycol)-succinyl-N-hydroxysuccinimidyl ester;
mAbMB: a kind of anti-mesothelin antibody;
DMA: dodecyl methacrylate;
MA: methacrylic acid;
mPEGMA: methyl ether methacrylate;
PEI: polyethyleneimine;
mPEG: poly (ethylene glycol) monomethyl ether;
DMSA: meso-2,3-dimercaptosuccinic acid.

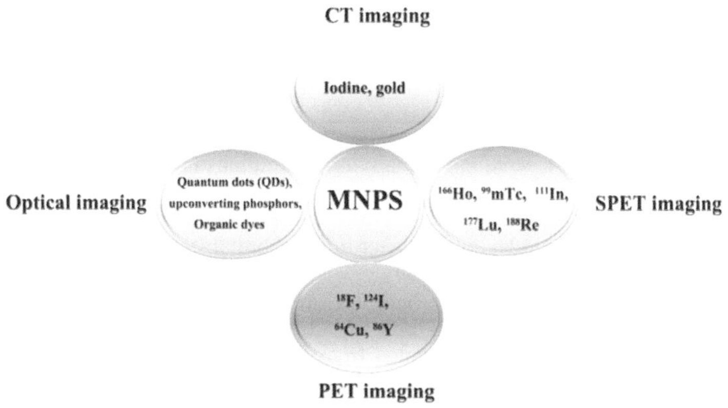

Fig. 6.10 Incorporation of various contrast agents for multimodality imaging.

Recently, trimodality imaging has also been successfully demonstrated by some research groups. Trimodal imaging provides more information than single- or dual-modality imaging. Weissleder and co-workers (Nahrendorf et al. 2008) decorated NIR fluorescent dyes on the crosslinked iron oxide nanoparticles (CLIO) which were coated with aminated dextran and then conjugated DTPA to the hybrids for the chelation of PET tracer [64]Cu. This CLIO hybrid provides three-modal imaging agents for PET, MRI and optical imaging. Core-shell strategy is another simple method to reach multimodal imaging. Zhu et al. developed core-shell Fe_3O_4@$NaLuF_4$: Yb, Er/Tm nanoparticles (MUCNP) which have multimodal imaging properties (Zhu et al. 2012). In this work, core-shell Fe_3O_4@$NaLuF_4$: Yb, Er/Tm nanoparticles (MUCNP) with Fe_3O_4 as the core and $NaLuF_4$: Yb, Er/Tm as the shell layer have been designed and synthesized with one-pot method. T2 magnetic resonance properties were the result from the Fe_3O_4 cores, while the $NaLuF_4$: Yb, Er/Tm outer shell exhibited not only excellent UCL emission but also positive CT contrast owing to the large atomic number and high X-ray absorption coefficient of lutetium. This type of hybrid nanomaterial was successfully used for MR, CT and UCL multimodal imaging of tumor-bearing mice (Fig. 6.11).

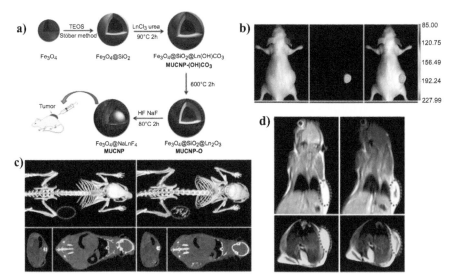

Fig. 6.11 (a) Schematic representation of the synthetic routine of the multifunctional Fe_3O_4@$NaLuF_4$: Yb, Er/Tm nanoparticle (MUCNP). Ln stands for lanthanide elements (Lu: Yb: Er = 78:20:2 or Lu : Yb : Tm = 79:20:1). (b) *In vivo* UCL images of the tumor-bearing nude mice in bright field, dark field and overlay images after intratumoral injection of MUCNP. (c) *In vivo* CT volume-rendered and maximum intensity projection of coronal, transversal images of the tumor-bearing mouse after intratumoral injection. (d) T2-weighted coronal and transversal cross-sectional images of the nude mice bearing a tumor pre-injection and at 30 min after intratumoral injection of MUCNP. (Reprinted from *Biomaterials*, 33, Zhu, X., Zhou, J., Chen, M., Shi, M., Feng, W. and Li, F., Core-shell Fe_3O_4@$NaLuF_4$: Yb, Er/Tm nanostructure for MRI, CT and upconversion luminescence tri-modality imaging, 4618-4627, Copyright 2012, with permission from Elsevier.)

Iron-oxide-based nanoparticles have been developed for theranostic applications with various formulations, in which the nanoparticle carries imaging, targeting and therapeutic capabilities. The main strategy for the design of multifunctional theranostic nanoparticles is the simultaneous incorporation of the dual modality of the diagnostic element and drug delivery motif into a particle. For instance, co-encapsulation of a lipophilic near infrared (NIR) dye and the chemotherapeutic drug (i.e. taxol) within hydrophobic pockets of the polymeric matrix of poly(acrylic acid) (PAA), which were coated on the surfaces of IONPs (PAA-IONPs), resulted in multifunctional hybrids for combinational optical imaging, magnetic resonance imaging (MRI) and cancer therapy. Covalent modification of the polymeric surfaces of the nanoparticles with folate allowed their specific uptake by A549 lung cancer cells, intracellular release of taxol and cell death (Santra et al. 2009). Since MNPs can generate heat under alternating magnetic fields due to their energy losses in the traversing of the magnetic hysteresis loop, MNPs themselves can serve as theranostic agents. Hayashi et al. reported the preparation of theranostic FA-PEG-SPION NCs using superparamagnetic nanoparticle clusters (SPION NCs) modified with PEG and folic acid (FA) (Hayashi et al. 2013). Clustering of SPION not only prevents their leaking from fenestrated capillaries, but also increases their relaxivity and the specific absorption rate. FA-PEG-SPION NCs were applied for cancer theranostics combining magnetic resonance imaging and hyperthermia (HT) treatment. 24 h after intravenous injection of the clusters into nude mice bearing multiple myeloma, FA-PEG-SPION NCs were found locally accumulating in cancer (not necrotic) tissues within the tumor with enhanced MRI. HT-treated mice were alive after 12 weeks, while control mice without HT treatment died within 8 weeks.

Quantum Dots (QDs)-Based Hybrids

QDs are one type of nanoparticles (NPs) with three characteristic properties: semiconductive, zero-dimensional and strong fluorescent. They are usually composed of semiconductor elements in group II-VI such as CdSe, CdS and CdTe, or group IV-VI such as PbS, PbSe, PbTe and SnTe, or group III-V such as InAs and InP. Their sizes typically range from 1 to 10 nm. QDs have broad absorption but narrow emission spectra and their emission maxima are tunable from 450 to 850 nm by changing their sizes. Their merits of high resistance to photobleaching and high quantum yields render QDs as one of the ideal materials for long-term cellular or deep-tissue imaging. For *in vivo* imaging, QDs must have adequate circulating lifetime, show minimal nonspecific deposition and retain their fluorescence for a sufficiently long time. Initially, chemically synthesized QDs are hydrophobic and could not be employed for biochemical applications. Thus, it is necessary to make them "water soluble" (i.e., well disperse in water) in order to facilitate their conjugation to biomolecules and to adapt them for biological imaging.

Approaches to prepare "water soluble QDs" include: (i) derivatizing their surface with mercaptoacetic acid; (ii) encapsulating them in phospholipid micelles; (iii) derivatizing their surface with silica over coat; and (iv) coating them with amphiphilic polymers such as poly(acrylic acid), PEG and polyanhydrides (Sharma et al. 2006). *In vivo* imaging of QDs has been applied to many avenues such as lymph node mapping (Kim et al. 2004), blood vessel imaging, liver imaging based on macrophage uptake and receptor-based specific tumor targeting (Cai et al. 2006).

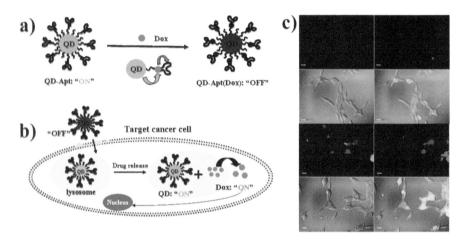

Fig. 6.12 (a) Schematic illustration of QD-Apt-Dox FRET system. (b) Schematic illustration of the specific uptake of QD-Apt-Dox conjugates into a target cancer cell through PSMA mediated endocytosis. (c) Confocal laser scanning microscopy images of PSMA expressing LNCaP cells after incubation with 100 nm QD-Apt-(Dox) conjugates for 0.5 h at 37°C, washed two times with PBS buffer and further incubated at 37°C for 0 h (upper two lines) and 1.5 h (lower two lines). Dox and QD are shown in red and green, respectively and the lower right images of each panel represents the overlay of Dox and QD fluorescent. The scale bar is 20 µm. (Reprinted with permission from Bagalkot, V., Zhang, L., Levy-Nissenbaum, E., Jon, S., Kantoff, P.W., Langer, R., et al., Quantum dot – Aptamer conjugates for synchronous cancer imaging, therapy and sensing of drug delivery based on Bi-fluorescence resonance energy transfer. *Nano Lett.* 7. 3065-3070. Copyright 2007 American Chemical Society.)

Since QDs possess a very broad excitation range extending to the UV region that allows them to be excited at any wavelength shorter than their excitation origin, QDs can be efficiently functionalized as FRET donors to various acceptors, with the possibility of selecting the wavelength at which the excitation of the acceptors does not occur. In 2007, Bagalkot et al. (Bagalkot et al. 2007) reported a novel quantum dot (QD)–aptamer(Apt)–doxorubicin (Dox) [QD–Apt(Dox)] bioconjugate system for targeted cancer imaging, therapy and sensing. As illustrated in Fig. 6.12, this conjugate is comprised of three components: (i) QDs, as fluorescent imaging vehicles, (ii) RNA aptamers covalently attached to the surface of QD, serving as a dual functional motif for cancer targeting and drug loading and (iii) doxorubicin

(Dox), a widely used anti-cancer drug with known fluorescent properties. A donor–acceptor model fluorescence resonance energy transfer (FRET) between QD and Dox and a donor–quencher model FRET between Dox and aptamer simultaneously existed when Dox was intercalated within the A10 aptamer. This is the "OFF" state, in which the fluorescence of the QDs is transferred to DOX while, simultaneously, the fluorescence of DOX is self-quenched among the molecules intercalated within the A10 PSMA aptamer. After the QD hybrids were uptaken by the targeting cancer cells, Dox was gradually released from the conjugate, turning the fluorescence of QD and Dox to the "ON" state.

QD-based imaging has advantages over those of organic fluorophores, but lacks anatomic resolution and is hampered by a limited tissue penetration depth even in the NIR region. Integration of QDs with targeting ligands or other imaging contrast agents renders a possible multimodal imaging with complementary advantages of each imaging modality. For examples, direct coating QDs with paramagnetic metal chelates (e.g., Gd chelates) using strategies of self-assembly or covalent coupling yields magnetic QDs with their sizes not being largely increased. By coating fluorescent CdSe QDs with paramagnetic BSA-coupled DTPA-Gd (GdDTPA·BSA) and further functionalization of the hybrid with a targeting warhead anti-Glut1 polyclonal antibody, Xing et al. constructed multifunctional paramagnetic quantum dots (GdDTPA·BSA@QDs-PcAb) for colorectal tumor reconfirmation (Xing et al. 2015). This GdDTPA·BSA@QDs hybrid exhibits strong fluorescence and much higher longitudinal and transverse relaxivities than those of commercial Gd-DTPA in water. *In vivo* results indicated that this GdDTPA·BSA@QDs-PcAb hybrid could be employed for real-time MR imaging of colorectal cancer (CRC) tumors on nude mice after intravenous injection of the nanoprobes. Doping QDs with paramagnetic metal ions is another method to achieve intrinsically paramagnetic QDs (MDQDs) with small size. In the last two decades, a large amount of MDQDs, such as Mn-doped CdTeSe/CdS QDs (Mn-QDs) (Yong 2009), Mn-doped ZnSe/ZnS QDs (Zhou et al. 2014), Mn-doped CdS/ZnS QDs (Santra et al. 2005), Ni-doped CdTeSe/CdS QDs (Singh et al. 2012), Fe-doped CdTeS QDs (Saha et al. 2013), Eu-doped GdS QDs (Jung et al. 2012), Gd-doped ZnO QDs (Liu et al. 2011c), Gd-doped CdSe QDs (Li and Yeh 2010) and Gd-doped CdTe QDs (Zhang et al. 2014), have been established as dual-modality imaging agents. Since the toxicity induced by Cd^{2+} is an unavoidable problem for transferring Gd-based imaging probes to clinic, scientists gradually pay more attention to Cd-free quantum dots. Ding et al. (Ding et al. 2014) developed dual-modality probes of PEGylated $CuInS_2$@ZnS: MnQDs for fluorescence/magnetic resonance imaging of tumors. Synthesis of this QDs hybrid includes four steps, (i) the preparation of fluorescent $CuInS_2$ seeds, (ii) particle surface coating of ZnS, (iii) Mn-doping of the ZnS shells and (iv) PEGylation of $CuInS_2$@ZnS: Mn QDs. Surface functionalization greatly improved the imaging properties of as-prepared QDs hybrid. On the one hand, the fluorescence QY of the

CuInS$_2$ core was greatly increased after surface coating. On the other hand, Mn-doping well balanced the optical and magnetic properties of the resultant QDs hybrid. *In vivo* preliminary experimental results indicated that both subcutaneous and intraperitoneal tumor xenografts can be visualized via fluorescence and MR imaging using this QDs hybrid.

Lanthanide Upconversion Nanoparticles (UCNPs)-Based Hybrids

UCNPs exhibit excellent luminescence properties (e.g., sharp emission lines and long lifetimes) through a unique upconversion luminescence (UCL) process. In this process, continuous-wave (CW) low-energy light in the near-infrared region (such as 980 nm) is converted into higher-energy visible light through multiple photon absorption or energy transfer (Yu et al. 2009). NIR-excited UCNPs have been known as low toxic and stable fluorescent probes, whose capability of deep tissue penetration greatly reduces their photodamage to the cells and avoids the auto-fluorescence from the cells. In recent years, lanthanide-doped UCNPs have been developed as a new class of luminescent optical labels and have emerged as a serious alternative to organic fluorophores and quantum dots for biomedical applications (Wang and Liu 2009). Typically, UCNPs are composed of three components: a host matrix, a sensitizer and an activator (Wang and Zhang 2014). Owing to their low photon energies and minimum quenching effects to the excited states of the lanthanide ions, fluorides are considered as the ideal host materials for the doping of lanthanide ions to achieve intense UC emissions. Yb^{3+} is always chosen as the sensitizer due to its large absorption cross-section around 980 nm. Er^{3+}, Tm^{3+} and Ho^{3+}, featured with ladder-like arranged energy levels which are in favor of a multiphoton process, are frequently used as activators. NaYF$_4$ co-doped with ytterbium (Yb^{3+}), erbium (Er^{3+}), or terbium (Tb^{3+}) have been developed as potential contrast agents for optical imaging (Vetrone et al. 2010). After subcutaneous injection, 50-nm NaYF$_4$: Yb/Er nanoparticles can be detected up to 10 mm beneath the skin of Wistar rats, which is far deeper than the depths managed through the use of quantum dots (Chatteriee et al. 2008). Using sub-10 nm β-NaLuF$_4$: Gd, Yb, Tm nanocrystals as a UCL probe, detection limits as low as 50 or 1000 nanocrystal-labeled cells for subcutaneous or intravenous injection could be reached, respectively. Particularly, high-contrast UCL imaging of a whole-body black mouse with a penetration depth of ~2 cm could be achieved with this probe (Liu et al. 2011b).

In order to improve their biocompatibility for *in vivo* applications, surface coating of the UCNPs is required. Both assembly and conjugation of coating polymers on the nanoparticles surfaces are promising strategies. Maji et al. reported the application of α-cyclodextrin-covered NaYF$_4$: Yb^{3+}, Er^{3+} upconversion nanoparticles (UC-α-CD) for *in vivo* photoacoustic imaging (PAI) of living mice (Maji et al. 2014). They employed the heating property of NaYF$_4$: Yb^{3+}, Er^{3+} UCNPs as well as their luminescence quenching effect to generate enhanced photoacoustic (PA) signal in aqueous conditions for *in vivo* PA imaging (PAI) of mice under 980 nm excitation (Fig. 6.13). Using photoacoustic tomography, *in vivo* localization of UC-α-CD in live mice was successfully demonstrated, suggesting that UC-α-CD could be employed as an efficient imaging contrast agent for diagnostic purposes.

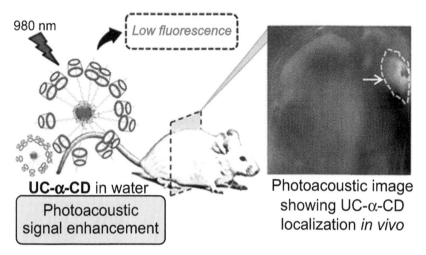

Fig. 6.13 (Right) Schematic illustration of luminescence quenching effect and subsequent photoacoustic signal enhancement from UC-α-CD in water. (Left) Individual anatomy sections recorded after 35 min post intravenous injection of UC-α-CD. Pointed areas indicate the localization of UC-α-CD. (From Maji, S.K. et al., Upconversion nanoparticles as a contrast agent for photoacoustic imaging in live mice. *Adv. Mater.* 2014. 26. 5633-5638. Copyright Wiley-VCH Verlag GmbH & Co. KGaA. Reproduced with permission.)

UCNPs can be integrated with other imaging agents of different modalities to form a multimodal imaging platform. Xing et al. (Xing et al. 2012) developed a trimodal imaging agent for upconversion fluorescent, MR and CT imaging by constructing a PEGylated NaY/GdF$_4$: Yb, Er, Tm @ SiO$_2$-Au@PEG5000 UCNPs hybrid. The feasibilies of this UCNPs hybrid for trimodal imaging were clearly evidenced by increased SNR of tumors after local injection of the nanoprobes *in vivo*. Gadolinium ion (Gd^{3+}), having seven unpaired electrons, provides high paramagnetic relaxivity. Nanoparticles based on NaGdF$_4$ co-doped with Yb^{3+}, Er^{3+}/Tm^{3+} or NaYF$_4$ co-doped with Gd^{3+}, possessing both upconversion luminescence (UCL) and magnetic resonance properties, have been reported for *in vivo* T$_1$-positive MR and UCL dual-modality imaging of small animals (Kumar et al. 2009; Zhou et al. 2010). Besides co-dopping strategy, surface modification can also be a good choice for constructing multifunctional nanoparticles. Liu et al. reported a multifunctional UCNPs hybrid ^{18}F-AA-Gd-UCNPs that integrates UCL (NaYF$_4$: Yb, Er) with magnetism (Gd^{3+}), radioactivity (^{18}F) and targeted recognition (folic acid) (Liu et al. 2011a). They claimed that Gd^{3+} is not doped within the host materials, but is distributed on the surface of the nanocrystals by cation exchange with Y^{3+}. Versatility of this ^{18}F-AA-Gd-UCNPs UCNPs hybrid was demonstrated by cellular targeted imaging, *in vivo* MRI, upconversion luminescence imaging and PET imaging of a small animal (Fig. 6.14).

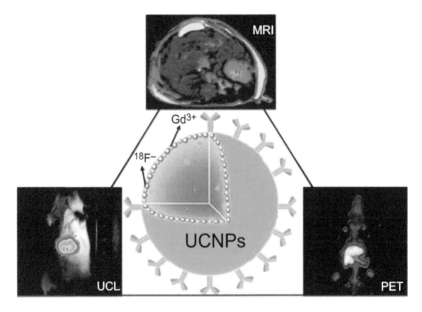

Fig. 6.14 Upconversion luminescence properties, MR, PET imaging of Multifunctional nanoparticles that integrate UCL (NaYF$_4$: Yb, Er), magnetism (Gd^{3+}), radioactivity (^{18}F) and targeted recognition (folic acid). (Reprinted with permission from Liu, Q., Sun, Y., Li, C.G., Zhou, J., Li, C.Y., Yang, T.S., et al., ^{18}F-labeled magnetic-upconversion nanophosphors via rare-earth cation-assisted ligand assembly. *ACS Nano* 5. 3146-3157. Copyright 2011 American Chemical Society.)

Gold Nanoparticles (AuNPs)-Based Hybrids

Gold nanoparticles (AuNPs) are metal NPs which are able to scatter incident light at their localized surface plasmon resonance (LSPR) frequency. Thus, AuNPs provide a simple method for tumor imaging (e.g., using an optical microscope equipped with a dark-field condenser). It is possible to tune the optical properties of AuNPs into near infrared (NIR, 650-1000 nm) region because their SPR effect is very sensitive to the changes of their size, shape and the dielectric constant of surrounding media. Thus, AuNPs are also one type of NPs for promising deep tissue imaging. For example, two-photon luminescence imaging of cancer cells in a 3D tissue phantom down to the 75 μm depth has been achieved using gold nanorods as imaging agents (Durr et al. 2007). Based on their size, shape and physical properties, AuNPs can be classified into many subtypes, as shown in Fig. 6.15. Gold nanocubes are applicable for cancer imaging since they possess the highest photoluminescence quantum yield among the AuNPs. Ding et al. reported novel protein–gold hybrid nanocubes (PGHNs) which were hybridized with gold nanoclusters, bovine serum albumin and tryptophan building blocks (Ding et al. 2015). The PGHNs are cube-shaped with a size of approximately 100 nm and emit blue fluorescence upon UV excitation. The quantum yield of PGHNs was 7.13% using rhodamine 6G as the reference, which is the highest among those of reported fluorescent gold NPs/NCs. These PGHNs were biocompatible and can serve as a dual-purpose tool: as a blue-emitting cell marker for bioimaging and as a nanocarrier for drug delivery.

Fig. 6.15 Different types of gold nanoparticles. Different shaped NPs exhibit different properties that could be useful for different applications. (From Cai, W. et al., *Nanotechnol. Sci. Appl.*, 1, 17-32, 2008. With permission.)

X-ray-based CT imaging is one of the most convenient imaging/diagnostic tools in hospitals today in terms of its availability, efficiency and cost. AuNPs have been tried as new CT contrast agents to improve the limitations of conventional iodine-based contrast agents, including restricted imaging acquisition time, the need of catheterization in many cases, poor contrast in large patients and renal toxicity. Compared with iodine-based X-CT contrast agents, AuNPs have several intrinsic merits such as easy control of sizes and shapes, biocompatibility, low toxicity, prolonged blood circulation time and a high X-ray absorption coefficient. Due to their larger surface area, smaller AuNPs exhibit higher X-ray attenuation. The use of gold NPs as CT contrast agents for various imaging applications, such as cancer imaging and blood-pool imaging, has been demonstrated. Similar to the above-mentioned MNPs and QDs, AuNPs can be additionally conjugated with other imaging reagents for multi-modal imaging. For examples, [125]I (Kim et al. 2011; Su et al. 2015) and [99m]Tc (Morales-Avila et al. 2011) were labeled onto AuNPs in order to realize SPECT/CT dual-modality imaging. [64]Cu was integrated with AuNPs for PET/CT dual-modality imaging (Sun et al. 2014; Xie et al. 2010). When this ^{64}Cu-AuNPs hybrid was simultaneously functionalized with active targeting ligands (e.g., cyclic RGD to target $\alpha_v\beta_3$ integrin), the radioactive hybrid could be used for tumor-targeted dual-modality PET/CT imaging.

When tissues are heated above 42°C, irreversible cellular damage occurs, resulting in eventual cell death (Cobley et al. 2010). AuNPs can generate heat upon long-wavelength (800-1300 nm) laser irradiation, which makes AuNPs applicable for photothermal therapy (PTT). Therefore, AuNPs can act as theranostic agents which combine diagnostic modality (i.e., CT) with the therapeutic application (i.e., PTT). Deng et al. reported AuNPs-polymer hybrids with different sizes and shapes for NIR photothermal therapy and X-Ray computed tomography imaging (Deng et al. 2014). Biodegradable hydrophobic poly(ϵ-caprolactone) (PCL) and hydrophilic poly 2-(2-methoxyethoxy) ethyl methacrylate (PMEO$_2$MA) were anchored onto 14 nm AuNPs to offer amphiphilic-driven force for self-assembling AuNPs-polymer hybrids with different sizes and shapes (Fig. 6.16). By optimizing the hydrophobic/hydrophilic ratio, large micelles of this AuNPs-polymer hybrid could be obtained which showed a commendable effect for tumor ablating and CT imaging and exhibited negligible toxicity to cells or animals and good biodegradability after treatment. These results suggest this AuNPs-polymer hybrid system could be a potential theranostic reagent candidate for clinical PTT and CT imaging.

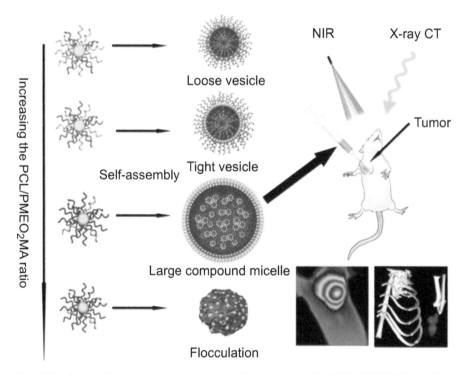

Fig. 6.16 Schematic representation of assemblies composed by PCL/PMEO2MA grafted AuNPs and their potential application in photothermal therapy and CT imaging of cancer. Bottom right shows thermal images of MCF-7 tumor-bearing mice exposed to an 808 nm laser for 5 min after the injection GA7 (2 mg/mL, 50 µL) (left) and three-dimensional *in vivo* CT image of tumor tissues after the injection of GA7 (C). Arrows indicate the location of tumor. (From Deng, H., et al., *Theranostics*, 2014, 4(9), 904-918. With permission.)

Structural Nanoparticle-Based Hybrids

In order to generate signal from tumor tissue with high signal to noise (S/N) ratio, the signal needs to be effectively amplified. Signal amplification strategies include carrying large numbers of contrast entities on one carrier or targeting receptors (or antigens) overexpressed in tumor cells. Typically, the structural nanoparticles have a lot of delivery sites due to their specific construction, which results in a high payload of functional agents. For instance, imaging entities can be located into the interior, the phospholipid bilayer or onto the outer surface of a liposome. As-prepared nanoparticles containing high payloads of contrast agents are able to provide similar quality of imaging but with lower concentration. In addition, since these structural nanoparticles are mainly made of biocompatible materials or biomaterials such as natural starch, chitosan, virus, or protein, their degradation products or themselves are safe to biosystems. Lastly, when the imaging agents are placed inside a nanocarrier, significant advantages of the imaging agents can be acquired: (i) decrease their interactions with the biological environment; (ii) change their hydrophobic property into "water soluble" by being encapsulated in hydrophilic nanocarriers; (iii) increase stability against photobleaching or oxidative degradation.

Polymer-Based Hybrid Nanoparticles

Based on their structural features, polymeric NPs can be classified into nano-capsules, polymeric micelles and nanospheres. For nanocapsules, imaging or therapeutic agents are usually loaded with the "tanker" method. The use of these amphiphilic polymers results in the formation of nanoparticles with a hydrophobic core and a hydrophilic shell. This structure favors encapsulation of inorganic nanoparticles together with anticancer drugs into the core of the polymeric nanoparticle. The polymeric NPs hybrids are superior in physical and chemical properties to commercial imaging agents. They provide localized contrast enhancement in different medical visualization techniques, rendering precise diagnosis or evaluation of therapeutic efficiency. Magnetic nanoparticles, quantum dots were encapsulated by poly(lactic-co-glycolic acid) (PLGA) to construct inorganic-organic polymeric hybrids for cancer imaging (Kim et al. 2008; Yang et al. 2007). After further encapsulation of drugs followed by modification with targeting molecules on their surfaces, these polymer nanomedical hybrids were applied for simultaneous cancer-targeted MR imaging, optical imaging and magnetically guided drug delivery. Jin et al. reported the preparation of PLA microcapsules with ultrasound contrast, which encapsulated AuNPs for X-CT imaging and were functionalized with graphene oxide (GO) for photothermal therapy (PTT) (Jin et al. 2013). This hybrid microcapsule was applied for effective US imaging, X-ray CT imaging and photothermal therapy of cancer. The combination of real-time ultrasound with 3-D computed tomography through a single microcapsule agent is very helpful for accurate interpretation of the obtained images, identifying the size and location of the tumor, as well as guiding and monitoring the photothermal therapy. Simultaneously, the effectiveness of photothermal therapy could be evaluated by the combinational US and CT imaging of the tumors with this microcapsule hybrid. Ye et al. (Ye et al. 2014) developed biodegradable polymeric vesicles containing magnetic nanoparticles, quantum dots and anticancer drugs for drug delivery and imaging (Fig. 6.17). The poly (lactic-co-glycolic acid) (PLGA) vesicle hybrids were fabricated by encapsulating inorganic imaging agents of superparamagnetic iron oxide nanoparticles (SPION), manganese-doped zinc sulfide (Mn: ZnS) quantum dots (QDs) and the anticancer drug busulfan with PLGA nanoparticles via an emulsion-evaporation method. Encapsulated quantum dots were used for enhanced fluorescence visualization of the cell uptake of the vesicles while the biodistribution of the vesicles in the rat model was doubly elucidated by *in vivo* MR imaging and histological studies of the animal.

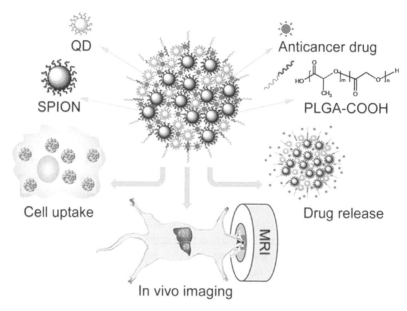

Fig. 6.17 Schematic diagram showing the composition of PLGAeSPIONeMn: ZnS nanoparticles and their multiple applications. (Reprinted from *Biomaterials*, 35, Ye, F., Barrefelt, A., Asem, H., Abedi-Valugerdi, M., El-Serafi, I., Saghafian, M., et al., Biodegradable polymeric vesicles containing magnetic nanoparticles, quantum dots and anticancer drugs for drug delivery and imaging, 3885-3894, Copyright 2014, with permission from Elsevier.)

Nanospheres consist of a solid mass and an outer surface which is always used for loading functional agents. Key et al. developed optical and MR dual-modality nanoparticle hybrid for tumor imaging using glycol chitosan nanoparticles (HGC), SPIOs and Cy5.5 dyes (Key et al. 2011). Monodispersive

SPIOs were physically loaded into HGC-Cy5.5 nanoparticles to form SPIO-HGC-Cy5.5 nanoparticle hybrid. The nanoparticle hybrid was injected into MBT-2 tumor-bearing mice through the tail vein. *In vivo* MR imaging results indicated that MR signals from the tumor were two times lower than those in the control group, suggesting that this hybrid could be a potential T2 MR contrast agent. Strong NIRF signals of Cy5.5 from tumors, kidneys and bladder were detected, indicating that many nanoparticles were accumulated in tumors and some of them were filtered in kidneys and excreted through urine. These results showed the possibility of this SPIO-HGC-Cy5.5 nanoparticle hybrid as an optical and MR dual-modality imaging agent for early cancer detection. Lipid–polymer hybrid nanoparticles (LPNs) are particular nanospheres with core–shell structures comprised of polymer cores and lipid/lipid–PEG shells (Fig. 6.18). They possess complementary characteristics of both polymer nanoparticles and liposomes: (i) high structural integrity and stability attributed to the polymer core and (ii) high biocompatibility and bioavailability owed to the lipid and lipid–PEG layers (Hadinoto et al. 2013). Owing to these features, LPNs have been predominantly applied for anticancer therapy. Additionally, LPNs have also been used as delivery vehicles to load contrast agents for biomedical imaging (especially for cancer diagnostics). Aryal et al. (Aryal et al. 2013) developed one type of lipid-polymer hybrid nanoparticles comprised of three distinct compartments with specific functions: (i) a solid PLGA polymeric core, acting as a cytoskeleton and providing mechanical stability and encapsulating poorly water-soluble payloads such as the hydrophobic USPIOs or drug molecules; (ii) a lipid shell wrapped around the core, acting as a cell membrane, where the Gd-DOTA molecules were anchored; and (iii) a hydrophilic polymer stealth layer outside the lipid shell, enhancing nanoparticle stability and circulation lifetime. The hybrid nanoparticles showed a remarkable T2 MRI behavior with a transversal relaxivity at least 3 times larger than that of superparamagnetic iron oxide nanoparticles used in clinic, which attributes to the significant darkening of the tumor tissue in T2-weighted MR imaging.

Polymeric micelles consist of amphiphilic block copolymers which self-assemble into micelles in aqueous solutions. The hydrophobic fragments of the amphiphilic copolymers form the core of a micelle, while the hydrophilic fragments form the shell of the micelle. Thus, poorly water-soluble, nonpolar imaging agents can be solubilized within the core of a micelle, while polar molecules can be adsorbed on the micelle surface and substances with intermediate polarity can be distributed along the intermediate position of a micelle (Torchilin 2007). Mathews et al. encapsulated lipophilic cyanine fluorescent dye DiI within the core of poly (ethylene oxide)-b-poly(caprolactone) (PEO-b-PCL) diblock polymeric micelles and modified the micelle surface with peptide 11 (RGDPAYQGRFL) and peptide 18 (WXEAAYQRFL) (Mathews et al. 2013). After incubation of the micelle hybrid with breast cancer cells, a stronger fluorescent signal of DiI from the cells as compared to that of conventional p160, c-RGD modified, or naked micelles was observed, indicating higher cellular uptake of the dye after micelle modification. This micelle hybrid seems to be a promising carrier for targeting delivery of drugs or diagnostic moieties to the breast cancer site. Zhang et al. used polyethylene glycol-coated, core-crosslinked polymeric micelles (CCPM) to trap hydrophobic Cy-7 in the core of the micelle and conjugated radioisotope chelator diethylene triamine pentaacetic acid (DTPA) on the surface of CCPM for the following labeling with [111]In (Zhang et al. 2011). Thus, each micellar nanoparticle was loaded with multiple Cy7 dye molecules and ^{111}In ions, providing a huge boost of signal intensity. Follow up conjugation of EphB4-binding peptide TNYL-FSPNGPIARAW (TNYL-RAW) to CCPM resulted in tumor-targeted TNYL-RAW-CCPM micelle hybrid with dual imaging modalities. These hybrid micelles not only had significantly prolonged blood half-life of TNYLRAW and reduced volume of distribution in mice, but also allowed clear visualization of EphB4-overexpressing tumors in mice with both SPECT and near-infrared fluorescence imaging.

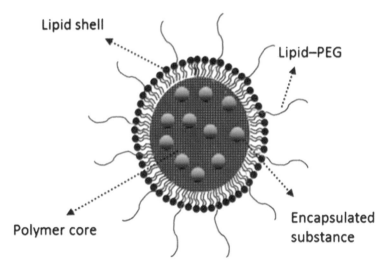

Lipid-Polymer Hybrid Nanoparticles (LPN)

Fig. 6.18 Structure of lipid–polymer hybrid nanoparticles (LPNs). They comprised (i) a polymer core in which the imaging substances are encapsulated, (ii) an inner lipid layer enveloping the polymer core, the main function of which is to confer biocompatibility to the polymer core and (iii) an outer lipid–PEG layer, which functions as a stealth coating that prolongs *in vivo* circulation time of the LPNs, along with providing steric stabilization. In addition, the inner lipid layer also functions as a molecular fence that minimizes leakage of the encapsulated content during the LPNs preparation. (Reprinted from *Eur. J. Pharm. Biopharm.*, 85, Hadinoto, K., Sundaresan, A. and Cheow, W.S., Lipid-polymer hybrid nanoparticles as a new generation therapeutic delivery platform: A review, 427-443, Copyright 2013, with permission from Elsevier.)

Silica-Based Hybrids

As compared to polymer matrix, silica nanoparticles have several advantages (Jain et al. 1998). Firstly, during their preparation, surface modification and other solution treatment, silica nanoparticles are easy to separate via centrifugation due to their high density (e.g., 1.96 g/cm^3 for silica vs 1.05 g/cm^3 for polystyrene). Secondly, silica nanoparticles are hydrophilic and biocompatible not subjected to microbial attack and do not swell or change their porosity with the changes of pH.

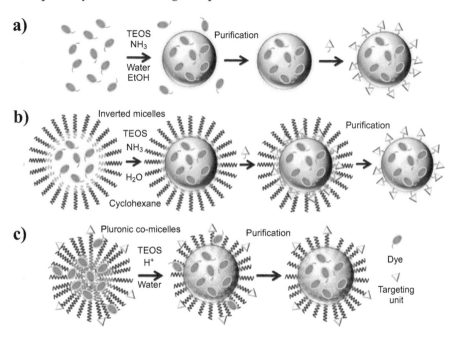

Fig. 6.19   Schematic representation of different synthetic strategies to obtain dye-doped silica nanoparticles: Stöber-Van Blaaderen method (a) reverse microemulsion method (b) and direct micelle assisted method (c). (From Montalti, M., Prodi, L., Rampazzo, E. and Zaccheroni, N., Dye-doped silica nanoparticles as luminescent organized systems for nanomedicine, Chem. Soc. Rev., 43, 4243-4268, 2014. Reprinted with permission of The Royal Society of Chemistry.)

Dye-doped SiNPs are very common silica-based hybrid nanoparticles. Doping of thousands of fluorescent dye molecules in one SiNP could result in an intense fluorescence signal up to 10,000 times that of organic fluorophores (Wang et al. 2006). Furthermore, the silica matrix serves as a protective shell, or dye isolator to limit the effect of the outside environment (such as oxygen, solvents and soluble species in buffer solutions) on the fluorescent dye doped in the nanoparticles. The main synthetic approaches to incorporate dyes into silica matrix include Stöber–Van Blaaderen method, reverse microemulsion (water-in-oil) and direct micelle-assisted methods (Fig. 6.19) (Montalti et al. 2014). The excellent optical characteristics, high biocompatibility and size-controllable properties of the dye-doped SiNPs ensure their usage as fluorescence agents for tumor imaging both *in vitro* and *in vivo*. Rampazzo et al. (Rampazzo et al. 2012) developed NIR-emitting hybrid nanoparticles doped with a Cy7 heptamethine dye alone or together with a trialkoxysilane derivative of rhodamine B. Thus, these hybrid nanoparticles have two separate emissions (NIR + Red) and exhibit suitable characteristics for both *in vivo* imaging and *ex vivo* microscopic imaging with the same NPs, a feature unattainable by QDs. Furthermore, imaging or therapeutic cargoes can be either directly incorporated in the silica matrix or grafted onto the outer surface of the solid silica particles.

Using other imaging entities (e.g., organic fluorophores, iron oxide nanoparticles, gold nanoparticles and quantum dots) as the core and dye-doped silica as the shell, we could construct core/shell silica nanoparticles (C/S-SiNPs) with multimodal imaging properties. For example, Organic fluorophores (CyN-12, (Wu et al. 2013)), Magnetic cobalt ferrite ($CoFe_2O_4$) nanoparticles (Yoon et al. 2006), gold nanoparticles (NPs) (Cao et al. 2011), Fe_3O_4 nanoparticles (Wang et al. 2011) have been successfully employed to construct C/S-SiNPs. Ean-Luc Bridot and co-workers encapsulated Gd_2O_3 core with a polysiloxane shell which carried organic fluorophores to fabricate a dual-imaging modal hybrid nanoplatform for *in vivo* magnetic resonance and fluorescence imaging (Bridot et al. 2007). Instead of being embedded into silica shell, imaging entities can also be anchored to the outer surface of silica nanoparticles. Lee et al. constructed "core–satellite" structured dual functional nanoparticles comprised of a dye-doped silica "core" and multiple "satellites" of magnetic nanoparticles (Lee et al. 2006). Rhodamine-dye-doped silica ($DySiO_2$) nanoparticles were covalently linked with Fe_3O_4 NPs through a coupling agent sulfosuccinimidyl-(4-N-maleimidomethyl) cyclohexane-1-carboxylate. The authors showed an interesting synergistic MR enhancement effect with the fluorescence of the hybrid $DySiO_2$–(Fe_3O_4) nanoparticles. Actually, besides dye-doped silica, the shell of the hybrid silica nanoparticles can be silica doped with other functional moieties (e.g., drugs). Lee et al. reported dual-functional silica–iron oxide hybrid NPs composed of a Fe_3O_4 core, a mesoporous silica shell and GSH-responsive (cyclodextrin) CD gatekeepers as an efficient carrier for anticancer drug delivery and MR imaging (Lee et al. 2012a). *In vivo* MR imaging indicated that the Fe@Si–DOX–CD–PEG hybrid nanoparticles were accumulated in the tumors and the growth of the tumors was effectively suppressed by anticancer drug DOX controllably released from the hybrid nanoparticles.

Mesoporous silica nanoparticles (MSNs) are mesoporous materials which contain hundreds of empty channels arranged along the 2D networks in the honeycomb-like porous structures. MSNs possess several merits such as uniform and tunable pore structure, high surface area, simple surface functionalization, good dispersibility, excellent biocompatibility and cellular membrane-penetrating capacity. Their porous structure enables the nanoparticles to load imaging agents with a high capacity and effectively protect the payload molecules from leaking. Zhang et al. prepared AIE-MSNs via one-pot encapsulation of AIE dye (An18) into MSNs using CTAB as both surfactant template and cell killer agent (Zhang et al. 2013). Such novel fluorescent MSNs showed promising properties for both cell imaging and cancer therapy. MSNs could be further modified through their outer surface modification with polymers or dendrimers, normally resulting in enhancement of their suspension stability. Using surface hyper branching polymerization, Rosenholm et al. (Rosenholm et al. 2009) functionalized mesoporous silica nanoparticles with poly (ethylene imine) (PEI) (Fig. 6.20a). Further modification of the polymer outer surface with fluorophores and tumor-targeting moieties resulted in hybrid MSNs for tumor-targeted imaging. Like solid silica nanoparticles, MSNs are also potential candidates for multimodal imaging. Lin et al. (Lin et al. 2006) proposed multifunctional hybrid MSNs which incorporate luminescent with magnetic resonance imaging (Fig. 6.20b). In this hybrid, Fe_3O_4@silica nanoparticles, potential T2 agents with high relaxivity at 0.47T, were merged with fluorescent mesoporous silica nanoparticles via silica deposition.

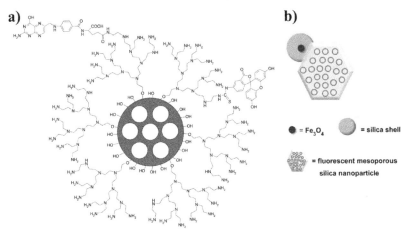

Fig. 6.20 (a) Schematic representation of the hybrid silica-dendrimer nanoparticle structure. (Reprinted with permission from Rosenholm, J.M., Meinander, A., Peuhu, E., Niemi, R., Eriksson, J.E., Sahlgren, C., et al., Targeting of porous hybrid silica nanoparticles to cancer cells, *ACS Nano.*, 3, 197-206. Copyright 2009, American Chemical Society.); (b) schematic and electron microscopy depictions of hybrid nanoparticles comprising of a Fe_3O_4@silica body with a fluorescent mesoporous silica nanoparticle appendage. (Adapted with permission from Lin, Y.S., Wu, S.H., Hung, Y., Chou, Y.H., Chang, C., Lin, M.L., et al., Multifunctional composite nanoparticles: Magnetic, luminescent and mesoporous, *Chem. Mater.*, 18, 5170-5172. Copyright 2006, American Chemical Society.)

Dendrimer-Based Hybrid Nanoparticles

Dendrimers are synthetic, branched macromolecules with tree-like structures consisting of a vast array of types, branched backbones and functional groups. These dendrimers, based on polyamidoamines, polyamines and polyamides, are commonly used for biological applications (Lee et al. 2005). The three areas of a dendrimer, namely, the encapsulated core, the interior branches and the surface functional groups, can be utilized for loading (or attaching) imaging agents or drugs (Fig. 6.21) (Dykes 2001). The high degree of functionality of dendrimers endows them with several remarkable merits such as the following (Ghobril et al. 2012): (i) sharply increasing the binding ratio of the nano-object (herein dendrimer) to the target tissue by increasing the number of biological effectors within the same size-controlled nano-object; (ii) allowing multi-modal imaging (MRI, PET, SPECT) through complexation of diverse metallic ions; (iii) having favorable biodistribution properties (elimination of the non-targeted complexes by a renal way).

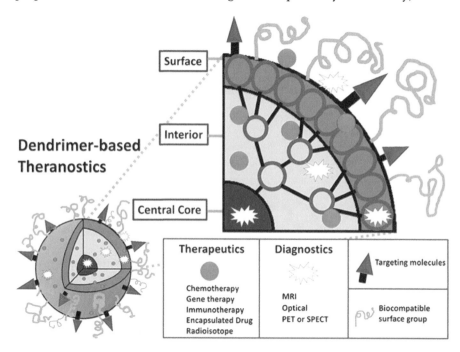

Fig. 6.21 Illustration of the areas in the dendrimer for the loading of therapeutics and diagnostics. (Reprinted with permission from Lo, S.T., Kumar, A., Hsieh, J.T. and Sun, X.K., Dendrimer nanoscaffolds for potential theranostics of prostate cancer with a focus on radiochemistry, *Mol. Pharm.* 10, 793-812. Copyright 2013, American Chemical Society.)

Dendrimers were first used as carriers for loading magnetic resonance imaging contrast reagents for *in vivo* applications. The dendritic structure is in favor of anchoring or incorporating different imaging entities. Wiener and Margerum both used polyamidoamine (PAMAM) dendrimers to prepare dendritic gadolinium poly-chelates which showed enhanced MR contrast as compared to that of conventional contrast agents (Margerum et al. 1997; Wiener et al. 1994). Alcala et al. (Alcala et al. 2011) created a luminescent dendrimer complex by substituting thirty-two naphthalimide fluorophores on the surface of the dendrimer and incorporating eight europium cations within the interior of generation-3-PAMAM-(glycine-4-amino-1, 8-naphthalimide) 32 (G3P4A18N) dendrimer. This dendrimer complex showed preferential accumulation within liver tumors and emitted persistent luminescence *in vivo* which was strongly resistant against photobleaching. The unique structure of dendrimer offers an ideal platform for fluorescence resonance energy transfer (FRET). For example, Adronov et al. (Adronov et al. 2000) placed one molecule of coumarin-343 dye in the core of Poly (aryl ether) dendrimer and conjugated coumarin-2 dye molecules on its surface, which resulted in energy transfer from the surface dyes to the core. For multifunctional applications, different imaging agents or drugs can be simultaneously integrated into one single dendrimer. Shi and co-workers (Wen et al. 2013) reported a gadolinium-loaded, gold nanoparticle-entrapped dendrimer complex (GdeAu DENPs) (Fig. 6.22) for dual-modality computed tomography (CT)/magnetic resonance (MR) imaging applications. Amine-terminated generation five poly (amidoamine) dendrimers (G5.NH₂) modified with gadolinium (Gd) chelator and PEG monomethyl ether were used as templates to synthesize and trap gold nanoparticles (AuNPs). Subsequent chelation of Gd(III) and acetylation of the remaining dendrimer terminal amine groups conferred the multifunctional GdeAu DENPs dendrimer complex. This GdeAu DENPs dendrimer complex showed extended blood circulation time and could be cleared from the major organs within 24 h. After that, they used folic acid (FA) terminated PEG to prepare another GdeAu DENPs-FA dendrimer complex for FA receptor-mediated KB tumor-targeted imaging *in vivo* (Chen et al. 2013).

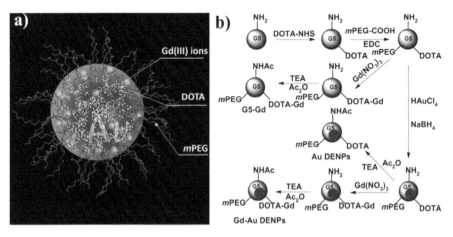

Fig. 6.22 (a) Schematic illustration of the designed nanostructure; (b) the synthesis procedure of the GdeAu DENPs. TEA and Ac₂O represent triethylamine and acetic anhydride, respectively. (Reprinted from *Biomaterials*, 34, Wen, S., et al., Multifunctional dendrimer-entrapped gold nanoparticles for dual mode CT/MR imaging applications, 1570-1580, Copyright 2013, with permission from Elsevier.)

Liposome-Based Hybrid Nanoparticles

Liposomes are spherical vesicles that consist of one or more phospholipid bilayers encapsulating water in their interior. These phospholipid bilayered membrane vesicles (Fig. 6.8c), with sizes ranging from 100 nm up to 5 μm, have been utilized for the delivery of small molecules, proteins and peptides, DNA and imaging contrast agents. Since the early 1980s when Vescan™ was developed and envisioned as a broadly usable imaging agent for *in vivo* tumor diagnostic (Proffitt et al. 1983), liposomes have been widely studied as radionuclide carriers for tumor imaging for many years. Methods for incorporating radionuclides into liposomes can be roughly divided into four categories (Petersen et al. 2012): passive encapsulation, membrane labeling, surface chelation and remote loading (Fig. 6.23). In passive encapsulation, the radionuclide labeling is performed during the manufacture of the liposomes. In membrane labeling, the radionuclides are conjugated to the surface or embedded within the bilayer of the liposomes. Surface chelation can be achieved by incorporating the lipid chelator with the radionuclides to form the liposomal membrane during liposome preparation. Remote loading is an active loading of the radionuclide into the aqueous phase of liposomes, where an ionophore or a lipophilic chelator transports the radionuclide over the membrane of the preformed liposomes and the radionuclide is delivered to a pre-encapsulated chelator. For example, [64]Cu (II) was traditionally loaded into liposomes by surface chelation using effective copper (II) macrocyclic chelators such as DOTA, NOTA, TETA, TE2A and BAT (Seo et al. 2008; Smith 2004). [111]In was remote loaded into liposomes by the use of the lipophilic chelator oxine with deferoxamine (Gabizon et al. 1991) or DTPA (Harrington et al. 2000). [99m]Tc and ^{111}In-labeled liposomes have been used as SPECT imaging agents in clinical studies for cancer diagnosis (Harrington et al. 2001). PEGylated ^{111}In-liposomes have been reported for effective targeting imaging of solid tumors in patients with locally advanced cancers. Pegylated liposomes have prolonged circulation half-life times, which benefit or favor their accumulation in the tumors. Harrington et al. 2001 reported the use of ^{111}In-DTPA-labeled pegylated liposomes (IDLPL) for whole body imaging of a patient with stage T1I1S1 AIDS-related Kaposi's sarcoma. Clear Kaposi's sarcoma lesions were visualized in the left foot and leg, right arm and face by the nuclear medicine whole body gamma camera imaging due to the uptake of radiolabeled liposomes. 7 d after injection, signals of IDLPL could still be found in the lesions, which indicated prolonged retention of the radiolabel.

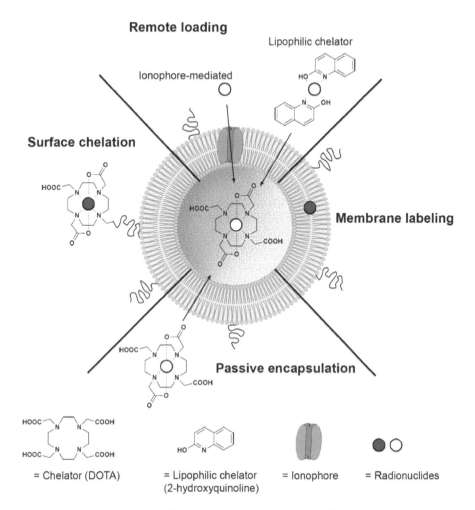

Fig. 6.23 Schematic diagram of the remote loading, membrane labeling, passive encapsulation and surface chelation methods for preparing radioactive liposomes. Radionuclides can be associated with the lipid membrane by hydrophobic interaction, through membrane conjugation or surface chelation using chelator-lipid conjugates in preformed liposomes (solid radionuclides). Radionuclides can alternatively be encapsulated inside liposomes during lipid hydration or can be transported through the lipid membrane of preformed liposomes by ionophores or lipophilic chelators (hollow radionuclides). In the latter case, the radionuclides are trapped inside the aqueous lumen by a hydrophilic chelator with high affinity for the radionuclide. (Reprinted from *Adv. Drug. Deliv. Rev.*, 64, Petersen, A.L., et al., Liposome imaging agents in personalized medicine, 1417-1435, Copyright 2012 with permission from Elsevier.)

Besides radionuclides, other imaging agents can also be loaded into liposomes. Quantum dot-lipid hybrid nanoparticle is a typical construct in which functionalized or non-functionalized quantum dots are encapsulated in the hollow part of a liposome or in the lipid bilayer of such vesicle, respectively. This type of structure not only retains the fluorescent properties of QDs, but also turns the hydrophobic QDs more compatible with aqueous phases. It has been shown that cationic liposomes can electrostatically interact with the tumor cell membrane, bind to the negatively charged plasma membrane and thus mediate their endocytosis. This was widely employed for the intracellular delivery of QDs to cancer cells (Al-Jamal et al. 2008). The huge storage space of liposomes favors the loading of different imaging agents for multimodal imaging. Ding et al. reported a fluorescent and paramagnetic bimodal liposome complex (Ding et al. 2011). In this work, Gd-DTPA-BSA was incorporated into the bilayer of the liposomes as a paramagnetic agent, whereas calcein, a fluorescent dye was encapsulated in the aqueous core of the liposomes. Folate-polyethylene glycol3350-cholesteryl hemisuccinate (folate-PEG3350-CHEMS) on the liposome surface was used as a ligand for FR targeting. Both MR and fluorescence cell imaging showed FR-specific uptake of these liposomes in FR-positive but not in FR-negative cells. Kim et al. reported a trimodal liposome comprising Gd-DOAT-lipid and ^{124}I radionuclide-labeled hexadecyl-4-iodo benzoate (^{124}I-HIB) for optical, nuclear and MR tumor imaging (Kim et al. 2014). Since the radiotracer ^{124}I-HIB simultaneously emits γ-rays and strong Cerenkov luminescence light, optical and PET images could be obtained from the single imaging component (Fig. 6.24a, b). The liposome-carried radiotracer showed longer retention in tumor but quick clearance from RES organs, resulting in vivid tumor imaging with minimum background (Fig. 6.24c).

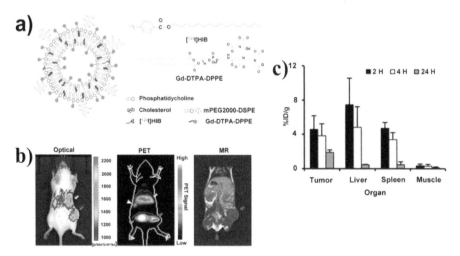

Fig. 6.24 (a) Schematic diagram of trimodal liposome and its imaging components. (b) Typical optical luminescence, PET and MR image of CT26 tumor-bearing mouse at 4 h post injection of [^{124}I]HIB-Gd-liposome (n = 3). Tumor, red arrow; liver, yellow arrow; spleen, white arrowhead; bladder, green arrowhead. (c) Radioactivity uptakes in various organs at 2, 4 and 24 h (n = 3). (Reprinted with permission from Kim, J., Pandya, D.N., Lee, W., Park, J.W., Kim, Y.J., Kwak, W., et al., Vivid tumor imaging utilizing liposome-carried bimodal radiotracer, *ACS Med. Chem. Lett.*, 5, 390-394. Copyright 2014, American Chemical Society.)

Conclusions and Future Prospects

Hybrid materials are emerging as powerful platforms for cancer imaging applications. The inclusion of organic and inorganic components within a single system allows these hybrid materials to be modified with miscellaneous agents for multiple purposes. In this chapter, we have summarized the various types of organic–inorganic hybrid nanomaterials for cancer imaging. Besides the hybrid NPs mentioned in this chapter, several other hybrid materials at nanoscale have also been reported for cancer imaging. For example, viral hybrid nanoparticles, peptide nanoparticles, nanoscale proteins and carbon nanotubes (CNTs) (single- and multi-walled) are all potential nanocarriers for loading imaging agents or drugs due to their excellent biocompatibility, low cytotoxicity and ease of surface modification. Recently, Liang's group has reported a series of peptide-based hybrid nanoparticles for cancer imaging (Cao et al. 2013; Huang et al. 2013; Liang et al. 2011; Miao et al. 2012; Yuan et al. 2015; Yuan et al. 2014). In these works, they employed a biocompatible click condensation reaction between 2-cyanobenzothiazole (CBT) and cysteine (Cys) to self-assemble various nanoparticles for cancer imaging. Properties of the assembled nanoparticles (e.g., molecular weight, size and morphology) could be controlled by adjusting the chemical structures of the monomers or the reaction conditions (i.e., pH, reduction or enzymatic cleavage). Since the ultimate goal of cancer imaging is to help doctors treat cancer more effectively and efficiently, in this chapter, we have also introduced some examples of theranostic hybrid nanoparticles which might allow the clinicians to more effectively monitor the progress of cancer and treat it with more efficacy. Although significant advances have been made in cancer diagnosis by organic–inorganic hybrid NPs, there still exist a lot of concerns which need to be addressed before they are formally moved to the clinical stage (e.g., their biocompatibility, *in vivo* targeting efficacy, long term stability and safety profile). Therefore, more efforts should be made to understand the pharmacokinetics of these hybrid NPs (absorption, uptake, distribution, metabolism and excretion). Herein, we envision that the continuous development of novel inorganic-organic hybrid nanomaterials and their new applications *in vivo* will open new avenues for their diagnostic and therapeutic applications in the clinic.

Acknowledgements

This work was supported by the Collaborative Innovation Center of Suzhou Nano Science and Technology, the Major Program of Development Foundation of Hefei Center for Physical Science and Technology, the National Natural Science Foundation of China (Grants 21175122, 21375121) and Anhui Provincial Natural Science Foundation (Grant 1508085JGD06).

References

Adronov, A., Gilat, S.L., Frechet, J.M.J., Ohta, K., Neuwahl, F.V.R. and Fleming, G.R. 2000. Light harvesting and energy transfer in laser-dye-labeled poly(aryl ether) dendrimers. J Am Chem Soc 122: 1175-1185.

Al-Jamal, W.T., Al-Jamal, K.T., Bomans, P.H., Frederik, P.M. and Kostarelos, K. 2008. Functionalized-quantum-dot-liposome hybrids as multimodal nanoparticles for cancer. Small 4: 1406-1415.

Alcala, M.A., Shade, C.M., Uh, H., Kwan, S.Y., Bischof, M., Thompson, Z.P., et al. 2011. Preferential accumulation within tumors and *in vivo* imaging by functionalized luminescent dendrimer lanthanide complexes. Biomaterials 32: 9343-9352.

Aryal, S., Key, J., Stigliano, C., Ananta, J.S., Zhong, M. and Decuzzi, P. 2013. Engineered magnetic hybrid nanoparticles with enhanced relaxivity for tumor imaging. Biomaterials 34: 7725-7732.

Bagalkot, V., Zhang, L., Levy-Nissenbaum, E., Jon, S., Kantoff, P.W., Langer, R., et al. 2007. Quantum dot – aptamer conjugates for synchronous cancer imaging, therapy and sensing of drug delivery based on Bi-fluorescence resonance energy transfer. Nano Lett 7: 3065-3070.

Bonnemain, B. 1998. Superparamagnetic agents in magnetic resonance imaging: physicochemical characteristics and clinical applications – A review. J Drug Target 6: 167-174.

Bridot, J.L., Faure, A.C., Laurent, S., Riviere, C., Billotey, C., Hiba, B., et al. 2007. Hybrid gadolinium oxide nanoparticles: multimodal contrast agents for *in vivo* imaging. J Am Chem Soc 129: 5076-5084.

Cai, W.B., Shin, D.W., Chen, K., Gheysens, O., Cao, Q.Z., Wang, S.X., et al. 2006. Peptide-labeled near-infrared quantum dots for imaging tumor vasculature in living subjects. Nano Lett 6: 669-676.

Cao, C.Y., Shen, Y.Y., Wang, J.D., Li, L. and Liang, G.L. 2013. Controlled intracellular self-assembly of gadolinium nanoparticles as smart molecular MR contrast agents. Sci Rep-Uk. 3.

Cao, F., Deng, R.P., Liu, D.P., Song, S.Y., Wang, S., Su, S.Q., et al. 2011. Fabrication of fluorescent silica-Au hybrid nanostructures for targeted imaging of tumor cells. Dalton Trans 40: 4800-4802.

Chakravarty, R., Valdovinos, H.F., Chen, F., Lewis, C.M., Ellison, P.A., Luo, H., et al. 2014. Intrinsically germanium-69-labeled iron oxide nanoparticles: synthesis and in-vivo dual-modality PET/MR imaging. Adv Mater 26: 5119-5123.

Chatterjee, D.K., Rufaihah, A.J. and Zhang, Y. 2008. Upconversion fluorescence imaging of cells and small animals using lanthanide doped nanocrystals. Biomaterials 29: 937-943.

Chen, K., Xie, J., Xu, H., Behera, D., Michalski, M.H., Biswal, S., et al. 2009. Triblock copolymer coated iron oxide nanoparticle conjugate for tumor integrin targeting. Biomaterials 30: 6912-6919.

Chen, Q., Li, K., Wen, S., Liu, H., Peng, C., Cai, H., et al. 2013. Targeted CT/MR dual mode imaging of tumors using multifunctional dendrimer-entrapped gold nanoparticles. Biomaterials 34: 5200-5209.

Choi, J.S., Park, J.C., Nah, H., Woo, S., Oh, J., Kim, K.M., et al. 2008. A hybrid nanoparticle probe for dual-modality positron emission tomography and magnetic resonance imaging. Angew Chem Int Ed 47: 6259-6262.

Choi, S.W., Kim, W.S. and Kim, J.H. 2003. Surface modification of functional nanoparticles for controlled drug delivery. J Disper Sci Technol 24: 475-487.

Cobley, C.M., Au, L., Chen, J.Y. and Xia, Y.N. 2010. Targeting gold nanocages to cancer cells for photothermal destruction and drug delivery. Expert Opin Drug Del 7: 577-587.

Corot, C., Robert, P., Idee, J.M. and Port, M. 2006. Recent advances in iron oxide nanocrystal technology for medical imaging. Adv Drug Deliv Rev 58: 1471-1504.

de Rosales, R.T.M., Tavare, R., Paul, R.L., Jauregui-Osoro, M., Protti, A., Glaria, A., et al. 2011. Synthesis of Cu-64 (II)-bis (dithiocarbamatebisphosphonate) and its conjugation with superparamagnetic iron oxide nanoparticles: *in vivo* evaluation as dual-modality PET-MRI agent. Angew Chem Int Ed 50: 5509-5513.

Debergh, I., Vanhove, C. and Ceelen, W. 2012. Innovation in cancer imaging. Eur Surg Res 48: 121-130.

Deng, H., Zhong, Y., Du, M., Liu, Q., Fan, Z., Dai, F., et al. 2014. Theranostic self-assembly structure of gold nanoparticles for NIR photothermal therapy and X-ray computed tomography imaging. Theranostics 4: 904-918.

Ding, H., Yang, D.Y., Zhao, C., Song, Z.K., Liu, P.C., Wang, Y., et al. 2015. Protein-gold hybrid nanocubes for cell imaging and drug delivery. ACS Appl Mater Inter 7: 4713-4719.

Ding, K., Jing, L., Liu, C., Hou, Y. and Gao, M. 2014. Magnetically engineered Cd-free quantum dots as dual-modality probes for fluorescence/magnetic resonance imaging of tumors. Biomaterials 35: 1608-1617.

Ding, N., Lu, Y., Lee, R.J., Yang, C., Huang, L., Liu, J., et al. 2011. Folate receptor-targeted fluorescent paramagnetic bimodal liposomes for tumor imaging. Int J Nanomedicine 6: 2513-2520.

Durr, N.J., Larson, T., Smith, D.K., Korgel, B.A., Sokolov, K. and Ben-Yakar, A. 2007. Two-photon luminescence imaging of cancer cells using molecularly targeted gold nanorods. Nano Lett 7: 941-945.

Dykes, G.M. 2001. Dendrimers: a review of their appeal and applications. J Chem Technol Biot 76: 903-918.

Enochs, W.S., Harsh, G., Hochberg, F. and Weissleder, R. 1999. Improved delineation of human brain tumors on MR images using a long-circulating, superparamagnetic iron oxide agent. J Magn Reson Imaging 9: 228-232.

Gabizon, A., Chisin, R., Amselem, S., Druckmann, S., Cohen, R., Goren, D., et al. 1991. Pharmacokinetic and imaging studies in patients receiving a formulation of liposome-associated adriamycin. Br J Cancer 64: 1125-1132.

Ghobril, C., Lamanna, G., Kueny-Stotz, M., Garofalo, A., Billotey, C. and Felder-Flesch, D. 2012. Dendrimers in nuclear medical imaging. New J Chem 36: 310-323.

Ghosh, P., Han, G., De, M., Kim, C.K. and Rotello, V.M. 2008. Gold nanoparticles in delivery applications. Adv Drug Deliv Rev 60: 1307-1315.

Hadinoto, K., Sundaresan, A. and Cheow, W.S. 2013. Lipid-polymer hybrid nanoparticles as a new generation therapeutic delivery platform: a review. Eur J Pharm Biopharm 85: 427-443.

Harrington, K.J., Rowlinson-Busza, G., Syrigos, K.N., Abra, R.M., Uster, P.S., Peters, A.M., et al. 2000. Influence of tumour size on uptake of ^{111}In-DTPA-labelled pegylated liposomes in a human tumour xenograft model. Br J Cancer 83: 684-688.

Harrington, K.J., Mohammadtaghi, S., Uster, P.S., Glass, D., Peters, A.M., Vile, R.G., et al. 2001. Effective targeting of solid tumors in patients with locally advanced cancers by radiolabeled pegylated liposomes. Clin Cancer Res 7: 243-254.

Hayashi, K., Nakamura, M., Sakamoto, W., Yogo, T., Miki, H., Ozaki, S., et al. 2013. Superparamagnetic nanoparticle clusters for cancer theranostics combining magnetic resonance imaging and hyperthermia treatment. Theranostics 3: 366-376.

Huang, J., Zhong, X.D., Wang, L.Y., Yang, L.L. and Mao, H. 2012. Improving the magnetic resonance imaging contrast and detection methods with engineered magnetic nanoparticles. Theranostics 2: 86-102.

Huang, R., Wang, X.J., Wang, D.L., Liu, F., Mei, B., Tang, A.M., et al. 2013. Multifunctional fluorescent probe for sequential detections of glutathione and caspase-3 *in vitro* and in cells. Anal Chem 85: 6203-6207.

Jain, P.K., Huang, X.H., El-Sayed, I.H. and El-Sayed, M.A. 2008. Noble metals on the-nanoscale: optical and photothermal properties and some applications inimaging, sensing, biology and medicine. Accounts Chem Res 41: 1578-1586.

Jain, T.K., Roy, I., De, T.K. and Maitra, A. 1998. Nanometer silica particles encapsulating active compounds: a novel ceramic drug carrier. J Am Chem Soc 120: 11092-11095.

Jin, Y., Wang, J., Ke, H., Wang, S. and Dai, Z. 2013. Graphene oxide modified PLA microcapsules containing gold nanoparticles for ultrasonic/CT bimodal imaging guided photothermal tumor therapy. Biomaterials 34: 4794-4802.

Josephson, L., Tung, C.H., Moore, A. and Weissleder, R. 1999. High-efficiency intracellular magnetic labeling with novel superparamagnetic-tat peptide conjugates. Bioconjug Chem 10: 186-191.

Jung, J., Kim, M.A., Cho, J.H., Lee, S.J., Yang, I., Cho, J., et al. 2012. Europium-doped gadolinium sulfide nanoparticles as a dual-mode imaging agent for T-1-weighted MR and photoluminescence imaging. Biomaterials 33: 5865-5874.

Key, J. and Leary, J.F. 2014. Nanoparticles for multimodal *in vivo* imaging in nanomedicine. Int J Nanomed 9: 711-726.

Key, J.H., Kim, K., Dhawan, D., Knapp, D.W., Kwon, I.C., Choi, K., et al. 2011. Dual-modality in vivo imaging for MRI detection of tumors and NIRF-guided surgery using multi-component nanoparticles. Nanoscale Imaging, Sensing and Actuation for Biomedical Applications Vii 7908: 1-8.

Kim, J., Lee, J.E., Lee, S.H., Yu, J.H., Lee, J.H., Park, T.G., et al. 2008. Designed fabrication of a multifunctional polymer nanomedical platform for simultaneous cancer-targeted imaging and magnetically guided drug delivery. Adv Mater 20: 478-483.

Kim, J., Piao, Y. and Hyeon, T. 2009. Multifunctional nanostructured materials for multimodal imaging and simultaneous imaging and therapy. Chem Soc Rev 38: 372-390.

Kim, J., Pandya, D.N., Lee, W., Park, J.W., Kim, Y.J., Kwak, W., et al. 2014. Vivid tumor imaging utilizing liposome-carried bimodal radiotracer. ACS Med Chem Lett 5: 390-394.

Kim, S., Lim, Y.T., Soltesz, E.G., De Grand, A.M., Lee, J., Nakayama, A., et al. 2004. Near-infrared fluorescent type II quantum dots for sentinel lymph node mapping. Nat Biotechnol 22: 93-97.

Kim, Y.H., Jeon, J., Hong, S.H., Rhim, W.K., Lee, Y.S., Youn, H., et al. 2011. Tumor targeting and imaging using cyclic RGD-PEGylated gold nanoparticle probes with directly conjugated iodine-125. Small 7: 2052-2060.

Kumar, R., Nyk, M., Ohulchanskyy, T.Y., Flask, C.A. and Prasad, P.N. 2009. Combined optical and MR bioimaging using rare earth ion doped $NaYF_4$ nanocrystals. Adv Funct Mater 19: 853-859.

Lee, C.C., MacKay, J.A., Frechet, J.M.J. and Szoka, F.C. 2005. Designing dendrimers for biological applications. Nat Biotechnol 23: 1517-1526.

Lee, J., Kim, H., Kim, S., Lee, H., Kim, J., Kim, N., et al. 2012a. A multifunctional mesoporous nanocontainer with an iron oxide core and a cyclodextrin gatekeeper for an efficient theranostic platform. J Mater Chem 22: 14061.

Lee, J.H., Jun, Y.W., Yeon, S.I., Shin, J.S. and Cheon, J. 2006. Dual-mode nanoparticle probes for high-performance magnetic resonance and fluorescence imaging of neuroblastoma. Angew Chem Int Ed 45: 8160-8162.

Lee, S.Y., Lim, J.S. and Harris, M.T. 2012b. Synthesis and application of virus-based hybrid nanomaterials. Biotechnol Bioeng 109: 16-30.

Li, I.F. and Yeh, C.S. 2010. Synthesis of Gd doped CdSe nanoparticles for potential optical and MR imaging applications. J Mater Chem 20: 2079-2081.

Li, J.C., Zheng, L.F., Cai, H.D., Sun, W.J., Shen, M.W., Zhang, G.X., et al. 2013. Facile one-pot synthesis of Fe_3O_4@Au composite nanoparticles for dual-mode MR/CT imaging applications. ACS Appl Mater Inter 5: 10357-10366.

Liang, G.L., Ronald, J., Chen, Y.X., Ye, D.J., Pandit, P., Ma, M.L., et al. 2011. Controlled self-assembling of gadolinium nanoparticles as smart molecular magnetic resonance imaging contrast agents. Angew Chem Int Ed 50: 6283-6286.

Lin, A.W.H., Yen Ang, C., Patra, P.K., Han, Y., Gu, H., Le Breton, J.-M., et al. 2011. Seed-mediated synthesis, properties and application of γ-Fe_2O_3–CdSe magnetic quantum dots. J Solid State Chem 184: 2150-2158.

Lin, Y.S., Wu, S.H., Hung, Y., Chou, Y.H., Chang, C., Lin, M.L., et al. 2006. Multifunctional composite nanoparticles: magnetic, luminescent and mesoporous. Chem Mater 18: 5170-5172.

Liu, Q., Sun, Y., Li, C.G., Zhou, J., Li, C.Y., Yang, T.S., et al. 2011a. ^{18}F-labeled magnetic-upconversion nanophosphors via rare-earth cation-assisted ligand assembly. ACS Nano 5: 3146-3157.

Liu, Q., Sun, Y., Yang, T.S., Feng, W., Li, C.G. and Li, F.Y. 2011b. Sub-10 nm hexagonal lanthanide-doped $NaLuF_4$ upconversion nanocrystals for sensitive bioimaging in vivo. J Am Chem Soc 133: 17122-17125.

Liu, Y.L., Ai, K.L., Yuan, Q.H. and Lu, L.H. 2011c. Fluorescence-enhanced gadolinium-doped zinc oxide quantum dots for magnetic resonance and fluorescence imaging. Biomaterials 32: 1185-1192.

Lu, A.H., Salabas, E.L. and Schuth, F. 2007. Magnetic nanoparticles: synthesis, protection, functionalization and application. Angew Chem Int Ed 46: 1222-1244.

Madru, R., Kjellman, P., Olsson, F., Wingardh, K., Ingvar, C., Stahlberg, F., et al. 2012. Tc-99m-labeled superparamagnetic iron oxide nanoparticles for multimodality SPECT/MRI of sentinel lymph nodes. J Nucl Med 53: 459-463.

Maji, S.K., Sreejith, S., Joseph, J., Lin, M.J., He, T.C., Tong, Y., et al. 2014. Upconversion nanoparticles as a contrast agent for photoacoustic imaging in live mice. Adv Mater 26: 5633-5638.

Margerum, L.D., Campion, B.K., Koo, M., Shargill, N., Lai, J.J., Marumoto, A., et al. 1997. Gadolinium (III) DO3A macrocycles and polyethylene glycol coupled to dendrimers – effect of molecular weight on physical and biological properties of macromolecular magnetic resonance imaging contrast agents. J Alloy Compd 249: 185-190.

Mathews, A.S., Ahmed, S., Shahin, M., Lavasanifar, A. and Kaur, K. 2013. Peptide modified polymeric micelles specific for breast cancer cells. Bioconjug Chem 24: 560-570.

McNeil, S.E. 2005. Nanotechnology for the biologist. J Leukocyte Biol 78: 585-594.

Miao, Q.Q., Bai, X.Y., Shen, Y.Y., Mei, B., Gao, J.H., Li, L., et al. 2012. Intracellular self-assembly of nanoparticles for enhancing cell uptake. Chem Commun 48: 9738-9740.

Misri, R., Meier, D., Yung, A.C., Kozlowski, P. and Hafeli, U.O. 2012. Development and evaluation of a dual-modality (MRI/SPECT) molecular imaging bioprobe. Nanomedicine: NBM 8: 1007-1016.

Montalti, M., Prodi, L., Rampazzo, E. and Zaccheroni, N. 2014. Dye-doped silica nanoparticles as luminescent organized systems for nanomedicine. Chem Soc Rev 43: 4243-4268.

Morales-Avila, E., Ferro-Flores, G., Ocampo-Garcia, B.E., De Leon-Rodriguez, L.M., Santos-Cuevas, C.L., Garcia-Becerra, R., et al. 2011. Multimeric system of ^{99m}Tc-labeled gold nanoparticles conjugated to c[RGDfK(C)] for molecular imaging of tumor alpha (v) beta (3) expression. Bioconjug Chem 22: 913-922.

Nahrendorf, M., Zhang, H., Hembrador, S., Panizzi, P., Sosnovik, D.E., Aikawa, E., et al. 2008. Nanoparticle PET-CT imaging of macrophages in inflammatory atherosclerosis. Circulation 117: 379-387.

Nguyen, X.P., Tran, D.L., Ha, P.T., Pham, H.N., Mai, T.T., Pham, H.L., et al. 2012. Iron oxide-based conjugates for cancer theragnostics. Adv Nat Sci: Nanosci Nanotechnol 3: 1-13.

Pansare, V., Hejazi, S., Faenza, W. and Prud'homme, R.K. 2012. Review of long-wavelength optical and NIR imaging materials: contrast agents, fluorophores and multifunctional nano carriers. Chem Mater 24: 812-827.

Peer, D., Karp, J.M., Hong, S., FaroKHzad, O.C., Margalit, R. and Langer, R. 2007. Nanocarriers as an emerging platform for cancer therapy. Nat Nanotechnol 2: 751-760.

Petersen, A.L., Hansen, A.E., Gabizon, A. and Andresen, T.L. 2012. Liposome imaging agents in personalized medicine. Adv Drug Deliv Rev 64: 1417-1435.

Proffitt, R.T., Williams, L.E., Presant, C.A., Tin, G.W., Uliana, J.A., Gamble, R.C., et al. 1983. Tumor-imaging potential of liposomes loaded with In-111-NTA – biodistribution in mice. J Nucl Med 24: 45-51.

Rampazzo, E., Boschi, F., Bonacchi, S., Juris, R., Montalti, M., Zaccheroni, N., et al. 2012. Multicolor core/shell silica nanoparticles for *in vivo* and *ex vivo* imaging. Nanoscale 4: 824-830.

Rosenholm, J.M., Meinander, A., Peuhu, E., Niemi, R., Eriksson, J.E., Sahlgren, C., et al. 2009. Targeting of porous hybrid silica nanoparticles to cancer cells. ACS Nano 3: 197-206.

Saha, A.K., Sharma, P., Sohn, H.B., Ghosh, S., Das, R.K., Hebard, A.F., et al. 2013. Fe doped CdTeS magnetic quantum dots for bioimaging. J Mater Chem B 1: 6312-6320.

Sailor, M.J. and Park, J.H. 2012. Hybrid nanoparticles for detection and treatment of cancer. Adv Mater 24: 3779-3802.

Sanchez, C., Julian, B., Belleville, P. and Popall, M. 2005. Applications of hybrid organic-inorganic nanocomposites. J Mater Chem 15: 3559-3592.

Santra, S., Yang, H.S., Holloway, P.H., Stanley, J.T. and Mericle, R.A. 2005. Synthesis of water-dispersible fluorescent, radio-opaque and paramagnetic CdS: Mn/ZnS quantum dots: a multifunctional probe for bioimaging. J Am Chem Soc 127: 1656-1657.

Santra, S., Kaittanis, C., Grimm, J. and Perez, J.M. 2009. Drug/dye-loaded, multifunctional iron oxide nanoparticles for combined targeted cancer therapy and dual optical/magnetic resonance imaging. Small 5: 1862-1868.

Schellenberger, E.A., Hogemann, D., Josephson, L. and Weissleder, R. 2002. Annexin V-CLIO: a nanoparticle for detecting apoptosis by MRI. Acad Radiol 9: S310-S311.

Sekhon, B.S. and Kamboj, S.R. 2010. Inorganic nanomedicine-Part 1. Nanomedicine: NBM 6: 516-522.

Seo, J.W., Zhang, H., Kukis, D.L., Meares, C.F. and Ferrara, K.W. 2008. A novel method to label preformed liposomes with ^{64}Cu for positron emission tomography (PET) imaging. Bioconjug Chem 19: 2577-2584.

Sharma, P., Brown, S., Walter, G., Santra, S. and Moudgil, B. 2006. Nanoparticles for bioimaging. Adv Colloid Interfac 123-126: 471-485.

Shen, T., Weissleder, R., Papisov, M., Bogdanov, A. and Brady, T.J. 1993. Monocrystalline iron-oxide nanocompounds (Mion) – physicochemical properties. Magn Reson Med 29: 599-604.

Singh, N., Charan, S., Sanjiv, K., Huang, S.H., Hsiao, Y.C., Kuo, C.W., et al. 2012. Synthesis of tunable and multifunctional Ni-doped near-infrared QDs for cancer cell targeting and cellular sorting. Bioconjug Chem 23: 421-430.

Smith, S.V. 2004. Molecular imaging with copper-64. J Inorg Biochem 98: 1874-1901.

Su, N., Dang, Y.J., Liang, G.L. and Liu, G.Z. 2015. Iodine-125-labeled cRGD-gold nanoparticles as tumor-targeted radiosensitizer and imaging agent. Nanoscale Res Lett 10: 1-9.

Sun, C., Veiseh, O., Gunn, J., Fang, C., Hansen, S., Lee, D., et al. 2008. *In vivo* MRI detection of gliomas by chlorotoxin-conjugated superparamagnetic nanoprobes. Small 4: 372-379.

Sun, E.Y., Josephson, L. and Weissleder, R. 2006. "Clickable" nanoparticles for targeted imaging. Mol Imaging 5: 122-128.

Sun, X.L., Huang, X.L., Yan, X.F., Wang, Y., Guo, J.X., Jacobson, O., et al. 2014. Chelatorfree ^{64}Cu-Integrated gold nanomaterials for positron emission tomography imaging guided photothermal cancer therapy. ACS Nano 8: 8438-8446.

Sunderland, C.J., Steiert, M., Talmadge, J.E., Derfus, A.M. and Barry, S.E. 2006. Targeted nanoparticles for detecting and treating cancer. Drug Dev Res 67: 70-93.

Taylor-Pashow, K.M.L., Della Rocca, J., Huxford, R.C. and Lin, W.B. 2010. Hybrid nanomaterials for biomedical applications. Chem Commun 46: 5832-5849.

Torchilin, V.P. 2007. Micellar nanocarriers: pharmaceutical perspectives. Pharm Res 24: 1-16.

Toy, R., Bauer, L., Hoimes, C., Ghaghada, K.B. and Karathanasis, E. 2014. Targeted nanotechnology for cancer imaging. Adv Drug Deliv Rev 76: 79-97.

Vetrone, F., Naccache, R., Juarranz de la Fuente, A., Sanz-Rodriguez, F., Blazquez-Castro, A., Rodriguez, E.M., et al. 2010. Intracellular imaging of HeLa cells by non-functionalized $NaYF_4$: Er^{3+}, Yb^{3+} upconverting nanoparticles. Nanoscale 2: 495-498.

Wang, F. and Liu, X.G. 2009. Recent advances in the chemistry of lanthanide-doped upconversion nanocrystals. Chem Soc Rev 38: 976-989.

Wang, L., Wang, K.M., Santra, S., Zhao, X.J., Hilliard, L.R., Smith, J.E., et al. 2006. Watching silica nanoparticles glow in the biological world. Anal Chem 78: 646-654.

Wang, L.A., Neoh, K.G., Kang, E.T. and Shuter, B. 2011. Multifunctional polyglycerol-grafted Fe_3O_4@SiO_2 nanoparticles for targeting ovarian cancer cells. Biomaterials 32: 2166-2173.

Wang, R. and Zhang, F. 2014. NIR luminescent nanomaterials for biomedical imaging. J Mater Chem B 2: 2422-2443.

Wen, S., Li, K., Cai, H., Chen, Q., Shen, M., Huang, Y., et al. 2013. Multifunctional dendrimer-entrapped gold nanoparticles for dual mode CT/MR imaging applications. Biomaterials 34: 1570-1580.

Wiener, E.C., Brechbiel, M.W., Brothers, H., Magin, R.L., Gansow, O.A., Tomalia, D.A., et al. 1994. Dendrimer-based metal-chelates – a new class of magnetic-resonance-imaging contrast agents. Magn Reson Med 31: 1-8.

Wu, X.M., Chang, S., Sun, X.R., Guo, Z.Q., Li, Y.S., Tang, J.B., et al. 2013. Constructing NIR silica-cyanine hybrid nanocomposite for bioimaging *in vivo*: a breakthrough in photo-stability and bright fluorescence with large stokes shift. Chem Sci 4: 1221-1228.

Wunderbaldinger, P., Josephson, L. and Weissleder, R. 2002. Tat peptide directs enhanced clearance and hepatic permeability of magnetic nanoparticles. Bioconjug Chem 13: 264-268.

Xie, H.A., Wang, Z.J., Bao, A.D., Goins, B. and Phillips, W.T. 2010. *In vivo* PET imaging and biodistribution of radiolabeled gold nanoshells in rats with tumor xenografts. Int J Pharm 395: 324-330.

Xing, H.Y., Bu, W.B., Zhang, S.J., Zheng, X.P., Li, M., Chen, F., et al. 2012. Multifunctional nanoprobes for upconversion fluorescence, MR and CT trimodal imaging. Biomaterials 33: 1079-1089.

Xing, X., Zhang, B., Wang, X., Liu, F., Shi, D. and Cheng, Y. 2015. An "imaging-biopsy" strategy for colorectal tumor reconfirmation by multipurpose paramagnetic quantum dots. Biomaterials 48: 16-25.

Xu, C., Xie, J., Ho, D., Wang, C., Kohler, N., Walsh, E.G., et al. 2008. Au-Fe_3O_4 dumbbell nanoparticles as dual-functional probes. Angew Chem Int Ed 47: 173-176.

Yang, J., Lee, C.H., Ko, H.J., Suh, J.S., Yoon, H.G., Lee, K., et al. 2007. Multifunctional magneto-polymeric nanohybrids for targeted detection and synergistic therapeutic effects on breast cancer. Angew Chem Int Ed 46: 8836-8839.

Ye, F., Barrefelt, A., Asem, H., Abedi-Valugerdi, M., El-Serafi, I., Saghafian, M., et al. 2014. Biodegradable polymeric vesicles containing magnetic nanoparticles, quantum dots and anticancer drugs for drug delivery and imaging. Biomaterials 35: 3885-3894.

Yong, K.T. 2009. Mn-doped near-infrared quantum dots as multimodal targeted probes for pancreatic cancer imaging. Nanotechnology 20: 015102.

Yoon, T.J., Yu, K.N., Kim, E., Kim, J.S., Kim, B.G., Yun, S.H., et al. 2006. Specific targeting, cell sorting and bioimaging with smart magnetic silica core-shell nanomaterials. Small 2: 209-215.

Yu, M.X., Li, F.Y., Chen, Z.G., Hu, H., Zhan, C., Yang, H., et al. 2009. Laser scanning up-conversion luminescence microscopy for imaging cells labeled with rare-earth nanophosphors. Anal Chem 81: 930-935.

Yuan, Y., Zhang, J., An, L.N., Cao, Q.J.W., Deng, Y. and Liang, G.L. 2014. Oligomeric nanoparticles functionalized with NIR-emitting CdTe/CdS QDs and folate for tumor-targeted imaging. Biomaterials 35: 7881-7886.

Yuan, Y., Ge, S., Sun, H., Dong, X., Zhao, H., An, L., et al. 2015. Intracellular self-assembly and disassembly of ^{19}F nanoparticles confer respective "off" and "on" ^{19}F NMR/MRI signals for legumain activity detection in zebrafish. ACS Nano 9: 5117-5124.

Zhang, F. 2015. Photon Upconversion Nanomaterials. Springer Press, Berlin Heidelberg, Germany.

Zhang, F., Sun, T.T., Zhang, Y., Li, Q., Chai, C., Lu, L., et al. 2014. Facile synthesis of functional gadolinium-doped CdTe quantum dots for tumor-targeted fluorescence and magnetic resonance dual-modality imaging. J Mater Chem B 2: 7201-7209.

Zhang, R., Xiong, C.Y., Huang, M., Zhou, M., Huang, Q., Wen, X.X., et al. 2011. Peptide-conjugated polymeric micellar nanoparticles for dual SPECT and optical imaging of EphB4 receptors in prostate cancer xenografts. Biomaterials 32: 5872-5879.

Zhang, X.Y., Zhang, X.Q., Wang, S.Q., Liu, M.Y., Zhang, Y., Tao, L., et al. 2013. Facile incorporation of aggregation-induced emission materials into mesoporous silica nanoparticles for intracellular imaging and cancer therapy. ACS Appl Mater Inter 5: 1943-1947.

Zhao, H.Y., Liu, S., He, J., Pan, C.C., Li, H., Zhou, Z.Y., et al. 2015. Synthesis and application of strawberry-like Fe_3O_4-Au nanoparticles as CT-MR dual-modality contrast agents in accurate detection of the progressive liver disease. Biomaterials 51: 194-207.

Zhou, J., Sun, Y., Du, X., Xiong, L., Hu, H. and Li, F. 2010. Dual-modality in vivo imaging using rare-earth nanocrystals with near-infrared to near-infrared (NIR-to-NIR) upconversion luminescence and magnetic resonance properties. Biomaterials 31: 3287-3295.

Zhou, R.H., Li, M., Wang, S.L., Wu, P., Wu, L. and Hou, X.D. 2014. Low-toxic Mn-doped ZnSe@ZnS quantum dots conjugated with nano-hydroxyapatite for cell imaging. Nanoscale 6: 14319-14325.

Zhu, X., Zhou, J., Chen, M., Shi, M., Feng, W. and Li, F. 2012. Core-shell Fe_3O_4@ $NaLuF_4$: Yb, Er/Tm nanostructure for MRI, CT and upconversion luminescence tri-modality imaging. Biomaterials 33: 4618-4627.

7

CHAPTER

Combinatorial Nanoparticles and Their Applications in Biomedicine

Lina Wu,[1,a] *Kai Cheng,*[2,b] *Baozhong Shen*[1,*] *and Zhen Cheng*[2,*]

INTRODUCTION

Combinatorial nanoparticles (NPs) or hybrid NPs are literally considered to be the combination of different constitute mixed together with the purpose of taking advantage of the effect of the synergetic properties of individual components for biological applications. Such a nanosystem provides a powerful strategy for constructing the so-called all-in-one vehicle by combining both imaging and therapeutic agents together, which is capable of diagnosis, targeted therapy and therapeutic response monitoring. By means of well-developed surface chemistry, multi-functional nanoparticles can be anchored with functional groups to endow specific targeting (Albanese et al. 2012). Recently, these kinds of nanoparticles have proved to be of great research interest because of their extensive applications in biomedical engineering, molecular imaging and disease therapy (Bigall et al. 2012; Taylor-Pashow et al. 2010).

Selection of the building blocks to be assembled represents the initial step for the rational design of discrete groups of combinatorial NPs. Colloid chemistry has become a reliable approach for obtaining high-quality hybrid NPs. Successful efforts have led to plenty of novel synthesis approaches, capable of producing exquisite control over a wide range of sizes, shapes and compositions of NPs (Mukerjee et al. 2012). A wide range of shapes are available, which show some anisotropic shapes, including cubes (Sun and Xia 2002), triangular plates (Millstone et al. 2005), octahedrons (Tao et al.

[1] Department of Medical Imaging and Nuclear Medicine, the Fourth Affiliated Hospital of Harbin Medical University, Harbin 150001, China.

[a] E-mail: LinaWuHMU@hotmail.com

[2] Molecular Imaging Program at Stanford (MIPS), Canary Center at Stanford for Cancer Early Detection, Department of Radiology and Bio-X Program, School of Medicine, Stanford University, Stanford, CA 94305-5484, USA.

[b] E-mail: kaicheng@stanford.edu

[*] Corresponding authors: shenbzh@vip.sina.com and zcheng@stanford.edu

2006), pentagonal bipyramids (Sánchez-Iglesias et al. 2006), rods with sharp-tips (Carbó-Argibay et al. 2007). The ability to combine nanosized domains of metallic, organic and magnetic materials into hybrid nanostructures with intriguing shapes and varied compositions provides a powerful way to engineer nanomaterials with multiple functionalities for biomedicine.

A series of recent review articles have covered several aspects of the effect of NPs with complex structures and compositions, from the basic concepts to specific applications (Louie 2010; Mukerjee et al. 2012; Romo-Herrera et al. 2011; Sun et al. 2015). In this chapter, the recent developments in multifunctional probes (which are the combinatorial nanoparticles that support the effect) are discussed, with our focus on their biomedical applications. The discussions will start with a description of the different kinds of combinatorial nanoparticles, followed by an overview of the biomedical applications developed for targeted imaging, cell separation or therapy.

Classification of Combinatorial Nanoparticles

Core/Shell Nanoparticles

Core/shell nanoparticles are gradually attracting more and more attention, since these nanoparticles have emerged at the frontier between materials chemistry and many other fields, such as electronics, biomedicine, pharmacy, optics and catalysis (Chaudhuri and Paria 2012). Core/shell nanoparticles are highly functional materials with modified properties. Sometimes the properties arising from either the core or shell materials can be quite different. The properties can be modified by changing either the constituting materials or the core to shell ratio (Oldenburg et al. 1998).

Core/Shell nanoparticles of various classes, properties, synthesis mechanisms, characterization and applications have been summarized in a recent review article by Santanu Paria et al. (Ghosh Chaudhuri and Paria 2011). Current applications of different core/shell nanoparticles have been summarized in a review article by Kalele et al. (Kalele et al. 2006). In spite of much focus and great efforts on core/shell nanocomposites, this field is still very much in its developing stage, making the classification of these materials difficult and quite arbitrary (Corr et al. 2008). Consulting some studies (Corr et al. 2008; Jin et al. 2010; Pan et al. 2010), we can identify five main types of combinatorial core/shell nanoparticles (Fig. 7.1).

A Nanoparticle Core Coated with a Silica Shell Containing a Second Nanocomponent

Silica shell is broadly adopted in the fabrication of core/shell NPs for it is relatively inert and optically transparent allowing incorporation of the second NPs constituent directly into the shell. Besides, it may reduce the potential toxic effects of the bare nanoparticles. It also helps to prevent particle aggregation and increases their stability in solution. Importantly, the silica surface can be easily functionalized, enabling chemical bonding of various second nanospecies to the surface. Finally, when compared to the traditional surfactant coatings such as lauric acid and oleic acid, silica coating has no risk of desorption because of the strong covalent bonding (Seo et al. 2014; Wang et al. 2011).

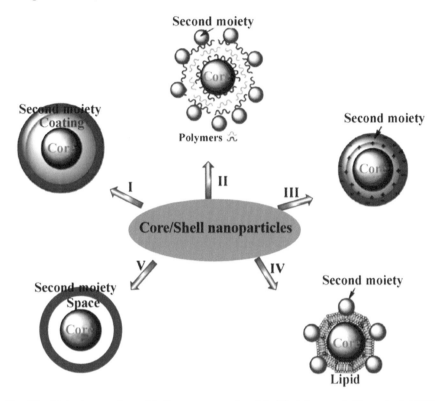

Fig. 7.1 Main types of core/shell nanocomposites. Modified from ref. (Corr et al. 2008). Copyright 2008 Springer.

Rieter et al. reported luminescent and paramagnetic hybrid silica NPs with a magnetic layer (Kim et al. 2007; Rieter et al. 2007). They incorporated a ruthenium complex within a silica NP that acts as the luminescent core, while the paramagnetic component is provided by a monolayer coating of different silyated Gd complexes. These particles were prepared by a water-in-oil reverse microemulsion procedure from $[Ru(bpy)_3]Cl_2$ and tetraethyl orthosilicate by adding ammonia. Similarly, there is a report on the silica-coated CdS Mn/ZnS quantum dots (QDs) treated with the chelating silane coupling agent triacetic acid trisodium salt and the subsequent coordination of Gd ions (Yang et al. 2006).

Up-converting fluorescent magnetic nanoparticles with a silica shell have been synthesized using ytterbium and erbium co-doped sodium yttrium fluoride (NaYF$_4$: Yb, Er), which were deposited on iron oxide nanoparticles by the co-precipitation of the rare-earth metal salts in the presence of a chelator, ethylenediaminetetraacetic acid (EDTA) (Lu et al. 2004).

Polymer-Coated Nanoparticles Core Functionalized with a Second NP Moiety

Various self-assembly techniques utilizing polymers or polyelectrolytes have received considerable interest. The use of several charged layers to provide a coating around the nanoparticle core has been termed the layer-by-layer (LBL) technique. The method has several advantages including the possibility of tuning the thickness of the polymer coating allowing the deposition of a monolayer of charged particles or molecules. By employing this approach, QD/polyelectrolyte/Fe$_3$O$_4$ multifunctional magnetic-fluorescent nanocomposites were fabricated (Fang et al. 2002; Hong et al. 2004). The thickness of the polyelectrolyte coating can be tuned by successive additions of oppositely charged polyelectrolytes, as shown in Fig. 7.2. The fluorescence intensity of the composites was found to vary according to the distance between the core and the QD layer (Okamoto et al. 2007).

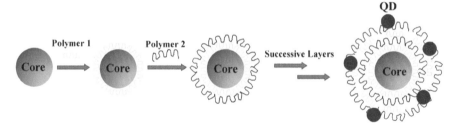

Fig. 7.2 LBL treatment of magnetite nanoparticles with multi-layers of polyelectrolytes, modified from ref. (Corr et al. 2008). Copyright 2008 Springer.

Sometimes the core and the shell nanocomponents cannot dissolve in the same medium, for example some QDs are oil soluble, when designing how to couple these QDs to water-soluble magnetic beads, scholars carried out such reactions in mixed solvents, such as in a 10:5:1 mixture of chloroform/methanol/water, to proceed such kind of reaction (Wang et al. 2004).

Ionic Aggregates Consisting of a Nanoparticle Core and Shell Through Charge Interactions

Electrostatic interactions have also been utilized to fabricate core/shell nanoparticles. You et al. (You et al. 2007) synthesized positively charged magnetite NPs and negatively charged QDs respectively. In order to improve the attachment of the QDs to the magnetic nanoparticles and to maintain these charges, the pH of the reaction medium was adjusted to 3. This lowered pH caused the QDs to flocculate and once the magnetic NPs were added to the suspension of these QDs, they associated via strong electrostatic attractions.

NPs Labeled Lipid-Coated NPs

Lipid layers are frequently used to improve the stability and biocompatibility of nanoparticles. This technique is based on the coating of the nanoparticle surface by amphiphilic lipid molecules, which could then be linked to various species. A novel T_1-weighted MR molecular imaging agent, based on Mn-Gd nanoparticles, has been reported by Lanza et al. (Wang et al. 2012). The 134 nm nanoparticles were synthesized by a bivalent manganese oleate (0.49 ± 0.02 mg Mn/ml)/polysorbate core encapsulated with phospholipid surfactant enriched with 1.25% Gd-DOTA-cholesterol (percentage of phospholipid) and 0.3% of quinolone-derived peptidomimetic $\alpha_v\beta_3$-integrin antagonist-coupled to phosphatidylethanolamine through a polyethylene glycol (PEG) (2000) spacer.

NPs Core Covalently Bonded to a NP Entity via a Spacer

Combining multiple discrete components into a single multifunctional nanoparticle could be useful in a variety of applications. Gao and his group developed a method to create a gap between the core material and gold shell in a compact, uniform format. They proved the success of this strategy on quantum dot core (Jin and Gao 2009) and magnetic NP (MNP) core (Jin et al. 2010), both coated by gold shell through LBL assembly. The wet-chemistry approach provides a general route for the deposition of ultra-thin gold layers onto virtually any discrete nanostructure or continuous surface. The corresponding schematic process was shown in Fig. 7.3.

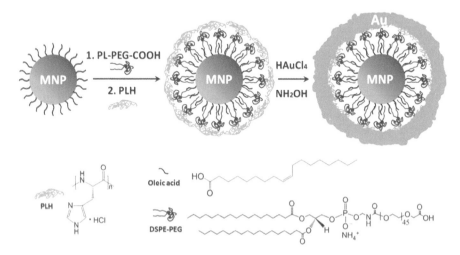

Fig. 7.3 Key steps involved in hybrid NP synthesis of a gold shell encapsulated magnetic nanoparticle by creating a gap between the core and the shell by means of LBL assembly. The molecular structures of oleic acid, phospholipid-PEG terminated with carboxylic acid (PL-PEG-COOH) and poly-L-histidine (PLH) are shown below the reaction scheme. Copyright 2010 Nature communications.

Dumbbell NPs

On the basis of core/shell nanostructure, the dumbbell nanoparticles ('Janus' NPs) have been vigorously developed (Avvisati et al. 2015; Munaò et al. 2015). In such a dumbbell heterostructure, one nanoparticle is linked to another and electronic communication across the junction can drastically change the local electronic structure, leading to an additional dimension of control in catalytic, magnetic and optical properties (Li et al. 2005). Moreover, the dumbbell structure offers two functional surfaces for the attachment of different kinds of molecules, making such species especially attractive as multifunctional probes (Gu et al. 2005). Synthesis of dumbbell-shaped nanoparticles containing different functionalities has attracted much attention (Cheng et al. 2014b). The dumbbell heterostructure usually requires firstly fabricate core/shell nanoparticles, then after a step of heating, for the mismatch of the lattices between the core and the shell, plus the surface tension can force these core/shell nanoparticles to evolve into heterodimers. There are several reports on the fabrication of dumbbell nanocomposites based on this strategy (Corr et al. 2008; Xu et al. 2008). In one study, a CdS shell was deposited on the surface of FePt nanoparticles to form a fluorescent-magnetic core/shell nanostructure (Gu et al. 2004). This was achieved by a relatively simple one-pot synthesis, which involved the following steps: (1) the thermolysis of $Fe(CO)_5$ and the reduction of $Pt(AcAc)_2$ by hexadecane-1, 2 diol, resulting in FePt magnetic nanoparticles in dioctyl ether; (2) the deposition of a sulphur layer on the particles by the addition of elemental sulphur to the dioctyl ether solution at 100°C; (3) the addition of trioctylphosphine oxide (TOPO), hexadecane-1, 2-diol and $Cd(AcAc)_2$ at 100°C yielding a metastable core/shell FePt@CdS structure and finally (4) the transformation of the CdS layer from amorphous to crystalline by heating to 280°C. Then the heterodimers consisting of CdS and FePt nanocrystals with size < 10 nm can finally be achieved.

Shim and co-workers used this approach to prepare a series of maghemite-metal sulphide (ZnS, CdS and HgS) hetero-nanostructures (Kwon and Shim 2005). These nanocomposites were prepared by the direct addition of sulphur and the appropriate metal-organic precursors to pre-formed γ-Fe_2O_3 nanoparticles, followed by high-temperature treatment. Similar to the above heterodimers, the large lattice mismatch between γ-Fe_2O_3 and metal sulphide nanocrystals resulted in the formation of non-centrosymmetric nanostructures, as shown in Fig. 7.4. Preferential formation of trimers and higher oligomers was observed for ZnS and dimers or isolated particles for CdS and HgS nanocomposites. However, the fluorescent and magnetic properties of these nanocomposites have not yet been investigated.

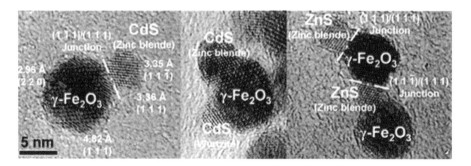

Fig. 7.4 Transmission electron microscope (TEM) images of QD-Fe$_2$O$_3$ nanoparticles. Reproduced with permission from (Kwon and Shim 2005). Copyright 2005 American Chemical Society.

Some other interesting shapes of hetero nanoparticles have also been successfully prepared and reported (Zeng et al. 2010). They were synthesized by employing pre-synthesized nanocrystals as seeds for heterogeneous nucleation and growth of Au. The seeds can include CdSe (Costi et al. 2008), PbSe (Shevchenko et al. 2006), FePt (Seki et al. 2008), Cu$_2$O (Hua et al. 2011) or FePt–CdS nanocrystals (Zeng et al. 2010). The sites of heterogeneous nucleation can be accurately controlled for nanocrystal seeds with a nonspherical shape. Another strategy to access several types of Au-tipped dumbbell-like nanocrystal heterostructures is also disclosed (Carbone et al. 2006), which involved the selective oxidation of either PbSe or CdTe sacrificial domains, initially grown on CdSe and CdS nanorods, with an Au (III) surfactant complex. This approach has allowed the growth of Au domains on specific locations of anisotropically shaped nanocrystals for which direct metal deposition is unfeasible, as in the case of CdS nanorods. It is believed that this strategy may be of general utility to create other types of complex colloidal nanoheterostructures, provided that a suitable sacrificial material can be grown on top of the starting nanocrystal seeds.

In spite of the fast development on the fabrication of some metallic nanostructures with complex shapes, great challenges still remain in the precise control of their geometries and monodispersity. Our group recently developed a controlled, stepwise strategy to build novel, anisotropic, branched, gold nanoarchitectures (Au-tripods) with pre-determined composition and morphology for bioimaging, their structure diagram is shown in Fig. 7.5. The resultant Au-tripods with size less than 20 nm showed great promise as contrast agents for *in vivo* photoacoustic imaging (Cheng et al. 2014a).

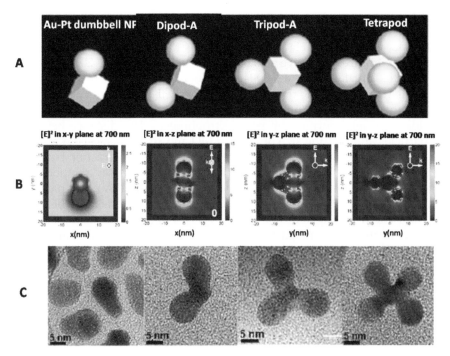

Fig. 7.5 (A) Construction of gold multipods (including Au–Pt dumbbell NPs, Au-dipods, Au-tripods and Au-tetrapods); (B) Electric field intensity contours in different planes of representative gold multipods from the finite-difference time-domain FDTD calculations; (C) high-resolution transmission electron microscopy (HRTEM) images of representative gold multipods. Reproduced with permission from (Cheng et al. 2014a). Copyright 2014 American Chemical Society.

NP/NP Combinatorial Structure

FRET Nanoparticles

Förster or fluorescence resonance energy transfer (FRET) technology is very convenient and appealing for bioanalysis, which is a nonradiative process whereby an excited state donor (D) transfers energy to a proximal ground state acceptor (A) (Sapsford et al. 2006; Wang et al. 2005b). FRET possesses intrinsic sensitivity to nanoscale changes in D/A separation distance (proportional to r^6). This property is exploited in FRET techniques ranging from the assay of interactions of an antigen with an antibody *in vitro* to the real-time imaging of protein folding *in vivo* (Lillo et al. 1997; Schobel et al. 1999). It can be applied routinely at the single molecule detection limit, so it should be an ideal method to detect the trace amount of molecules. In the FRET system, QDs-Au based FRET nanoparticles have attracted broad attention. QDs have several unique intrinsic photophysical properties which make them attractive biolabels, including relatively high quantum yields, high resistance to photobleaching and chemical degradation, as well as molar extinction coefficients 10 to 100 times those of organic dyes (He et al.

2010; Medintz et al. 2005; Michalet et al. 2005). QDs consisting of various other binary and ternary semiconductor materials including ZnS, CdTe, PbSe, CdS, CdSe and CdHgTe with emissions ranging from the Ultraviolet (UV) to the Infrared (IR) have also been synthesized (Murphy 2002).

Alivisatos et al. (Fu et al. 2004) developed an approach to fabricate nanostructures of colloidal CdSe/ZnS core/shell QDs surrounded by a discrete number of Au nanoparticles, which were generated via DNA hybridization and purified by gel electrophoresis. The distance between Au particles and QD, the size of Au and QD and the content of Au around the central QD can be adjusted and they can affect the interaction between Au NPs and QDs (Wang et al. 2010d). An inhibition assay method was developed based on the FRET efficiency between QDs and Au nanoparticles in the presence of the molecules which inhibit the interactions between QD- and AuNP-conjugated biomolecules (Fig. 7.6) (Oh et al. 2005). For the functionalization, Au nanoparticles were first stabilized with *n*-alkanethiols and then linked with the first generation polyamidoamine dendrimers. By employing a streptavidin–biotin couple as a model system, avidin was quantitatively analyzed as an inhibitor by sensing the change in the photoluminescence quenching of SA–QDs by biotin–Au nanoparticles. The detection limit for avidin was about 10 nM.

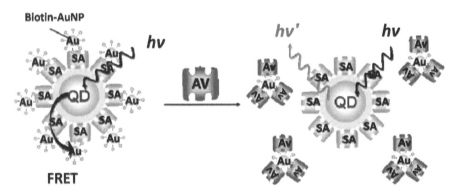

Fig. 7.6   Inhibition assay of biomolecules based on FRET between QDs and gold nanoparticles. Reproduced with permission from ref. (Oh et al. 2005). Copyright 2005 American Chemical Society.

Using a rigid variable-length polypeptide as a bifunctional biological linker, Hedi Mattoussi developed CdSe/ZnS QDs-Au FRET nanoparticles where nonradiative quenching of the QD emission by proximal Au nanoparticles is over a broad range of values (~50–200 Å), exceeding the distance limitations of traditional FRET applications within 100 Å (Fig. 7.7) (Pons et al. 2007). This is ascribed to long-distance dipole–metal interactions that extend significantly beyond the classical Förster range. Compared to Au nanoparticles, Au nanorods have more coupling choice with QDs because they present different localized surface plasmon resonance peaks, which can be adjusted by varying the ratio of diameter to length (Eustis and El-Sayed 2006). The quenching efficiency is directly related to the spectral overlap of Au nanorod-absorption and QD-emission, as well as the separation distance between these two nanoparticles. Au nanorods are advantageous for biomolecular conjugation and the high efficiency of Au nanorod-QD FRET nanoplatform can be utilized for sensitive DNA detection (Xin et al. 2009), homogeneous immunoassays (Zeng et al. 2012) and nanosensors (Xia et al. 2011b).

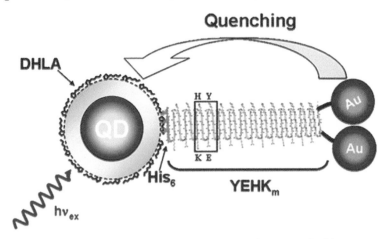

Fig. 7.7 On the quenching of QD photoluminescence by proximal gold nanoparticles. Reproduced with permission from ref. (Pons et al. 2007). Copyright 2007 American Chemical Society.

Although QDs have found plenty of applications, their intrinsic toxicity limits their extensive application in biology. Further, QDs are usually excited by UV or visible radiation. In the actual applicability, lots of impurities are present in biological, toxicological and environmental samples, which can also be excited under such conditions. The sensitivity and correctivity of FRET detection would hence be reduced. Upconversion (UC) nanophosphors are an alternative. UCs are excited in the infrared region instead of the UV and visible region to give emission in the visible domain (Auzel 2004; Wang et al. 2010a; Wang et al. 2010b). UC nanophosphores have the advantageous features of low toxicity, low autofluorescence background, long fluorescence lifetime, high quantum yields, low photobleaching, large Stokes shifts, narrow emission peak, high chemical stability and tunable optical property by varying lanthanide dopants and host matrix (Vetrone et al. 2010; Wang and Liu 2009; Wang and Li 2006; Wang et al. 2005b; Wang et al. 2009b; Yi and Chow 2007). In particular, the luminescence from backgrounds of biological, toxicological and environmental samples upon excitation with IR radiation is extremely low. Wang and coworkers (Wang et al. 2011) developed a facile LBL technology to synthesize a bifunctional water-soluble upconversion (UC)/Au nanocomposite in which the mercapto-silica shell is used as the functional layer coating on the central UC nanocrystals (shown in Fig. 7.8). By adjusting the mole ratio of two kinds of nanoparticles, control of the gold loading on the surface of UC nanocrystal is achieved. The combinatorial nanoparticles simultaneously present UC luminescence and high X-ray attenuation. Under excitation with a 970 nm laser, UC/Au nanocomposite gives emission peak at ~540 nm. In the presence of target molecules, the nanocomposites are aggregated and FRET enhanced between UC nanocrystals and Au nanoparticles. The relative intensity of luminescence at ~540 nm has a linear relationship with the concentration of target in the solution (Wang et al. 2013a). Based on this kind of heterostructure, the DNA, IgG and other biomolecules can be determined (Wang and Li 2006; Wang et al. 2009a; Wang et al. 2009b).

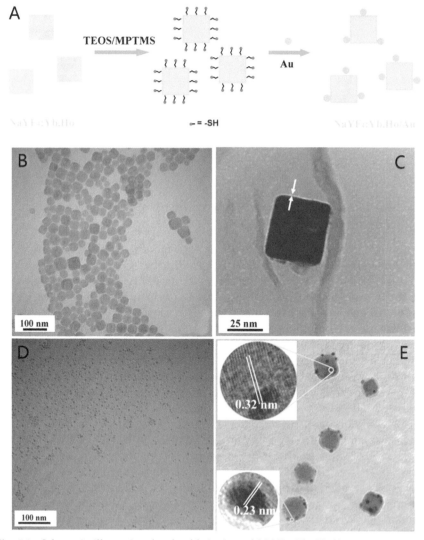

Fig. 7.8 Schematic illustration for the fabrication of NaYF$_4$: Yb, Ho/Au nanocomposites (A) TEM image of NaYF$_4$: Yb, Ho nanocrystals. (B) mercapto-functionalized NaYF$_4$: Yb, Ho nanocrystals. (C) Au nanoparticles. (D) NaYF$_4$: Yb, Ho/Au nanocomposites (E). inset being the HRTEM images. Reproduced with permission (Wang et al. 2013a). Copyright 2013, Elsevier.

Carbon Nanotube Conjugating with NPs

Carbon materials can display intrinsic near-infrared (NIR) luminescence, making them promising candidates for various applications in nanobio-technology (Baughman et al. 2002; Wu et al. 2013a; Wu et al. 2013b). Lots of approaches have been reported in the literature to immobilize nanoparticles on carbon nanotubes (CNTs) to endow them with multifunctional properties and they can be divided into two main pathways: (a) *in situ* growth of metal or oxides nanoparticles directly on the surface of carbon nanotube and (b) the connection of chemically modified nanoparticles to carbon nanotubes or to modified CNTs. A selection of representative examples of the synthesis of CNT based combinatorial nanoparticles is reported and discussed here. In the CNT-nanohybrids, weak interactions between the CNT and nanoparticles will lead to rather non-uniform distributions of the nanoparticles (Kongkan and et al. 2007; Wang et al. 2005a). Endeavors to conquer this problem have been pursued. Wang developed a simple hydrothermal synthesis where a high-density coating of monodisperse Fe_3O_4 nanocrystals with a core size of around 8 nm can be grown *in situ* on the multi-walled carbon nanotube (MWCNT) (Fig. 7.9). Fe_3O_4 nanocrystals are immobilized on the CNT surface by a strong chemical interaction of Fe-O-C bonding. Charges in the CNT/Fe_3O_4 hybrids transfer from the conduction band in Fe_3O_4 to C $2p$-derived states in the CNT substrate (Wang et al. 2013b). By a sol-gel method, RuO_2 (Zhou et al. 2009) SnO_2 (Wen et al. 2007) TiO_2 (Yao et al. 2008) and ZnO (Khanderi et al. 2009) can also be tightly bonded with CNT by a chemical bonding.

CNT-inorganic hybrid architectures can be non-covalently constructed by chemically modified nanoparticles or CNT. To attach Au or quantum dots on CNT, thiol-modification on CNT is the most commonly adopted approach where thiol-terminated molecules can be linked with gold by Au-S tightly bonding or with quantum dots by metal-S bonds. Liu et al. (Liu et al. 2003) used a thiol-terminated pyrene derivative as the inter-linker to assemble Au NPs onto MWCNTs. The pyrene units strongly linked with the MWCNT surface, the thiol group on the other end anchored the as-synthesized Au NPs, thus forming Au-MWCNT hybrids. Han et al. (Han et al. 2007) functionalized the single-wall nanotubes (SWNTs) by thiolated DNA strands. Then, the DNA wrapped SWNTs were mixed with the pre-synthesized Au NPs, thus forming Au-SWNT hybrids. To attach Fe_3O_4 on the CNT, carboxylation function process was adopted to modify MWCNT (Georgakilas et al. 2005). The means of electrostatic interactions to connect CNT and nanoparticles has been demonstrated to be an effective and universal approach. Metals including Pt, Pd, Ru, Rh, Ir, Au and Ag (Zhang and Cui 2009) oxides such as ZnO, SnO_2 and TiO_2 (Chu et al. 2010; Park et al. 2009a) have been decorated on MWCNT.

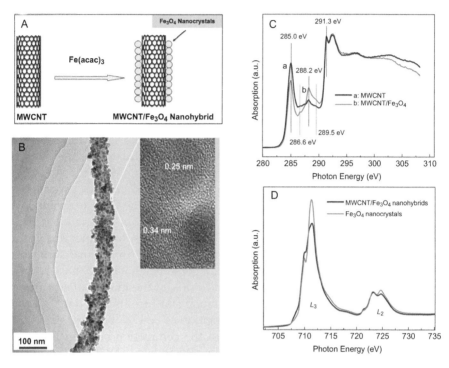

Fig. 7.9 Monodisperse Fe_3O_4 nanocrystals with a core size of around 8 nm grown *in situ* on the MWCNT. (A) Schematic illustration for fabrication of MWCNT/Fe_3O_4 nanohybrid; (B) TEM images of the MWCNT/Fe_3O_4 nanohybrid; (C) XANES C K-edge of MWCNT and MWCNT/Fe_3O_4 nanohybrid; (D) XANES Fe $L_{3,2}$-edge of Fe_3O_4 and MWCNT/Fe_3O_4 nanohybrid.

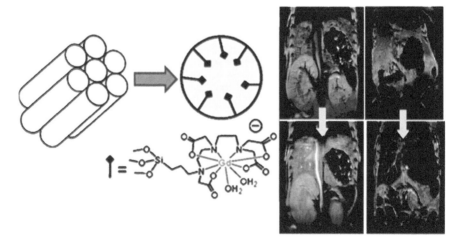

Fig. 7.10 MSN–Gd particles with high magnetic resonance (MR) relaxivities and its proof in disease diagnosis. Reproduced with permission from ref (Taylor et al. 2008). Copyright 2008 American Chemical Society.

Besides, there is a report using 3-Aminopropyl (trimethoxysilyl) diethylenetriamine tetraacetic acid (Si-DTTA) and its Gd complex to coat mesoporous silica nanoparticles (MSNs) containing a hexagonal array of one-dimensional channels with diameters of 2.4 nm, as shown in Fig. 7.10. Owing to the ready access of water molecules through the nanochannels of the MSN-Gd particles and a high payload of Gd centers these MSN-Gd particles exhibit very high MR relaxivities. This work demonstrates the design and synthesis of highly efficient nanoparticulates as contrast-enhanced magnetic resonance imaging (MRI) agents based on MSNs and suggests the potential of using these new hybrid nanomaterials for early disease diagnosis.

Colloidal Nanoparticle Cluster (CNC)

Colloidal nanoparticle cluster (CNC) is one kind of material which is obtained by assembly of colloidal nanoparticles into secondary structures. CNCs not only allow the combination of physicochemistry properties of individual particles, but also produce coupling interactions between neighboring nanoparticles which can yield new properties (Lu and Yin 2012). The present work is specifically centered on Au and Fe_3O_4 NPs as building blocks. Three main assembly approaches (linkers, asymmetric functionalization and electrostatic interactions) summarize the currently available tools to fabricate colloidal nanoparticle clusters.

Pioneering works by Alivisatos et al. (Alivisatos et al. 1996) and Feldheim et al. (Brousseau III et al. 1999) demonstrated the formation of Au NP dimers and trimers, through assembly with molecular linkers, such as DNA or organic molecules, respectively. The DNA approach comprised NP assembly using molecular linkers. DNA complementary recognition features were exploited, by using single-stranded DNA molecules with complementary sequences bound at the surfaces of the NPs, to induce their linear assembly into dimers or trimers. By using phenylacetylene oligomers as 'molecular wire' linkers between the nanoparticles, discrete assemblies of dimer and trimer nanoclusters were fabricated allowing the production of well-defined, rigid arrays of gold clusters with a variety of geometries. As it is known that the distance and the medium between clusters can influence the optical absorption and electron transport (Brousseau III et al. 1999). The same group subsequently reported the first traces of plasmonic coupling from discrete clusters of plasmonic NPs (Novak and Feldheim 2000). Olson et al. used host–guest organic complexes (CBPQT[4+]/TTF-DEG) to reversibly assemble dimers and trimers, using non-covalent bonding interactions. They were able to assemble/disassemble the discrete clusters by oxidizing/reducing the anchoring functional groups (Olson et al. 2009).

A completely different approach, which is starting to be used for plasmonic NPs, is based on the control over the aggregation kinetics of a colloid by fine tuning of the electrostatic interactions. Wang et al. reported formation of dimers and trimers from Au NPs (15 nm in diameter) by inducing aggregation with HCl (Wang et al. 2008). The $Fe(NO_3)_3$ they used to induce the aggregation plays a double role of inducing assembly and etching the initial cubic shaped NPs into spheres. The approach proved efficient with building blocks within a range of particle sizes, in a yield of dimers as high as 60% and with interparticle gaps smaller than 1 nm.

There are two basic strategies for the fabrication of CNCs other than dimers or trimers: (1) one-step approach which combines the preparation of nanoparticles and their aggregation into clusters in a single step (Song et al. 2010); and (2) multi-step processes which firstly obtain nanoparticles with the desired surface functionality and then assemble them into clusters by methods, such as electrostatic attraction, solvent evaporation, chemical interaction or interfacial tension (Kuhlicke et al. 2015; Peng et al. 2013; Tsunoyama et al. 2009).

One-step approach is efficient to produce CNC structures. The techniques of thermolysis and solvothermal are mostly employed (Peng et al. 2013). In a typical thermolysis process, the reactions are typically composed of three critical components: precursors, capping ligands and solvents. Among them, the capping ligands anchor on the nanoparticle surfaces, inhibit their growth and limit interparticle agglomeration by steric interactions. By reducing the degree of ligand protection, complex 3D nanostructures are produced by the oriented attachment of primary nanoparticles. Yin has developed a one-pot polyol process for the synthesis of polyelectrolyte-capped superparamagnetic CNCs of magnetite (Fe_3O_4) (Ge et al. 2007). Fe_3O_4 CNCs were prepared by hydrolyzing $FeCl_3$ with NaOH in a diethylene glycol (DEG) solution with short chain polyacrylic acid (PAA) as a surfactant. The CNC size can be tuned from 30 to 180 nm by varying the amount of NaOH. The CNCs were grown in two stages in which the primary nanoparticles nucleated first in the super saturated DEG solution and then aggregated into larger secondary particles. Solvothermal synthesis is processed in a closed reaction vessel (autoclave) at temperatures higher than the boiling point of the solvent, which has the advantages of relatively simple appliance, easy steps involved and reduced energy requirements. Liang et al. have successfully prepared Fe_3O_4 CNCs by solvothermal synthesis in which polyvinylpyrrolidone was utilized as the surfactant (Liang et al. 2013).

Multi-step approaches are flexible and universal for assembling nanoparticles into CNCs with highly configurable structures. LBL assembly technique was a versatile route for the creation of various CNCs. Wang et al. (Wang et al. 2013c) developed a facile LBL approach to prepare Fe_3O_4 and Fe_3O_4-Au CNCs (Fig. 7.11). The monodisperse Fe_3O_4 nanoparticles capped by triethylene glycol were firstly prepared. Then, Stöber sol-gel method was used to produce thiol-functionalized Fe_3O_4 CNCs with core size of around 80 nm in which 3-mercaptopropyltrimethoxysilane was adopted as the silicon source. Lastly, the thiol-modified Fe_3O_4 CNCs were incubated with citrate-stabilized Au nanoparticles and durian-like multi-functional Fe_3O_4-Au

CNCs were produced. Evaporation of solvents in the presence of block-copolymers as structure directing templates can induce the assembly of nanoparticles to form CNCs. Niederberger (Ba et al. 2005) presents a typical process, monodisperse tin oxide nanoparticles nanometers were synthesized first and subsequently dispersed in tetrahydrofuran (THF), forming a transparent and stable dispersion with the addition of polybutadiene-block-poly (ethylene oxide) (PB-PEO) block copolymer. The evaporation of THF solvent led to the assembly of nanoparticles, finally forming CNCs.

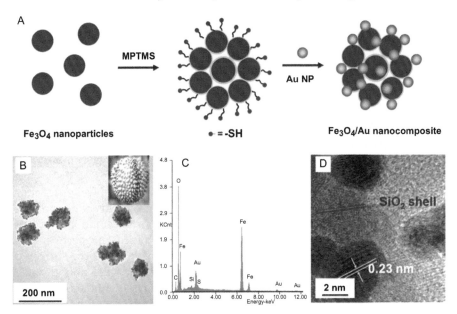

Fig. 7.11 Schematic illustration for the fabrication of durian-like Fe_3O_4/Au nanocomposite (A) TEM image. (B) Energy-dispersive X-ray spectroscopy(EDS) spectrum. (C) and HRTEM image (D) of Fe_3O_4/Au nanocomposites. The image at bottom left of Fig. 7.11b is a photo of a durian. Reproduced with permission from ref. (Wang et al. 2013c). Copyright 2013 Royal Society of Chemistry.

Biomedical Application

Combinatorial nanoparticles are highly functional materials with modified properties that consist of more than two functions combined in one probe. A number of recent studies have highlighted a myriad of applications based on the use of combinatorial NPs, such as molecular imaging, cell tracking and sorting, bioseparation, drug delivery the theranostic in nanomedicine (Hoskins et al. 2012; Jarzyna et al. 2010). These kinds of agents can vigorously develop as promising probes because of their extensive applications in multi-modality imaging which can combine the complementary strengths of different imaging modalities to enable comprehensive diagnostic function.

The present part aims to give an extensive account of the progress of combinatorial nanoparticles in terms of the blooming research with regards to their application in biomedicine.

Multi-Modality Imaging

The development of multi-modality bioimaging is an important advance in molecular imaging, because multi-modality bioimaging not only validates its accuracy, but also enables additional characteristics to be obtained (Fan et al. 2014). Common imaging techniques include magnetic resonance imaging, X-ray computer tomography (CT), Ultrasound imaging, Optical imaging, etc. In multi-modality bioimaging methods, combinatorial nanocomposites can be designed to introduce other functions in addition to the single imaging ability.

Gold-Based Combinatorial Nanoparticles

Gold nanoparticles have found applications in biomedicine since their first colloidal syntheses more than three centuries ago. Their unique combination of properties is beginning to be fully realized in medical diagnostic and therapeutic applications. Due to the advantages of multiple approaches on regulation of their size and surface chemistry, gold nanoparticles preferentially accumulate at sites of tumor growth, inflammation and can be encapsulated into cells. Their unique photophysical properties give make them of use in highly specific thermal ablation of diseased or infected tissues (Huang et al. 2007; Seo et al. 2014). Well-ordered gold nanoparticle superlattices can be used as an active substrate for surface-enhanced Raman scattering (SERS) that can quantitatively detect markers with very low detection limits (Im et al. 2014; Wu et al. 2013c). Choo and Lee developed a highly sensitive optical imaging technology using SERS-fluorescence dual-modality nanoprobes, which can be applied as a tool to image and determine the co-localization of CD24 and CD44 in the cell (Lee et al. 2012). The ability of gold to attenuate X-ray radiation can be also utilized to enhance cancer radiation therapy or improve imaging contrast in diagnostic CT scans (Schirra et al. 2012; Wang et al. 2010e).

Because of the fantastic properties of gold NPs, they are widely employed as important components in combinatorial nanoparticle systems. For example, Minati (Minati et al. 2012) used CNT/gold hybrids for the delivery of an anticancer drug doxorubicin hydrochloride into A549 lung cancer cell line. The hybrids presented a broad absorption band in the red and NIR regions allowing their use for imaging applications. *In vitro* cellular tests showed that the nanostructures can efficiently transport and deliver doxorubicin inside the cells. Currently, highly-integrated multifunctional "all-in-one" nanoparticles combining diagnostic, targeting and therapeutic functions together are attracting the scientists' attention, which enable to perform multiple parallel tasks and provide a platform onto which image-guided therapeutic technologies can be developed for the targeted imaging or therapy of cancers simultaneously (Li 2014). Gold-manganese oxide (Au@MnO) hybrid nanoflowers are capable of imaging the same tissue field with both MRI and an optical source without the fast signal loss observed in the common fluorescent labeling (Schladt et al. 2010). Meanwhile, these hybrid nanoflowers can also be easily functionalized to target cancer cells and capture adenosine triphosphate (Ocsoy et al. 2012).

Fe_3O_4-Based Combinatorial Nanoparticles

Superparamagnetic iron oxide (SPIO) nanoparticles with appropriate surface chemistry can be utilized for numerous *in vivo* applications (Weissleder et al. 2014), such as MRI contrast enhancement (Yang et al. 2011), immunoassay, detoxification of biological fluids, tissue repair and cell separation (Gupta and Gupta 2005; Laurent et al. 2008). The combination of a magnetic and a fluorescent entity provides us with a new two-in-one multifunctional nanomaterial with a broad range of potential applications, which is very beneficial for *in vitro* and *in vivo* bioimaging applications. Ying and co-workers (Jiang et al. 2008) synthesized Fe_3O_4-Ag heterodimer nanoparticles, which combined two-photon fluorescence and MRI imaging in one system. Ding et al. (Fan et al. 2010) developed a new class of magnetic-fluorescent nanoprobes in which cationic polyethylenimine capped QDs were grafted into negatively charged magnetite nanorings (Fig. 7.12). The obtained hybrids exhibit a much stronger magnetic resonance T_2^* effect where the r_2^* relaxivity and r_2^*/r_1 ratio are 4 times and 110 times respectively larger than those of a commercial superparamagnetic iron oxide. These hybrids can escape from endosomes and can be released into the cytoplasm, which suggests that these hybrids could be a useful tool as dual-modality nanoprobes for intracellular imaging and therapeutic applications.

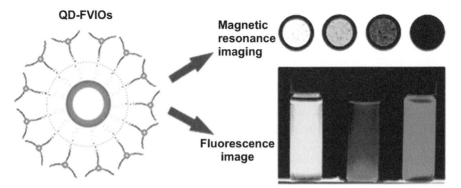

Fig. 7.12 QD capped magnetite nanorings as high-performance nanoprobes for multiphoton fluorescence and magnetic resonance imaging. Reproduced with permission from ref. (Fan et al. 2010). Copyright 2010 American Chemical Society.

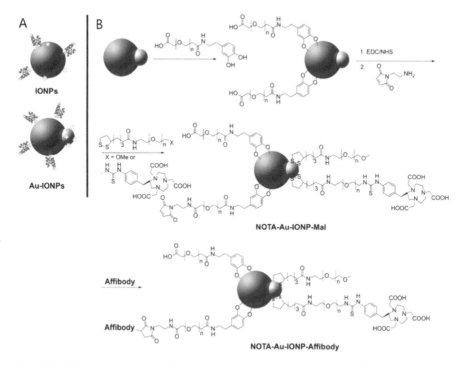

Fig. 7.13 (A) Schematic of the affibody binding domain with conventional single-component iron oxide nanoparticle (IONP) (top) and dumbbell Au-IONP (bottom). (B) Schematic illustration of Au-IONPs surface functionalization and conjugation with affibody. Reproduced with permission from ref. (Yang et al. 2013a). Copyright 2013 Elsevier.

Significantly advanced functionality has been achieved in dual-functional imaging probes based on dumbbell Au-Fe$_3$O$_4$ NPs as introduced by Sun and co-workers (Xu et al. 2008). The structure contained both a Fe$_3$O$_4$ dumbbell-side useful for MRI imaging and an optically active plasmonic Au part for simultaneous optical detection. These hybrid nanoparticles were toposelectively modified with an epidermal growth factor receptor antibody linked through a PEG chain to the Fe$_3$O$_4$ part and PEGylated with a SH-PEG-NH$_2$ polymer on the other Au side. This renders the probe biocompatible and active to target A431 human epithelial carcinoma cell line that overexpresses EGFR. The magnetically and optically active materials are useful for simultaneous magnetic and optical detection, which shows Au–Fe$_3$O$_4$ nanoparticles to be promising new types of multifunctional probes for diagnostic and therapeutic applications. Our group (Yang et al. 2013a) developed an affibody based trimodality nanoprobe (positron emission tomography, PET; optical imaging; and MRI) for imaging of epidermal growth factor receptor positive tumors, which had the capability to combine several imaging modalities together to provide complementary information (Fig. 7.13). *In vitro* and *in vivo* study indicates that the prepared nanoprobe provided high specificity, sensitivity and excellent tumor contrast for both PET and MRI imaging in the human EGFR expressing cells and tumors. Based on this heteronanostructure, we further built new and powerful hybrid combinatorial NPs by fusing distinct nanocrystals, Au, Pt, Gd and IONP, together via solid-state interfaces. The resultant nanotrimers combine both T_1 and T_2-weighted contrast enhancements together with the desirable shape and size for dual MRI, which can enhance the accuracy and reliability of the images achieved from MRI (Cheng et al. 2014b). Xie et al. (Xie et al. 2011) have used an Au-Fe$_3$O$_4$ nanoparticle as a template to construct an optical probe on the iron oxide and SH-PEG$_{5000}$ on the gold surface that can be specifically activated by matrix metallo proteinases expressed in tumors. By a combination of the two kinds of materials' properties, this nanostructure showed performance that is unachievable with the separate components. This architecture of nanoparticles can be tailored to enhance their function as molecular imaging and therapeutic agents.

CNT-Based Combinatorial Nanoparticles

The natural intensity of NIR luminescence emitted by CNT made it a notable component for multimodality imaging technologies. While it is believed that the intensity of luminescence emitted by CNT might be too weak to achieve a whole-body imaging *in vivo*. Shi and co-workers developed luminescent multi-walled carbon nanotubes labeled with QDs in an effort to overcome this limitation (Shi et al. 2007). They adopted plasma surface activation technique to achieve polymerization. The composite quantum dot and carbon nanotubes exhibited intense visible light emissions in both fluorescent spectroscopy and *in vivo* imaging and this kind of combination is believed to be an effective biomarker for applications in cancer diagnosis and treatment. Further, Chen et al. developed carbon nanotube-based magnetic-fluorescent nanohybrids as highly efficient contrast agents for multimodal cellular imaging (Chen et al. 2010a). SPIO nanoparticles and NIR fluorescent CdTe quantum dots were covalently coupled on the surface of CNTs in sequence via LBL assembly. The multifunctional CNT-based magnetic-fluorescent nanohybrids showed an enhanced MRI signal as contrast agents for detecting 293T cells in comparison with the pure SPIO. Moreover, the multifunctional CNT-based magnetic-fluorescent nanohybrids exhibited the higher intracellular labeling efficiency due to the greater ability of CNTs for penetrating into cells in comparison with pure SPIO-CdTe nanoparticles. Combinatorial nanoparticles for this kind of probes are not only limited for CNT, but also extended to single-walled carbon nanohorns (SWNHs). SWNHs are new carbonaceous materials with conical structures and a particularly sharp apical angle. Zhang et al. report the first successful preparation of SWNHs encapsulating trimetallic nitride template endohedral metallofullerenes (Zhang et al. 2010). The resultant materials were functionalized with CdSe/ZnS quantum dots. This kind of combinatorial nanoparticles provide a dual diagnostic platform for *in vitro* and *in vivo* biomedical applications.

Upconversion Nanoparticles (UCNP)-Based Combinatorial Nanoplatforms

Upconversion nanophosphors have the ability to generate visible or NIR emissions under continuous-wave NIR excitation (Chen et al. 2014b; Guo and Wong 2014). Utilizing these special photoluminescent properties, upconversion nanophosphors can be used as key components in combinatorial nanoparticles for a wide range of applications (Yang et al. 2012a).

The paramagnetic constituent locates differently in the crystal lattice will greatly influence the property of their corresponding MRI performance (Chen et al. 2011). (Feng et al. 2013). Because magnetic resonance signals mainly originate from the H_2O molecules, the component contributed for MRI, located on the surface of the combinatorial nanoparticles will provide a greater contribution to the MRI contrast. For example, it is reported that the r_1 values of $NaGdF_4$: Yb, Er (Park et al. 2009b) and $NaYF_4$: 60% Gd, Yb, Er (Zhou et al. 2011) were measured to be 1.40 and 0.41 $s^{-1}mM^{-1}$, respectively. These low r_1 values could be attributed to the location of Gd^{3+} ions inside the UCNPs. Li et al. used Gd^{3+} to modify $NaYF_4$: Yb, Ervia a surface exchange process, achieving a high r_1 value of 28.39 s^{-1} (mg/mL)$^{-1}$ (Liu et al. 2011). T_2-weighted contrast in the nanocomposite is usually provided by Fe_3O_4 nanocrystals. Several effective strategies of fabricating combinatorial nanoparticles as stated before had been found to design nanoprobes for multi-modality imaging with UCL imaging and T_2-weighted MRI, such as core/shell structure (Xia et al. 2011a), composites fabricated by electrostatic attraction (Cheng et al. 2011) and Shi et al. used the so-called "neck formation" to construct a nanocomposite by linking SiO_2-coated UCNPs and Fe_3O_4 with Si–O–Si covalent bonds (Chen et al. 2010b).

X-ray imaging is a traditional and powerful tool used to obtain tissue information with high spatial resolution. An ideal X-ray CT contrast agent would have a high atomic number and electron density to achieve high attenuation of the X-rays. Because lanthanide elements have large atomic numbers, lanthanide ions themselves can act as X-ray CT contrast agents. To date, sole upconversion nanophosphors, such as $NaGdF_4$: Yb, Er (He et al. 2011), $NaYbF_4$: Gd, Er (Liu et al. 2012), $NaYbF_4$: Tm (Xing et al. 2012) and $NaLuF_4$: Yb, Tm (Yang et al. 2013b), have been directly used as effective X-ray CT contrast agents and have further application in dual modality X-ray CT and upconversion luminesce (UCL) imaging. The design of combinatorial nanoparticles for X-ray CT/UCL dual modality imaging requires linking a conventional contrast agent to the UCNPs. To date, Au (Wang et al. 2011),

TaO$_x$ (Xiao et al. 2012) and 5-amino-2, 4, 6-triiodoisophthalic acid (Zhang et al. 2011) nanoparticles have been incorporated with UCNPs.

QD-Based Combinatorial Nanoparticles

The success of detecting cancer at early stages relies greatly on the specificity and sensitivity of *in vivo* molecular imaging. The unique intrinsic photophysical properties of QDs make them fascinating for application as attractive biolabels in optical imaging (Cheng and Cheng 2012; Gao et al. 2011; Su et al. 2013). A research team from the University of Maastricht developed a biocompatible molecular structure capable of binding strongly to eight gadolinium atoms and then linked multiple carriers to each fluorescent QD. By such an assembly that combines a quantum dot with a novel carrier of the MRI agent gadolinium, shown in Fig. 7.14(a), it can spot apoptosis or programmed cell death, using both MRI and fluorescence imaging. Tests in animals showed that this combinatorial nanoparticle can provide anatomical information using MRI and cellular level information using fluorescence imaging. Imaging programmed cell death in the body could provide an early indication that an antitumor therapy is indeed killing cancer cells. The investigators also attached one molecule of annexin A5, a molecule that binds to the surface of cells undergoing apoptosis. The resulting nanoparticle contained enough gadolinium atoms to produce a strong MRI signal that would be detectable even if only a few of the nanoparticles were able to bind themselves to an apoptotic cell. MRI experiments from their research showed that the nanoparticle produced an imaging signal that was approximately 40 times stronger than that produced by the gadolinium carrier alone. Subsequent imaging experiments were able to detect injury-induced apoptosis in mice (Prinzen et al. 2007).

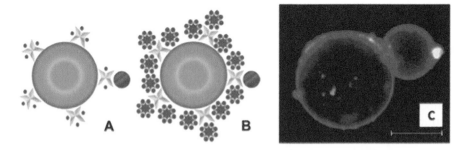

Fig. 7.14 Schematic representation of the two nanoparticles. (A) Nanoparticle with single biotinylated Gd-DTPA (AnxA5-QD-Gd); (B) Nanoparticle with the biotinylated Gd-wedge, containing eight Gd-DTPA complexes each (AnxA5-QD-Gd-wedge). Green: QD, yellow: streptavidin, red dot: Gd-DTPA, red star: lysine-wedge, blue: AnxA5; (C) TPLSM image of a Jurkat cell, incubated with AnxA5-QD525-Gd (green), while simultaneously inducing apoptosis by anti-Fas antibodies. Reproduced with permission from ref. (Prinzen et al. 2007). Copyright 2007 American Chemical Society.

Cell Separation and Purification

Cell separation and purification is a critical step in most applied cell science, from basic research to cell therapy. Using combinatorial NPs can simultaneously achieve separation and monitoring of this process. Wang et al. (Wang et al. 2004) describes the synthesis of new nanocomposites that consist of polymer coated γ-Fe$_2$O$_3$ superparamagnetic cores and CdSe/ZnS quantum dots to form luminescent/magnetic nanocomposite particles. The luminescent/magnetic nanoparticles were easily separated from solution using a permanent magnet and the particles were bright and were easily observed using a conventional fluorescence microscope. The kind of particles could be used in a variety of bioanalytical assays involving luminescence detection and magnetic separation, as shown in Fig. 7.15. The anticycline E antibodies were conjugated to the surface of the nanocomposites and were used to separate MCF-7 breast cancer cells from serum solutions. The separated breast cells were easily observed by fluorescence imaging microscopy due to the strong luminescence of the luminescent/magnetic nanocomposite particles.

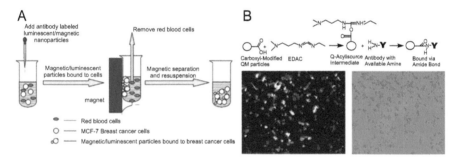

Fig. 7.15 (A) Schematic of cell separation based on luminescent/magnetic particles; (B) Covalent attachment of anticycline E to the luminescent/magnetic particles (top), fluorescence (left) and transmission (right) microscopy images of anticycline E labeled luminescent/ magnetic nanoparticles bound to MCF-7 breast cancer cells. Reproduced with permission from ref. (Wang et al. 2004). Copyright 2004 American Chemical Society.

Jackie Y. Ying et al. (Jiang et al. 2008) synthesized Fe$_3$O$_4$-Ag heterodimer nanocomposites. A ligand-exchange scheme is devised utilizing the bifunctionality of the heterodimer system to render the nanocomposite soluble in aqueous solutions. Cells labeled with the heterodimer nanoparticles can be magnetically manipulated and imaged with two-photon fluorescence microscopy. An interesting study was done by Nicholas A. Kotov using gold nanorods which were selectively modified either on the sides or ends using complementary microcystin (MC-LR) antibody and antigen (blue) (Wang et al. 2010c). Fast detection of MC-LR (green) was successfully achieved with these assemblies and both sensitivity and detection ranges were markedly better for the end-to-end motif (right) than the side-to-side variant (left).

Theranostics

The fast development of nanobiotechnology makes the medical application of highly specific medical intervention at the nanoscale for diagnosing, curing or preventing diseases accessible. The application of nanoparticles for the delivery and targeting of pharmaceutical, therapeutic and diagnostic agents in cancer therapy has received significant attention in recent years (Chen et al. 2014a). There is a professional journal Theranostics to report the creation and applications of nanobiomaterials and devices at the molecular level for personalized diagnosis, imaging and therapy. Nanoparticles may be constructed from a wide range of materials and used for improved delivery *in vivo* or to provide unique optical, magnetic and electrical properties for imaging and therapy (Huang et al. 2012; Serganova et al. 2006). Nowadays, the performance of theranostic systems has improved enormously because of the development of nanotechnology. Simultaneously, very specific drug delivery is now possible to a particular location inside the body or to an organ in what is called "targeted delivery" or "image guided drug delivery", which can achieve controlled release of drugs over the traditional uncontrolled release. By this way, treatments can be individualized through visualizing pathology and controlling the local delivery of therapy. In spite of the fast developments in this field, numerous challenges still face all new technologies, materializing the opportunities presented by theranostics continues to require addressing the significant interdisciplinary challenges and biological barriers (Lanza et al. 2014). Among the candidates for theranostics, the core/shell structure is one kind of typical and classical structure in this direction. While considering that most of the drugs refer to chemotherapeutic molecules, we mainly covered combinatorial NPs in this chapter, which mainly emphasize the scope of NPs and their applications in biomedicine. So we will only give several examples as following. Some relevant content on theranostics NPs can be found in recent reviews. (Frima et al. 2012; Gindy and Prud'homme 2009; Iyer et al. 2012; Ruenraroengsak et al. 2010; Shi et al. 2011; Singh and Lillard Jr 2009).

A reduced graphene oxide-iron oxide nanoparticle (RGO-IONP) complex which is non-covalently functionalized with PEG (RGO–IONP–PEG) has been developed and for the first time used as a novel probe for *in vivo* multimodal tumor imaging and image-guided highly-effective photothermal therapy. Using external radioactive or fluorescent labels and intrinsic optical and magnetic properties, biodistribution and triple-modal fluorescence/MR/photoacoustic tomography (PAT) *in vivo* tumor imaging with RGO–IONP–PEG were carried out, revealing high passive tumor accumulation of the nanoprobe owing to the EPR effect (Yang et al. 2012a). Beyond that, they further used RGO–IONP–PEG as a theranostic probe for *in vivo* MR imaging guided treatment and achieved ultra-efficient tumor ablation using a rather lower laser power density at 0.5 W cm^{-2}, which is much lower than

1-4 W cm^{-2} that is usually applied using gold nanomaterials (Yang et al. 2012b). The dual scattering/fluorescence and absorption properties of gold nanoparticles enable simultaneous cancer detection and therapy. The studies using gold in the combinatorial NPs for imaging and therapy are always a hot topic in the theranostic field (Huang et al. 2006; Loo et al. 2005). Ross's group has been actively pursuing light-activable theranostic nanoparticles for the imaging and photodynamic therapy of brain tumors (Reddy et al. 2006). In this construct, both iron oxide nanoparticles and photofrin (a potent photosensitizer) was incorporated within PEG-modified polyacrylamide nanoparticles. Further, the particles were tagged with a vascular homing (F3) peptide that binds selectively to nucleolin on angiogenic endothelial cells and tumor cell surfaces. These particles demonstrated selective targeting and pronounced antitumor effect in rat brain animal tumor models. In another study, Santra et al. have used multimodal optical and MRI agents containing biocompatible nanoparticles that could demonstrate targeted optical/MR imaging and cell killing towards folate receptors expressing cancer cells (Santra et al. 2009).

Due to the difficulties in large-scale preparation with high reproductivity, good biocompatibility without toxicity, most of the combinatorial nanosystems were limited only to the level of cells and experimental animals. Therefore, the development of quality stable, biocompatible, appropriate size of particles and surface-coated combinatorial nanoparticles, will be an urgent need for future research. In addition, the combinatorial strategy also includes using nanocago space to encapsulate different drugs to affect the biochemical networks of cancer cells, enabling the directed design of treatment regimens that can better thwart the resistance of cancer (Fang and Zhang 2014).

For the exploration of the biomedical applications, the nano-platforms not only refer to the tangible metallic particles, but also to the "soft" nanoscopic objects, with prerequisite features for different imaging modalities with a potential for image-guided drug delivery. The newly published book "Nanomedicine: A Soft Matter Perspective" takes a balanced and in-depth look at the potential and challenges faced in the field (Pan 2014). The reader interested in this aspect can find proofs in this book on the soft matter varieties developed for biomedical imaging and how they can be translated into nanomedicine technologies.

Summary and Outlook

Functionally combinatorial nanoparticle is an active and promising area for noninvasive molecular imaging. This chapter summarizes the progress of combinatorial nanoparticles from design and synthesis to their biomedical application, including contrast enhancement for multi-modality diagnostics, cell separation or theranostic capabilities. It is not hard to see combinatorial nanoparticles open up the opportunity to provide new nanocomposites which could act as multi-targeting, multi-modality, multi-functional and multi-treating tools. It is no wonder that these kinds of NPs have already received a lot of interest, we envision that this field will see significant progress within the next decade for applications *in vitro, in vivo* and in adjuvant settings. Combinatorial nanoparticles applied in molecular imaging should ultimately allow us to simultaneously visualize the disease sites with complementary modalities and probably cure them on-site. Combinatorial nanoparticle is truly emblematic of the evolution of multidisciplinary nanoscience, as a gradual convergence of multiple disciplines including molecular imaging, chemistry, material science, electromagnetics, biology, medical physics and oncology. We can predict high-impact advances in this field by researchers' pioneer approaches to develop nanoscale platforms with multiple functionalities. However, several aspects must be evaluated before combinatorial NPs can advance, including nanoparticle size and surface characteristics, appropriate dosage of both diagnostic and therapeutic functions, toxicity and biocompatibility, the detailed tracking of combinatorial nanoparticles and their interactions *in vivo*. Pursuing these important aspects will enable combinatorial NPs to transit from the laboratory to clinical settings. With great effort in this field, there is no doubt that more combinatorial nanoparticles will emerge not only for imaging applications, but also for target-specific diagnosis and treatment of diseases.

Acknowledgements

Authors would like to acknowledge the support from the Major State Basic Research Development Program of China (973 Program) (No. 2015CB931801, 2015CB931803), the project of International Cooperation and Exchanges NSFC (No. 31210103913) and Major Program of NSFC (No. 81130028). We also acknowledge the support from NSFC (81101087), Heilongjiang Province Foundation for Returnees (LC2016034), Heilongjiang Postdoctoral Foundation (LBH-Q15090) and WuLiande Foundation of Harbin Medical University (WLD-QN1404).

References

Albanese, A., Tang, P.S. and Chan, W.C.W. 2012. The effect of nanoparticle size, shape and surface chemistry on biological systems. Annu Rev Biomed Eng 14: 1-16.
Alivisatos, A.P., Johnsson, K.P., Peng, X., Wilson, T.E., Loweth, C.J., Bruchez, M.P., et al. 1996. Organization of 'nanocrystal molecules' using DNA. Nature 382: 609-611.
Auzel, F. 2004. Upconversion and anti-stokes processes with f and d ions in solids. Chem Rev 104: 139-173.
Avvisati, G., Vissers, T. and Dijkstra, M. 2015. Self-assembly of patchy colloidal dumb-bells. J Chem Phys 142: 084905.

Ba, J., Polleux, J., Antonietti, M. and Niederberger, M. 2005. Non-aqueous synthesis of tin oxide nanocrystals and their assembly into ordered porous mesostructures. Adv Mater 17: 2509-2512.

Baughman, R.H., Zakhidov, A.A. and de Heer, W.A. 2002. Carbon nanotubes–the route toward applications. Science 297: 787-792.

Bigall, N.C., Parak, W.J. and Dorfs, D. 2012. Fluorescent, magnetic and plasmonic-hybrid multifunctional colloidal nano objects. Nano Today 7: 282-296.

Brousseau III, L.C., Novak, J.P., Marinakos, S.M. and Feldheim, D.L. 1999. Assembly of phenylacetylene–bridged gold nanocluster dimers and trimers. Adv Mater 11: 447-449.

Carbó-Argibay, E., Rodríguez-González, B., Pacifico, J., Pastoriza-Santos, I., Pérez-Juste, J. and Liz-Marzán, L.M. 2007. Chemical sharpening of gold nanorods: the rod-to-octahedron transition. Angew Chem Int Ed 46: 8983-8987.

Carbone, L., Kudera, S., Giannini, C., Ciccarella, G., Cingolani, R., Cozzoli, P.D., et al. 2006. Selective reactions on the tips of colloidal semiconductor nanorods. J Mater Chem 16: 3952-3956.

Chaudhuri, R.G. and Paria, S. 2012. Core/shell nanoparticles: classes, properties, synthesis mechanisms, characterization and applications. Chem Rev 112: 2373-2433.

Chen, B., Zhang, H., Zhai, C., Du, N., Sun, C., Xue, J., et al. 2010a. Carbon nanotube-based magnetic-fluorescent nanohybrids as highly efficient contrast agents for multimodal cellular imaging. J Mater Chem 20: 9895-9902.

Chen, F., Zhang, S., Bu, W., Liu, X., Chen, Y., He, Q., et al. 2010b. A "Neck-Formation" strategy for an antiquenching magnetic/upconversion fluorescent bimodal cancer probe. Chem Eur J 16: 11254-11260.

Chen, F., Bu, W., Zhang, S., Liu, X., Liu, J., Xing, H., et al. 2011. Positive and negative lattice shielding effects co-existing in Gd (III) ion doped bifunctional upconversion nanoprobes. Adv Funct Mater 21: 4285-4294.

Chen, F., Ehlerding, E.B. and Cai, W. 2014a. Theranostic nanoparticles. J Nucl Med 55: 1919-1922.

Chen, G., Qiu, H., Prasad, P.N. and Chen, X. 2014b. Upconversion nanoparticles: design, nanochemistry and applications in theranostics. Chem Rev 114: 5161-5214.

Cheng, K. and Cheng, Z. 2012. Near infrared receptor-targeted nanoprobes for early diagnosis of cancers. Curr Med Chem 19: 4767-4785.

Cheng, K., Kothapalli, S.-R., Liu, H., Koh, A.L., Jokerst, J.V., Jiang, H., et al. 2014a. Construction and validation of nano gold tripods for molecular imaging of living subjects. J Am Chem Soc 136: 3560-3571.

Cheng, K., Yang, M., Zhang, R., Qin, C., Su, X. and Cheng, Z. 2014b. Hybrid nanotrimers for dual T1 and T2-weighted magnetic resonance imaging. ACS Nano 8: 9884-9896.

Cheng, L., Yang, K., Li, Y., Chen, J., Wang, C., Shao, M., et al. 2011. Facile preparation of multifunctional upconversion nanoprobes for multimodal imaging and dual-targeted photothermal therapy. Angew Chem Int Ed 50: 7385-7390.

Chu, H., Shen, Y., Lin, L., Qin, X., Feng, G., Lin, Z., et al. 2010. Ionic-liquid-assisted preparation of carbon nanotube-supported uniform noble metal nanoparticles and their enhanced catalytic performance. Adv Funct Mater 20: 3747-3752.

Corr, S.A., Rakovich, Y.P. and Gun'ko, Y.K. 2008. Multifunctional magnetic-fluorescent nanocomposites for biomedical applications. Nanoscale Res Lett 3: 87-104.

Costi, R., Saunders, A.E., Elmalem, E., Salant, A. and Banin, U. 2008. Visible light-induced charge retention and photocatalysis with hybrid CdSe–Au nanodumb-bells. Nano Lett 8: 637-641.

Eustis, S. and El-Sayed, M.A. 2006. Why gold nanoparticles are more precious than pretty gold: noble metal surface plasmon resonance and its enhancement of the radiative and nonradiative properties of nanocrystals of different shapes. Chem Soc Rev 35: 209-217.

Fan, H.-M., Olivo, M., Shuter, B., Yi, J.-B., Bhuvaneswari, R., Tan, H.-R., et al. 2010. Quantum dot capped magnetite nanorings as high performance nanoprobe for multiphoton fluorescence and magnetic resonance imaging. J Am Chem Soc 132: 14803-14811.

Fan, Q., Cheng, K., Hu, X., Ma, X., Zhang, R., Yang, M., et al. 2014. Transferring bio-marker into molecular probe: melanin nanoparticle as a naturally active plat-form for multimodality imaging. J Am Chem Soc 136: 15185-15194.

Fang, M., Grant, P.S., McShane, M.J., Sukhorukov, G.B., Golub, V.O. and Lvov, Y.M. 2002. Magnetic bio/nanoreactor with multilayer shells of glucose oxidase and inorganic nanoparticles. Langmuir 18: 6338-6344.

Fang, R.H. and Zhang, L. 2014. Combinatorial nanotherapeutics: rewiring, then kill-ing, cancer cells. Sci Signal 7: pe13.

Feng, W., Han, C. and Li, F. 2013. Upconversion-nanophosphor-based functional nanocomposites. Adv Mater 25: 5287-5303.

Frima, H.J., Gabellieri, C. and Nilsson, M.-I. 2012. Drug delivery research in the European Union's seventh framework programme for research. J Control Release 161: 409-415.

Fu, A., Micheel, C.M., Cha, J., Chang, H., Yang, H. and Alivisatos, A.P. 2004. Discrete nanostructures of quantum dots/Au with DNA. J Am Chem Soc 126: 10832-10833.

Gao, J., Chen, K., Luong, R., Bouley, D.M., Mao, H., Qiao, T., et al. 2011. A novel clini-cally translatable fluorescent nanoparticle for targeted molecular imaging of tumors in living subjects. Nano Lett 12: 281-286.

Ge, J.P., Hu, Y.X., Biasini, M., Beyermann, W.P. and Yin, Y.D. 2007. Superparamagnetic magnetite colloidal nanocrystal clusters. Angew Chem Int Ed 46: 4342-4345.

Georgakilas, V., Tzitzios, V., Gournis, D. and Petridis, D. 2005. Attachment of mag-netic nanoparticles on carbon nanotubes and their soluble derivatives. Chem Mater 17: 1613-1617.

Ghosh Chaudhuri, R. and Paria, S. 2011. Core/shell nanoparticles: classes, proper-ties, synthesis mechanisms, characterization and applications. Chem Rev 112: 2373-2433.

Gindy, M.E. and Prud'homme, R.K. 2009. Multifunctional nanoparticles for imaging, delivery and targeting in cancer therapy. Expert Opin Drug Deliv 6: 865-878.

Gu, H., Zheng, R., Zhang, X. and Xu, B. 2004. Facile one-pot synthesis of bifunc-tional heterodimers of nanoparticles: a conjugate of quantum dot and magnetic nanoparticles. J Am Chem Soc 126: 5664-5665.

Gu, H., Yang, Z., Gao, J., Chang, C. and Xu, B. 2005. Heterodimers of nanoparticles: formation at a liquid-liquid interface and particle-specific surface modification by functional molecules. J Am Chem Soc 127: 34-35.

Guo, L. and Wong, M.S. 2014. Multiphoton excited fluorescent materials for frequency upconversion emission and fluorescent probes. Adv Mater 26: 5400-5428.

Gupta, A.K. and Gupta, M. 2005. Synthesis and surface engineering of iron oxide nanoparticles for biomedical applications. Biomaterials 26: 3995-4021.

Han, X.G., Li, Y.L. and Deng, Z.X. 2007. DNA-wrapped single-walled carbon nanotubes as rigid templates for assembling linear gold nanoparticle arrays. Adv Mater 19: 1518-1522.

He, M., Huang, P., Zhang, C., Hu, H., Bao, C., Gao, G., et al. 2011. Dual phase-controlled synthesis of uniform lanthanide-doped NaGdF$_4$ upconversion nanocrystals via an OA/ionic liquid two-phase system for in vivo dual-modality imaging. Adv Funct Mater 21: 4470-4477.

He, X., Wang, K. and Cheng, Z. 2010. In vivo near-infrared fluorescence imaging of cancer with nanoparticle-based probes. WIREs Nanomed Nanobiotechnol 2: 349-366.

Hong, X., Li, J., Wang, M., Xu, J., Guo, W., Li, J., et al. 2004. Fabrication of magnetic luminescent nanocomposites by a layer-by-layer self-assembly approach. Chem Mater 16: 4022-4027.

Hoskins, C., Min, Y., Gueorguieva, M., McDougall, C., Volovick, A., Prentice, P., et al. 2012. Hybrid gold-iron oxide nanoparticles as a multifunctional platform for biomedical application. J Nanobiotechnol 10: 27.

Hua, Q., Shi, F., Chen, K., Chang, S., Ma, Y., Jiang, Z., et al. 2011. Cu$_2$O-Au nanocomposites with novel structures and remarkable chemisorption capacity and photocatalytic activity. Nano Res 4: 948-962.

Huang, T., Civelek, A.C., Li, J., Jiang, H., Ng, C.K., Postel, G.C., et al. 2012. Tumor microenvironment–dependent ^{18}F-FDG, ^{18}F-fluorothymidine and ^{18}F-misonidazole uptake: a pilot study in mouse models of human non–small cell lung cancer. J Nucl Med 53: 1262-1268.

Huang, X., El-Sayed, I.H., Qian, W. and El-Sayed, M.A. 2006. Cancer cell imaging and photothermal therapy in the near-infrared region by using gold nanorods. J Am Chem Soc 128: 2115-2120.

Huang, X., Qian, W., El-Sayed, I.H. and El-Sayed, M.A. 2007. The potential use of the enhanced nonlinear properties of gold nanospheres in photothermal cancer therapy. Laser Surg Med 39: 747-753.

Im, H., Shao, H., Park, Y.I., Peterson, V.M., Castro, C.M., Weissleder, R., et al. 2014. Label-free detection and molecular profiling of exosomes with a nano-plasmonic sensor. Nature Biotechnol 32: 490-495.

Iyer, A.K., He, J. and Amiji, M. 2012. Image-guided nanosystems for targeted delivery in cancer therapy. Curr Med Chem 19: 3230-3240.

Jarzyna, P.A., Gianella, A., Skajaa, T., Knudsen, G., Deddens, L.H., Cormode, D.P., et al. 2010. Multifunctional imaging nanoprobes. WIREs Nanomed Nanobiotechnol 2: 138-150.

Jiang, J., Gu, H., Shao, H., Devlin, E., Papaefthymiou, G.C. and Ying, J.Y. 2008. Bifunctional Fe$_3$O$_4$–Ag heterodimer nanoparticles for two-photon fluorescence imaging and magnetic manipulation. Adv Mater 20: 4403-4407.

Jin, Y. and Gao, X. 2009. Plasmonic fluorescent quantum dots. Nat Nanotechnol 4: 571-576.

Jin, Y., Jia, C., Huang, S.-W., O'Donnell, M. and Gao, X. 2010. Multifunctional nanoparticles as coupled contrast agents. Nat Commun 1: 41.

Kalele, S., Gosavi, S., Urban, J. and Kulkarni, S. 2006. Nanoshell particles: synthesis, properties and applications. Curr Sci 91: 1038-1052.

Khanderi, J., Hoffmann, R.C., Gurlo, A. and Schneider, J.J. 2009. Synthesis and sensoric response of ZnO decorated carbon nanotubes. J Mater Chem 19: 5039-5046.

Kim, J.S., Rieter, W.J., Taylor, K.M.L., An, H., Lin, W. and Lin, W. 2007. Self-assembled hybrid nanoparticles for cancer-specific multimodal imaging. J Am Chem Soc 129: 8962-8963.

Kongkanand, A., Martínez Domínguez, R. and Kamat, P.V. 2007. Single wall carbon nanotube scaffolds for photoelectrochemical solar cells. Capture and transport of photogenerated electrons. Nano Lett 7: 676-680.

Kuhlicke, A., Rylke, A. and Benson, O. 2015. On-demand electrostatic coupling of individual precharacterized nano- and microparticles in a segmented paul trap. Nano Lett 15: 1993-2000.

Kwon, K.-W. and Shim, M. 2005. γ-Fe_2O_3/II-VI sulfide nanocrystal heterojunctions. J Am Chem Soc 127: 10269-10275.

Lanza, G.M., Moonen, C., Baker, J.R., Chang, E., Cheng, Z., Grodzinski, P., et al. 2014. Assessing the barriers to image-guided drug delivery. WIREs Nanomed Nanobiotechnol 6: 1-14.

Laurent, S., Forge, D., Port, M., Roch, A., Robic, C., Elst, L.V., et al. 2008. Magnetic iron oxide nanoparticles: synthesis, stabilization, vectorization, physicochemical characterizations and biological applications. Chem Rev 108: 2064-2110.

Lee, S.M., Park, H.Y., Yoo, K-H. 2010. Synergistic cancer therapeutic effects of locally delivered drug using multifunctional nanoparticles. Adv Mater 22: 4049-4053

Lee, S., Chon, H., Yoon, S.-Y., Lee, E.K., Chang, S.-I., Lim, D.W., et al. 2012. Fabrication of SERS-fluorescence dual modal nanoprobes and application to multiplex cancer cell imaging. Nanoscale 4: 124-129.

Li, C. 2014. A targeted approach to cancer imaging and therapy. Nat Mater 13: 110-115.

Li, Y., Zhang, Q., Nurmikko, A.V. and Sun, S. 2005. Enhanced magnetooptical response in dumbbell-like Ag-$CoFe_2O_4$ nanoparticle pairs. Nano Lett 5: 1689-1692.

Liang, J., Ma, H., Luo, W. and Wang, S. 2013. Synthesis of magnetite submicrospheres with tunable size and superparamagnetism by a facile polyol process. Mater Chem Phys 139: 383-388.

Lillo, M.P., Szpikowska, B.K., Mas, M.T., Sutin, J.D. and Beechem, J.M. 1997. Real-time measurement of multiple intramolecular distances during protein folding reactions: a multisite stopped-flow fluorescence energy-transfer study of yeast phosphoglycerate kinase. Biochemistry 36: 11273-11281.

Liu, L., Wang, T.X., Li, J.X., Guo, Z.X., Dai, L.M., Zhang, D.Q., et al. 2003. Self-assembly of gold nanoparticles to carbon nanotubes using a thiol-terminated pyrene as interlinker. Chem Phys Lett 367: 747-752.

Liu, Q., Sun, Y., Li, C., Zhou, J., Li, C., Yang, T., et al. 2011. ^{18}F-Labeled magnetic-upconversion nanophosphors via rare-earth cation-assisted ligand assembly. Acs Nano 5: 3146-3157.

Liu, Y., Ai, K., Liu, J., Yuan, Q., He, Y. and Lu, L. 2012. A high-performance ytterbium-based nanoparticulate contrast agent for in vivo X-ray computed tomography imaging. Angew Chem Int Ed 51: 1437-1442.

Loo, C., Lowery, A., Halas, N., West, J. and Drezek, R. 2005. Immunotargeted nanoshells for integrated cancer imaging and therapy. Nano Lett 5: 709-711.

Louie, A. 2010. Multimodality imaging probes: design and challenges. Chem Rev 110: 3146.

Lu, H., Yi, G., Zhao, S., Chen, D., Guo, L.-H. and Cheng, J. 2004. Synthesis and characterization of multi-functional nanoparticles possessing magnetic, up-conversion fluorescence and bio-affinity properties. J Mater Chem 14: 1336-1341.

Lu, Z. and Yin, Y. 2012. Colloidal nanoparticle clusters: functional materials by design. Chem Soc Rev 41: 6874-6887.

Medintz, I.L., Uyeda, H.T., Goldman, E.R. and Mattoussi, H. 2005. Quantum dot bioconjugates for imaging, labelling and sensing. Nat Mater 4: 435-446.

Michalet, X., Pinaud, F.F., Bentolila, L.A., Tsay, J.M., Doose, S., Li, J.J., et al. 2005. Quantum dots for live cells, in vivo imaging and diagnostics. Science 307: 538-544.

Millstone, J.E., Park, S., Shuford, K.L., Qin, L., Schatz, G.C. and Mirkin, C.A. 2005. Observation of a quadrupole plasmon mode for a colloidal solution of gold nanoprisms. J Am Chem Soc 127: 5312-5313.

Minati, L., Antonini, V., Dalla Serra, M. and Speranza, G. 2012. Multifunctional branched gold–carbon nanotube hybrid for cell imaging and drug delivery. Langmuir 28: 15900-15906.

Mukerjee, A., P Ranjan, A. and K Vishwanatha, J. 2012. Combinatorial nanoparticles for cancer diagnosis and therapy. Curr Med Chem 19: 3714-3721.

Munaò, G., O'Toole, P., Hudson, T.S., Costa, D., Caccamo, C., Sciortino, F., et al. 2015. Cluster formation and phase separation in heteronuclear janus dumbbells. J Phys Condens Mat 27: 234101.

Murphy, C.J. 2002. Peer reviewed: optical sensing with quantum dots. Anal Chem 74: 520 A-526 A.

Novak, J.P. and Feldheim, D.L. 2000. Assembly of phenylacetylene-bridged silver and gold nanoparticle arrays. J Am Chem Soc 122: 3979-3980.

Ocsoy, I., Gulbakan, B., Shukoor, M.I., Xiong, X., Chen, T., Powell, D.H., et al. 2012. Aptamer-conjugated multifunctional nanoflowers as a platform for targeting, capture and detection in laser desorption ionization mass spectrometry. Acs Nano 7: 417-427.

Oh, E., Hong, M.-Y., Lee, D., Nam, S.-H., Yoon, H.C. and Kim, H.-S. 2005. Inhibition assay of biomolecules based on fluorescence resonance energy transfer (FRET) between quantum dots and gold nanoparticles. J Am Chem Soc 127: 3270-3271.

Okamoto, Y., Kitagawa, F. and Otsuka, K. 2007. Online concentration and affinity separation of biomolecules using multifunctional particles in capillary electrophoresis under magnetic field. Anal Chem 79: 3041-3047.

Oldenburg, S., Averitt, R., Westcott, S. and Halas, N. 1998. Nanoengineering of optical resonances. Chem Phys Lett 288: 243-247.

Olson, M.A., Coskun, A., Klajn, R., Fang, L., Dey, S.K., Browne, K.P., et al. 2009. Assembly of polygonal nanoparticle clusters directed by reversible noncovalent bonding interactions. Nano Lett 9: 3185-3190.

Pan, D. (Ed.) 2014. Nanomedicine: A Soft Matter Perspective. CRC Press Inc, New York.

Pan, D., Caruthers, S.D., Chen, J., Winter, P.M., SenPan, A., Schmieder, A.H., et al. 2010. Nanomedicine strategies for molecular targets with MRI and optical imaging. Future Med Chem Mar 2: 471-490.

Park, H.S., Choi, B.G., Yang, S.H., Shin, W.H., Kang, J.K., Jung, D., et al. 2009a. Ionic-liquid-assisted sonochemical synthesis of carbon-nanotube-based nanohybrids: control in the structures and interfacial characteristics. Small 5: 1754-1760.

Park, Y.I., Kim, J.H., Lee, K.T., Jeon, K.S., Na, H.B., Yu, J.H., et al. 2009b. Nonblinking and nonbleaching upconverting nanoparticles as an optical imaging nanoprobe and T1 magnetic resonance imaging contrast agent. Adv Mater 21: 4467-4471.

Peng, E., Choo, E.S., Tan, C.S., Tang, X., Sheng, Y. and Xue, J. 2013. Multifunctional PEGylated nanoclusters for biomedical applications. Nanoscale 5: 5994-6005.

Pons, T., Medintz, I.L., Sapsford, K.E., Higashiya, S., Grimes, A.F., English, D.S., et al. 2007. On the quenching of semiconductor quantum dot photoluminescence by proximal gold nanoparticles. Nano Lett 7: 3157-3164.

Prinzen, L., Miserus, R.-J.J., Dirksen, A., Hackeng, T.M., Deckers, N., Bitsch, N.J., et al. 2007. Optical and magnetic resonance imaging of cell death and platelet activation using annexin A5-functionalized quantum dots. Nano Lett 7: 93-100.

Reddy, G.R., Bhojani, M.S., McConville, P., Moody, J., Moffat, B.A., Hall, D.E., et al. 2006. Vascular targeted nanoparticles for imaging and treatment of brain tumors. Clin Cancer Res 12: 6677-6686.

Rieter, W.J., Kim, J.S., Taylor, K.M., An, H., Lin, W., Tarrant, T., et al. 2007. Hybrid silica nanoparticles for multimodal imaging. Angew Chem Int Ed 46: 3680-3682.

Romo-Herrera, J.M., Alvarez-Puebla, R.A. and Liz-Marzán, L.M. 2011. Controlled assembly of plasmonic colloidal nanoparticle clusters. Nanoscale 3: 1304-1315.

Ruenraroengsak, P., Cook, J.M. and Florence, A.T. 2010. Nanosystem drug targeting: facing up to complex realities. J Control Release 141: 265-276.

Sánchez-Iglesias, A., Pastoriza-Santos, I., Pérez-Juste, J., Rodríguez-González, B., García de Abajo, F.J. and Liz-Marzán, L.M. 2006. Synthesis and optical properties of gold nanodecahedra with size control. Adv Mater 18: 2529-2534.

Santra, S., Kaittanis, C., Grimm, J. and Perez, J.M. 2009. Drug/dye-loaded, multifunctional iron oxide nanoparticles for combined targeted cancer therapy and dual optical/magnetic resonance imaging. Small 5: 1862-1868.

Sapsford, K.E., Berti, L. and Medintz, I.L. 2006. Materials for fluorescence resonance energy transfer analysis: beyond traditional donor–acceptor combinations. Angew Chem Int Ed 45: 4562-4589.

Schirra, C.O., Senpan, A., Roessl, E., Thran, A., Stacy, A.J., Wu, L., et al. 2012. Second generation gold nanobeacons for robust K-edge imaging with multi-energy CT. J Mater Chem 22: 23071-23077.

Schladt, T.D., Shukoor, M.I., Schneider, K., Tahir, M.N., Natalio, F., Ament, I., et al. 2010. Au@MnO nanoflowers: hybrid nanocomposites for selective dual functionalization and imaging. Angew Chem Int Ed 49: 3976-3980.

Schobel, U., Egelhaaf, H.J., Brecht, A., Oelkrug, D. and Gauglitz, G. 1999. New donor-acceptor pair for fluorescent immunoassays by energy transfer. Bioconjugate Chem 10: 1107-1114.

Seki, T., Hasegawa, Y., Mitani, S., Takahashi, S., Imamura, H., Maekawa, S., et al. 2008. Giant spin hall effect in perpendicularly spin-polarized FePt/Au devices. Nat Mater 7: 125-129.

Seo, S.-H., Kim, B.-M., Joe, A., Han, H.-W., Chen, X., Cheng, Z., et al. 2014. NIR-light-induced surface-enhanced Raman scattering for detection and photothermal/photodynamic therapy of cancer cells using methylene blue-embedded gold nanorod@SiO₂ nanocomposites. Biomaterials 35: 3309-3318.

Serganova, I., Humm, J., Ling, C. and Blasberg, R. 2006. Tumor hypoxia imaging. Clin Cancer Res 12: 5260-5264.

Shevchenko, E., Talapin, D., Murray, C. and Obrien, S. 2006. Structural characterization of self-assembled multifunctional binary nanoparticle superlattices. J Am Chem Soc 128: 3620-3637.

Shi, D., Guo, Y., Dong, Z., Lian, J., Wang, W., Liu, G., et al. 2007. Quantum-dot-activated luminescent carbon nanotubes via a nano scale surface functionalization for in vivo imaging. Adv Mater 19: 4033-4037.

Shi, D., Bedford, N.M. and Cho, H.S. 2011. Engineered multifunctional nanocarriers for cancer diagnosis and therapeutics. Small 7: 2549-2567.

Singh, R. and Lillard Jr, J.W. 2009. Nanoparticle-based targeted drug delivery. Exp Mol Pathol 86: 215-223.

Song, J., Xia, Y. and Song, H.-S. 2010. One-step generation of cluster state by adiabatic passage in coupled cavities. Appl Phys Lett 96: 071102.

Su, X., Cheng, K., Wang, C., Xing, L., Wu, H. and Cheng, Z. 2013. Image-guided resection of malignant gliomas using fluorescent nanoparticles. WIRES Nanomed Nanobi 5: 219-232.

Sun, X., Cai, W. and Chen, X. 2015. Positron emission tomography imaging using radiolabeled inorganic nanomaterials. Accounts Chem Res 48: 286-294.

Sun, Y. and Xia, Y. 2002. Shape-controlled synthesis of gold and silver nanoparticles. Science 298: 2176-2179.

Tao, A., Sinsermsuksakul, P. and Yang, P. 2006. Polyhedral silver nanocrystals with distinct scattering signatures. Angew Chem Int Ed 45: 4597-4601.

Taylor, K.M.L., Kim, J.S., Rieter, W.J., An, H., Lin, W. and Lin, W. 2008. Mesoporous silica nanospheres as highly efficient MRI contrast agents. J Am Chem Soc 130: 2154-2155.

Taylor-Pashow, K.M.L., Della Rocca, J., Huxford, R.C. and Lin, W. 2010. Hybrid nanomaterials for biomedical applications. Chem Commun 46: 5832-5849.

Tsunoyama, H., Ichikuni, N., Sakurai, H. and Tsukuda, T. 2009. Effect of electronic structures of Au clusters stabilized by poly(N-vinyl-2-pyrrolidone) on aerobic oxidation catalysis. J Am Chem Soc 131: 7086-7093.

Vetrone, F., Naccache, R., Zamarron, A., de la Fuente, A.J., Sanz-Rodriguez, F., Maestro, L.M., et al. 2010. Temperature sensing using fluorescent nanothermometers. Acs Nano 4: 3254-3258.

Wang, D., He, J., Rosenzweig, N. and Rosenzweig, Z. 2004. Superparamagnetic Fe_2O_3 beads-CdSe/ZnS quantum dots core-shell nanocomposite particles for cell separation. Nano Lett 4: 409-413.

Wang, F. and Liu, X.G. 2009. Recent advances in the chemistry of lanthanide-doped upconversion nanocrystals. Chem Soc Rev 38: 976-989.

Wang, F., Han, Y., Lim, C.S., Lu, Y.H., Wang, J., Xu, J., et al. 2010a. Simultaneous phase and size control of upconversion nanocrystals through lanthanide doping. Nature 463: 1061-1065.

Wang, G.F., Peng, Q. and Li, Y.D. 2010b. Luminescence tuning of upconversion nanocrystals. Chem Eur J 16: 4923-4931.

Wang, G.X., Zhang, B.L., Yu, Z.L. and Qu, M.Z. 2005a. Manganese oxide/MWNTs composite electrodes for supercapacitors. Solid State Ionics 176: 1169-1174.

Wang, K., Pan, D., Schmieder, A.H., Zhang, H., SenPan, A., Williams, T.A., et al. 2012. Probing atherosclerotic angiogenesis with new manganese-based nanocolloid for T1-weighted MRI. J Cardiovasc Magn R 14: O11.

Wang, L., Zhu, Y., Xu, L., Chen, W., Kuang, H., Liu, L., et al. 2010c. Side-by-side and end-to-end gold nanorod assemblies for environmental toxin sensing. Angew Chem Int Ed 49: 5472-5475.

Wang, L.Y. and Li, Y.D. 2006. Green upconversion nanocrystals for DNA detection. Chem Commun 2557-2559.

Wang, L.Y., Yan, R.X., Hao, Z.Y., Wang, L., Zeng, J.H., Bao, H., et al. 2005b. Fluorescence resonant energy transfer biosensor based on upconversion-luminescent nanoparticles. Angew Chem Int Ed 44: 6054-6057.

Wang, M., Hou, W., Mi, C.C., Wang, W.X., Xu, Z.R., Teng, H.H., et al. 2009a. Immunoassay of goat antihuman immunoglobulin G antibody based on luminescence resonance energy transfer between near-infrared responsive NaYF$_4$: Yb, Er upconversion fluorescent nanoparticles and gold nanoparticles. Anal Chem 81: 8783-8789.

Wang, M., Mi, C.C., Wang, W.X., Liu, C.H., Wu, Y.F., Xu, Z.R., et al. 2009b. Immunolabeling and NIR-excited fluorescent imaging of HeLa cells by using NaYF$_4$: Yb, Er upconversion nanoparticles. Acs Nano 3: 1580-1586.

Wang, Q., Wang, H., Lin, C., Sharma, J., Zou, S. and Liu, Y. 2010d. Photonic interaction between quantum dots and gold nanoparticles in discrete nanostructures through DNA directed self-assembly. Chem Commun 46: 240-242.

Wang, X., Li, G., Chen, T., Yang, M., Zhang, Z., Wu, T., et al. 2008. Polymer-encapsulated gold-nanoparticle dimers: facile preparation and catalytical application in guided growth of dimeric ZnO-nanowires. Nano Lett 8: 2643-2647.

Wang, Z., Wu, L., Shen, B. and Jiang, Z. 2013a. Highly sensitive and selective cartap nanosensor based on luminescence resonance energy transfer between NaYF$_4$: Yb, Ho nanocrystals and gold nanoparticles. Talanta 114: 124-130.

Wang, Z., Wu, L., Zhou, J., Cai, W., Shen, B. and Jiang, Z. 2013b. Magnetite nanocrystals on multiwalled carbon nanotubes as a synergistic microwave absorber. J Phys Chem C 117: 5446-5452.

Wang, Z.J., Wu, L.N. and Cai, W. 2010e. Size-tunable synthesis of monodisperse water-soluble gold nanoparticles with high X-ray attenuation. Chem Eur J 16: 1459-1463.

Wang, Z.J., Wu, L.N., Liang, H.J., Cai, W., Zhang, Z.G. and Jiang, Z.H. 2011. Controllable synthesis of bifunctional NaYF$_4$: Yb^{3+}/Ho^{3+}@SiO$_2$/Au nanoparticles with upconversion luminescence and high X-ray attenuation. J Alloy Compd 509: 9144-9149.

Wang, Z.J., Wu, L.N., Wang, F.P., Jiang, Z.H. and Shen, B.Z. 2013c. Durian-like multifunctional Fe$_3$O$_4$-Au nanoparticles: synthesis, characterization and selective detection of benzidine. J Mater Chem A 1: 9746-9751.

Weissleder, R., Nahrendorf, M. and Pittet, M.J. 2014. Imaging macrophages with nanoparticles. Nat Mater 13: 125-138.

Wen, Z., Wang, Q., Zhang, Q. and Li, J. 2007. In situ growth of mesoporous SnO$_2$ on multiwalled carbon nanotubes: a novel composite with porous-tube structure as anode for lithium batteries. Adv Funct Mater 17: 2772-2778.

Wu, L., Cai, X., Nelson, K., Xing, W., Xia, J., Zhang, R., et al. 2013a. A green synthesis of carbon nanoparticles from honey and their use in real-time photoacoustic imaging. Nano Res 6: 312-325.

Wu, L., Luderer, M., Yang, X., Swain, C., Zhang, H., Nelson, K., et al. 2013b. Surface passivation of carbon nanoparticles with branched macromolecules influences near infrared bioimaging. Theranostics 3: 677-686.

Wu, L., Wang, Z. and Shen, B. 2013c. Large-scale gold nanoparticle superlattice and its SERS properties for the quantitative detection of toxic carbaryl. Nanoscale 5: 5274-5278.

Xia, A., Gao, Y., Zhou, J., Li, C., Yang, T., Wu, D., et al. 2011a. Core-shell $NaYF_4$: Yb^{3+}, Tm^{3+}@Fe_xO_y nanocrystals for dual-modality T2-enhanced magnetic resonance and NIR-to-NIR upconversion luminescent imaging of small-animal lymphatic node. Biomaterials 32: 7200-7208.

Xia, Y., Song, L. and Zhu, C. 2011b. Turn-on and near-infrared fluorescent sensing for 2, 4, 6-trinitrotoluene based on hybrid (gold nanorod)–(quantum dots) assembly. Anal Chem 83: 1401-1407.

Xiao, Q., Bu, W., Ren, Q., Zhang, S., Xing, H., Chen, F., et al. 2012. Radiopaque fluorescence-transparent TaO_x decorated upconversion nanophosphors for in vivo CT/MR/UCL trimodal imaging. Biomaterials 33: 7530-7539.

Xie, J., Zhang, F., Aronova, M., Zhu, L., Lin, X., Quan, Q., et al. 2011. Manipulating the power of an additional phase: a flower-like $Au–Fe_3O_4$ optical nanosensor for imaging protease expressions in vivo. Acs Nano 5: 3043-3051.

Xin, L., Jun, Q., Li, J. and Sailing, H. 2009. Fluorescence quenching of quantum dots by gold nanorods and its application to DNA detection. Appl Phys Lett 94: 063111-063111-063113.

Xing, H., Bu, W., Ren, Q., Zheng, X., Li, M., Zhang, S., et al. 2012. A $NaYbF_4$: Tm^{3+} nanoprobe for CT and NIR-to-NIR fluorescent bimodal imaging. Biomaterials 33: 5384-5393.

Xu, C., Xie, J., Ho, D., Wang, C., Kohler, N., Walsh, E.G., et al. 2008. $Au–Fe_3O_4$ dumbbell nanoparticles as dual-functional probes. Angew Chem Int Ed 47: 173-176.

Yang, H., Santra, S., Walter, G.A., Holloway, P.H. 2006. Gd^{III}-functionalized fluorescent quantum dots as multimodal imaging probes. Adv Mater 18: 2890-2894.

Yang, K., Hu, L., Ma, X., Ye, S., Cheng, L., Shi, X., et al. 2012b. Multimodal imaging guided photothermal therapy using functionalized graphene nanosheets anchored with magnetic nanoparticles. Adv Mater 24: 1868-1872.

Yang, M., Cheng, K., Qi, S., Liu, H., Jiang, Y., Jiang, H., et al. 2013a. Affibody modified and radiolabeled gold–iron oxide hetero-nanostructures for tumor PET, optical and MR imaging. Biomaterials 34: 2796-2806.

Yang, Y., Aw, J., Chen, K., Liu, F., Padmanabhan, P., Hou, Y., et al. 2011. Enzyme-responsive multifunctional magnetic nanoparticles for tumor intracellular drug delivery and imaging. Chem Asian J 6: 1381-1389.

Yang, Y., Shao, Q., Deng, R., Wang, C., Teng, X., Cheng, K., et al. 2012a. In vitro and in vivo uncaging and bioluminescence imaging by using photocaged upconversion nanoparticles. Angew Chem Int Ed 51: 3125-3129.

Yang, Y., Sun, Y., Cao, T., Peng, J., Liu, Y., Wu, Y., et al. 2013b. Hydrothermal synthesis of $NaLuF_4$: ^{153}Sm, Yb, Tm nanoparticles and their application in dual-modality upconversion luminescence and SPECT bioimaging. Biomaterials 34: 774-783.

Yao, Y., Li, G., Ciston, S., Lueptow, R.M. and Gray, K.A. 2008. Photoreactive TiO_2/carbon nanotube composites: synthesis and reactivity. Environ Sci Technol 42: 4952-4957.

Yi, G.S. and Chow, G.M. 2007. Water-soluble $NaYF_4$: Yb, Er(Tm)/$NaYF_4$/polymer core/shell/shell nanoparticles with significant enhancement of upconversion fluorescence. Chem Mater 19: 341-343.

You, X., He, R., Gao, F., Shao, J., Pan, B. and Cui, D. 2007. Hydrophilic high-luminescent magnetic nanocomposites. Nanotechnology 18: 035701.

Zeng, J., Huang, J., Liu, C., Wu, C.H., Lin, Y., Wang, X., et al. 2010. Gold-based hybrid nanocrystals through heterogeneous nucleation and growth. Adv Mater 22: 1936-1940.

Zeng, Q., Zhang, Y., Liu, X., Tu, L., Kong, X. and Zhang, H. 2012. Multiple homogeneous immunoassays based on a quantum dots-gold nanorods FRET nanoplatform. Chem Commun 48: 1781-1783.

Zhang, G., Liu, Y., Yuan, Q., Zong, C., Liu, J. and Lu, L. 2011. Dual modal in vivo imaging using upconversion luminescence and enhanced computed tomography properties. Nanoscale 3: 4365-4371.

Zhang, H. and Cui, H. 2009. Synthesis and characterization of functionalized ionic liquid-stabilized metal (gold and platinum) nanoparticles and metal nanoparticle/carbon nanotube hybrids. Langmuir 25: 2604-2612.

Zhang, J., Ge, J., Shultz, M.D., Chung, E., Singh, G., Shu, C., et al. 2010. In vitro and in vivo studies of single-walled carbon nanohorns with encapsulated metallofullerenes and exohedrally functionalized quantum dots. Nano Lett 10: 2843-2848.

Zhou, J., Yu, M., Sun, Y., Zhang, X., Zhu, X., Wu, Z., et al. 2011. Fluorine-18-labeled $Gd^{3+}/Yb^{3+}/Er^{3+}$ co-doped $NaYF_4$ nanophosphors for multimodality PET/MR/UCL imaging. Biomaterials 32: 1148-1156.

Zhou, J.G., Fang, H.T., Hu, Y.F., Sham, T.K., Wu, C.X., Liu, M., et al. 2009. Immobilization of RuO_2 on carbon nanotube: an X-ray absorption near-edge structure study. J Phys Chem C 113: 10747-10750.

8

Engineering of Hybrid Upconversion Nanoparticles for Biodetection and Cancer Imaging

Dalong Ni,[1,a] *Feng Chen,*[2,c] *Wenbo Bu*[1,]* and *Jianlin Shi*[1,b]

INTRODUCTION

Upconversion nanoparticles (UCNPs), which utilize the sequential absorption and energy transfer steps to emit a higher energy photon after absorbing two or more low-energy excitation photons, have evoked great attention for bio-applications in areas such as biodetection and cancer imaging (Chen et al. 2014a; Zhou et al. 2015). This conversion from long-wavelength near-infrared (NIR) stimulation to short-wavelength outcome (e.g., ultraviolet, visible and NIR) endows UCNPs with attractive optical features, such as large anti-Stokes shifts (Haase and Schafer 2011), high detection sensitivity (Cheng et al. 2010), non-blinking and non-bleaching properties (Park et al. 2009; Wu et al. 2009), low autofluorescence background (Kobayashi et al. 2009; Xiong et al. 2009a), minimal photodamage (Nam et al. 2011) and extremely high tissue penetration depth (Liu et al. 2011f). In general, UCNPs comprise an inorganic host, a sensitizer and an activator. Fluorides including $NaYF_4$ (Ni et al. 2014b; Wang et al. 2010), $NaYbF_4$ (Liu et al. 2012b; Xing et al. 2012a), $NaGdF_4$ (Qiao et al. 2015) and $NaLuF_4$ (Liu et al. 2011f), which exhibit low phonon energies

[1] State Key Laboratory of High Performance Ceramics and Superfine Microstructure, Shanghai Institute of Ceramics, Chinese Academy of Sciences, Shanghai 200050, People's Republic of China.
[a] E-mail: dlni2011@126.com
[b] E-mail: jlshi@mail.sic.ac.cn
[2] Department of Radiology, University of Wisconsin-Madison, Wisconsin 53705, USA.
[c] E-mail: chenf@mskcc.org
* Corresponding author: wbbu@mail.sic.ac.cn

(~350 cm^{-1}) and high chemical stability, have become the most popular host materials. The doped lanthanide (Ln^{3+}) ions such as Er^{3+}, Tm^{3+} and Ho^{3+} are frequently used as activators to generate upconversion luminescence (UCL) emission under NIR excitation. To enhance the UCL efficiency, Yb^{3+}, which has a larger NIR absorption strength than other Ln^{3+} ions, is often co-doped as a sensitizer (Gai et al. 2014).

Surface functionalization of hydrophobic UCNPs to engineer the hybrid nanomaterials with good dispersibility in biological buffers is the first and vital step before any bio-application of UCNPs (Zhou et al. 2012). These hybrid UCNPs with multicolor emission favorable for multiplexed biodetection can measure the presence or concentration of biomolecules (DNA, glucose, ATP, etc.), metal ions (Zn^{2+}, Ca^{2+}, Hg^{2+}, etc.) and gas molecules (ammonia, O$_2$, CO$_2$, etc.) by luminescence resonance energy transfer (LRET) detection (Zheng et al. 2015). Moreover, by doping with Gd^{3+} in UCNPs or by combination with iron oxide nanoparticles, magnetic resonance (MR) and UCL imaging could be achieved. With the presence of heavy metal ions like Yb^{3+}, Gd^{3+} and Ba^{3+} that possess high X-ray attenuation, hybrid UCNPs have been designed for computed tomography (CT) contrast enhancement. Furthermore, by incorporating the radionuclide (^{18}F) or by doping radioactive ^{153}Sm^{3+} ions into the matrix, hybrid UCNPs have also been recently considered as intrinsic radio-nuclear imaging (RNI) contrast agents for positron emission tomography (PET) or single-photon emission computed tomography (SPECT) imaging (Liu et al. 2011e; Sun et al. 2013). By combining one or more of the above-mentioned imaging modalities with UCL, UCNPs possesses the capability to obtain more anatomical and physiological details for future cancer imaging (Park et al. 2015). All of these have made hybrid UCNPs one of the most promising nanoparticles for future biodetection and cancer imaging.

Here, we focus on the recent developments in the design and applications of hybrid UCNPs with emphasis on the fields of biodetection and cancer imaging. Firstly, we summarize the surface functionalization (including hydrophilic modification and bioconjugation) techniques for hybrid UCNPs. Secondly, the strategies to sense biomolecules using the UCL signal of LRET-based UCNPs, such as LRET-on and LRET-off detection are discussed. Finally, we will discuss how to use hybrid UCNPs as an efficient nanoplatform to combine UCL imaging with other imaging modalities such as MRI, CT and RNI (e.g., PET or SPECT) for multimodality cancer imaging.

Surface Modification and Hybrid UCNPs

The optical properties of UCNPs are highly dependent on their composition, structure and crystallinity. Suitable size and uniform morphology of UCNPs are the primary prerequisites for the bio-applications of nanoparticles. In the last few decades, the precise control of the synthesis of UCNPs with tunable size (from sub-10 nm to sub-μm) and various nano-shapes (e.g., spheres, cubes, rods, etc.) has been successfully achieved by the thermal decomposition and hydro (solvo) thermal reaction methods, which have become the two of the most popular synthetic methods (Wang et al. 2005b; Zhang et al. 2006b). These synthesized UCNPs are often capped with hydrophobic organic ligands such as oleic acid (OA) or oleylamine (OM). Therefore, in order to be successfully used in biodetection and cancer imaging, hydrophobic UCNPs should be made water-soluble via surface hybrid engineering, which includes ligand engineering, polymer self-assembly and one-pot hydrophilic synthesis. Ligand engineering, which involves the removal of the surface ligand, exchange or oxidation, has been widely used to realize hydrophilic modification. Capobianco et al. and colleagues reported that the surface OA ligands from $NaYF_4$: Yb, Er UCNPs could be removed through a facile stirring treatment of the OA-capped nanoparticles (NPs) in HCl solution (Bogdan et al. 2011). The ligand-free UCNPs are suitable for direct conjugation with electronegative groups such as $-SH$, $-COOH$ and $-NH_2$. Ligand exchange is an important and effective approach for converting OA/OM-capped UCNPs into aqueous solutions by using various hydrophilic organic molecules to replace the original hydrophobic ligands. For instance, Li et al. has utilized citrate to exchange the OA ligand of hydrophobic UCNPs (Zhou et al. 2011), while Capobianco et al. acquired dispersible $NaGdF_4$: Yb, Ho in water by replacing the UCNPs surface's OA ligands with polyacrylic acid (PAA) (Naccache et al. 2009). Gao's group found that two phosphate groups of asymmetric polyethylene glycols (PEGs) have a higher binding affinity to Gd^{3+} of $NaGdF_4$: Yb, Er UCNPs than the carboxyl group from OA, leading to enhanced colloidal stability of UCNPs in phosphate buffer solution (Liu et al. 2013a). Other hydrophilic organic molecules also have been used in the ligand exchange process, which include 6-aminohexanoic acid (Meiser et al. 2004), PEG-diacid (Yi and Chow 2006), 3-mercaptopropionic acid (Kumar et al. 2009), poly (amidoamine) dendrimers (Bogdan et al. 2010), polyvinylpyrrolidone (Johnson et al. 2011) and meso-2, 3-dimercaptosuccinic acid (Chen et al. 2011b). Ligand oxidation, in which the surface OA ligands were directly oxidized into azelaic acids with free $-COOH$ groups on the surface by Lemieux-von Rudloff reagent (Chen et al. 2008; Hu et al. 2008) or ozone under specific conditions, as a simple and versatile strategy has been developed (Zhou et al. 2009) (Figs. 8.1a-b).

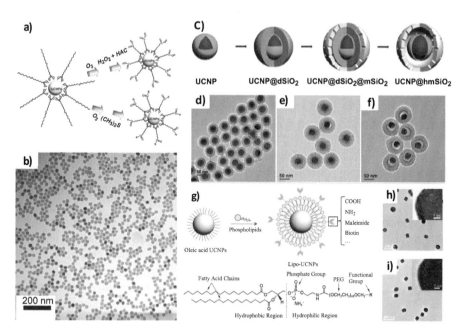

Fig. 8.1 (a) Schematic of surface engineering of ligands oxidation. (b) The –COOH groups modified NaYF$_4$: Yb, Er dispersed in water. (c) The formation process of UCNP@hmSiO$_2$. d-f) TEM images of (d) UCNP@dSiO$_2$, (e) UCNP@dSiO$_2$@mSiO$_2$ and (f) UCNP@hmSiO$_2$. (g) Illustration of the polymer assembly of the water-soluble UCNPs through hydrophobic-hydrophobic interaction in which the fatty acid chains of the phospholipids were embedded in the hydrophobic surface of the UCNPs, while the hydrophilic part assembled toward the aqueous environment. The molecular structure of the designed phospholipid was shown at the bottom. TEM images of the (h) OA-capped UCNPs and (i) UCNPs coated with phospholipids DSPE-PEG-COOH. Inset shows the high-resolution TEM image of the respective sample.

Apart from the ligand engineering technique, coating processes like surface silanization and polymer self-assembly are other effective ways to obtain hydrophilic hybrid UCNPs. Silanization is a popular method for hydrophilic modification. Usually, a dense or mesoporous silica shell can be grown on the UCNPs by hydrolysis and condensation of tetraethoxysilane. We have previously reported that the silica shell can be precisely controlled from sub-10 nm to 100 nm. The surface of silica coated UCNPs can be further modified with functional groups, such as –COOH, –NH$_2$ or –SH (Chen et al. 2013; Xing et al. 2012b). To form dense silica (or dSiO$_2$) coated UCNPs structure (i.e., UCNPs@dSiO$_2$), both reverse microemulsion (Chen et al. 2010b; Mahalingam et al. 2009) and classic Stöber methods (Li and Zhang 2006) have been reported. The mesoporous silica coated UCNPs may offer more space for the accommodation of functional molecules, such as drugs and photosensitizers (Liu et al. 2013b; Idris et al. 2012). Notably, we have showed that the mesoporous silica (or mSiO$_2$) layer with a thickness ranging from 15 nm to 45 nm can be directly coated on the cetyltrimethylammonium bromide (CTAB)-stabilized UCNPs using ultrasonication treatment (Liu et al. 2012a). Furthermore, core-multiple-shell structure, such as UCNPs@dSiO$_2$@ mSiO$_2$, can also be easily achieved by further coating a mesoporous silica shell onto pre-synthesized UCNPs@dSiO$_2$ (Kang et al. 2011; Li et al. 2014; Qian et al. 2009). By etching away the dense silica of UCNPs@dSiO$_2$@mSiO$_2$ based on a "surface-protected hot water etching" strategy, we have recently fabricated rattle-structured UCNPs@hmSiO$_2$ (hmSiO$_2$ = hollow mesoporous silica) (Figs. 8.1c-f) (Liu et al. 2014b), leaving a large cavity between the UCNPs core and the mSiO$_2$ shell. The presence of the cavity is highly desirable for significantly enhanced drug delivery (Fan et al. 2014; Liu et al. 2015). Similar structure (Gd$_2$O$_3$: Eu@hmSiO$_2$) has also reported by Lin's group for drug delivery (Kang et al. 2012).

Polymer self-assembly is another coating method used to achieve hydrophilic UCNPs through electrostatic absorption (layer-by-layer method) or surface interaction (hydrophobic-hydrophobic interaction and host-guest interaction). The layer-by-layer method based on the oppositely charged species has been developed by Wang et al. (2005a) to prepare hydrophilic UCNPs@PAH@PSS@PAH NPs ((PAH = poly(allylamine hydrochloride); PSS = poly(styrene sulfonate)) with –NH$_2$ groups. This method permits the preparation of different shapes and sizes with controllable shell thickness. However, before the layer-by-layer modification, the UCNPs should be hydrophilic. In contrast, surface interaction can directly make the hydrophobic UCNPs soluble in aqueous solutions through hydrophobic-hydrophobic or host-guest interaction. Hyeon et al. reported that the PEG-phospholipids can be coated onto the surface of the UCNPs by hydrophobic-hydrophobic (also called van der Waals) interaction in which the fatty acid chains of the phospholipids were embedded in the hydrophobic surface of the UCNPs, while the hydrophilic part assembled towards the aqueous environment (Park et al. 2012). Li et al. also demonstrated the feasibility of this method nearly at the same time and they found that various functional groups (e.g., carboxylic acid, amine, maleimide, biotin) at the end of the PEG segment could be used for biomimetic surface engineering of UCNPs (Figs. 8.1g-i) (Li et al. 2012). Cheng et al. (2010) have adopted an amphiphilic polymer Octylamine-PAA-PEG to coat on the surface of OA-capped UCNPs based on the hydrophobic–hydrophobic interactions between the OA and the hydrocarbon chains of the polymer. Host-guest interaction developed by Li's group is a rapid (< 20s) and high conversion yield (> 95%) approach to transfer hydrophobic nanoparticles ones, based on the robust interaction of the host molecule alpha (or beta)-cyclodextrin and guest molecule OA (or adamantine-acetic acid) on the surface of UCNPs (Liu et al. 2011c; Liu et al. 2010).

Recently, one-step synthetic strategies for acquiring hydrophilic UCNPs have attracted increasing attention due to their simple reaction procedure and reduced post-treatment procedure. Li's group has developed various one-pot methods to directly obtain hydrophilic UCNPs, including polyol process (Zhou et al. 2010), hydrothermal route assisted by binary cooperative ligands (Cao et al. 2011) and hydrothermal microemulsion synthesis (Xiong et al. 2009b). Other one-pot methods such as the synthesis assisted by hydrophilic ligands and ionic liquid-based synthesis have also been employed. For example, Lin and colleagues reported the one-pot synthesis of hydrophilic LnF_3 NPs by hydrothermal techniques in the presence of hydrophilic citrate (Li et al. 2008). Ionic liquid-based synthesis reported by groups of Yan and Cui is an efficient and environmental method to directly synthesize the hydrophilic UCNPs, in which the green solvents of ionic liquid have been used as nonvolatile, nonflammable and thermally stable organic salts to replace conventional organic solvents (Chen et al. 2010a; He et al. 2011). As compared with the two-step synthetic method, one-step strategy can simplify the reaction procedure, but the size, shape and crystallinity of hydrophilic UCNPs can be hardly obtained, which are easily controllable in two-step methods.

For further biodetection and cancer imaging, hydrophilic UCNPs are required to be conjugated with biomolecules such as RGD, antibody, FA, avidin (or biotin), protein and DNA and so on. The amido bond between – COOH and $-NH_2$ by a typical EDC chemistry is the most used reaction for bioconjugation of UCNPs. Based on this principle, the hybrid UCNPs with $-NH_2$ groups on their surface are always linked with –COOH groups containing biomolecules and vice versa. For example, our group recently reported that Angiopep-2 as a crossing blood brain barrier (BBB) peptide can be successfully conjugated with amine-functionalized PEG-UCNPs (Ni et al. 2014b). Prasad et al. demonstrated the –COOH modified UCNPs could be reacted with the $-NH_2$ from antibodies (anti-claudin-4) for targeted cancer cells delivery (Kumar et al. 2009). Another classic interaction between –SH groups and maleimide (MA) groups is always used for UCNPs' bioconjugation. Gao's group reported the bioconjugation of MA-modified UCNPs with – SH groups of MGb_2 antibody via "click" reaction for ultrasensitive *in vivo* detection of primary gastric tumor and lymphatic metastasis (Qiao et al. 2015). By liking thiolated RGD (arginine-glycine-aspartic acid) peptides on the MA-UCNPs, we also acquired the RGD-UCNPs for tumor targeting CT imaging and radio-enhanced therapy (Xing et al. 2013).

Engineering of Hybrid UCNPs for Biodetection

Biodetection that measures the presence or concentration of biomolecules (e.g., DNA, protein, cancer biomarkers, etc.) plays an important role for a variety of bio-applications. Recent years have witnessed the numerous applications of hybrid UCNPs in biodetection due to their special optical performance, such as little photodamage to biological species, free from background autofluorescence and large anti-Stokes shifts without interference from NIR excitation. These features make hybrid UCNPs new generation luminescent bio-probes that possess higher sensitivity and circumvent the limitations of traditional organic dyes and QDs. In general, UCNPs are labeled with capture molecules (or analytes) to combine with analytes (or capture molecules) based on the high binding affinity between the capture molecules and analytes. After washing off the excess unbound UCNPs, the concentration of the analytes can be quantified by directly measuring the UCL signal. This "mix-separate-read" method involves multi-step separations or washing operations and thus is not a convenient way. To realize fast detection in practical application, another "mix-read" method has been developed based on the luminescence resonance energy transfer process (LRET, also called FRET) which utilizes an energy donor (UCNPs) and a nearby energy acceptor. When they are placed at a close distance and the absorption of the acceptor and the emission UCNPs are overlapped, the excitation of UCNPs can excite the energy acceptor. Therefore, turning on or off the LRET process by the interaction with the analytes in solutions will lead to a UCL light-off or light-on biodetection. It should be noted that the multi-wavelength emissions of the hybrid UCNPs allow the acceptors to selectively absorb on emission band and leave the others nearly constant for ratiometric detection which is independent of the excitation intensity and endows the biodetection with high stability and flexibility (Liu et al. 2011b; Liu et al. 2011d). In this section, we mainly focus on the recent development of UCNPs-based LRET biodetection from a new horizon of LRET-on or -off detection. Moreover, how to improve the LRET efficiency and achieve high detection sensitivity is also discussed.

LRET-On Detection

LRET-on detection means the LRET process is turned on to produce detectable signal relevant with analytes in solution. The hybrid UCNPs are firstly excited upon NIR excitation and then transfer their emission energy to the nearby acceptors (normally within 10 nm or smaller), quenching the luminescence of UCNPs or exciting the acceptors to emit another detectable fluorescence. As illustrated in Fig. 8.2a, LRET-on detection consists of two strategies: i) LRET-on sandwich assays. LRET-on sandwich assay is performed in a way that the UCNPs and acceptors are separated at the beginning and assemble in close proximity only in the presence of analytes for quantifying changeable optical signal. ii) LRET-on sustain assays. Unlike LRET-on sandwich assay, LRET-on sustain assay is carried out by initially linking the UCNPs and acceptors together and then the UCNPs' emission excites the acceptors to emit the detectable fluorescence that is highly sensitive to the analytes for quantitative detection. Since the first report of UCNPs in LRET-on detection in 2005 (Wang et al. 2005a), a variety of LRET-on biodetections including sandwich assay and sustain assay formats have been developed.

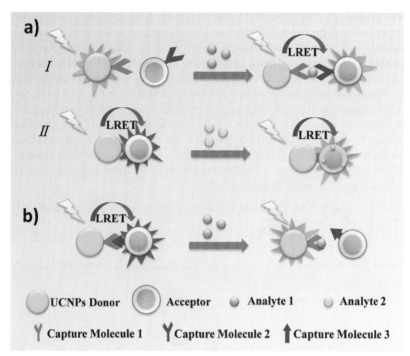

Fig. 8.2 Schematic illustration of typical LRET detection based on the hybrid UCNPs: (a) LRET-on detection which includes (i) LERT-on sandwich detection and (ii) LERT-on sustain detection. (b) LRET-off detection.

LRET-on sandwich assay is a popular and effective method to detect the analytes in which both the hybrid UCNPs and the acceptors are labeled with the capture molecules that can specifically recognize the target analytes. After the addition of analytes, the capture molecule labeled UCNPs and the acceptors, which are initially far apart from each other in solution, are brought together through the specific binding between the capture molecules and target analytes, forming a LRET pair and giving the binding relevant signals. As a result, the concentration of the analytes can be quantified by directly reading the optical signals of the UCNPs or acceptors. To date, numerous LRET-on assays have been applied by utilizing different donor-acceptor binding to detect the biotin (Kuningas et al. 2005), avidin (Wang et al. 2005b), DNA (Chen et al. 2008; Zhang et al. 2006a), siRNA (Jiang and Zhang 2010), antibody (Wang et al. 2009a), thrombin (Yuan et al. 2014) and gene (Ye et al. 2014). For instance, Kuningas et al. (2005) reported an LRET biosensor for the detection of biotin with streptavidin labeled UCNPs as donors and biotinylated phycobiliproteins as the acceptors, with a limit of detection (LOD) as low as subnanomolar concentration range for biotin biodetection. Later, the DNA sensor in LRET-on sandwich assay was reported by Zhang et al. (2006a) who successfully detected the target-DNA utilizing the oligonucleotide-labeled UCNPs, reaching a LOD down to 1.3 nM and subsequently by Li and co-workers using streptavidin-modified

UCNPs as donors and -DNA-TAMRA as the acceptor to detect concentration of target-DNA (Chen et al. 2008). Besides using the molecules as acceptors, nanoparticles such as Au with larger absorption coefficient at the wavelength matchable to the emission from UCNPs are also widely adopted as acceptors for LRET-on sandwich assay. Wang et al. (2005a) demonstrated the feasibility of Au NPs as the acceptors for the detection of avidin, for the first time, where an LOD as low as 0.5 nM was acquired under non-optimized conditions. Other systems using Au NPs as the acceptors have also been developed, including human IgG modified Au NPs for convenient goat anti-human IgG detection (Fig. 8.3a) (Wang et al. 2009a) and oligonucleotide-labeled Au NPs for the rapid detection of disease markers of H7 hemagglutinin genes of avian influenza viruses (Ye et al. 2014).

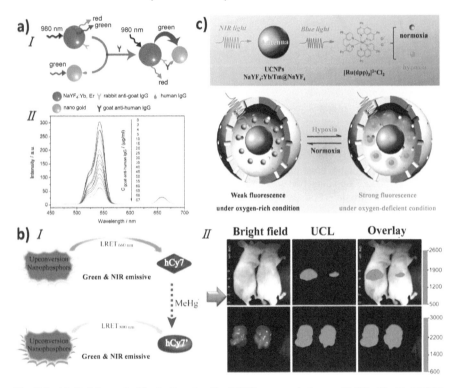

Fig. 8.3 (a) (i) Schematic illustration for the LRET process between NaYF$_4$: Yb, Er UCNPs (donor) and Au NPs (acceptor); UCL spectra of the LRET system at different concentrations of goat antihuman IgG. (b) (i) Schematic illustration of the upconversion LRET process from the upconversion emission of UCNPs to the absorption of the Dye hCy7 or hCy7′; (ii) Upper: *in vivo* UCL images of 40 μg hCy7-UCNPs-pretreated living mice injected intravenously with 0.2 mL normal saline (left mouse) or 0.1 mM MeHg+ solution (right mouse); Down: the corresponding UCL images of the livers which were isolated from the above-dissected mice. The UCL emission was collected at 800 ± 12 nm upon irradiation at 980 nm. (c) Schematic illustration of the structure of nanosensors and their sensing to oxygen with a change in luminescence emission.

LRET-on sustain assays where the hybrid UCNPs act as energy transfer machines and the LRET lasts during the whole detection have also been performed recently. In this method, the acceptor should be deliberately selected, not only should their excitation match the UCNPs' emission, but they should also be sensitive to the target analytes. This strategy is widely used to carry out photodynamic therapy (PDT) in which upon NIR shine, the UCNPs emit visible light to excite photo-sensitizers producing singlet oxygen for PDT (Idris et al. 2012; Zhang et al. 2007). Li and co-workers reported a novel LRET-on sustain sensor for $MeHg^+$ detection by assembling the $MeHg^+$ sensitive dyes (cyanine dye hCy7) on the surface of UCNPs, in which the presence of $MeHg^+$ induced a significant red shift of the maximum absorption peaks of hCy7 dye from 670 to 845 nm that matched the UCL emissions (Liu et al. 2013c). This transform from $LRET_{660\,nm}$ to $LRET_{800\,nm}$ will endow the UCNPs ratiometric UCL detection of $MeHg^+$ with high sensitivity where an LOD as low as 0.18 ppb was achieved, much lower than the blood level. Importantly, this nanosystem was capable of monitoring $MeHg^+$ in living cells and small animals by UCL imaging (Fig. 8.3b). Recently, our group developed a oxygen sensor by using the acceptor $[Ru(dpp)_3]^{2+}Cl_2$, the 613 nm red luminescence of which can be strongly sensitive to oxygen upon 463 nm excitation that strongly overlaps with the two emission wavelengths (at 450 and 475 nm) of UCNPs (Fig. 8.3c). These LRET-on sustain assays are attractive for the ratiometric quantitative measurement of hypoxic level, which is vitally important and necessary for hypoxic tumor imaging and therapy (Liu et al. 2014b). Another important application of LRET-on sustain assay was for cancer drug detection including drug release and drug monitoring. Lin's group found that the quenching of the UCL of $NaYF_4$: Yb^{3+}/Tm^+ UCNPs by DOX due to LRET mechanism can be applied as an optical probe to confirm the DOX conjunction and monitor the release of DOX (Dai et al. 2012). Our group has recently developed a NIR light-triggered drug release system by integrating photo-responsive azobenzene into mesopore of $UCNPs@mSiO_2$ and the UCNPs' UV/Vis emission matches well with the absorption of azo molecules, thereby leading to the continuous rotation-inversion movements of azo molecules in the mesopore channels to release cancer drugs (Liu et al. 2013b). Based on LRET-on sustain methods, the precise control and quantification of drug release by UCL can be achieved using the UCNP@ $hmSiO_2$ nanosystem (Liu et al. 2014a).

LRET-Off Detection

LRET-off detection means the LRET process is turned off to recover either UCL or fluorescence of acceptor molecules only in the presence of analytes that cleaved the linkage (Fig. 8.2b). This method is also known as "quenching-to-recovery" assay since the optical signals undergo a quenching and recovery process. To effectively quench the UCL emission, the acceptors should possess a large extinction coefficient at wavelength matchable to the emissions of UCNPs. Organic dyes that are sensitive to the target analytes are usually adopted as the acceptors to quench emissions of UCNPs. In the presence of analytes, organic dyes are transformed and turn off the LRET process to recover the detectable optical signal that is relevant to the analytes' concentration, such as the N719 to detect Hg^{2+} ions (Liu et al. 2011d), Zn^{2+} responsive dyes to sensor Zn^{2+} ions (Peng et al. 2015), Fluo-4 to monitor Ca^{2+} ions (Li et al. 2015), rhodamine B derivatives to detect Fe^{3+} ions (Ding et al. 2013) and complex Ir1 to test CN^- (Liu et al. 2011b). For instance, Peng et al. recently developed novel Zn^{2+} sensors, in which LRET was turned off due to the presence of Zn^{2+} that resulted in an energy mismatch between the absorption of chromophore and the blue emission of UCNPs (Peng et al. 2015). This variation in UCL intensity allows one to achieve the quantitative monitoring of Zn^{2+} with high sensitivity down to 0.78 µM and quick response within 5s (Fig. 8.4a). More importantly, their nanosystem was capable of implementing an efficient *in vitro* and *in vivo* detection of Zn^{2+} in mouse brain slice with Alzheimer's disease and zebrafish, respectively. It is worth noting that the organic dyes are always used as acceptors for detection of different kinds of metal or anion ions in living organisms.

Nanoparticles with unique electronic and absorption properties are widely used as other kinds of acceptors that are attached to the UCNPs to quench UCL and are detached from the UCNPs with the addition of analytes to recover the UCL emission for detection. These nanoparticles include Au nanorods (Chen et al. 2014b), Ag nanoclusters (Xiao et al. 2015), graphene oxide (GO) (Zhang et al. 2011), manganese dioxide (MnO_2) nanosheets (Deng et al. 2011) and carbon NPs (Wang et al. 2011b). For example, Liu and co-workers constructed a novel UCNP-GO sensor for detection of glucose in human serum samples without any background interference where the various biomolecules in the serum do not have a significant influence on either the LRET process or the optical measurements (Zhang et al. 2011). As an efficient platform, UCNP-GO sensors, with different functional molecules modification, are also applied for sensitive and selective determination of human immunodeficiency virus (HIV) antibodies (Wu et al. 2014) or adenosine triphosphate (Liu et al. 2011a). To rapidly and selectively detect glutathione in aqueous solutions and living cells, Liu's group reported a novel design, based on a combination of hybrid UCNPs and MnO_2 nanosheets (Deng et al. 2011), where MnO_2 formed on the surface of UCNPs served as an efficient UCL quencher. The LRET can be turned off and then UCL turned on by introducing glutathione that reduces MnO_2 into Mn^{2+} (Figs. 8.4b-c). Due to the rich electrical and catalytic properties of MnO_2 nanomaterial, this strategy also provides a sensor platform useful for photocatalytic and photoelectric studies. Other nanoparticles are also used to detect biomolecules such as Au nanorods for cancer markers sensing (Wang et al. 2011b) and carbon NPs for thrombin detection in human plasma (Chen et al. 2014b).

Improving LRET Efficiency

Despite the attractive development and great success with UCNPs-based LRET bioassay, there is an intractable problem with UCNPs as the LRET platform which is the limited low energy transfer efficiency from UCNPs to energy acceptors. A higher efficiency will undoubtedly improve the sensitivity of LRET-based biodetection and decrease the LOD. There are mainly two factors affecting the LRET efficiency: the donor-to-acceptor distance and the acceptors' quenching ability. As known, LRET efficiency is highly dependent on the donor-to-acceptor distance and the nonradiative energy transfer is suggested to be effective within a distance of 10 nm. However, since the size of commonly used UCNPs is normally at tens of nanometers, only those emitting ions near the surface of the UCNPs are acting as the energy donors. Quite a large part of the emitters are out of the energy-transfer distance and are non-functional. To address this issue, sub- 10 nm UCNPs can shorten the LRET distance due to their smaller diameter (Zheng et al. 2013). Recently, Liu and co-workers fabricated a sandwich UCNP (SWUCNPs: $NaYF_4$@$NaYF_4$: Yb, Er@$NaYF_4$), in which the outer layer thickness of $NaYF_4$ could be turned to control the donor-to-acceptor distance (Li et al. 2015). Furthermore, to decrease the distance from the emitters to the quenchers due to surface ligands, Fluo-4 is directly linked to bared SWUCNPs via the coordination between carboxyl of Fluo-4 and the exposed Ln^{3+} ions on SWUCNPs. A > 7-fold enhancement of LRET efficiency with SWUCNPs-2.5 than that of homogeneous UCNPs was achieved (Fig. 8.4d). On the other hand, the acceptors' quenching ability also determinates the LRET efficiency, the organic molecules as energy acceptors always fall short on energy-transfer efficiency, whereas inorganic energy acceptors are able to provide higher energy-transfer efficiencies possibly due to the surface energy-transfer mechanism, but they tend to be less flexible and less biocompatible. Xiao et al. (2015) discovered that the ultrasmall silver nanoclusters (Ag NCs) exhibit pronounced energy-accepting capability to quench a luminophore and they applied core/satellite structure in which each UCNP core is surrounded by a few Ag NCs. The molecule-like properties of the few-atom Ag NCs endowed the nanoprobe with both high LRET efficiency and good water solubility and biocompatibility for intracellular and *in vivo* biothiol levels detection with high discrimination.

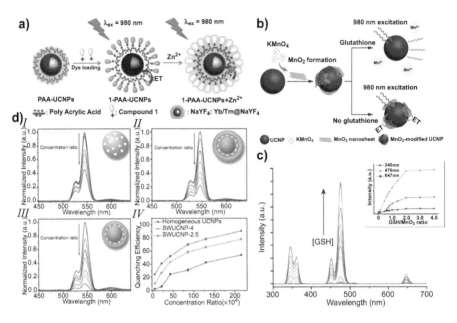

Fig. 8.4 (a) Schematic illustration showing the synthesis of chromophore-assembled UCNPs and their response to Zn^{2+}. (b) Experimental design for GSH detection using MnO_2-nanosheet-modified UCNPs. (c) Photoluminescence response of MnO_2-modified UCNPs as a function of GSH concentration (0–10 mM) in an aqueous solution. Inset: Plot of luminescence intensity at 345, 476 and 647 nm, respectively, against the GSH/MnO_2 molar ratio. (d) (i–iii) Emission spectra of the three energy donors (with Er^{3+} as the emitting ions) at varying concentration ratios of dyes over UCNPs in MES buffer: (i) homogeneous UCNPs, (ii) SWUCNP-4 and (iii) SWUCNP-2.5; (iv) Comparison on the quenching degrees of the three energy donors.

Engineering of Hybrid UCNPs for Cancer Imaging

Currently, significant attention has been focused on early cancer diagnosis, using single or combined imaging modalities, such as optical, MR, CT, ultrasound and radio-nuclear imaging (RNI, e.g., PET or SPECT). Each imaging modality has its own merits for acquiring anatomic, physiologic or molecular information, but none of them provides comprehensive medical imaging. Therefore, multimodal imaging has become an attractive strategy for *in vivo* cancer imaging to obtain high sensitivity, resolution and deep penetration depth, enabling excellent visualization from the cellular scale to whole-body. For example, the PET/CT and PET/MRI dual-modality scanners have become commercially available. The rapid development of UCNPs provides an efficient platform for acquiring multimodal imaging. In this section, we firstly focus on the upconversion luminescence imaging (UCL) of UCNPs for cellular imaging and deep body tissue imaging. Then UCL imaging combined with one or more imaging modalities such as MR, CT or RNI for multimodal imaging are also summarized.

Upconversion Luminescence Imaging

Optical imaging has a high imaging resolution and sensitivity at cellular level and plays an important role in early cancer detection and screening. However, the commonly used fluorescent probes such as organic dyes or quantum dots are not ideal for bioimaging because the excitation in UV or visible range is almost non-penetrable to tissues and evokes strong imaging background. As previously described, the NIR-excited UCNPs own prominent advantages because the "optical transparency window" of biological tissues are in the NIR range (750-1100 nm), which allows for deeper light penetration depth and lower auto-fluorescence. Since the first bio-application of UCNPs to detect cell and tissue surface antigens in 1999 (Zijlmans et al. 1999), high contrast cellular (e.g., HeLa, U87MG, KB, etc.) imaging of the hybrid UCNPs has been reported in recent years. Due to their inherent non-blinking emission and photostable behavior, the hybrid UNCPs have been acquired for reliable single molecular imaging with long-time tracking capability (Wu et al. 2009). Suh et al. for the first time, obtained real-time images of endocytosed UCNPs at the single vesicle level for 6 h continuously, in which the dynamics of UCNPs transport were composed of multiple phases within a single trajectory (Nam et al. 2011). Later, Wong et al. found that the cellular uptake depended on the surface charge of UCNPs and the positively charged PEI-UCNPs have greatly enhanced cellular uptake than their neutral PVP-UCNPs and negative PAA-UCNPs, mainly through clathrin-mediated endocytosis (Jin et al. 2011). In contrast, another strategy for efficient cellular uptake and enhancing imaging contrast was developed by using targeting molecules for specific binding. These targeting molecules include folic acid for targeting folate receptors over-expressed on various cancer cells (Chien et al. 2013; Wang et al. 2011a; Xiong et al. 2009b), the peptides such as RGD for $\alpha_v\beta_3$ integrin receptor (Xiong et al. 2009a) and angiopep-2 peptide for LRP receptor (Ni et al. 2014b) of U87MG cells, the antibodies such as anti-CEA8 for targeting antigen receptor of HeLa cells (Wang et al. 2009b), anti-Claudin 4 for antigen receptor on Panc 1 cells (Kumar et al. 2009), anti-EGFR for EGFR receptor on LS180 cells (Liu et al. 2013a), MGb$_2$ for TRAK1 antigen on gastric carcinomas (Qiao et al. 2015), etc. Importantly, the targeting capability of these hybrid UCNPs provides opportunities to diagnose tumors inside the body.

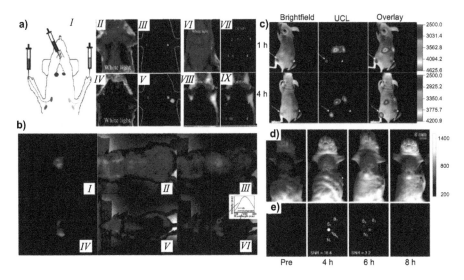

Fig. 8.5 (a) *In vivo* multiplexed lymphangiography with UCNPs: (I) schematic illustration of UCNPs-based lymph node mapping; (II) white light image; (III) a three-color spectrally resolved *in vivo* UCL image; (IV) white light image of the same mouse after dissection; (V) UCL image of the dissected mouse. (VI-IX) comparison of imaging sensitivities between UCNPs and QDs: (VI) white light image; (VII) UCL image; (VIII) white light images of mice subcutaneously injected with QDs, (IX) red and green colors represent QD fluorescence and autofluorescence, respectively. (b) Whole-animal imaging of a BALB/c mouse injected with the water-soluble UCNPs@CaF₂ NPs: (I, IV) UCL images; (II, V) bright-field images; and (III, VI) merged bright-field and UCL images, Inset in (VI) shows the spectra of the NIR UCL and background taken from the circled area. (c) Time-dependent *in vivo* UCL imaging of subcutaneous U87MG tumor (left hind leg, indicated by short arrows) and MCF-7 tumor (right hind leg, indicated by long arrows) after intravenous injection of RGD-UCNPs. (d-e) *In vivo* optical images obtained before and at different time points after intravenous injection of the MGb₂-UCNPs nanoprobes: bright-field images superimposed with color-coded UCL images (d). Luminescence signals of different sites labeled S for signal, N for noise and B for background to determine SNR (e).

Besides the various types of cellular imaging, *in vivo* UCL imaging (e.g., elegans, mice and zebrafish) is also obtained using hybrid UCNPs due to their intrinsic multiplexed imaging and deep tissue penetration. Since the precursory work of UCL imaging in live elegans by Lim et al. (Lim et al. 2006), Zhang and colleagues first reported the advantages of using NIR-excited hybrid UCNPs in small animal imaging, in which UCNPs showed a higher imaging depth than QDs (Chatterjee et al. 2008). Then *in vivo* multiplexed biomedical imaging including lymphatic imaging using three different kinds of UCNPs with multicolor emission (red, green and blue) has been realized by (Cheng et al. 2010) (Fig. 8.5a). They also demonstrated that the *in vivo* imaging detection sensitivity of UCNPs is at least one order of magnitude higher than that of QDs. By doping 800 nm emitting Tm³⁺ in the host, the NIR-to-NIR UCL imaging can be acquired with a deeper penetration depth.

Li and co-workers have developed higher efficient 800 nm-emission UCNPs based on $NaLuF_4$ matrix and high contrast UCL imaging of deeper body tissue (~2 cm) was achieved (Liu et al. 2011f; Yang et al. 2012). More recently, by coating CaF_2 on Tm^{3+} -doped $NaYbF_4$ UCNPs, Han's group developed novel and efficient 800 nm-emitting probes for deep tissue body imaging (~3.2 cm) with a high signal-to-background ratio (SBR) of 310 under low imaging power density (0.3 W/cm^2) at 980 nm (Fig. 8.5b) (Chen et al. 2012). In addition, *in vivo* targeting UCL imaging is another attractive application for hybrid UCNPs due to their high sensitivity. Zhang et al. (2012a) has synthesized a novel mesoporous nanorattle structure (Fe_3O_4@SiO_2@$NaYF_4$: Yb, Er) for magnetic targeted *in vivo* UCL imaging and chemotherapy. Furthermore, by conjugating targeting molecules, *in vivo* tumor targeting UCL imaging has been achieved by the ligand-receptor or antigen-antibody interaction. The FA and RGD are the most used targeting peptides for *in vivo* UCL imaging of tumors-bearing mice with the over-expression of folate receptor or $\alpha_v\beta_3$ integrin receptor (Fig. 8.5c) (Xing et al. 2013; Xiong et al. 2009a; Xiong et al. 2009b). In addition, neurotoxin-mediated UCL imaging for direct visualization of tumors *in vivo* has also been achieved using the highly efficient red light-emitting CTX-UCNPs (Yu et al. 2010). Very recently, Gao's group have linked different kinds of antibodies (anti-EGFR or MGb_2) to the surface of UCNPs for *in vivo* early cancer imaging (Liu et al. 2013a; Qiao et al. 2015). For example, they constructed a highly specific gastric tumor probe by linking the antibody MGb_2 on $NaGdF_4$: Yb, Er@$NaGdF_4$ UCNPs, which showed outstanding UCL imaging ability for detecting very small lesions at the primary gastric tumor and lymphatic metastatic sites due to their excellent tumor-targeting specificity and imaging sensitivity (Figs. 8.5d-e).

Magnetic Resonance Imaging

To obtain the merits of UCL and MR imaging, hybrid UCNPs have been widely engineered as nanoprobes for *in vivo* fluorescent and MR imaging. The Gd^{3+}-based UCNPs are promising probes for use as T_1-weighted MRI due to their seven unpaired inner 4f electrons, which can influence the longitudinal relaxation of water protons. Since the first reported T_1-MR imaging of Gd^{3+}-based UCNPs ($NaYF_4$: 2% Er, 20% Yb, 10% Gd, r_1 = 0.14 $mM^{-1}s^{-1}$) by Kumar et al. (2009), numerous works have focused on developing high performance MRI contrast agents with efficient UCL imaging, as the Gd^{3+} ions in UCNPs matrix can efficiently induce crystal phase transformations to increase the UCL emission (Wang et al. 2010). To enhance Gd^{3+} doping amount to 60%, Zhou et al. (2011) also demonstrated T_1-MRI capability of Gd^{3+}-based UCNPs ($NaYF_4$: 2% Er, 20% Yb, 60% Gd), but the imaging performance was still poor as the r_1 was only 0. 405 $mM^{-1}s^{-1}$, as compared to the clinically used Gd-DTPA (r_1 = 4-5 $mM^{-1}s^{-1}$). Instead of Gd^{3+} doping, UCNPs of $NaGdF_4$ matrix ($NaGdF_4$: 2% Er, 18% Yb) were also used for MR imaging, which exhibited a higher r_1 value of 3.33 $mM^{-1}s^{-1}$ than that of Gd^{3+} co-doped UCNPs (Liu et al. 2013a). After coating a $NaGdF_4$ layer ($NaGdF_4$: 2% Er, 18% Yb@$NaGdF_4$), the r_1 value decreased to 2.25 $mM^{-1}s^{-1}$, reported by the same group recently (Qiao et al. 2015). These results strongly suggest that the MRI performance of Gd^{3+} ions depends on their location in UCNPs. Our group found that the surface Gd^{3+} ions are mainly responsible for MR imaging, while a nearly 100% loss of relaxivity of Gd^{3+} ions buried deeply within the crystal lattice (> 4 nm) was found. Namely, a Gd^{3+} free UCNP core with a $NaGdF_4$ layer structure exhibits high MRI performance and the r_1 value can be regulated to 6.18 $mM^{-1}s^{-1}$ by controlling the $NaGdF_4$ shell (Chen et al. 2011a). Moreover, to guide the MRI sensitivity optimization, we revealed the detailed longitudinal relaxation mechanism of Gd^{3+}-based UCNPs, in which the inner- and outer-sphere relaxivity mechanisms have been demonstrated to be coexisting in ligand-free probes (resulting in the highest r_1 value of 14.73 $mM^{-1}s^{-1}$), while the outer-sphere mechanism has been suggested to be the main contribution to that of silica-shielded probes (Fig. 8.6a) (Chen et al. 2013). Our new findings provided not only a deeper insight into the origin of contrast enhancements of Gd^{3+}-based UCNPs, but also provided useful strategies for achieving high MR imaging performance. Hyeon and group demonstrated the feasibility of Gd^{3+}-based UCNPs ($NaYF_4$: Yb, Er@$NaGdF_4$) for *in vivo* MR/UCL dual-modal imaging of tumors in the mouse model by the enhanced permeability and retention (EPR) effect (Park et al. 2012). Later, the successful *in vivo* targeting tiny tumors (1.7 mm × 1.9 mm) for UCL/MR imaging using anti-EGFR antibody linked $NaGdF_4$: Er, Yb UCNPs was reported by Liu et al. (2013a). We recently have developed the dual-targeting angipep-2 modified UCNPs ($NaYF_4$: Yb/Tm/Gd@$NaGdF_4$) nanoprobes to cross the BBB, target the glioblastoma and then function as preoperative MR diagnosing and intraoperative agents positioning the brain tumors by NIR-to-NIR UCL imaging (Ni et al. 2014b), which showed a much enhanced imaging performance in comparison with the clinically used single MRI contrast (Gd-DTPA) and fluorescent dye of 5-ALA (Fig. 8.6b).

Another main class of MRI contrast agents is based on shortening the transversal relation time for T_2-weighted imaging to darken the image. Superparamagnetic iron oxide NPs are also introduced to combine with UCNPs for T_2-weighted MR and UCL dual-modal imaging. Typically, the construction includes layer-by-layer UCNPs/Fe_3O_4/Au (Fig. 8.6c) (Cheng et al. 2011), core@shell UCNPs@Fe_xO_y (Xia et al. 2011) or Fe_4O_3@UCNPs (Zhang et al. 2012b), core/satellite UCNPs/Fe_4O_3 (Shen et al. 2010) or Fe_4O_3/UCNPs (Gai et al. 2010) nanostructures. These prepared superparamagnetic UCNPs suffer from the drawback that emission intensity is significantly reduced because both the excitation and emission light are absorbed by the black iron oxide NPs. Our groups have developed a neck-formation strategy for construction of the antiquenching MR/UCL imaging nanoprobes using dense silica shell to avoid the direct interaction of UCNPs and Fe_3O_4 NPs (Chen et al. 2010b). However, these T_2-MR/UCL dual-modal imaging nanoprobes still suffer from various intractable issues such as complicated multi-step syntheses for composite nanostructures and the possible detachment of different imaging agents from the heterogeneous nanostructure remains an unsolved issue. Interestingly, we have synthesized Ho^{3+}-doped $NaYbF_4$ UCNPs for excellent T_2-MR/UCL imaging. The efficient T_2-MR performance of activators (Ho^{3+}) and sensitizers (Yb^{3+}) has been investigated for the first time (Ni et al. 2014a). More importantly, the Ho^{3+}-doped UCNPs have further been used to successfully diagnose glioblastoma, which is the most lethal brain tumor and remains highly challenging for detection because of its easy evasiveness of diagnosis (Fig. 8.6d).

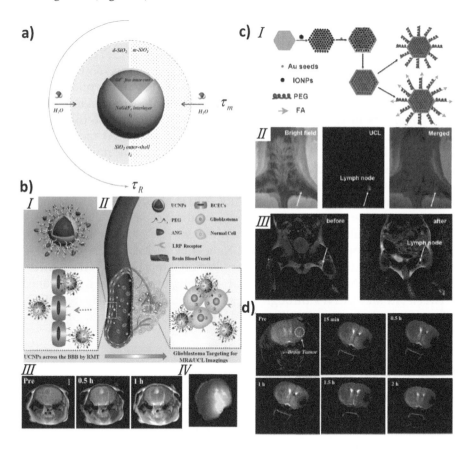

Fig. 8.6 (a) Well-designed silica protected water soluble core@NaGdF$_4$ nanoprobes for investigating the MR relaxivity mechanism and sensitivity optimization. (b) Design of the dual-targeting ANG-UCNPs (I). (II) schematic diagram of the ANG-UCNPs as the dual targeting system to cross the BBB and target the glioblastoma, enabling MR (III) and UCL (IV) imaging of intracranial glioblastoma. (c) Strategy for layer-by-layer UCNPs/Fe$_3$O$_4$/Au synthesis and functionalization (I). Multimodal UCL (II) and MR (III) imaging for *in vivo* lymphangiography mapping using UCNPs/Fe$_3$O$_4$/Au NPs. (d) *In vivo* T$_2$-MR imaging of glioblastoma-bearing mice before and at various time points after intravenous injection of Ho^{3+}-doped NaYbF$_4$ UCNPs.

Computed Tomography Imaging

Lanthanide elements have a higher X-ray attenuation and K-edge values within the X-ray spectrum in medical use, thus making UCNPs inherent CT contrast agents. The sensitizers Yb^{3+} ions have been regarded as ideal candidates for CT/UCL imaging due to their higher X-ray absorption coefficient and larger NIR absorption strength. Both the results from Lu and our group demonstrated that $NaYbF_4$-based UCNPs own higher CT imaging ability than clinical iodinated agents with efficient UCL imaging simultaneously (Fig. 8.7a) (Liu et al. 2012b; Xing et al. 2012a). Integrating Ba^{3+} and Yb^{3+} with differential K-edge value in single NPs, Liu et al. (2012c) performed the first example of binary CT contrast agents ($BaYbF_5$ NPs) with a much higher CT performance than the iodinated agent at different voltages for various patients. Doping 2% Er (or Tm) in $BaYbF_5$ NPs, our group has demonstrated the feasibility of *in vivo* targeting CT/UCL dual-modality imaging to diagnose tumors by conjugating RGD peptide on the surface of UCNPs. Moreover, these UCNPs can also function as irradiation dose enhancers in tumors during radiotherapy, providing the theranostic UCNPs by the precise positioning of tumors and enhanced radiotherapy (Fig. 8.7b) (Xing et al. 2013). Lu^{3+} ions have the highest atomic number among lanthanide elements and $NaLuF_4$-based UCNPs have also been reported by Li's group for UCL/CT imaging combined with Gd^{3+} complex or Fe_3O_4 for simultaneous MR imaging (Xia et al. 2012; Zhu et al. 2012). Our groups have developed another strategy for adding CT contrast capability by decorating CT NPs, such as ultrasmall Au or tantalum oxide (TaO_x) NPs on the surface of the hybrid UCNPs (Xiao et al. 2012; Xing et al. 2012b). Importantly, these nanostructures can realize a win-win multimodal imaging (CT/UCL/MRI). For example, Au NPs will enhance UCL imaging by adjusting the distance between Au and UCNPs through the silica shell and the transparent TaO_x NPs improve the biocompatibility and do not quench the UCL imaging (Figs. 8.7c-d).

Radio Nuclear Imaging

Radio nuclear imaging (RNI) is an extremely sensitive and quantitative tool, including positron emission tomography (PET) or single-photon emission computed tomography (SPECT) imaging, but is limited in resolution and provides no anatomical information. Therefore, RNI is usually combined with other imaging methods to obtain images with high sensitivity and resolution as PET/MRI and PET/CT scanners are commercial available, promoting researchers to synthesize the multimodal contrast agents. Li et al. developed a simple, rapid and efficient way to acquire ^{18}F-labeled UCNPs by incorporating [18]F$^-$ into a $NaYF_4$ matrix due to the strong binding between

[18]F− and lanthanide ions and the spleen and liver could be visualized by PET and MRI after injecting ^{18}F-labeled UCNPs (Fig. 8.8a) (Liu et al. 2011e). They also found that these ^{18}F-labeled UCNPs could be used for PET monitoring of the sentinel lymph node (Figs. 8.8b-c) (Sun et al. 2011). Moreover, by doping [153]Sm^{3+} in NaLuF$_4$: Yb, Tm@NaGd([153]Sm)F$_4$, four-modal imaging nanoprobes have been developed for tumor angiogenesis detection (Fig. 8.8d), using UCL(Yb^{3+}/Tm^{3+}), CT (Lu^{3+}, Yb^{3+}), T$_1$-MRI (Gd^{3+}) and SPECT ([153]Sm^{3+}) multimodal imaging (Sun et al. 2013).

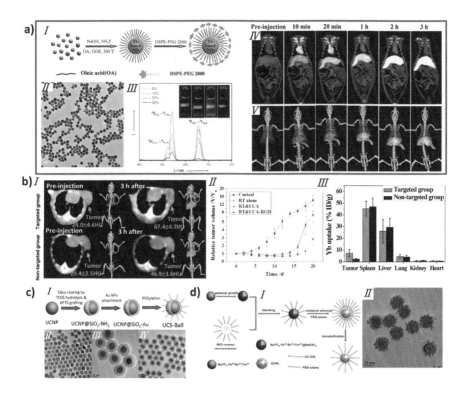

Fig. 8.7 (a) Schematic illustration of the synthesis and surface modification of OA-capped UCNPs (I); (II) TEM image of PEG-UCNPs dispersed in water; (III) UCL spectra of UCNPS with different concentrations of Gd, Inset shows the UCL photograph of the UCNPs with different concentrations of Gd^{3+} under excitation at 980 nm; *In vivo* CT coronal view images of a rat after intravenous injection of PEG-UCNPs solution at timed intervals; (IV) Heart and liver. (V) The corresponding 3D renderings of *in vivo* CT images. (b) *In vivo* transverse slices and 3D volume rendering CT images of U87MG tumour-bearing mice at pre-injection and 3 h after intravenous injection of RGD-UCNPs or PEG-UCNPs (I); (II) Radio-enhanced therapy of RGD-UCNPs (growth of U87MG tumors after various treatments); (III) Tissue distributions of NPs *in vivo*. (c) Schematic illustration for the synthesis of UCNPs@SiO$_2$@Au (I); TEM images of UCNPs (II), UCNPS@SiO$_2$ (III) and UCNPs@SiO$_2$@Au (IV). (d) Schematic illustration of the synthetic procedures of UCNPs@TaOx (I), TEM image of UCNPs@TaOx (II).

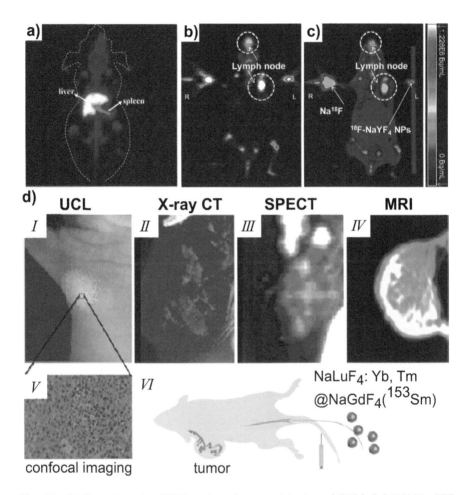

Fig. 8.8 (a) Kunming mice PET imaging after post-injection of ^{18}F-labeled UCNPs. PET imaging (b) and PET/CT imaging (c) of lymph node 30 min after subcutaneous injection of ^{18}F-labeled UCNPs, 30 min after subcutaneous injection of ^{18}F-labeled UCNPs into the left paw footpad, signal in lymph node reached the peak intensity and maintained it to 60 min post-injection. While as control free 18F$^-$ ions injected into the right paw showed no lymphatic imaging ability. d) Four-modal imaging of the focused tumor from the tumor-bearing nude mouse after intravenous injection of NaLuF$_4$: Yb, Tm@NaGd([153]Sm)F$_4$: (I) *In vivo* UCL image, (II) CT image, (III) SPECT image, (IV) MR imaging of tumor. (V) UCL confocal image of the paraffin section of tumor tissue. (VI) Schematic illustration of tumor angiogenesis imaging using NaLuF$_4$: Yb, Tm@NaGd([153]Sm)F$_4$ as the nanoprobe.

Conclusions

In summary, we have surveyed the recent advancements in the engineering of hybrid UCNPs in the field of surface hydrophilic modification and conjugation, biodetection and multimodal cancer imaging such as UCL, MRI,

CT, PET and SPECT imaging. Engineering of hybrid UCNPs with excellent water solubility and stability is a key precondition for their further bio-applications both *in vitro* and *vivo*. By attaching special functional molecules, organic dyes or inorganic NPs, these UCNPs have developed a great potential for application in numerous fields such as the biodetection of metal or anion ions (e.g., Zn^{2+}, Ca^{2+}, Fe^{2+}, Hg^{2+}, $-SH$, CN^-, etc.), biomolecules (e.g., DNA, RNA, gene, glucose, etc.), antibodies and cancer markers involved in the vital process of the organism. The utilization of unique optical advantages did make the LOD of these nanosystems much lower than that of traditional organic dyes. In addition, multiplexed detections of many analytes or multicolor sensing of biochemical entities like the ratiometric biodetection are emerging as rapid, convenient and effective ways to accelerate their bio-applications in clinical analysis and detection. Furthermore, multimodal UCNPs involving UCL, MRI, CT, PET and SPECT imaging are regarded as the new generation contrast agents for early cancer diagnosis as well as imaging guided therapy. The efficient multimodal diagnostics of early tumors with UCNPs, which offer a more convenient platform than other hybrid materials to reduce the patient discomfort, is highly promising for versatile cancer screening and monitoring on clinically used machines.

One of the major problems of the current hybrid UCNPs is their biosafety. Recent works regarding the cellular internalization, biodistribution, excretion, cytotoxicity and *in vivo* toxic effects of UCNPs have been reported. However, the toxicity of these hybrid UCNPs involving organism process *in vivo* is very complicated and remains to be deeper investigated. In particular, the gold standards for assessing the biocompatibility of UCNPs in detail (such as *in vitro* and *in vivo* evaluation parameters: surface modification, experiment time, the using dose, laser density, etc.) need to be urgently established. Meanwhile, low UCL quantum yields (QY) of UCNPs also limit their further commercial bio-applications. To date, the highest obtained QY of UCNPs (~7.6%) is lower than that of organic dyes or quantum dots. It is even not acceptable for some *in vivo* applications. And increasing the QY of UCNPs relies on a comprehensive study of the optical and structural properties of UCNPs, which will guide the synthesis and optimization of these hybrid materials.

Acknowledgements

This work has been financially supported by the National Natural Science Foundation of China (Grant No. 51372260, 51132009).

References

Bogdan, N., Vetrone, F., Roy, R. and Capobianco, J.A. 2010. Carbohydrate-coated lan-thanide-doped upconverting nanoparticles for lectin recognition. J Mater Chem 20: 7543.

Bogdan, N., Vetrone, F., Ozin, G.A. and Capobianco, J.A. 2011. Synthesis of ligand-free colloidally stable water dispersible brightly luminescent lanthanide-doped upconverting nanoparticles. Nano Lett 11: 835-840.

Cao, T., Yang, Y., Gao, Y., Zhou, J., Li, Z. and Li, F. 2011. High-quality water-soluble and surface-functionalized upconversion nanocrystals as luminescent probes for bioimaging. Biomaterials 32: 2959-2968.

Chatterjee, D.K., Rufaihah, A.J. and Zhang, Y. 2008. Upconversion fluorescence imaging of cells and small animals using lanthanide doped nanocrystals. Biomaterials 29: 937-943.

Chen, C., Sun, L., Li, Z., Li, L., Zhang, J., Zhang, Y., et al. 2010a. Ionic liquid-based route to spherical NaYF$_4$ nanoclusters with the assistance of microwave radiation and their multicolor upconversion luminescence. Langmuir 26: 8797-8803.

Chen, F., Zhang, S., Bu, W., Liu, X., Chen, Y., He, Q., et al. 2010b. A "neck-formation" strategy for an antiquenching magnetic/upconversion fluorescent bimodal cancer probe. Chem Eur J 16: 11254-11260.

Chen, F., Bu, W., Zhang, S., Liu, X., Liu, J., Xing, H., et al. 2011a. Positive and negative lattice shielding effects co-existing in Gd (III) ion doped bifunctional upconversion nanoprobes. Adv Funct Mater 21: 4285-4294.

Chen, F., Bu, W., Zhang, S., Liu, J., Fan, W., Zhou, L., et al. 2013. Gd^{3+}-doped upconversion nanoprobes: relaxivity mechanism probing and sensitivity optimization. Adv Funct Mater 23: 298-307.

Chen, G., Shen, J., Ohulchanskyy, T.Y., Patel, N.J., Kutikov, A., Li, Z., et al. 2012. (α-NaYbF$_4$: Tm^{3+})/CaF$_2$ core/shell nanoparticles with efficient near-infrared to near-infrared upconversion for high-contrast deep tissue bioimaging. ACS Nano 6: 8280-8287.

Chen, G., Qiu, H., Prasad, P.N. and Chen, X. 2014a. Upconversion nanoparticles: design, nanochemistry and applications in theranostics. Chem Rev 114: 5161-5214.

Chen, H., Guan, Y., Wang, S., Ji, Y., Gong, M. and Wang, L. 2014b. Turn-on detection of a cancer marker based on near-infrared luminescence energy transfer from NaYF$_4$: Yb, Tm/NaGdF$_4$ core-shell upconverting nanoparticles to gold nanorods. Langmuir 30: 13085-13091.

Chen, Q., Wang, X., Chen, F., Zhang, Q., Dong, B., Yang, H., et al. 2011b. Functionalization of upconverted luminescent NaYF$_4$:Yb/Er nanocrystals by folic acid–chitosan conjugates for targeted lung cancer cell imaging. J Mater Chem 21: 7661.

Chen, Z., Chen, H., Hu, H., Yu, M., Li, F., Zhang, Q., et al. 2008. Versatile synthesis strategy for carboxylic acid-functionalized upconverting nanophosphors as biological labels. J Am Chem Soc 130: 3023-3029.

Cheng, L., Yang, K., Zhang, S., Shao, M., Lee, S. and Liu, Z. 2010. Highly-sensitive multiplexed in vivo imaging using pegylated upconversion nanoparticles. Nano Res 3: 722-732.

Cheng, L., Yang, K., Li, Y., Chen, J., Wang, C., Shao, M., et al. 2011. Facile preparation of multifunctional upconversion nanoprobes for multimodal imaging and dual-targeted photothermal therapy. Angew Chem Int Ed 50: 7385-7390.

Chien, Y.H., Chou, Y., Wang, S., Hung, S., Liau, M., Chao, Y., et al. 2013. Near-infrared light photocontrolled targeting, bioimaging and chemotherapy with caged upconversion nanoparticles in vitro and in vivo. ACS Nano 7: 8516-8528.

Dai, Y., Yang, D., Ma, P., Kang, X., Zhang, X., Li, C., et al. 2012. Doxorubicin conjugated $NaYF_4$: Yb^{3+}/Tm^{3+} nanoparticles for therapy and sensing of drug delivery by luminescence resonance energy transfer. Biomaterials 33: 8704-8713.

Deng, R., Xie, X., Vendrell, M., Chang, Y.T. and Liu, X. 2011. Intracellular glutathione detection using MnO(2)-nanosheet-modified upconversion nanoparticles. J Am Chem Soc 133: 20168-20171.

Ding, Y., Zhu, H., Zhang, X., Zhu, J.J. and Burda, C. 2013. Rhodamine B derivative-functionalized upconversion nanoparticles for FRET-based Fe^{3+}-sensing. Chem Commun 49: 7797-7799.

Fan, W., Shen, B., Bu, W., Chen, F., He, Q., Zhao, K., et al. 2014. A smart upconversion-based mesoporous silica nanotheranostic system for synergetic chemo-/radio-/photodynamic therapy and simultaneous MR/UCL imaging. Biomaterials 35: 8992-9002.

Gai, S., Yang, P., Li, C., Wang, W., Dai, Y.L., Niu, N., et al. 2010. Synthesis of magnetic, up-conversion luminescent and mesoporous core-shell-structured nanocomposites as drug carriers. Adv Funct Mater 20: 1166-1172.

Gai, S., Li, C., Yang, P. and Lin, J. 2014. Recent progress in rare earth micro/nanocrystals: soft chemical synthesis, luminescent properties and biomedical applications. Chem Rev 114: 2343-2389.

Haase, M. and Schafer, H. 2011. Upconverting nanoparticles. Angew Chem Int Ed 50: 5808-5829.

He, M., Huang, P., Zhang, C., Hu, H., Bao, C., Gao, G., et al. 2011. Dual phase-controlled synthesis of uniform lanthanide-doped $NaGdF_4$ upconversion nanocrystals via an OA/ionic liquid two-phase system for in vivo dual-modality imaging. Adv Funct Mater 21: 4470-4477.

Hu, H., Yu, M., Li, F., Chen, Z., Gao, X., Xiong, L., et al. 2008. Facile epoxidation strategy for producing amphiphilic up-converting rare-earth nanophosphors as biological labels. Chem Mater 20: 7003-7009.

Idris, N.M., Gnanasammandhan, M.K., Zhang, J., Ho, P.C., Mahendran, R. and Zhang, Y. 2012. In vivo photodynamic therapy using upconversion nanoparticles as remote-controlled nanotransducers. Nat Med 18: 1580-1585.

Jiang, S. and Zhang, Y. 2010. Upconversion nanoparticle-based FRET system for study of siRNA in live cells. Langmuir 26: 6689-6694.

Jin, J., Gu, Y.J., Man, C.W., Cheng, J., Xu, Z., Zhang, Y., et al. 2011. Polymer-coated $NaYF_4$: Yb^{3+}, Er^{3+} upconversion nanoparticles for charge-dependent cellular imaging. ACS Nano 5: 7838-7847.

Johnson, N.J.J., Oakden, W., Stanisz, G.J., Scott Prosser, R. and van Veggel, F.C.J.M. 2011. Size-tunable, ultrasmall $NaGdF_4$ nanoparticles: insights into their T_1 MRI contrast enhancement. Chem Mater 23: 3714-3722.

Kang, X., Cheng, Z., Li, C., Yang, D., Shang, M., Ma, P.A., et al. 2011. Core–shell structured up-conversion luminescent and mesoporous $NaYF_4$: Yb^{3+}/Er^{3+}@ $nSiO_2$@$mSiO_2$ nanospheres as carriers for drug delivery. J Phys Chem C 115: 15801-15811.

Kang, X., Cheng, Z., Yang, D., Ma, P.A., Shang, M., Peng, C., et al. 2012. Design and synthesis of multifunctional drug carriers based on luminescent rattle-type mesoporous silica microspheres with a thermosensitive hydrogel as a controlled switch. Adv Funct Mater 22: 1470-1481.

Kobayashi, H., Kosaka, N., Ogawa, M., Morgan, N.Y., Smith, P.D., Murray, C.B., et al. 2009. In vivo multiple color lymphatic imaging using upconverting nanocrystals. J Mater Chem 19: 6481.

Kumar, R., Nyk, M., Ohulchanskyy, T.Y., Flask, C.A. and Prasad, P.N. 2009. Combined optical and MR bioimaging using rare earth ion doped $NaYF_4$ nanocrystals. Adv Funct Mater 19: 853-859.

Kuningas, K., Rantanen, T., Ukonaho, T., Lovgren, T. and Soukka, T. 2005. Homogeneous assay technology based on upconverting phosphors. Anal Chem 77: 7348-7355.

Li, C., Yang, J., Yang, P., Lian, H. and Lin, J. 2008. Hydrothermal synthesis of lanthanide fluorides LnF_3 (Ln = La to Lu) nano-/microcrystals with multiform structures and morphologies. Chem Mater 20: 4317-4326.

Li, L.L., Zhang, R., Yin, L., Zheng, K., Qin, W., Selvin, P.R., et al. 2012. Biomimetic surface engineering of lanthanide-doped upconversion nanoparticles as versatile bioprobes. Angew Chem Int Ed 51: 6121-6125.

Li, X., Zhou, L., Wei, Y., El-Toni, A.M., Zhang, F. and Zhao, D. 2014. Anisotropic growth-induced synthesis of dual-compartment janus mesoporous silica nanoparticles for bimodal triggered drugs delivery. J Am Chem Soc 136: 15086-15092.

Li, Z. and Zhang, Y. 2006. Monodisperse silica-coated polyvinylpyrrolidone/$NaYF_4$ nanocrystals with multicolor upconversion fluorescence emission. Angew Chem Int Ed 45: 7732-7735.

Li, Z., Lv, S., Wang, Y., Chen, S. and Liu, Z. 2015. Construction of LRET-Based nanoprobe using upconversion nanoparticles with confined emitters and bared surface as luminophore. J Am Chem Soc 137: 3421-3427.

Lim, S.F., Riehn, R., Ryu, W.S., Khanarian, N., Tung, C.K., Tank, D., et al. 2006. In vivo and scanning electron microscopy imaging of up-converting nanophosphors in caenorhabditis elegans. Nano Lett 6: 169-174.

Liu, C., Wang, Z., Jia, H. and Li, Z. 2011a. Efficient fluorescence resonance energy transfer between upconversion nanophosphors and graphene oxide: a highly sensitive biosensing platform. Chem Commun 47: 4661-4663.

Liu, C., Gao, Z., Zeng, J., Hou, Y., Fang, F., Li, Y., et al. 2013a. Magnetic/upconversion fluorescent $NaGdF_4$: Yb, Er nanoparticle-based dual-modal molecular probes for imaging tiny tumors in vivo. ACS Nano 7: 7227-7240.

Liu, J., Liu, Y., Liu, Q., Li, C., Sun, L. and Li, F. 2011b. Iridium (III) complex-coated nanosystem for ratiometric upconversion luminescence bioimaging of cyanide anions. J Am Chem Soc 133: 15276-15279.

Liu, J., Bu, W., Zhang, S., Chen, F., Xing, H., Pan, L., et al. 2012a. Controlled synthesis of uniform and monodisperse upconversion core/mesoporous silica shell nanocomposites for bimodal imaging. Chem Eur J 18: 2335-2341.

Liu, J., Bu, W., Pan, L. and Shi, J. 2013b. NIR-triggered anticancer drug delivery by upconverting nanoparticles with integrated azobenzene-modified mesoporous silica. Angew Chem Int Ed 52: 4375-4379.

Liu, J., Bu, J., Bu, W., Zhang, S., Pan, L., Fan, W., et al. 2014a. Real-Time in vivo quantitative monitoring of drug release by dual-mode magnetic resonance and upconverted luminescence imaging. Angew Chem Int Ed 53: 4551-4555.

Liu, J., Liu, Y., Bu, W., Bu, J., Sun, Y., Du, J., et al. 2014b. Ultrasensitive nanosensors based on upconversion nanoparticles for selective hypoxia imaging in vivo upon near-infrared excitation. J Am Chem Soc 136: 9701-9709.

Liu, Q., Li, C., Yang, T., Yi, T. and Li, F. 2010. "Drawing" upconversion nanophosphors into water through host-guest interaction. Chem Commun 46: 5551-5553.

Liu, Q., Chen, M., Sun, Y., Chen, G., Yang, T., Gao, Y., et al. 2011c. Multifunctional rare-earth self-assembled nanosystem for tri-modal upconversion luminescencefluorescence/positron emission tomography imaging. Biomaterials 32: 8243-8253.

Liu, Q., Peng, J., Sun, L. and Li, F. 2011d. High-efficiency upconversion luminescent sensing and bioimaging of Hg (II) by chromophoric ruthenium complex-assembled nanophosphors. ACS Nano 5: 8040-8048.

Liu, Q., Sun, Y., Li, C.G., Zhou, J., Li, C.Y., Yang, T.S., et al. 2011e. ^{18}F-labeled magnetic-upconversion nanophosphors via rare-earth cation-assistant ligand assembly. ACS Nano 5: 3146-3157.

Liu, Q., Sun, Y., Yang, T., Feng, W., Li, C. and Li, F. 2011f. Sub-10 nm hexagonal lanthanide-doped NaLuF4 upconversion nanocrystals for sensitive bioimaging in vivo. J Am Chem Soc 133: 17122-17125.

Liu, Y., Ai, K., Liu, J., Yuan, Q., He, Y. and Lu, L. 2012b. A high-performance ytterbium-based nanoparticulate contrast agent for in vivo X-ray computed tomography imaging. Angew Chem Int Ed 51: 1437-1442.

Liu, Y., Ai, K., Liu, J., Yuan, Q., He, Y. and Lu, L. 2012c. Hybrid BaYbF$_5$ nanoparticles: novel binary contrast agent for high-resolution in vivo X-ray computed tomography angiography. Adv Healthc Mater 1: 461-466.

Liu, Y., Chen, M., Cao, T., Sun, Y., Li, C., Liu, Q., et al. 2013c. A cyanine-modified nanosystem for in vivo upconversion luminescence bioimaging of methylmercury. J Am Chem Soc 135: 9869-9876.

Liu, Y., Bu, W., Xiao, Q., Sun, Y., Zhao, K., Fan, W., et al. 2015. Radiation-/hypoxia-induced solid tumor metastasis and regrowth inhibited by hypoxia-specific upconversion nanoradiosensitizer. Biomaterials 49: 1-8.

Mahalingam, V., Vetrone, F., Naccache, R., Speghini, A. and Capobianco, J.A. 2009. Colloidal Tm^{3+}/Yb^{3+}-doped LiYF$_4$ nanocrystals: multiple luminescence spanning the UV to NIR regions via low-energy excitation. Adv Mater 21: 4025-4028.

Meiser, F., Cortez, C. and Caruso, F. 2004. Biofunctionalization of fluorescent rare-earth-doped lanthanum phosphate colloidal nanoparticles. Angew Chem Int Ed 43: 5954-5957.

Naccache, R., Vetrone, F., Mahalingam, V., Cuccia, L.A. and Capobianco, J.A. 2009. Controlled synthesis and water dispersibility of hexagonal phase NaGdF$_4$: Ho^{3+}/Yb^{3+} nanoparticles. Chem Mater 21: 717-723.

Nam, S.H., Bae, Y.M., Park, Y.I., Kim, J.H., Kim, H.M., Choi, J.S., et al. 2011. Long-term real-time tracking of lanthanide ion doped upconverting nanoparticles in living cells. Angew Chem Int Ed 50: 6093-6097.

Ni, D., Bu, W., Zhang, S., Zheng, X., Li, M., Xing, H., et al. 2014a. Single Ho^{3+}-doped upconversion nanoparticles for high-performance T$_2$-weighted brain tumor diagnosis and MR/UCL/CT multimodal imaging. Adv Funct Mater 24: 6613-6620.

Ni, D., Zhang, J., Bu, W., Xing, H., Han, F., Xiao, Q., et al. 2014b. Dual-targeting upconversion nanoprobes across the blood-brain barrier for magnetic resonance/fluorescence imaging of intracranial glioblastoma. ACS Nano 8: 1231-1242.

Park, Y.I., Kim, J.H., Lee, K.T., Jeon, K.-S., Na, H.B., Yu, J.H., et al. 2009. Nonblinking and nonbleaching upconverting nanoparticles as an optical imaging nanoprobe and T$_1$ magnetic resonance imaging contrast agent. Adv Mater 21: 4467-4471.

Park, Y.I., Kim, H.M., Kim, J.H., Moon, K.C., Yoo, B., Lee, K.T., et al. 2012. Theranostic probe based on lanthanide-doped nanoparticles for simultaneous in vivo dual-modal imaging and photodynamic therapy. Adv Mater 24: 5755-5761.

Park, Y.I., Lee, K.T., Suh, Y.D. and Hyeon, T. 2015. Upconverting nanoparticles: a versatile platform for wide-field two-photon microscopy and multi-modal in vivo imaging. Chem Soc Rev 44: 1302-1317.

Peng, J., Xu, W., Teoh, C.L., Han, S., Kim, B., Samanta, A., et al. 2015. High-efficiency in vitro and in vivo detection of zn(2+) by dye-assembled upconversion nanoparticles. J Am Chem Soc 137: 2336-2342.

Qian, H.S., Guo, H.C., Ho, P.C.L., Mahendran, R. and Zhang, Y. 2009. Mesoporous-silica-coated up-conversion fluorescent nanoparticles for photodynamic therapy. Small 5: 2285-2290.

Qiao, R., Liu, C., Liu, M., Hu, H., Hou, Y., Wu, K., et al. 2015. Ultrasensitive in vivo detection of primary gastric tumor and lymphatic metastasis using upconversion nanoparticles. ACS Nano 9: 2120-2129.

Shen, J., Sun, L.D., Wen, Z.Y., Yan, C.H. and 2010. Superparamagnetic and upconversion emitting Fe3O4/NaYF4: Yb, Er hetero-nanoparticles via a crosslinker anchoring strategy. Chem Commun 46: 5731-5733.

Sun, Y., Yu, M., Liang, S., Zhang, Y., Li, C., Mou, T., et al. 2011. Fluorine-18 labeled rare-earth nanoparticles for positron emission tomography (PET) imaging of sentinel lymph node. Biomaterials 32: 2999-3007.

Sun, Y., Zhu, X., Peng, J. and Li, F. 2013. Core-shell lanthanide upconversion nanophosphors as four-modal probes for tumor angiogenesis imaging. ACS Nano 7: 11290-11300.

Wang, C., Cheng, L. and Liu, Z. 2011a. Drug delivery with upconversion nanoparticles for multi-functional targeted cancer cell imaging and therapy. Biomaterials 32: 1110-1120.

Wang, F., Han, Y., Lim, C.S., Lu, Y., Wang, J., Xu, J., et al. 2010. Simultaneous phase and size control of upconversion nanocrystals through lanthanide doping. Nature 463: 1061-1065.

Wang, L., Yan, R., Huo, Z., Zeng, J., Bao, J., Wang, X., et al. 2005a. Fluorescence resonant energy transfer biosensor based on upconversion-luminescent nanoparticles. Angew Chem Int Ed 44: 6054-6057.

Wang, M., Hou, W., Mi, C., Wang, W., Xu, Z., Teng, H., et al. 2009a. Immunoassay of goat antihuman immunoglobulin G antibody based on luminescence resonance energy transfer between near-infrared responsive NaYF4: Yb, Er upconversion fluorescent nanoparticles and gold nanoparticles. Anal Chem 81: 8783-8789.

Wang, M., Mi, C.C., Wang, W.X., Liu, C.H., Wu, Y.F., Xu, Z.R., et al. 2009b. Immunolabeling and NIR-excited fluorescent imaging of HeLa cells by using NaYF$_4$: Yb, Er upconversion nanoparticles. ACS Nano 3: 1580-1586.

Wang, X., Zhuang, J., Peng, Q. and Li, Y., 2005b. A general strategy for nanocrystal synthesis. Nature 437: 121-124.

Wang, Y., Bao, L., Liu, Z. and Pang, D.W. 2011b. Aptamer biosensor based on fluorescence resonance energy transfer from upconverting phosphors to carbon nanoparticles for thrombin detection in human plasma. Anal Chem 83: 8130-8137.

Wu, S., Han, G., Milliron, D.J., Aloni, S., Altoe, V., Talapin, D.V., et al. 2009. Non-blinking and photostable upconverted luminescence from single lanthanide-doped nanocrystals. Proc Natl Acad Sci USA 106: 10917-10921.

Wu, Y.M., Cen, Y., Huang, L.J., Yu, R.Q. and Chu, X. 2014. Upconversion fluorescence resonance energy transfer biosensor for sensitive detection of human immunodeficiency virus antibodies in human serum. Chem Commun 50: 4759-4762.

Xia, A., Gao, Y., Zhou, J., Li, C., Yang, T., Wu, D., et al. 2011. Core-shell $NaYF_4$: Yb^{3+}, Tm^{3+}@Fe_xO_y nanocrystals for dual-modality T_2-enhanced magnetic resonance and NIR-to-NIR upconversion luminescent imaging of small-animal lymphatic node. Biomaterials 32: 7200-7208.

Xia, A., Chen, M., Gao, Y., Wu, D., Feng, W. and Li, F. 2012. Gd^{3+} complex-modified $NaLuF_4$-based upconversion nanophosphors for trimodality imaging of NIR-to-NIR upconversion luminescence, X-ray computed tomography and magnetic resonance. Biomaterials 33: 5394-5405.

Xiao, Q., Bu, W., Ren, Q., Zhang, S., Xing, H., Chen, F., et al. 2012. Radiopaque fluorescence-transparent TaO_x decorated upconversion nanophosphors for in vivo CT/MR/UCL trimodal imaging. Biomaterials 33: 7530-7539.

Xiao, Y., Zeng, L., Xia, T., Wu, Z. and Liu, Z. 2015. Construction of an upconversion nanoprobe with few-atom silver nanoclusters as the energy acceptor. Angew Chem Int Ed 54: 5323-5327.

Xing, H., Bu, W., Ren, Q., Zheng, X., Li, M., Zhang, S., et al. 2012a. A $NaYbF_4$: Tm^{3+} nanoprobe for CT and NIR-to-NIR fluorescent bimodal imaging. Biomaterials 33: 5384-5393.

Xing, H., Bu, W., Zhang, S., Zheng, X., Li, M., Chen, F., et al. 2012b. Multifunctional nanoprobes for upconversion fluorescence, MR and CT trimodal imaging. Biomaterials 33: 1079-1089.

Xing, H., Zheng, X., Ren, Q., Bu, W., Ge, W., Xiao, Q., et al. 2013. Computed tomography imaging-guided radiotherapy by targeting upconversion nanocubes with significant imaging and radiosensitization enhancements. Sci Rep 3: 1751.

Xiong, L., Chen, Z., Tian, Q., Cao, T., Xu, C. and Li, F. 2009a. High contrast upconversion luminescence targeted imaging in vivo using peptide-labeled nanophosphors. Anal Chem 81: 8687-8694.

Xiong, L., Chen, Z., Yu, M., Li, F., Liu, C. and Huang, C., 2009b. Synthesis, characterization and in vivo targeted imaging of amine-functionalized rare-earth upconverting nanophosphors. Biomaterials 30: 5592-5600.

Yang, T., Sun, Y., Liu, Q., Feng, W., Yang, P. and Li, F. 2012. Cubic sub-20 nm $NaLuF_4$-based upconversion nanophosphors for high-contrast bioimaging in different animal species. Biomaterials 33: 3733-3742.

Ye, W., Tsang, M.K., Liu, X., Yang, M. and Hao, J. 2014. Upconversion luminescence resonance energy transfer (LRET)-based biosensor for rapid and ultrasensitive detection of avian influenza virus H7 subtype. Small 10: 2390-2397.

Yi, G.S. and Chow, G.M. 2006. Synthesis of hexagonal-phase $NaYF_4$: Yb, Er and $NaYF_4$: Yb, Tm nanocrystals with efficient up-conversion fluorescence. Adv Funct Mater 16: 2324-2329.

Yu, X., Sun, Z., Li, M., Xiang, Y., Wang, Q., Tang, F., et al. 2010. Neurotoxin-conjugated upconversion nanoprobes for direct visualization of tumors under near-infrared irradiation. Biomaterials 31: 8724-8731.

Yuan, F., Chen, H., Xu, J., Zhang, Y., Wu, Y. and Wang, L. 2014. Aptamer-based luminescence energy transfer from near-infrared-to-near-infrared upconverting nanoparticles to gold nanorods and its application for the detection of thrombin. Chem Eur J 20: 2888-2894.

Zhang, C., Yuan, Y., Zhang, S., Wang, Y. and Liu, Z. 2011. Biosensing platform based on fluorescence resonance energy transfer from upconverting nanocrystals to graphene oxide. Angew Chem Int Ed 50: 6851-6854.

Zhang, F., Braun, G.B., Pallaoro, A., Zhang, Y., Shi, Y., Cui, D., et al. 2012a. Mesoporous multifunctional upconversion luminescent and magnetic "nanorattle" materials for targeted chemotherapy. Nano Lett 12: 61-67.

Zhang, L., Wang, Y.S., Yang, Y., Zhang, F., Dong, W.F., Zhou, S.Y., et al. 2012b. Magnetic/upconversion luminescent mesoparticles of $Fe_3O_4@LaF_3$: Yb^{3+}, Er^{3+} for dual-modal bioimaging. Chem Commun 48: 11238-11240.

Zhang, P., Rogelj, S., Nguyen, K. and Wheeler, D. 2006a. Design of a highly sensitive and specific nucleotide sensor based on photon upconverting particles. J Am Chem Soc 128: 12410-12411.

Zhang, P., Steelant, W., Kumar, M. and Scholfield, M. 2007. Versatile photosensitizers for photodynamic therapy at infrared excitation. J Am Chem Soc 129: 4526-4527.

Zhang, Y., Mai, H., Si, R., Yan, Z., Sun, L., You, L., et al. 2006b. High-quality sodium rare-earth fluoride nanocrystals: controlled synthesis and optical properties. J Am Chem Soc 128: 6426-6436.

Zheng, W., Zhou, S., Chen, Z., Hu, P., Liu, Y., Tu, D., et al. 2013. Sub-10 nm lanthanide-doped CaF2 nanoprobes for time-resolved luminescent biodetection. Angew Chem Int Ed 52: 6671-6676.

Zheng, W., Huang, P., Tu, D., Ma, E., Zhu, H. and Chen, X. 2015. Lanthanide-doped upconversion nano-bioprobes: electronic structures, optical properties and bio-detection. Chem Soc Rev 44: 1379-1415.

Zhou, H., Xu, C., Sun, W. and Yan, C. 2009. Clean and flexible modification strategy for carboxyl/aldehyde-functionalized upconversion nanoparticles and their optical applications. Adv Funct Mater 19: 3892-3900.

Zhou, J., Yao, L., Li, C. and Li, F. 2010. A versatile fabrication of upconversion nano-phosphors with functional-surface tunable ligands. J Mater Chem 20: 8078.

Zhou, J., Yu, M., Sun, Y., Zhang, X., Zhu, X., Wu, Z., et al. 2011. Fluorine-18-labeled $Gd^{3+}/Yb^{3+}/Er^{3+}$ co-doped $NaYF_4$ nanophosphors for multimodality PET/MR/ UCL imaging. Biomaterials 32: 1148-1156.

Zhou, J., Liu, Z. and Li, F.Y. 2012. Upconversion nanophosphors for small-animal imaging. Chem Soc Rev 41: 1323-1349.

Zhou, J., Liu, Q., Feng, W., Sun, Y. and Li, F. 2015. Upconversion luminescent materials: advances and applications. Chem Rev 115: 395-465.

Zhu, X., Zhou, J., Chen, M., Shi, M., Feng, W. and Li, F. 2012. Core-shell $Fe_3O_4@$ $NaLuF_4$: Yb, Er/Tm nanostructure for MRI, CT and upconversion luminescence tri-modality imaging. Biomaterials 33: 4618-4627.

Zijlmans, H.J.M.A.A., Bonnet, J., Burton, J., Kardos, K., Vail, T., Niedbala, R.S., et al. 1999. Detection of cell and tissue surface antigens using up-converting phosphors: a new reporter technology. Anal Biochem 267: 30-36.

9

Radionanomedicine with Hybrid Nanomaterials

Yun-Sang Lee[1,2,a] and *Dong Soo Lee*[2,3,*]

INTRODUCTION

Nanomaterials can be modified on their surface or at the core (Kango et al. 2013; Sperling and Parak 2010). The modification of nanomaterials is important especially for the *in vivo* application because the hydrophobicity will render the nanomaterials being captured by the MPS (mononuclear phagocytic system) and the positive surface charge will give the cytotoxicity (Nam et al. 2013). Also the contents of the core materials will determine the traceability or capability of imaging. If we choose quantum dots or fluorescent silica particles, fluorescence gives us chance for sensing (*in vitro*) and imaging (*in vivo*) but the limitation is that the energy transferred by excitation laser usually kills the cells. Upconverting nanoparticles (UCNPs) are the solution to this problem in that they use near-infrared excitation laser and the live cells can be monitored for hours and days (Nam et al. 2011; Park et al. 2015). Multiplexing capability of nanomaterials are realized by SERS (Surface Enhanced Raman Spectroscopy) dots but these particles received attention as topical agents for the purpose of endoscopic characterization of tumor epitopes (Jeong et al. 2015). Some of the agents enable *in vivo*

[1] Department of Nuclear Medicine, Seoul National University College of Medicine, 103, Daehak-ro, Jongno-gu, Seoul, 110-799, Republic of Korea.
[2] Department of Molecular Medicine and Biopharmaceutical Sciences, Graduate School of Convergence Science and Technology, Seoul National University, Seoul, Republic of Korea.
[a] E-mail: wonza43@snu.ac.kr
[3] Department of Nuclear Medicine, Seoul National University Hospital, 101, Daehak-ro, Jongno-gu, Seoul, 110-744, Republic of Korea.
[*] Corresponding author: dsl@snu.ac.kr

imaging but if the core is ferromagnetic or paramagnetic we can combine MRI (magnetic resonance imaging) (Yin et al. 2015) and fluorescence or other optical imaging (Jun et al. 2010). MRI or PET (positron emission tomography) /SPECT (single photon emission computed tomography) enables depth imaging and even possible human application (Bouziotis et al. 2012; Garcia et al. 2015; Misri et al. 2012) (Fig. 9.1).

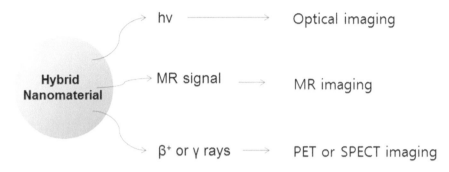

Fig. 9.1 Hybrid nanomaterials and imaging devices.

Among the various hybrid nanomaterials, optical and iron oxide or optical and radionuclide combinations are proposed to be optimal as *ex vivo* confirmation is done by optical characterization and *in vivo* (small animal or human) imaging is performed with iron oxide in the core or with positron or gamma emitters on the surface. After the synthesis of iron oxide core, the surface of iron oxide should also be modified for hydrophilization and ligand binding for biological applications. The core or the surface can be supplied for radiolabeling and the core can be irradiated in the nuclear reactor or the particle accelerator with neutrons or protons to yield radioactive core (Goel et al. 2014; Sun et al. 2012; Yang et al. 2013b; Zhao et al. 2014; Zhou et al. 2010) and the surface can be modified with the chelator to bind another radionuclide such as ^{68}Ga, ^{64}Cu, ^{89}Zr, ^{177}Lu or ^{90}Y, etc. (Anderson and Ferdani 2009; Fischer et al. 2013; Kam et al. 2012; Lee 2012; Sainz-Esteban et al. 2012; Velikyan 2013; Wadas et al. 2010; Xing et al. 2014) (Fig. 9.2).

The name of 'radionanomedicine' was given to these efforts and is, in one sentence, the combined effort of nanomedicine and nuclear medicine to realize targeted delivery with or without imaging (Lee et al. 2015a; Lee et al. 2015b). The major benefits of radionanomedicine are these two, 1) we can decrease the amount of nanomaterials, which will be delivered *in vivo*, down to microgram or nanomolar amount instead of milligram amount and we no longer have to worry about the toxicity of the pharmacological effect of nanomaterials and 2) using radio-tracing and tracer kinetic interpretation of the radiolabeled nanomaterials, we can build tracer kinetic models and using system identification on several mathematical software such as Matlab® or Mathematica® or specialized commercial software of PMOD®, we can find the mechanism of targeting and retention of radio-nanomaterials in the target tissues.

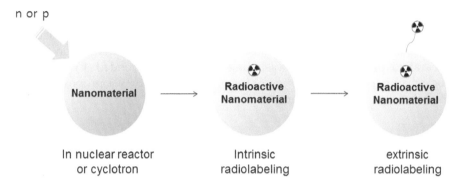

Fig. 9.2 Intrinsic and extrinsic radiolabeling for radionanomedicine.

Thus, the successful application of radionanomedicine is based on the optimal labeling of radionuclide mainly on the surface of the core nanomaterials. The surface modification of nanomaterials with chelators for radionuclide labeling uses all the technologies reported in the literature (Xing et al. 2014). New methods such as encapsulation methods or other simpler methods were proposed as the more appropriate solution to this problem (Lee et al. 2012).

Surface Modification

Surfaces of nanomaterials are going to be modified stepwise or in one step. The modification with targeting ligands, chelating agents, or polyethylene glycol (PEG) can be made by the trans-chelation method, the chemical bond formation using covalent or other bonds, physical interposition using intercalation or hydro-phobic/phobic (van der Waals') interactions. The surface should at least be hydrophilized from the original hydrophobic nature of these inorganic materials, which are sometimes metallic. Nano-surface modification was initially developed for hydrophilization for further engineering or *in vitro* diagnostics (Carion et al. 2007; Dubertret et al. 2002; Fan et al. 2004; Wu et al. 2008). It was later modified to yield the modified nanoparticles having several ligands on the surface which was made in mild condition in order to preserve the integrity of the ligands (Lee et al. 2012). Chelators should have been bound on the surface to label radionuclides later. And the investigators should develop the method of radiolabeling also in mild condition.

Cupper-free click chemistry was also proposed to preserve the integrity of the ligands on the surface (Chang et al. 2010; Gordon et al. 2012). These ligands include mononuclear antibodies, affibody, aptamer, aptide, avibody, nanobody, peptides and any kinds of 3-dimensional structures for the targeting. These ligands should be bound on the surface in the right direction, in very mild condition and without any degeneration. These bound ligands should also maintain their integrity in the further procedure of radiolabeling. The separation procedures should be developed and the eradication of the separation process after radiolabeling is also recommended. Cupper-free click chemistry for the adoption of the bioorthogonality was proposed to be the solution for these problems.

General Method: Stepwise

Once people tried to label nanoparticles simultaneously with a ligand for a target and with a chelator for further radionuclides labeling, they needed to label first the targeting ligands such as cyclic RGD, TOC or TATE and then the chelator later (Jimenez-Mancilla et al. 2013; Zhong et al. 2014). These stepwise or multi-step approaches have many difficulties. The multi-step chemical reactions can cause the decreasing of the total yield of modified nanoparticles. In addition, in each reaction step, they have to consider the reaction time, temperature, pH, impurities, solvents, reagent concentrations, etc. Finally, after all modification, they should check the targeting ability of those modified nanoparticles (Fig. 9.3).

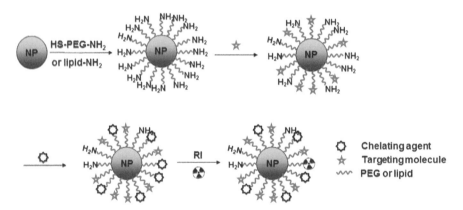

Fig. 9.3 The stepwise approach for the surface modification of nanomaterials.

Radiolabeling of chelator-bound nanoparticles adds another step of separation. The problem is that we need to examine the radiolabeling capability, radiolabeling yield and radiochemical purity of these chelator-bound nanoparticles using radio-chromatography or other separation and radioactive detection methods.

One-Pot Method

The simultaneous surface modification of nanoparticles with ligand and/or chelator and PEG, if feasible, will make the production of the ligand and/or chelator-bound and PEGylated nanoparticle easy. More important is that only when the chemical process is going to be mild, one-pot and able to be multiplexed, then the manufacturing and compounding will enable the radio-nanomaterials to be used for clinical or human purposes. For one-pot method, the micelle-encapsulation method was introduced and proven to be easily carried out for various nanomaterials (Lee et al. 2012). Once the mild and one-pot method was devised, the separation procedure was made simpler so that only the gel filtration could lead to the simple separation. Previously for other purposes, they used solvent evaporation method (Carion et al. 2007; Wu et al. 2008), but the micelle-encapsulation method using just mixing/sonication the pretreated-micelle from appropriate amphiphiles and nanomaterials, which have hydrophobic tails on the surface, needed just the size-exclusion gel filtration (Lee et al. 2012) (Fig. 9.4).

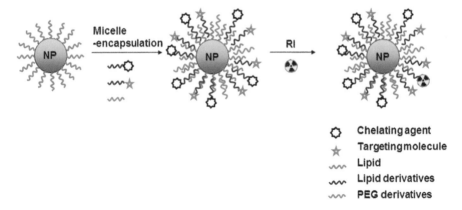

Fig. 9.4 One-pot approach for the surface modification of nanomaterials using micelle-encapsulation method.

The synthesis of amphiphile required that the meticulous recipe and the quality assurance in every production of the surface modified nanoparticles should be checked for their robustness in terms of the targeting affinity *in vitro*, the characteristic behavior of biodistribution *in vivo* and the targeting efficacy *in vivo* after systemic administration. One can optimize the composition of the amphiphile as needed and it will be up to the manufacturing when being considered for clinical trials and supply for clinical application, however, the further procedure of the surface modification and separation and further radiolabeling will be up to the compounding. Radiopharmacy will ensure once the vials for supply and ready-for-use after compounding are obtained from the registered suppliers or pharmaceutical companies.

Upconverting Nanoparticles (UCNPs)

Upconverting nanoparticles (UCNPs) are very well known to be an ideal nanoparticle for live cells used *in vitro* and possible *in vivo* and also can be used for photodynamic therapy (PDT) (Park et al. 2012) (Fig. 9.5). Near-infrared lasers of 980 nm or 800 nm are going to be used for excitation and the lights of visible range are emitted (Bae et al. 2012; Li et al. 2012; Li et al. 2013; Park et al. 2012; Yang et al. 2013c). The usual light-microscopes could be used for tracking the behaviors of the nanoparticles inside the alive cells after real-time or intermittent serial sequential imaging (Bae et al. 2012). The limitation of ordinary particles such as quantum dots is its high energy transfer which deteriorates the cellular viability during *in vivo* cell imaging (DaCosta et al. 2014; Park et al. 2015; Sedlmeier and Gorris 2015). UCNP can overcome this limitation and is free of the autofluorescence if used *in vivo*. Also the near-infrared ranged excitation ray can penetrate the inside of the body up to few centimeters; however, the emission light is in the visual (Y/Er/Yb) or infrared (Gd/Nd/Er/Yb) range and can penetrate only millimeters of the tissues. Scattering and absorption of the emission light and the excitation light (with the lesser amount) will require the injected amount to be higher. Thus, though the UCNP looks inert in the body, the amount of the UCNP needed for *in vivo* luminescence imaging will be several tens times more than the amount needed for PET imaging.

Radiolabeled nanoparticles were investigated for their biodistribution and found to be excreted as integral form (Kreyling et al. 2015). *In vivo* PET imaging and then the validation by *ex vivo* luminescence imaging cooperated to elucidate this behavior of the radiolabeled UCNP after systemic administration. Electron microscopy also helped to disclose mechanism of hepatobiliary excretion (Seo et al. 2015).

If the surface of UCNP is modified to have ligand(s) and/or chelator(s), these nanoparticles will target the tumor and/or microenvironment and will deliver the therapeutic radionuclides, such as ^{177}Lu and/or ^{90}Y, chelated with DOTA to the tumors. The multiplexed modification of the surface of UCNP will render the opportunity of radionuclide therapy using radiolabeled UCNP.

Surface-Enhanced Raman Scattering (SERS) Dots

Surface-Enhanced Raman Scattering (SERS) allows multiplexing in that SERS dots yield specific shift on spectroscopic spectrum according to the chemicals used to contact with the surface of the silver or gold nanoparticles (Samanta et al. 2014; Vendrell et al. 2013; Zavaleta et al. 2009; Zhang et al. 2011). This was the main advantage of SERS sensing and imaging. SERS dots, if labeled with antibodies, can be used for *in vitro* diagnostics to characterize the tumor epitopes depending on the specific antibodies against the epitopes according to the chemical-specific shift, which is the bar-code or certificate of each dot (Jokerst et al. 2011b; Jun et al. 2012b; Jun et al. 2011; Jun et al. 2010).

When the SERS dots were labeled with the fluorescent dye on the surface and were co-labeled with antibodies, the SERS dots could be used for endoscopic characterization of the suspected tumor markers on the surfaces of hollow viscus (Garai et al. 2015; Jeong et al. 2015; Zavaleta et al. 2013). In this case SERS dots are used for topical application and could successfully characterize EGFR and HER2 receptors of the implanted tumors. Topically applied SERS dots will be the most useful guide of *in vivo* application for multiplexed sensing and imaging. Intravesical cystoscopic, colonoscopic and gastroduodenal endoscopic application is expected.

Quantum Dot Square (QD²)

Quantum dots themselves were good examples of the multiplexing from the first report decades ago (Alivisatos et al. 2005) and size determined the emission spectrum in the visible range of wavelength. Thus, the mixing of the quantum dots with different sizes will yield multiple visible photons, however, though without flickering nor fading unlike fluorescent dyes, the range of emission rays are broad and the core contents of cadmium raise concerns (Geissler et al. 2014; Petryayeva et al. 2013). Thus one report proposed to use the silica coating over the core of silica and positioning the quantum dots between the coat and the core (Jun et al. 2012a). By making these types of quantum dots intermingled within the two layers of silica, the investigators could maximize the fluorescent intensity up to hundred times or more. When the surface was modified with the chelator and PEG and radiolabeled further with ^{68}Ga and ^{64}Cu, *in vivo* biodistribution could be traced.

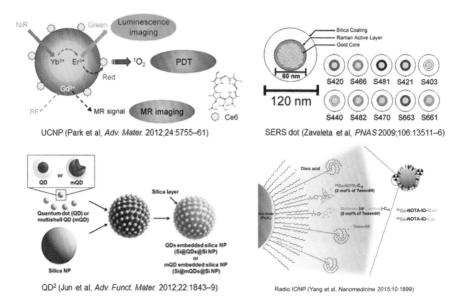

UCNP (Park et al, *Adv. Mater.* 2012;24:5755–61)

SERS dot (Zavaleta et al, *PNAS* 2009;106:13511–6)

QD² (Jun et al, *Adv. Funct. Mater.* 2012;22:1843–9)

Radio IONP (Yang et al, *Nanomedicine* 2015;10:1899)

Fig. 9.5 Hybrid nanomaterials for radionanomedicine.

Iron Oxide Nanoparticles (IONPs)

Iron oxide nanoparticle (IONP) is the one of the FDA approved metallic nanoparticles and could be used for magnetic resonance (MR) imaging contrast agent. Commercially available dextran coated IONP, Feridex I.V.®, was used for diagnosis of liver function of reticuloendothelial system (RES).

Iron oxide core was successfully modified with the micelle-encapsulation method and these surface modified IONPs could be used for *in vivo* in small animal model. One-pot method could make the surface of iron oxide simultaneously PEGylated, mannose-labeled and NOTA-labeled. After further labeling with ^{64}Cu, these radionanoparticles were engulfed by macrophages and could visualize the sentinel lymph node draining from the footpad (Yang et al. 2013a).

Material-Based *In Vivo* Stability

Gold nanoparticles were irradiated with neutrons in the nuclear reactor for the core to have ^{198}Au (Khan et al. 2008) and when the encapsulation was done with ^{111}In DOTA with PEGylation, the *in vivo* biodistribution data using radioactivity counting *ex vivo* after sacrifice showed that the capsules are detached and excreted via kidneys (Kreyling et al. 2015). This was presented to the scientific community to ponder upon the *in vivo* integrity of polymer-coated nanoparticles. The composition of the capsule and the bonds between the core and capsule, intracapsular bonds and radionuclide and the chelator might be the points of difference which determine the possibility of detachment or maintenance of the integrity.

Another report is coming that the integral form of radiolabeled encapsulated PEGylated UCNP was excreted via hepatobiliary route. Hepatobiliary excretion was not accompanied by any renal excretion and thus integral form was maintained until the excretion (Fig. 9.6) (Seo 2015).

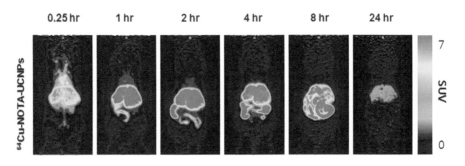

Fig. 9.6 Hepatobiliary excretion of the integral form of ^{64}Cu labeled UCNPs confirmed by 24 h PET images.

When ^{59}Fe, which has 44 days physical half-life, was labeled inside of the core of iron oxide nanoparticles and the biodistribution was monitored for several months, the PEGylated iron oxide was taken up by both hepatocytes and Kupffer cells and these iron oxide nanoparticles were degraded into ionic iron and these ions were recirculated after several months. The iron metabolism cycle was evidenced from the normal mouse by transient increasing in serum[59]Fe level from one day after administration and up to a month after administration (Fig. 9.7) (Yoo et al. 2014).

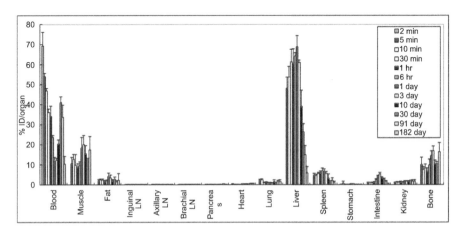

Fig. 9.7 Long-term biodistribution results of iron oxide nanoparticles by the core ^{59}Fe radiolabeling.

The integrity is determined by the constitution of the core material, the binding nature between the core and the capsules and the composition of the capsule and the chelator.

Hydrophilization

Commonly used nanomaterials have hydrophobic surfaces after the core synthesis and need the common use of polyethylene glycol (PEG) or natural and synthetic polymers including dextran for making the surface hydrophilic and retaining the material's stability in a biological environment (Jokerst et al. 2011a; Joralemon et al. 2010; Karakoti et al. 2011; Pasut and Veronese 2012). In addition, ligands and chelators need to be added to the termini of PEGs or the surface of dextran.

PEGylation or dextran was associated with so many water molecules and this has been considered to make the nanoparticles 'stealth' or 'cloaking', which means that the bodily reticuloendothelial system (RES) could not recognize the nanomaterials (Amoozgar and Yeo 2012; Noble et al. 2014).

The Sizes and Size Distributions

The size of the nanoparticle is well known to be a major predictive factor of nanoparticles for their biodistribution and excretion (Almeida et al. 2011; Khan et al. 2008; McNeil 2009; Tomalia 2009). With the final goal being the *in vivo* or clinical application, the fine tuning of nanoparticles is quite essential. And additionally, the sizes and size distributions of nanoparticles directly influence the *in vivo* distribution and excretion. With starting from the exact measuring of the sizes or size distributions of nanoparticles, we can speculate or understand well the *in vivo* distribution and excretion of nanoparticles.

The sizes and size distributions can be measured by several commercially available products using different kinds of methods, such as dynamic light scattering (DLS), nanoparticle tracking analysis (NTA), differential centrifugal sedimentation (DCS), size-exclusion chromatography (SEC), transmission electron microscopy (TEM), etc. Each method has pros and cons; we have to select a proper method for each nanoparticle to be measured. TEM directly shows the particle images, which can be used for the size of particle core and except for TEM, other methods can measure the hydrodynamic size of nanoparticles. Recently, one research group proposed a rapid and reliable method for determining the size and size distribution of iron oxide particles using matrix-assisted laser desorption/ionization time of flight (MALDI-TOF) mass spectrometry (MS) (Kim et al. 2013).

Larger-sized nanomaterials are not excreted through the kidneys while the nanomaterials less than the diameter of 5 nm are easily excreted through the kidneys (Choi et al. 2007; McNeil 2009; Tomalia 2009). The shape of nanoparticle was another factor to facilitate the renal filtration while the rod-shaped nanomaterials can pass through the glomerular basement membrane as if pricking the membranes (Ruggiero et al. 2010).

Nanoparticles of the size between 10 nm and 100 nm are considered to pass through the fenestra (windows) of the endothelial barriers to reach the tumor cells or to reach the hepatocytes to be excreted via the hepatobiliary route (Fig. 9.8). Enhanced Permeability and Retention (EPR) was proposed to take up nanoparticles in the tumor tissues and this was the passive targeting. And also the fenestra of the sinusoidal membranes pass the nanomaterials of the size between 10 nm and 100 nm and endocytose and exocytose the nanomaterials. However, in order to realize the proper excretion of the nanomaterials, the nanomaterials should already have been hydrophilized, such as PEGylation, to remove the aggregation factor *in vivo*.

Nanoparticles with the size of 100 nm or more are prone to be captured by the mononuclear phagocytes system (MPS) in the body, especially by the RES in liver. These are considered to be the characteristics of nanomaterials mimicking the viral particles or the debris of bacteria or fungi. Mostly the nanomaterials of this size are PEGylated with the hope of being undetected for an appropriately long time to reach the target without being excreted via hepatobiliary route or captured by the MPS or RES.

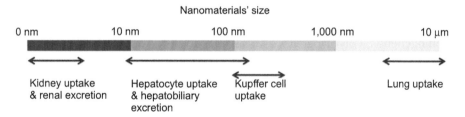

Fig. 9.8 The proposed *in vivo* distribution and excretion windows of nanomaterials.

Once the nanoparticles of the size over 10 nm are administered to the body systemically, they are supposed not to be excreted via kidneys. The examination of the circulation time and the measurement of the concentration of the metal contents after incineration of the mouse body organs using ICP-MS (inductively coupled plasma-mass spectrometry), were not sufficient to give insight about the final fate of the nanoparticles within the body and outside too.

Radionanomaterials could be examined for their biodistribution exactly regarding their transient and final fate of persistent retention and excretion from the body. PET/CT or SPECT/CT dedicated to mouse or rat work well for this purpose (Liu et al. 2014; Sun et al. 2014; Wang et al. 2012; Yang et al. 2013b). The ambiguity regarding renal excretion, hepatobiliary excretion and transient/prolonged/permanent persistence are going to be elucidated soon once the investigators adopt long half-life radionuclide, multiplex-modified radionanomaterials and microPET/CT or microSPECT/CT for this investigation (Rangger et al. 2013; Souris et al. 2010; Xiao et al. 2012; Xie et al. 2010).

Monodispersity or Polydispersity

Radiolabeled nanomaterials can be used for imaging with PET or SPECT to visualize the biodistribution of those *in vivo*. From these PET or SPECT images in literatures, we can easily find the radiolabeled nanoparticles distributed to the kidneys and the liver, simultaneously and sometimes even in the lung. As mentioned before, the kidney uptake came from 1~5 nm sized nanoparticles, the liver uptake from 10~100 nm sized nanoparticles for hepatocytes and more than 100 nm sized nanoparticles for RES. The lung uptake came from the *in vivo* aggregation of nanoparticles, which have higher than 5 μm size. To prevent these kinds of phenomenon, they should check the dispersity before/after the surface modification and before/after the radiolabeling.

Dispersity of the nanoparticles is evaluated with the same idea to measure the dispersity or heterogeneity of the polymer size or chain length. Polydispersity are measured by polydispersity index (PDI) defined by 'mass average molar mass divided by number average molar mass' (Gilbert et al. 2009). This index is 1 (unity) when the mass distribution is perfectly homogeneous. Polydispersity is measured by either DLS or NTA methods. The NTA methods are preferred as NTA traces all the particles using laser and generates the curve of the mass distribution over the size of the nanomaterials.

Finely tuned nanoparticles can have a monodispersity, which is quite difficult to achieve and is obtained from almost mono-sized core synthesis and mono-dispersed nanoparticles after the surface modification. Few labs have reached the state-of-the-art stage of nanoparticle handling procedure for this monodispersity issue in the world.

Finding the Fate to Choose Between Kupffer Cells vs Hepatocytes

Recently the size distribution of the nanomaterials can be examined exactly after all the preparation, just before injection into the body. The PEGylation helped nanoparticles keep hydrophilicity and to not be aggregated after systemic administration (Jokerst et al. 2011b; Joralemon et al. 2010; Karakoti et al. 2011; Pasut and Veronese 2012) and the fate of the radionanomaterials or un-radiolabeled (bare) nanomaterials is expected to choose between two choices of clearance from the body. We now know that the aggregated nanomaterials, if their size reaches twice, three times or more, are taken up by the Kupffer cells, however, if their size is maintained after injection despite the wrapping-up by soft or hard corona proteins (Kelly et al. 2015; Milani et al. 2012; Monopoli et al. 2012), these nanomaterials might choose to be endocytosed by the hepatocytes. Kupffer cells seem to keep the particles for a long time and will finally process the foreign materials like other living invaders such as bacteria and viruses but hepatocytes will choose between the excretion and recycling the materials within or encapsulates over the materials. If we monitor the fates until days or even weeks using ^{64}Cu (half-life: 12.7 h) or ^{89}Zr (half-life: 3.3 d), respectively we can say that the short-term fate of administered radio or bare nanomaterials are so and so.

Nano-Periodic Table

When the investigators chose dendrimers as their nanoplatform, they gave several advantages for examination of the size effects. The dendrimers of varying size with uniform dispersity were available easily and the spaces within the dendrimers were ample enough to contain enough amounts of metals such as Gadolinium (Tomalia 2009). The size effect over the biodistribution was once clearly reported using Gd-PAMAM and was concordant with the findings obtained using gold or silica nanoparticles (Khan et al. 2008). Smaller particles were excreted via kidneys and larger particles via liver. Much larger particles are expected to clog the pulmonary vasculature and lung uptake will clearly show these characteristics. Starting from this idea, we can expand and adopt other nanomaterials and evaluate the size effect over biodistribution using radiolabeled nanomaterials. Finally, we can establish the nano-periodic table from biodistribution images and the quantification data of each radiolabeled nanomaterial.

Particle Quantification

Through the recent introduction and ready availability of NTA whose commercial product is called NanoSight®, we can measure and quantify the nanomaterials (Fig. 9.9). This enabled the measuring of the number of nanomaterials and by just looking at the proportion of the materials with any size, we can even say that how many particles of such size are there in the specimen. This calculated number of nanomaterials can now be used for estimating how many ligands, antibodies or chelators are on the surface of the nanoparticles. As we already know the molar amount of ligands or others, we can design the sparseness or denseness of the ligands on the surface of nanomaterials. The radionuclide-binding specific activity of radionanomaterials can also be estimated.

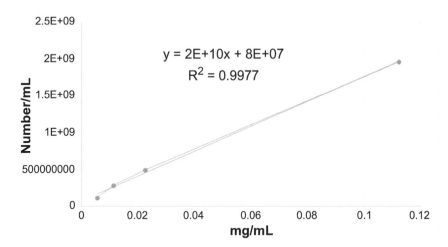

Fig. 9.9 Particle number measured by NTA method using PEGylated iron oxide nanoparticles.

The total amount of milligrams of nanoparticles is transformed to molar propensity and the molar propensity of composed amphiphile of ligands/chelators/PEG (or Tween®) will be used to understand the composition of the final nanomaterials and after another measurement of the number of the nanomaterials and measurement of the radioactivity of the radionuclides, we now know the amount of radionanomaterials, the number of ligands and other molecules per particle and the average number of radionuclides per particle. If the specific activity of radionanomaterials leads to the situation that each nanomaterial is bound to each radionuclide, we can say that we are now using trace amount of radionanomaterials.

This might be one of the possible breakthroughs for minimizing the concerns about toxicity *in vivo*. Trace amount of a substance does not cause any concern regarding the toxicity which was acquired from pharmacologic effect and side effects of this substance when administered systemically.

Quality Control

In vitro during and after preparation of the nanomaterials or radionanomaterials, at first we have to be concerned about the precipitation and aggregation which are quite visible on visual inspection. The second thing is the shelf-life of the authenticity of the nanomaterials after weeks and months. The recently introduced NTA should be used for the maintenance of the original composition and size.

The charge of nanoparticles can be measured by Zeta potential, which will also define the possible characteristics of these nanoparticles which mostly depend upon the modified surface. This will also be maintained in the aqueous media.

Monodispersity is desired to sustain and polydispersity index shall be measured during and after the production and compounding. After compounding, such as after mixing/vortexing and separation steps in micelle-encapsulation procedures, one should check the polydispersity again. After the radiolabeling, the radiochemical purity and also the polydispersity and zeta potential shall be measured again.

Before using the final radiolabeled product, the information should be given regarding the batch, number of particles, amount of radioactivity, the size and size distributions on NTA, zeta potential and its history during and after manufacturing/compounding, radiochemical purity and also the specific activity of the radionanomaterials. The issues of sterility and impurity should also be taken care of according to any pharmaceutical rules and regulations.

Dosing Effect

Dosing effect was once raised to be a factor for the *in vivo* side effects when we try to use nanomaterials as carriers or delivery vehicles to transport the cargos to the targets (mostly tumors). When we tried to use luminescence imaging of UCNP and PET imaging of surface labeled ^{64}Cu, we needed to increase the amount of radionanomaterials more than ten times in terms of the amount of nanomaterials. However, in terms of the representativeness of radiolabeled nanomaterials over bare nanomaterials, there have been no reports in the literature that the dosing effect will influence the measured (observed) behavior of radionanomaterials. As is well expected from the knowledge in the field of nuclear medicine, radiolabeled nanomaterial behaves the same *in vivo* and represents the behavior of the 'mother' substance of the nanomaterial. The only concern is that as the stability of radiolabeling was not good and the radionuclides themselves or radionuclide-chelator complex or even radionuclide-containing capsules are detached and could give the wrong information regarding *in vivo* behavior. This happened in one report (Panagi et al. 2001) but not in any others. This possibility should be paid sincere attention to by every investigator whenever they try to assume the observed behavior *in vivo* of radionanomaterials representing exactly the behavior of the un-radiolabeled nanomaterials.

In Vitro **Application**

Hybrid nanomaterials had shown the most promise in *in vitro* application while the variety of the ligands determined the specificity. The most elementary ones are the ones to process the images of the biochip which grew bacteria in the body fluids drawn from the patients with or without antibiotics (Choi et al. 2014). The combination of biochip, microfluidics and nano-surface technologies enabled the recognition of whether the microorganisms are proliferating in segmental fibrillary form or not. Antibiotic susceptibility results were produced within hours instead of 2 or 7 days according to the classic standard medium.

If the ligands consisted of the nucleic acid, the nucleotide sequence could yield the diversity of the representative targets, which has been called DNA barcodes (Lee et al. 2014; Nam et al. 2004; Nam et al. 2003). The hydrogen bonds of sizable sequence of nucleic acid could be used to find the complementary sequences and the microarray of these sequence-labeled nanoparticles will emit the signals when the target nucleic acids are encountered. The readout of the signals can be done automatically and if the signal is in the visual range, the test kits can become point-of-care test. Multiplexing and point-of-care capability could be the caveat to enhance the possibility of using these technologies to be used clinically. Proving the superiority or non-inferiority and incremental values in clinical terms is the mainstay of the clinical translation while no toxicity issues are problems for *in vitro* diagnostics.

The holy grail of the use of hybrid nanomaterials resides in tracing the biological processes in live cells *in vitro*. Quantum dots, organic-dye labeled silica or gold nanoparticles and quantum dot squares are not appropriate for this purpose. Excitation lasers deliver too high energy, while repeated imaging will heat, bake and kill the cells. Instead, UCNP composed by yttrium (Y) doped with erbium (Er) and ytterbium (Yb) will respond to the near-infrared light and excitation laser of 980 nm wavelength which will not harm or damage the object cells (Nam et al. 2011). Live cell imaging could disclose the movement of UCNPs within the HeLa cells *in vitro* and shed the light upon the *in vitro* live cell imaging (Bae et al. 2012). If we bind appropriate targeting ligands over the surface of these UCNPs, we can monitor the intracytoplasmic biological processes. These ligands include nucleic acids, peptides and other synthetic macromolecules, which can form the 3 dimensional structures. Another UCNP was proposed as the better candidate than the UCNP mentioned above. This UCNP contains neodymium (Nd) instead of Er/Yb and the excitation laser will be 800 nm (Li et al. 2012). As the water absorption is much less with Nd UCNP than Er/Yb UCNP, the investigators suggested least damage to the object cells while they are imaged alive *in vitro*.

In Vivo **Application**

Targeted delivery and imaging is the best-hoped application of nanomaterials *in vivo*. For *in vivo* application, we have to use the finely tuned nanoparticles, from which we can speculate or understand well the *in vivo* distribution and excretion. Finely tuned nanoparticles could maintain their sizes and size distributions as well as the integrity of targeting ligands. And furthermore, we have to clearly show the qualification and quantification data of each nanoparticle throughout the whole chemical process including radiolabeling.

To maintain the quality of nanoparticles for *in vivo* application, we should carefully consider and choose the proper surface modification method after the core synthesis. One-pot micelle-encapsulation method will be the solution for this concern and help maintaining the uniformity and the integrity of the surface of nanoparticles.

Recently, the cupper-free click chemistry (Baranyai et al. 2015; Zeng et al. 2012) was proposed and adopted in nanotechnology for *in vivo* radiolabeling using bioorthogonal reaction, which has proven to be reliable in the field of chemistry. This cupper-free click chemistry is carried out in the very mild reaction condition without any catalyst and also helps in maintaining the uniformity of nanoparticles for the surface modification. Clickable nanomaterials made from the micelle-encapsulation method can be used for further surface tagging, such as the targeting ligands such as antibodies (Alt et al. 2015; Kumar et al. 2015) and aptamers (Oishi et al. 2014; Zhang et al. 2013) or therapeutic drugs and even for radiolabeling both *in vitro* and *in vivo*.

Therefore, combining together the micelle-encapsulation method along with the cupper-free click chemistry for nano-surface modification, we could get the real meaning of a versatile nano-platform for clinical application. This new concept of nano-platform can be used in personalized medicine, which is the core concept of precision medicine and the clinical adaptability to the nanoparticles will be highly increased.

Even though we could solve the quality issue of nanomaterials, the toxicity issue still remains for *in vivo* application. The study on the toxicity of nanomaterials has been and is still being carried out by several research groups, however, it is still not clear whether the nanomaterials themselves, or the high dosing effect, surface-bound small or macromolecules, immune/inflammatory response to these surface-bound ligands, hydrophobic/hydrophilic surface characteristics or surface charges, are the critical determinants of the biomedical hazard to the recipient (Jenkins et al. 2013; Johnston et al. 2010). For nano-toxicity, radionanomedicine will be a solution, which has the capability of dose reduction of nanomaterials administered *in vivo* and can also help them to expect and prevent the damaged organs from the understanding of the exact biodistribution of nanomaterials.

Conclusion

A large number of hybrid nanomaterials have been introduced and used for *in vitro* or *in vivo* application in the biological or medical field. Especially for *in vivo* application, radionanomedicine merged with hybrid nanomaterials can eliminate the concerns about the possible hazards of pharmacologic amount of nanomaterials administered and the trace technology behind radionanomedicine will help the progress of the translation of nanomedicine. In addition, the qualification or quantification is another key factor for *in vivo* application of hybrid nanomaterials, thus, the dosimetric analysis using radionanomedicine can help the exact qualification and quantification of hybrid nanomaterials. Finally, we hope that radionanomedicine along with hybrid nanomaterials can help people who struggle with different diseases.

Acknowledgments

This chapter was funded by National Research Foundation of Korea (2014M2A2A7045043 and 2012R1A1A2008799).

References

Alivisatos, A.P., Gu, W.W. and Larabell, C. 2005. Quantum dots as cellular probes. Annu Rev Biomed Eng 7: 55-76.

Almeida, J.P., Chen, A.L., Foster, A. and Drezek, R. 2011. In vivo biodistribution of nanoparticles. Nanomedicine (Lond) 6: 815-835.

Alt, K., Paterson, B.M., Westein, E., Rudd, S.E., Poniger, S.S., Jagdale, S., et al. 2015. A versatile approach for the site-specific modification of recombinant antibodies using a combination of enzyme-mediated bioconjugation and click chemistry. Angew Chem Int Edit 54: 7515-7519.

Amoozgar, Z. and Yeo, Y. 2012. Recent advances in stealth coating of nanoparticle drug delivery systems. Wiley Interdiscip Rev Nanomed Nanobiotechnol 4: 219-233.

Anderson, C.J. and Ferdani, R. 2009. Copper-64 radiopharmaceuticals for PET imaging of cancer: advances in preclinical and clinical research. Cancer Biother Radiopharm 24: 379-393.

Bae, Y.M., Park, Y.I., Nam, S.H., Kim, J.H., Lee, K., Kim, H.M., et al. 2012. Endocytosis, intracellular transport and exocytosis of lanthanide-doped upconverting nanoparticles in single living cells. Biomaterials 33: 9080-9086.

Baranyai, Z., Reich, D., Vagner, A., Weineisen, M., Toth, I., Wester, H.J., et al. 2015. A shortcut to high-affinity Ga-68 and Cu-64 radiopharmaceuticals: one-pot click chemistry trimerisation on the TRAP platform. Dalton Trans 44: 11137-11146.

Bouziotis, P., Psimadas, D., Tsotakos, T., Stamopoulos, D. and Tsoukalas, C. 2012. Radiolabeled iron oxide nanoparticles as dual-modality SPECT/MRI and PET/MRI agents. Curr Top Med Chem 12: 2694-2702.

Carion, O., Mahler, B., Pons, T. and Dubertret, B. 2007. Synthesis, encapsulation, purification and coupling of single quantum dots in phospholipid micelles for their use in cellular and in vivo imaging. Nat Protoc 2: 2383-2390.

Chang, P.V., Prescher, J.A., Sletten, E.M., Baskin, J.M., Miller, I.A., Agard, N.J., et al. 2010. Copper-free click chemistry in living animals. Proc Natl Acad Sci USA 107: 1821-1826.

Choi, H.S., Liu, W., Misra, P., Tanaka, E., Zimmer, J.P., Itty Ipe, B., et al. 2007. Renal clearance of quantum dots. Nat Biotechnol 25: 1165-1170.

Choi, J., Yoo, J., Lee, M., Kim, E.G., Lee, J.S., Lee, S., et al. 2014. A rapid antimicrobial susceptibility test based on single-cell morphological analysis. Sci Transl Med 6: 267ra174.

DaCosta, M.V., Doughan, S., Han, Y. and Krull, U.J. 2014. Lanthanide upconversion nanoparticles and applications in bioassays and bioimaging: a review. Anal Chim Acta 832: 1-33.

Dubertret, B., Skourides, P., Norris, D.J., Noireaux, V., Brivanlou, A.H. and Libchaber, A. 2002. In vivo imaging of quantum dots encapsulated in phospholipid micelles. Science 298: 1759-1762.

Fan, H., Yang, K., Boye, D.M., Sigmon, T., Malloy, K.J., Xu, H., et al. 2004. Self-assembly of ordered, robust, three-dimensional gold nanocrystal/silica arrays. Science 304: 567-571.

Fischer, G., Seibold, U., Schirrmacher, R., Wangler, B. and Wangler, C. 2013. (89)Zr, a radiometal nuclide with high potential for molecular imaging with PET: chemistry, applications and remaining challenges. Molecules 18: 6469-6490.

Garai, E., Sensarn, S., Zavaleta, C.L., Loewke, N.O., Rogalla, S., Mandella, M.J., et al. 2015. A real-time clinical endoscopic system for intraluminal, multiplexed imaging of surface-enhanced Raman scattering nanoparticles. PLoS One 10: e0123185.

Garcia, J., Tang, T. and Louie, A.Y. 2015. Nanoparticle-based multimodal PET/MRI probes. Nanomedicine (Lond) 10: 1343-1359.

Geissler, D., Linden, S., Liermann, K., Wegner, K.D., Charbonniere, L.J. and Hildebrandt, N. 2014. Lanthanides and quantum dots as forster resonance energy transfer agents for diagnostics and cellular imaging. Inorg Chem 53: 1824-1838.

Gilbert, R.G., Hess, M., Jenkins, A.D., Jones, R.G., Kratochvil, R. and Stepto, R.F.T. 2009. Dispersity in polymer science (IUPAC Recommendations 2009) (vol 81, pg 351, 2009). Pure Appl Chem 81: 779-779.

Goel, S., Chen, F., Ehlerding, E.B. and Cai, W. 2014. Intrinsically radiolabeled nanoparticles: an emerging paradigm. Small 10: 3825-3830.

Gordon, C.G., Mackey, J.L., Jewett, J.C., Sletten, E.M., Houk, K.N. and Bertozzi, C.R. 2012. Reactivity of biarylazacyclooctynones in copper-free click chemistry. J Am Chem Soc 134: 9199-9208.

Jenkins, J.T., Halaney, D.L., Sokolov, K.V., Ma, L.L., Shipley, H.J., Mahajan, S., et al. 2013. Excretion and toxicity of gold-iron nanoparticles. Nanomedicine 9: 356-365.

Jeong, S., Kim, Y.I., Kang, H., Kim, G., Cha, M.G., Chang, H., et al. 2015. Fluorescence-Raman dual modal endoscopic system for multiplexed molecular diagnostics. Sci Rep-Uk 5: 9455.

Jimenez-Mancilla, N., Ferro-Flores, G., Santos-Cuevas, C., Ocampo-Garcia, B., Luna-Gutierrez, M., Azorin-Vega, E., et al. 2013. Multifunctional targeted therapy system based on (99m) Tc/(177) Lu-labeled gold nanoparticles-Tat (49-57)-Lys(3) -bombesin internalized in nuclei of prostate cancer cells. J Labelled Comp Radiopharm 56: 663-671.

Johnston, H.J., Hutchison, G., Christensen, F.M., Peters, S., Hankin, S. and Stone, V. 2010. A review of the in vivo and in vitro toxicity of silver and gold particulates: particle attributes and biological mechanisms responsible for the observed toxicity. Crit Rev Toxicol 40: 328-346.

Jokerst, J.V., Lobovkina, T., Zare, R.N. and Gambhir, S.S. 2011a. Nanoparticle PEGylation for imaging and therapy. Nanomedicine (Lond) 6: 715-728.

Jokerst, J.V., Miao, Z., Zavaleta, C., Cheng, Z. and Gambhir, S.S. 2011b. Affibody-functionalized gold-silica nanoparticles for Raman molecular imaging of the epidermal growth factor receptor. Small 7: 625-633.

Joralemon, M.J., McRae, S. and Emrick, T. 2010. PEGylated polymers for medicine: from conjugation to self-assembled systems. Chem Commun (Camb) 46: 1377-1393.

Jun, B.H., Noh, M.S., Kim, J., Kim, G., Kang, H., Kim, M.S., et al. 2010. Multifunctional silver-embedded magnetic nanoparticles as SERS nanoprobes and their applications. Small 6: 119-125.

Jun, B.H., Kim, G., Noh, M.S., Kang, H., Kim, Y.K., Cho, M.H., et al. 2011. Surface-enhanced Raman scattering-active nanostructures and strategies for bioassays. Nanomedicine (Lond) 6: 1463-1480.

Jun, B.H., Hwang, D.W., Jung, H.S., Jang, J., Kim, H., Kang, H., et al. 2012a. Ultrasensitive, biocompatible, quantum-dot-embedded silica nanoparticles for bioimaging. Adv Funct Mater 22: 1843-1849.

Jun, B.H., Kang, H., Lee, Y.S. and Jeong, D.H. 2012b. Fluorescence-based multiplex protein detection using optically encoded microbeads. Molecules 17: 2474-2490.

Kam, B.L., Teunissen, J.J., Krenning, E.P., de Herder, W.W., Khan, S., van Vliet, E.I., et al. 2012. Lutetium-labelled peptides for therapy of neuroendocrine tumours. Eur J Nucl Med Mol Imaging 39 Suppl 1: S103-112.

Kango, S., Kalia, S., Celli, A., Njuguna, J., Habibi, Y. and Kumar, R. 2013. Surface modification of inorganic nanoparticles for development of organic-inorganic nanocomposites-A review. Prog Polym Sci 38: 1232-1261.

Karakoti, A.S., Das, S., Thevuthasan, S. and Seal, S. 2011. PEGylated inorganic nanoparticles. Angew Chem Int Edit 50: 1980-1994.

Kelly, P.M., Aberg, C., Polo, E., O'Connell, A., Cookman, J., Fallon, J., et al. 2015. Mapping protein binding sites on the biomolecular corona of nanoparticles. Nat Nanotechnol 10: 472-479.

Khan, M.K., Minc, L.D., Nigavekar, S.S., Kariapper, M.S., Nair, B.M., Schipper, M., et al. 2008. Fabrication of {198Au0} radioactive composite nanodevices and their use for nanobrachytherapy. Nanomedicine 4: 57-69.

Kim, B.H., Shin, K., Kwon, S.G., Jang, Y., Lee, H.S., Lee, H., et al. 2013. Sizing by weighing: characterizing sizes of ultrasmall-sized iron oxide nanocrystals using MALDI-TOF mass spectrometry. J Am Chem Soc 135: 2407-2410.

Kreyling, W.G., Abdelmonem, A.M., Ali, Z., Alves, F., Geiser, M., Haberl, N., et al. 2015. In vivo integrity of polymer-coated gold nanoparticles. Nat Nanotechnol 10: 619-623.

Kumar, A., Hao, G., Liu, L., Ramezani, S., Hsieh, J.T., Oz, O.K., et al. 2015. Click-chemistry strategy for labeling antibodies with copper-64 via a cross-bridged tetraazamacrocyclic chelator scaffold. Bioconjug Chem 26: 782-789.

Lee, D.S., Im, H.J. and Lee, Y.S. 2015a. Radionanomedicine: widened perspectives of molecular theragnosis. Nanomedicine 11: 795-810.

Lee, H., Park, J.E. and Nam, J.M. 2014. Bio-barcode gel assay for microRNA. Nat Commun 5: 3367.

Lee, Y.-S. 2012. Radiopharmaceutical chemistry. pp. 21-51. In: E.E. Kim, D.S. Lee, U. Tateishi and R. Baum [ed.]. Handbook of Nuclear Medicine and Molecular Imaging. World Scientific Publishing Co. Pte. Ltd, NJ, USA.

Lee, Y.-S., Kim, Y.-I., Lee, D.S. 2015b. Future perspectives of the chemistry for radio-nanomedicine. Nucl Med Mol Imaging. (in press).

Lee, Y.K., Jeong, J.M., Hoigebazar, L., Yang, B.Y., Lee, Y.S., Lee, B.C., et al. 2012. Nanoparticles modified by encapsulation of ligands with a long alkyl chain to affect multispecific and multimodal imaging. J Nucl Med 53: 1462-1470.

Li, F., Li, C., Liu, X., Chen, Y., Bai, T., Wang, L., et al. 2012. Hydrophilic, upconvert-ing, multicolor, lanthanide-doped NaGdF4 nanocrystals as potential multifunc-tional bioprobes. Chemistry 18: 11641-11646.

Li, X., Wang, R., Zhang, F., Zhou, L., Shen, D., Yao, C., et al. 2013. Nd3+ sensitized up/down converting dual-mode nanomaterials for efficient in vitro and in vivo bioimaging excited at 800 nm. Sci Rep 3: 3536.

Liu, Y., Sun, Y., Cao, C., Yang, Y., Wu, Y., Ju, D., et al. 2014. Long-term biodistribu-tion in vivo and toxicity of radioactive/magnetic hydroxyapatite nanorods. Biomaterials 35: 3348-3355.

McNeil, S.E. 2009. Nanoparticle therapeutics: a personal perspective. Wiley Interdiscip Rev Nanomed Nanobiotechnol 1: 264-271.

Milani, S., Bombelli, F.B., Pitek, A.S., Dawson, K.A. and Radler, J. 2012. Reversible versus irreversible binding of transferrin to polystyrene nanoparticles: soft and hard corona. ACS Nano 6: 2532-2541.

Misri, R., Saatchi, K. and Hafeli, U.O. 2012. Nanoprobes for hybrid SPECT/MR molecular imaging. Nanomedicine (Lond) 7: 719-733.

Monopoli, M.P., Aberg, C., Salvati, A. and Dawson, K.A. 2012. Biomolecular coro-nas provide the biological identity of nanosized materials. Nat Nanotechnol 7: 779-786.

Nam, J., Won, N., Bang, J., Jin, H., Park, J., Jung, S., et al. 2013. Surface engineering of inorganic nanoparticles for imaging and therapy. Adv Drug Deliv Rev 65: 622-648.

Nam, J.M., Thaxton, C.S. and Mirkin, C.A. 2003. Nanoparticle-based bio-bar codes for the ultrasensitive detection of proteins. Science 301: 1884-1886.

Nam, J.M., Stoeva, S.I. and Mirkin, C.A. 2004. Bio-bar-code-based DNA detection with PCR-like sensitivity. J Am Chem Soc 126: 5932-5933.

Nam, S.H., Bae, Y.M., Park, Y.I., Kim, J.H., Kim, H.M., Choi, J.S., et al. 2011. Long-term real-time tracking of lanthanide ion doped upconverting nanoparticles in living cells. Angew Chem Int Edit 50: 6093-6097.

Noble, G.T., Stefanick, J.F., Ashley, J.D., Kiziltepe, T. and Bilgicer, B. 2014. Ligand-targeted liposome design: challenges and fundamental considerations. Trends Biotechnol 32: 32-45.

Oishi, M., Nakao, S. and Kato, D. 2014. Enzyme-free fluorescent-amplified aptasensors based on target-responsive DNA strand displacement via toehold-mediated click chemical ligation. Chem Commun 50: 991-993.

Panagi, Z., Beletsi, A., Evangelatos, G., Livaniou, E., Ithakissios, D.S. and Avgoustakis, K. 2001. Effect of dose on the biodistribution and pharmacokinetics of PLGA and PLGA-mPEG nanoparticles. Int J Pharm 221: 143-152.

Park, Y.I., Kim, H.M., Kim, J.H., Moon, K.C., Yoo, B., Lee, K.T., et al. 2012. Theranostic probe based on lanthanide-doped nanoparticles for simultaneous in vivo dual-modal imaging and photodynamic therapy. Adv Mater 24: 5755-5761.

Park, Y.I., Lee, K.T., Suh, Y.D. and Hyeon, T. 2015. Upconverting nanoparticles: a versatile platform for wide-field two-photon microscopy and multi-modal in vivo imaging. Chem Soc Rev 44: 1302-1317.

Pasut, G. and Veronese, F.M. 2012. State of the art in PEGylation: the great versatility achieved after forty years of research. J Control Release 161: 461-472.

Petryayeva, E., Algar, W.R. and Medintz, I.L. 2013. Quantum dots in bioanalysis: a review of applications across various platforms for fluorescence spectroscopy and imaging. Appl Spectrosc 67: 215-252.

Rangger, C., Helbok, A., Sosabowski, J., Kremser, C., Koehler, G., Prassl, R., et al. 2013. Tumor targeting and imaging with dual-peptide conjugated multifunctional liposomal nanoparticles. Int J Nanomed 8: 4659-4670.

Ruggiero, A., Villa, C.H., Bander, E., Rey, D.A., Bergkvist, M., Batt, C.A., et al. 2010. Paradoxical glomerular filtration of carbon nanotubes. Proc Natl Acad Sci USA 107: 12369-12374.

Sainz-Esteban, A., Prasad, V., Schuchardt, C., Zachert, C., Carril, J.M. and Baum, R.P. 2012. Comparison of sequential planar 177Lu-DOTA-TATE dosimetry scans with 68Ga-DOTA-TATE PET/CT images in patients with metastasized neuroendocrine tumours undergoing peptide receptor radionuclide therapy. Eur J Nucl Med Mol Imaging 39: 501-511.

Samanta, A., Jana, S., Das, R.K. and Chang, Y.T. 2014. Biocompatible surface-enhanced Raman scattering nanotags for in vivo cancer detection. Nanomedicine (Lond) 9: 523-535.

Sedlmeier, A. and Gorris, H.H. 2015. Surface modification and characterization of photon-upconverting nanoparticles for bioanalytical applications. Chem Soc Rev 44: 1526-1560.

Seo, H.J., Nam, S.H., Im, H.-J., Park, J.Y., Lee, J.Y., Yoo, B., et al. 2015. Rapid hepatobiliary excretion of micelle-encapsulated/radiolabeled upconverting nanoparticles as an integrated form. Sci Rep 5: 15685.

Souris, J.S., Lee, C.H., Cheng, S.H., Chen, C.T., Yang, C.S., Ho, J.A.A., et al. 2010. Surface charge-mediated rapid hepatobiliary excretion of mesoporous silica nanoparticles. Biomaterials 31: 5564-5574.

Sperling, R.A. and Parak, W.J. 2010. Surface modification, functionalization and bio-conjugation of colloidal inorganic nanoparticles. Philos T R Soc A 368: 1333-1383.

Sun, M., Hoffman, D., Sundaresan, G., Yang, L., Lamichhane, N. and Zweit, J. 2012. Synthesis and characterization of intrinsically radiolabeled quantum dots for bimodal detection. Am J Nucl Med Mol Imaging 2: 122-135.

Sun, X., Huang, X., Yan, X., Wang, Y., Guo, J., Jacobson, O., et al. 2014. Chelator-free (64)Cu-integrated gold nanomaterials for positron emission tomography imaging guided photothermal cancer therapy. ACS Nano 8: 8438-8446.

Tomalia, D.A. 2009. In quest of a systematic framework for unifying and defining nanoscience. J Nanopart Res 11: 1251-1310.

Velikyan, I. 2013. Prospective of (6)(8)Ga-radiopharmaceutical development. Theranostics 4: 47-80.

Vendrell, M., Maiti, K.K., Dhaliwal, K. and Chang, Y.T. 2013. Surface-enhanced Raman scattering in cancer detection and imaging. Trends Biotechnol 31: 249-257.

Wadas, T.J., Wong, E.H., Weisman, G.R. and Anderson, C.J. 2010. Coordinating radio-metals of copper, gallium, indium, yttrium and zirconium for PET and SPECT imaging of disease. Chem Rev 110: 2858-2902.

Wang, Y., Liu, Y., Luehmann, H., Xia, X., Brown, P., Jarreau, C., et al. 2012. Evaluating the pharmacokinetics and in vivo cancer targeting capability of Au nanocages by positron emission tomography imaging. ACS Nano 6: 5880-5888.

Wu, H., Zhu, H., Zhuang, J., Yang, S., Liu, C. and Cao, Y.C. 2008. Water-soluble nano-crystals through dual-interaction ligands. Angew Chem Int Edit 47: 3730-3734.

Xiao, Y.L., Hong, H., Javadi, A., Engle, J.W., Xu, W.J., Yang, Y.A., et al. 2012. Multifunctional unimolecular micelles for cancer-targeted drug delivery and positron emission tomography imaging. Biomaterials 33: 3071-3082.

Xie, J., Chen, K., Huang, J., Lee, S., Wang, J., Gao, J., et al. 2010. PET/NIRF/MRI triple functional iron oxide nanoparticles. Biomaterials 31: 3016-3022.

Xing, Y., Zhao, J., Conti, P.S. and Chen, K. 2014. Radiolabeled nanoparticles for multi-modality tumor imaging. Theranostics 4: 290-306.

Yang, B.Y., Seelam, S.R., Young, I.K., Lee, Y.S., Lee, D.S., Chung, J.K., et al. 2013a. Preparation of a multimodal iron oxide nanoparticle conjugated with mannose, NOTA and PEG by encapsulation with specific amphiphiles. J Labelled Compd Rad 56: S63-S63.

Yang, L.K., Sundaresan, G., Sun, M.H., Jose, P., Hoffman, D., McDonagh, P.R., et al. 2013b. Intrinsically radiolabeled multifunctional cerium oxide nanoparticles for in vivo studies. J Mater Chem B 1: 1421-1431.

Yang, Y., Velmurugan, B., Liu, X. and Xing, B. 2013c. NIR photoresponsive crosslinked upconverting nanocarriers toward selective intracellular drug release. Small 9: 2937-2944.

Yin, C., Hong, B., Gong, Z., Zhao, H., Hu, W., Lu, X., et al. 2015. Fluorescent oligo(p-phenyleneethynylene) contained amphiphiles-encapsulated magnetic nanoparticles for targeted magnetic resonance and two-photon optical imaging in vitro and in vivo. Nanoscale 7: 8907-8919.

Yoo, R.E., Choi, S.H., Cho, H.R., Jeon, B.S., Kwon, E., Kim, E.G., et al. 2014. Magnetic resonance imaging diagnosis of metastatic lymph nodes in a rabbit model: efficacy of PJY10, a new ultrasmall superparamagnetic iron oxide agent, with monodisperse iron oxide core and multiple-interaction ligands. PLoS One 9: e107583.

Zavaleta, C.L., Smith, B.R., Walton, I., Doering, W., Davis, G., Shojaei, B., et al. 2009. Multiplexed imaging of surface enhanced Raman scattering nanotags in living mice using noninvasive Raman spectroscopy. Proc Natl Acad Sci USA 106: 13511-13516.

Zavaleta, C.L., Garai, E., Liu, J.T., Sensarn, S., Mandella, M.J., Van de Sompel, D., et al. 2013. A Raman-based endoscopic strategy for multiplexed molecular imaging. Proc Natl Acad Sci USA 110: e2288-2297.

Zeng, D., Lee, N.S., Liu, Y., Zhou, D., Dence, C.S., Wooley, K.L., et al. 2012. 64Cu core-labeled nanoparticles with high specific activity via metal-free click chemistry. ACS Nano 6: 5209-5219.

Zhang, H., Feng, G., Guo, Y. and Zhou, D. 2013. Robust and specific ratiometric biosensing using a copper-free clicked quantum dot-DNA aptamer sensor. Nanoscale 5: 10307-10315.

Zhang, Y., Hong, H., Myklejord, D.V. and Cai, W. 2011. Molecular imaging with SERS-active nanoparticles. Small 7: 3261-3269.

Zhao, Y., Sultan, D., Detering, L., Cho, S., Sun, G., Pierce, R., et al. 2014. Copper-64-alloyed gold nanoparticles for cancer imaging: improved radiolabel stability and diagnostic accuracy. Angew Chem Int Edit 53: 156-159.

Zhong, Y.A., Meng, F.H., Deng, C. and Zhong, Z.Y. 2014. Ligand-directed active tumor-targeting polymeric nanoparticles for cancer chemotherapy. Biomacromolecules 15: 1955-1969.

Zhou, M., Zhang, R., Huang, M., Lu, W., Song, S., Melancon, M.P., et al. 2010. A chelator-free multifunctional [64Cu]CuS nanoparticle platform for simultaneous micro-PET/CT imaging and photothermal ablation therapy. J Am Chem Soc 132: 15351-15358.

10

Radionuclide Generators for Biomedical Applications: Advent of Nanotechnology

Rubel Chakravarty,* Ramu Ram[a] *and Ashutosh Dash*[b]

INTRODUCTION

Over the last 5 decades, radionuclide generators have attracted substantial attention, indeed even the close scrutiny of the nuclear medicine community owing to their ability to provide short-lived radioisotopes without the need for on-site nuclear reactor or accelerator facilities for the preparation of a myriad of diagnostic and therapeutic radiopharmaceuticals (Boschi et al. 2014; Chakravarty and Dash 2012a; Dash and Chakravarty 2014; Knapp and Baum 2012; Knapp and Mirzadeh 1994; Mushtaq 2004; Pillai et al. 2012). Not only do they provide no-carrier-added (NCA) radionuclides on-demand basis, but also in a cost-effective way where the payoff of benefits is substantial and invaluable (Boschi et al. 2014; Chakravarty and Dash 2012a; Dash and Chakravarty 2014; Knapp and Baum 2012; Mirzadeh and Knapp Jr 1996; Mushtaq 2004; Pillai et al. 2012). Nuclear medicine and radionuclide generator feed off of one another, propelling both forward. Utility of radionuclide generators has virtually pervaded most areas of activities in the field of nuclear medicine and their importance has been well demonstrated and recognized (Boschi et al. 2014; Chakravarty and Dash 2012a; Dash and Chakravarty 2014; Mushtaq 2004; Pillai et al. 2012). In fact, radionuclide generator technology has catalyzed the progress of nuclear medicine practice and has driven the field a quantum leap forward. The current importance and success of diagnostic imaging in nuclear medicine is

Isotope Production and Applications Division, Bhabha Atomic Research Centre, Trombay, Mumbai 400 085, India.
[a] E-mail: ramuram@barc.gov.in
[b] E-mail: adash@barc.gov.in
* Corresponding author: rubelc@barc.gov.in

primarily due to $^{99}Mo/^{99m}Tc$ generator (Eckelman 2009). A large number of the nuclear medicine procedures in remote areas far from the site of a cyclotron or reactor facility would not have been possible but for the availability of this radionuclide generator (Eckelman 2009).

The remarkable prospects associated with radionuclide generators in nuclear medicine along with the challenge of providing daughter radionuclides of requisite quality have led to a considerable amount of captivating research for the development of innovative separation strategies (Chakravarty and Dash 2012a; Chakravarty et al. 2012a; Dash and Chakravarty 2014; Dash and Knapp 2015; Dash et al. 2013; Knapp and Mirzadeh 1994). Spurred by the compelling need to obtain the daughter radionuclide in a chemical form of requisite purity amenable for formulation of radiopharmaceuticals, every conceivable separation strategy has been adequately exploited (Dash and Chakravarty 2014). Among the various separation methods that have been utilized for the development of radionuclide generators, the column chromatographic technique has dominated the field significantly due to its operational simplicity, flexibility and versatility (Chakravarty et al. 2012c; Dash and Chakravarty 2014; Mushtaq 2004). The survival, strength and success of the column chromatographic separation technique are driven by the sorbent, which remains the backbone of the process and sequesters the daughter radionuclide of interest in a desired chemical form. While the column chromatography separation technique represents a successful paradigm for developing radionuclide generators, the limited capacity of sorbents emerged as the main impediment which necessitated the use of high specific activity parent radioisotopes (Chakravarty and Dash 2013b, 2014; Chakravarty et al. 2012c). Since the production of high specific activity radioisotopes is not always economically viable, the prospect of using high capacity sorbent material was considered essential in order to enhance the scope of the column chromatographic technique in radionuclide generator technology (Chakravarty and Dash 2013b; Dash and Chakravarty 2014).

A great deal of effort has been expended in both academia and industries in an attempt to develop high capacity sorbents that satisfy the emerging needs of radionuclide generator technology (Dash and Chakravarty 2014; Mushtaq 2004). Although arduous efforts in this direction have culminated in the development of a wide variety of exotic high capacity sorbents, the majority of such overtures turned out to be futile as most of them exhibited only incremental improvements due to the limitations posed by the density of surface active sites, the activation energy of adsorptive bonds and the mass transfer rate to the sorbent surface (Chakravarty and Dash 2013b; Mushtaq 2004). Subsequently, there appeared to be enticing interest to consider the prospect of using nanomaterial-based sorbents (nanosorbents) in radionuclide generators as they differ from micron-sized and bulk materials not only in the scale of their characteristic dimensions, but also possess unique physical properties which offer new possibilities for developing high capacity sorbents (Fig. 10.1). This paradigm-changing approach using nanosorbents is poised to take column chromatographic generator technology to an exciting new stage.

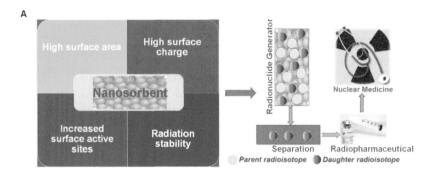

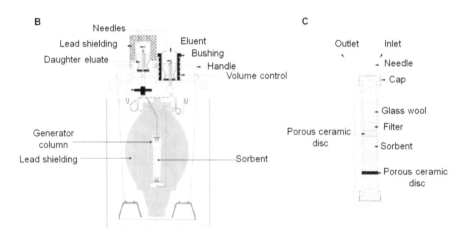

Fig. 10.1 Column chromatographic radionuclide generators using nanosorbents. (A) Utilization of nanosorbents for radiochemical separation of the daughter radioisotope for use in nuclear medicine. (B) Schematic diagram of a column chromatographic radionuclide generator. Adapted from (Dash and Chakravarty 2014) with permission from Royal Society of Chemistry. (C) Chromatographic column of a typical radionuclide generator. Adapted from (Dash and Chakravarty 2014) with permission from Royal Society of Chemistry.

To provide a comprehensive overview of the role of nanosorbents in the development of column chromatographic radionuclide generators, this chapter begins with a brief introduction to radionuclide generator principles and discusses the role of nanosorbents in the preparation of column chromatographic radionuclide generators. The present status of nanomaterial-based radionuclide generators is highlighted and the great potential and intriguing opportunities for future development which might aid in bringing this exciting research avenue closer to widespread clinical acceptance are discussed.

Concept of Radionuclide Generator

A radionuclide generator is a self-contained system housing an equilibrium mixture of a parent/daughter radionuclide pair (Knapp and Baum 2012; Knapp and Mirzadeh 1994). The system is designed to separate the daughter radionuclide formed by the decay of a parent radionuclide by virtue of the differences in their chemical properties (Knapp and Baum 2012) (Fig. 10.1). The parent-daughter nuclear relationships offer the possibility to separate the daughter radionuclide at suitable time intervals (Fig. 10.2). Before discussing the role of nanomaterials in radionuclide generator technology, it is important to throw some light on the basic principles of radionuclide generators, parent–daughter equilibrium and the intimate relationship that exists between them. A list of some medically important radionuclide generator systems is given in Table 10.1, along with the decay characteristics of the parent and daughter radionuclides.

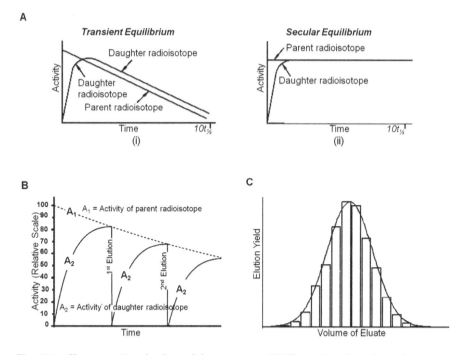

Fig. 10.2 Characteristics of radionuclide generators. (A) Illustration of secular and transient equilibrium. (i) Transient equilibrium is a condition when the t½ of the parent is approximately 10 times greater than the t½ of the daughter. (b) Secular equilibrium is a condition when the t½ of the parent is many times greater than the t½ of the daughter (100-1000 times greater or more). Adapted from (Dash and Chakravarty 2014) with permission from Royal Society of Chemistry. (B) Example of radionuclide daughter activity in-growth over repeated elutions. The radionuclide generator provides a reduced level of the daughter radionuclide activity by each subsequent elution. Adapted from (Dash and Chakravarty 2014) with permission from Royal Society of Chemistry. (C) Daughter activity eluted from a radionuclide generator generally follows a Gaussian distribution.

Table 10.1 Some medically important radionuclide generator systems.

Daughter radionuclide					Parent radionuclide		
Radio-nuclide	Half-life	Mode of decay*	β⁻, α energy in keV	Principal γ energy in keV (% abundance)	Radio-nuclide	Half-life	Mode of decay*
^{44}Sc	3.9 h	EC, β⁺	–	511.0 (188.0)	^{44}Ti	46 y	EC
^{62}Cu	9.7 min	β⁺	–	511.0 (194.9)	^{62}Zn	9.2 h	EC, β⁺
^{66}Cu	5.1 min	β⁻, γ	2642.4	1039.2 (8.0)	^{66}Ni	54.6 h	β⁻
^{68}Ga	67.6 min	β⁺	–	511.0 (178.2)	^{68}Ge	270.8 d	EC
^{72}As	26.0 h	β⁺	–	511.0 (150.0)	^{72}Se	8.4 d	EC
^{82}Rb	1.3 min	β⁺	–	511.0 (190.6)	^{82}Sr	25.6 d	EC
^{87m}Sr	2.8 h	IT	–	388.0 (80.0)	^{87}Y	3.3 d	EC, β⁺
^{90}Y	64.1 h	β⁻	2282.0	Nil	^{90}Sr	28.8 y	β⁻
^{99m}Tc	6.0 h	IT	–	140.5 (88.9)	^{99}Mo	65.9 h	β⁻, γ
^{103m}Rh	65.1 min	IT	–	39.7 (0.07)	^{103}Ru	39.4 d	β⁻, γ
^{109m}Ag	39.8 s	IT	–	88 (3.7)	^{109}Cd	453.0 d	EC
^{112}Ag	3.1 h	β⁻, γ	3956.0	616.8 (42.9)	^{112}Pd	21.0 h	β⁻, γ
^{113m}In	1.7 h	IT	–	391.7 (64.9)	^{113}Sn	115.1 d	EC
^{132}I	2.3 h	β⁻, γ	2118.1	667.7 (98.7)	^{132}Te	3.3 d	β⁻
^{137m}Ba	2.6 min	IT	–	661.6 (89.9)	^{137}Cs	30.1 y	β⁻
^{140}La	1.7 d	β⁻, γ	3761.9	1596.2 (95.4)	^{140}Ba	12.7 d	β⁻, γ
^{166}Ho	26.8 h	β⁻, γ	1854.5	80.6 (6.2)	^{166}Dy	81.6 h	β⁻, γ
^{178}Ta	9.3 min	EC	1910.0	325.8 (84.7)	^{178}W	21.6 d	EC
^{188}Re	16.9 h	β⁻, γ	2120.4	155.0 (14.9)	^{188}W	69.4 d	β⁻, γ
^{194}Ir	19.2 h	β⁻, γ	2246.9	328.4 (13.0)	^{194}Os	6.0 y	β⁻, γ
^{195m}Au	30.6 s	IT	–	261.7 (68.0)	^{195m}Hg	40.0 h	EC, γ
^{212}Bi	60.6 min	α, β⁻, γ	6207.1 (α) 2254.02 (β)	727.2 (11.8)	^{212}Pb	10.6 h	β⁻, γ
^{213}Bi	45.9 min	α, γ	5982.0	439.7 (27.3)	^{225}Ac	10.0 d	α, γ
^{225}Ac	10.0 d	α, γ	5935.1	99.7 (3.5)	^{229}Th	7430 y	α, γ

* Only principal decay mode is mentioned; EC indicates decay by electron capture; IT indicates isomeric transition.

In a radionuclide generator, a 'parent' radionuclide (A) decays to a 'daughter' radionuclide (B) which further decays to stable/nearly stable 'grand-daughter' nuclide (C). In this decay scheme λ_1 is the decay constant for A having N_1^0 initial number of atoms and λ_2 the decay constant for B. For the sake of simplicity, it is generally assumed that $N_2(t = 0) = N_2^0 = 0$ and $N_3(t = 0) = N_3^0 = 0$ and the grand-daughter product C, as a stable nuclide characterized by $\lambda_3 = 0$.

Then at any given time t, one can write the following differential equations:

$$dN_1(t) = -\lambda_1 N_1(t)dt \text{ or } \frac{dN_1(t)}{dt} = A_1(t) = -\lambda_1 N_1(t) \tag{1}$$

$$dN_2(t) = +\lambda_1 N_1(t)dt - \lambda_2 N_2(t)dt \text{ or } \frac{dN_2(t)}{dt} \lambda_1 N_1(t) - \lambda_2 N_2(t) \tag{2}$$

and $\quad dN_3(t) = +\lambda_2 N_2(t)dt \text{ or } \dfrac{dN_3(t)}{dt} = +\lambda_2 N_2(t) \tag{3}$

Equation 1 describes the rate of change of the number of atoms of type A, supposing that the only source of these atoms is from the initial supply N_1^0 at $t = 0$. Equation 2 describes the rate of change of the number of atoms of type B which is equal to the supply by the decay of atoms A corrected for the loss through its own decay. Equation 3 describes the rate of change of the number of atoms of type C fed by only the decay of atoms B (being stable, C is continuously accumulated). The solution of equation 1 leads to the well-known decay equation of a single radionuclide i.e.,:

$$N_1(t) = N_1^0 e^{-\lambda_1 t} \text{ and } A_1(t) = \lambda_1 N_1^0 e^{-\lambda_1 t} = A_1^0 e^{-\lambda_1 t} \tag{4}$$

The solution of equation 2 leads to the following expression (assuming $N_2(0) = N_2^0 = 0$):

$$N_2(t) = N_1^0 \frac{\lambda_1}{\lambda_2 - \lambda_1}\left(e^{-\lambda_1 t} - e^{-\lambda_2 t}\right) \tag{5}$$

Here, the activity $A_2(t)$ is given by the general definition of the activity. Therefore,

$$A_2(t) = \lambda_2 N_2(t) = N_1^0 \frac{\lambda_1 \lambda_2}{\lambda_2 - \lambda_1}\left(e^{-\lambda_1 t} - e^{-\lambda_2 t}\right) \tag{6}$$

or $\quad A_2(t) = A_1^0 \dfrac{\lambda_2}{\lambda_2 - \lambda_1}\left(e^{-\lambda_1 t} - e^{-\lambda_2 t}\right) \tag{7}$

where $\quad A_1^0 = \lambda_1 N_1^0$

Time Taken by the Daughter Radioisotope to Attain Maximum Radioactivity

The radioactivity of the daughter isotope reaches maxima when the feeding of the daughter atoms (B) exactly compensates their decay:

i.e., when $\quad A_1(t) = A_2(t) \quad$ or $\quad \lambda_1 N_1(t) = \lambda_2 N_2(t) \tag{8}$

or when $\quad \dfrac{dA_2(t)}{dt} = \dfrac{dN_2(t)}{dt} = 0$

Then one can write:

$$t_{max} = \frac{\ln(\lambda_2/\lambda_1)}{\lambda_2 - \lambda_1} = \left(\frac{1.44\,t_1\,t_2}{t_1 - t_2}\right)\ln\left(\frac{t_1}{t_2}\right) \tag{9}$$

where, t_1 and t_2 are the half-lives of the parent and daughter radionuclides, respectively and t_{max} is the time taken by the daughter radioisotope to attain maximum radioactivity.

Radioactive Equilibrium

When the 'daughter' radionuclide decays at the same rate at which it is being produced, it is said to have reached the state of 'radioactive equilibrium' (Fig. 10.2A). Radionuclide generators can be categorized on the basis of the ratio of the decay constants (λ) of the parent to that of the daughter radionuclide and two principal situations are considered: (1) If the ratio of $\lambda_1:\lambda_2$ is ~0.1, the parent-daughter system is said to exist in a state of transient equilibrium (2) If λ_1 is very much smaller than λ_2, the system is said to exist in secular equilibrium (Dash and Chakravarty 2014). Both the cases are discussed briefly, with suitable examples.

Case 1: Transient Equilibrium

In this case, $\lambda_1 < \lambda_2$ or $t_1 > t_2$ with t_1/t_2 ~10. The most suitable example for this case is the ^{99}Mo/$^{99\,m}$Tc radionuclide pair.

$$^{99}_{42}\text{Mo} \xrightarrow{\beta^- \, (t_1 = 65.94h)} {}^{99\,m}_{43}\text{Tc} \xrightarrow{\gamma(t_2 = 6.01h)} {}^{99}_{43}\text{Tc}\,(\text{ground state})$$

In this case, the transient equilibrium is reached at:

$$t_{max} = \left(\frac{1.44\,t_1\,t_2}{t_1 - t_2}\right)\ln\frac{t_1}{t_2} = 22.79 \text{ hours}$$

After this period of time, the daughter activity ($^{99\,m}$Tc) comes into equilibrium with the parent activity (^{99}Mo) and then it follows the same half-life as that of its parent.

Case 2: Secular Equilibrium

In this case, λ_1 is far less than λ_2 or t_1 is much greater than t_2. Some well-known examples of this case are ^{90}Sr/^{90}Y, ^{137}Cs/$^{137\,m}$Ba parent-daughter systems.

$$^{137}_{55}\text{CS} \xrightarrow{\beta(t_1 = 30.07\,y)} {}^{137\,m}_{56}\text{Ba} \xrightarrow{\beta(t_2 = 2.55m)} {}^{137}_{56}\text{Ba}$$

or

$$^{90}_{38}\text{Sr} \xrightarrow{\beta(t_1 = 28.8\,y)} {}^{90}_{39}\text{Y} \xrightarrow{\beta(t_2 = 2.55m)} {}^{90}_{40}\text{Zr}$$

For practical purposes, these situations could be considered to correspond to $\lambda_1 \to 0$ or $t_1 \to \infty$. Thus $A_1(t) \approx \lambda_1 N_1^0 = A_1^0$ and $N_1(t) = N_1^0$ due to the fact that $e^{-\lambda_1 t} = (1 - \lambda_1 t) \approx 1$. Consequently, with respect to the decay of the daughter activity, the radioactivity of the parent appears to be constant. The equation for the radioactivity of the daughter radioisotope becomes:

$$A_2(t) = A_1^0 \frac{\lambda_2}{\lambda_2 - \lambda_1}\left(e^{-\lambda_1 t} - e^{-\lambda_2 t}\right) \approx A_1^0 \left(1 - e^{-\lambda_2 t}\right) \tag{10}$$

Under this condition, the total radioactivity of the parent and daughter reaches the maximum and does not decrease appreciably for several half-lives of the daughter product.

In-Growth and Isolation of the Daughter Radionuclide from the Parent Radionuclide

For practical considerations, radionuclide generators are eluted at periodic intervals depending on the daughter activity requirements. Often the separation of the daughter from the parent may not occur at the time the daughter activity is at its maximum as calculated using Equation (9). Owing to the decrease of the parent radionuclide activity within the time between two elution steps, the radionuclide generator is expected to provide a reduced level of the daughter radionuclide activity by each subsequent elution as shown in Fig. 10.2B. The growth and separation of the daughter radionuclide can be continued as long as there are useful activity levels of the parent radionuclide available. Elution of the generator is performed either manually or with an automated system. The activity eluted from the generator typically follows a Gaussian distribution as shown in Fig. 10.2C with the maximum activity being eluted in the intermediate fractions.

The eluent can also be 'fractionated' by discarding those fractions with low activity if the activity concentration is a critical issue. Separation may be performed any time before equilibrium is reached and the activity levels of daughter recovered will depend on the time elapsed since the last separation. In-growth of the daughter species is continuous and once the activity of the daughter is recovered from the mixture, the daughter activity increases until its activity level reaches a maximum and is in equilibrium with the parent radionuclide (Fig. 10.2B). The growth of the daughter depends on the half-life of the daughter radionuclide which also governs the frequency of its separation from the parent radionuclide. When the daughter radionuclide is relatively long-lived, periodic elution will take place prior to reaching the maximum equilibrium daughter activity levels and it is normal to use generators in this way. As an example, 50% daughter activity in-growth is detected in one half-life, 75% in two half-lives and the daughter activity reaches the activity of the parent radionuclide in 5-6 half lives.

Criteria for Selection of Parent/Daughter Pairs

While a myriad of factors contribute to the development of column chromatographic radionuclide generators, selection of an appropriate parent/daughter pair is a key determinant that underpins its success in nuclear medicine. While selecting a parent/daughter pair for making a radionuclide generator, the following criteria need to be considered:

- *Availability of parent radionuclide:* The production route for the parent radionuclide should be cost-effective. Parent radionuclide which exhibits attractive characteristics but lacks a cost-effective production route might not be suitable for the preparation of clinically useful radionuclide generators.

- *Parent radionuclide half-life:* The physical half-life of the parent radionuclide should be long enough to ensure availability of the daughter nuclide for an extended period of time.

- *Parent specific activity:* With a view to adsorb required quantities of activity in a given mass of sorbent within the chromatographic column, the prospect of using high specific activity of the parent radionuclide is a trustworthy proposition.

- *Daughter radionuclide half-life:* The physical half-life of the daughter radionuclide should be compatible with the *in vivo* pharmacokinetics of the radiolabeled targeting molecule.

- *Emission and energy of the radiation of the daughter radionuclide:* The daughter product of a radionuclide generator will decay by any of the decay modes (isomeric transition, β^-, β^+, electron capture, α decay) or by a combination of two or more decay modes. Consequently, the applications of generators vary depending on the decay characteristics. Gamma emitters, with gamma energy within the range of 100-250 keV are suitable for single photon emission computed tomography (SPECT) imaging. Particle emitting radionuclides (α particle, β^- particle or Auger electron emitters) are suitable for therapy. Positron (β^+) emitting radionuclides are needed for positron emission tomography (PET) imaging.

- *Decay of daughter radionuclide:* In order to preclude radiation dose to the patient undergoing diagnosis or therapy, the daughter radionuclide should preferably decay to a stable or very long-lived product.

- *Purity of daughter radionuclide:* The success of a radionuclide generator resides on its ability to provide the daughter radionuclide of required purity (radionuclidic, radiochemical and elemental purity) amenable for the radiolabeling of a gamut of biological carriers for *in vivo* applications.

- *Chemical characteristics of daughter radionuclide*: The daughter radionuclide should have chemistry amenable to its attachment with a broad class of carrier molecules and binding must exhibit high *in vivo* stability when attached to the radiopharmaceutical.

Column Chromatographic Technique in Preparation of Radionuclide Generators

This is a solid-liquid separation technique in which the stationary phase, a solid sorbent, is packed in a chromatographic column and a liquid containing the parent/daughter radionuclide mixture, is added to the top and flows down through the column either by gravity or by external pressure (Mushtaq 2004). The stationary phase interacts with metal ions and selectively retains the parent metal ion by virtue of its affinity towards the same. The goal is to have a high binding [high distribution coefficient (K_D)] of the parent radionuclide to the sorbent with a low K_D for the decay product (daughter radionuclide), which can then be readily and reproducibly removed on demand by using a suitable eluent. The performance of column chromatographic separation is dictated by both adsorption equilibrium and kinetics. In general, the adsorption reaction is known to proceed through the following three steps:

- Transfer of metal ions from bulk solution to the sorbent surface known as diffusion.
- Migration of metal ions into pores.
- Interaction of metal ions with available sites on the interior surface of pores.

Although the mechanisms of metal ion retention on a sorbent are distinct, the characteristics of the sorbent that are responsible for the radiochemical separation are largely the same. Before discussing the parent/daughter radionuclide separation, a practical understanding of the adsorption process is necessary. Factors affecting adsorption as well as characteristics of the sorbent are the following:

- Surface area of sorbent: Larger surface area implies a greater adsorption capacity.
- Particle size of sorbent: Smaller particle sizes reduce internal diffusion and mass transfer limitation and facilitate penetration of metal ions through the pores of the sorbent. It offers scope for achieving theoretical adsorption capacity owing to the ready attainment of equilibrium.
- Kinetic of adsorption: The fast kinetic means short contact time or residence time to complete the adsorption.

- Affinity of the metal ion: Metal ions with high affinity for the sorbent (high K_D) confer very strong binding and are preferentially retained by the sorbent.
- pH of the external solution: The degree of ionization of a species is affected by the pH (e.g., a weak acid or a weak basis). This, in turn, affects adsorption.
- Concentration of metal ion: At high-level metal ion concentrations, the available adsorption sites become fewer.

The selection of an appropriate sorbent to separate the daughter radionuclide from the parent/daughter equilibrium mixture with requisite purity and appreciable yield constitutes the pillars for the success of column chromatographic technique in radionuclide generator technology. Selection of a sorbent is primarily aimed at simplification of column chromatographic procedure and maximizing the activity yield of the daughter radioisotope. While selecting a sorbent in the development of radionuclide generator, the following points should be taken into consideration.

- Possess sufficient specificity for a given oxidation state of the parent radionuclide and negligible specificity for the daughter radionuclide.
- High adsorption capacity: The sorbent should have high specific surface area, chemical reactive surface and suitable pore size to facilitate parent metal ion diffusion.
- The sorbent should have sufficient chemical and radiation stability.
- Sorbents need to be mechanically strong and robust enough to stand attrition, erosion and crushing during column operation.
- With a view to minimize decay losses of the daughter radionuclide, the kinetics of adsorption must be fast.
- Ability to sequester the daughter radionuclide in the desired chemical form (preferably with saline solution) with requisite purity, i.e., free from parent radionuclide breakthrough as well as trace levels of other metal ion impurities.
- Provide reproducible elution yield of the daughter radionuclide over an extended period of time.
- Reversibility of adsorption process to facilitate parent metal ion recovery.
- Simple synthesis procedure to undertake large-scale manufacturing in a cost-effective manner.

While performing separation, a number of aspects must be considered for efficient elution of the daughter radionuclide from the generator column, which include column type, column bed dimension, pretreatment of column bed, parent radioisotope loading technique, sample throughput and selection of a suitable eluent.

Fabrication of Column Chromatographic Radionuclide Generators

The important components of a typical column chromatographic radionuclide generator are as follows (Fig. 10.1B):

Separation System

The chromatographic column containing the sorbent to sequester the daughter radionuclide of interest from the parent with high efficiency constitutes the heart of a radionuclide generator (Fig. 10.1C). The column is usually constructed of glass or an appropriate plastic material into which the sorbent is packed. The parent radionuclide is taken up by the sorbent and the daughter radionuclide is separated readily and reproducibly on demand by using a suitable reagent.

Filter Systems

In order to preclude the presence of particulate impurities in the separated daughter radionuclide, filters in the form of porous frits are attached to the chromatographic column. Generally, in order to obtain the daughter radionuclide in a sterile form, it is filtered through commercially available membrane filters made of cellulose esters, available in various pore sizes as disposable units.

Tubing Components

The fluid inlet and outlet lines consist of radiation resistant or stable materials such as KYNAR (polyvinylidene fluoride), PFA (perfluoroalkoxy fluorocarbon), PEEK (polyetheretherketone) or stainless steel, which are attached to the chromatography column through connectors to allow passage of eluent through the generator column and collection of daughter eluate.

Shielding

The separation system, connectors and connection tubing are integrated within a small portable lead shielded unit throughout the use of the generator for the purpose of radioprotection. Only the output vial is accessible externally. Most generators use an evacuated vial for collection of the daughter radionuclide. Additional shielding is used during the collection process. The separated daughter radionuclide must also be shielded once it is collected from the generator.

Handle

The generator is provided with handles to allow manual or mechanical lifting and positioning.

The important steps for preparation of column chromatographic radionuclide generators are as follows:

- An accurately weighed amount of sorbent is taken into a sterile generator column and a sintered disc is put on the top to prevent disturbance of the column bed during flow of liquid through it.
- The top and bottom of the generator column are sealed, connected with the tubing and kept inside the lead shielded assembly.
- The column matrix is conditioned by passing the desired solution through the column bed.
- The pH of the loading solution is adjusted to the desired value and passed through the generator column.
- The column is washed with the eluent solution to remove trace level of unadsorbed or loosely held parent radioisotope.
- After allowing time for in-growth of the daughter activity, the generator is eluted using desired volume of a suitable eluent.

Quality Control of Radionuclide Generators Prior to Clinical Use

Since generator derived radionuclides are intended for clinical use, it is imperative that the radionuclide generators undergo strict quality control procedures before being handed over to the nuclear medicine physicians (Chakravarty and Dash 2013a; Knapp and Baum 2012; Knapp and Mirzadeh 1994). Quality control involves specific tests and measurements that ensure the elution efficiency of the generator, product purity, biological safety and the efficacy of the radionuclide for the preparation of radiopharmaceuticals. These methods are briefly described below.

Elution Efficiency of the Radionuclide Generator

The elution efficiency of the radionuclide generator is defined as the proportion of the daughter radioisotope present in the generator system that is separated during the elution process. Theoretically, the activity (A_2) of the daughter radioisotope present in the generator system at the time of elution is given by the equation (7). In practice, the activity of the daughter radioisotope eluted is less than that predicted by the theory. If the measured activity of the separated daughter radioisotope after allowing time 't' for its growth, is denoted by A_s, then the elution efficiency can be defined by the equation:

$$\% \text{ Elution efficiency} = \frac{A_s}{A_2} \times 100 \qquad (11)$$

For a radionuclide generator to be cost-effective, it is essential that its elution efficiency should be fairly high (not < 60%) and it should remain consistent during the stipulated period of utilization of the generator.

Radionuclidic Purity of the Separated Daughter Radioisotope

Radionuclidic purity is defined as the fraction of the total radioactivity in the form of the desired radionuclide. In the separated daughter radioisotope obtained from the radionuclide generator, the primary radionuclidic impurity that may be expected is the long-lived parent radioisotope. Sometimes, the parent radioisotope may be associated with other radionuclidic impurities which may also come in the separated daughter activity. Understandably, radionuclidic impurities are undesirable in the daughter product, as these can have various implications for its use in nuclear medicine, such as interference in the reaction for preparation of the radiopharmaceutical leading to poor yields or unwanted compounds, increase in the unwanted radiation exposure to the patient, obscure scintigraphic images and can possibly lead to radiotoxicity (Chakravarty and Dash 2013a). Hence the radionuclidic purity of the intended radionuclide needs to be determined and the impurities ascertained to be well within the stipulated limits (Saha 2010). Determination of radionuclidic impurities in the separated daughter radioisotope is generally done by γ-ray spectrometry using high purity germanium (HPGe) detector coupled with a multi-channel analyzer (Chakravarty and Dash 2013a). This technique can be used for both qualitative as well as quantitative estimation of radionuclidic impurities in the daughter product obtained from the radionuclide generator.

Radiochemical Purity of the Separated Daughter Radioisotope

The radiochemical purity of a generator produced radionuclide may be defined as the fraction of the total radioactivity present in the desired chemical form (Chakravarty and Dash 2013a). Radiochemical impurities may arise in the daughter radionuclide during its separation from the parent or its subsequent storage due to several factors such as the action of the solvent and the effect of radiolysis, change in temperature or pH, presence of oxidizing or reducing agents (Saha 2010). The radiochemical impurities present in the daughter radionuclide may not be suitable for labeling with ligands and biomolecules. This may also affect the biological behavior of the radiopharmaceutical as the agent may not be selectively taken up by the target organs (Chakravarty and Dash 2013a). The presence of radiochemical impurities in generator produced radioisotopes can be detected and determined by various analytical methods. These include, paper chromatography, thin layer chromatography, paper electrophoresis, high performance liquid chromatography, gel filtration, gel chromatography, ion exchange chromatography, solvent extraction, inverse dilution and precipitation (Chakravarty and Dash 2013a).

Chemical Purity

Any unwanted chemical species (organic or inorganic) present in the daughter product is considered to be chemical impurity. The presence of these chemical impurities may affect the chemistry of the radionuclide for the preparation of radiopharmaceuticals. The chemical impurities may be introduced in the daughter radionuclide by a variety of ways including the use of impure chemicals and the use of radioactive parent solutions containing undesired chemicals introduced during its radiochemical processing. Also, in case of column chromatographic generators, the radiolytic or chemical degradation of the column matrix may lead to the addition of chemical impurities to the daughter radionuclide. The presence of these chemical impurities can be avoided by the use of highly pure chemicals and adoption of appropriate separation methodology. The level of these chemical impurities may be detected and determined by various analytical techniques like colorimetry, spot-tests, spectrophotometry, inductively coupled plasma atomic emission spectroscopy (ICP-AES) etc. (Saha 2010).

Labeling Efficacy

Generally, generator produced radionuclides are used for clinical application, only after radiolabeling a suitable ligand or a biomolecule with it. The suitability of a generator produced radionuclide for radiopharmaceutical applications can be demonstrated by its efficacy to prepare standard radiolabeled agents. This is also an indirect test of the chemical purity of the radionuclide as high chemical purity is required for the preparation of the radiolabeled agent with the NCA radionuclide.

Biological Tests

Biological quality control tests are carried out to examine the sterility and apyrogenicity of the generator-produced radionuclide before the preparation of radiopharmaceuticals (Saha 2010). Sterility indicates the absence of any viable bacteria or microorganism in a radiochemical. It is essential to avail the daughter radionuclide from the generator in sterile form in order to prepare clinical-grade radiopharmaceuticals. Administration of non-sterile radiopharmaceutical can cause a wide variety of infections leading to several physiological problems including death (Saha 2010).

Pyrogens are either polysaccharides or proteins produced by the metabolism of microorganisms and if present in a radiopharmaceutical can cause a wide variety of physiological problems, such as, fever, chills, malaise, leucopenia, flushing, sweating, headache and pain in joints (Saha 2010). It is mandatory that all products intended for injection into humans, including radiopharmaceuticals should have pyrogens below the stipulated limits. Pyrogen free radionuclides can be obtained from the generator without much difficulty by using high-quality chemicals and taking particular care during the preparation and storage of the generator.

Shelf-Life of a Radionuclide Generator

The shelf-life of a radionuclide generator is the period for which the generator can safely be used for the designated clinical applications (Chakravarty and Dash 2013a). The shelf-life of a typical radionuclide generator is influenced by the following factors:

- Physical half-life of the parent radioisotope. The parent radioisotopes having longer physical half-lives are generally expected to have a longer shelf-life.
- Generator performance which is monitored by measuring the elution yield, radioactive concentration and the purity of the daughter radioisotope.

However, the shelf-life of a radionuclide generator is also influenced by the procedure adopted for the separation of the parent-daughter pairs. The suitability of a separation procedure to withstand the effects of radiolysis and chemical degradation over a prolonged period of time enhances the shelf-life of the generator. Overall, the economics of production of short-lived radioisotopes via a radionuclide generator is decided by the shelf-life of the generator which in turn determines the cost of treatment using radiopharmaceuticals prepared using generator produced radionuclides.

Nanomaterials as Sorbents in Column Chromatographic Radionuclide Generators

A nanomaterial can be defined as a material that has a structure in which at least one of its phases has one or more dimensions in the nanometer size range (1-100 nm). Such materials hold promise as they take advantage of unique physiochemical properties realized at nanoscale that cannot be anticipated from bulk counterparts of the same chemical composition (Bushra et al. 2012; Cao and Wang 2011; Chakravarty and Dash 2013a; Chakravarty and Dash 2013b; Chakravarty et al. 2012c; Chang et al. 2012). The defining feature of a nanomaterial is that its properties depend not only on composition but also on its size and shape. One of the specific properties of this class of material is that a high percent of the atoms lie on the surface (Cao and Wang 2011). These surface atoms are unsaturated and possess high chemical activity. From the perspective of realizing the scope of using nanomaterials as sorbents in radionuclide generators, there are a few key characteristics that contribute to the uniqueness of such materials, which include:

- Large fraction of surface atoms;
- High surface energy;
- Spatial confinement;
- Reduced imperfections.

These exciting properties of nanomaterials offer the ability to interact with the metal ions in the aqueous solution in new ways. Use of nanomaterial-based sorbent offers an exciting platform into the realm of column chromatography that is difficult to achieve using bulk materials.

Considering the aforementioned properties of nanomaterials, it is imperative to provide a broad snapshot of the characteristics of nanomaterial-based sorbents in the development of radionuclide generators. Some key attributes include:

- Grain size on the order of 1-100 nm: Smaller particle size enables packing a larger amount of sorbent in a specified volume, which can be effective for making high activity generators commensurate with centralized radiopharmacy requirement.

- Extremely large specific surface area: When the size of the particles is reduced to the nanometric scale, the proportion of atoms located at a particular surface is considerably high in relation to the total volume of atoms in the material. Surface area determines the number of sites for target radionuclides to attach to the solid sorbent surface. With the enormous increase in available surface area per unit mass, this class of materials provides a remarkable capacity for the target radionuclide. This will provide scope for using low specific activity parent radionuclide without altering the design of current generation radionuclide generators.

- Nanomaterials possess unique sorption properties due to different distributions of reactive surface sites and disordered surface regions. It is reported that the surface or interfacial tension decreases with decreasing particle size because of the increase in the potential energy of the bulk atoms of the particles (Cao and Wang 2011). Smaller particles with increased molar free energy are more prone to adsorption per unit area of molecules or ions onto their surfaces in order to decrease the total free energy and to become more stable and therefore, smaller particles have a higher adsorption coefficient. Thus, adsorption takes place relatively easily on the surface of nanomaterials.

- Presence of a high percentage of the atoms on the surface: One of the specific properties of nanomaterials is that a high percent of the atoms reside on the surface. Atoms at surfaces have fewer neighbors than atoms in the bulk. Because of this lower coordination and unsatisfied bonds, surface atoms are less stabilized than bulk atoms. The unsaturated surface atoms can therefore bind with other atoms/ions that come in contact with them.
- Strongly influence the thermodynamics and kinetics of metal ion adsorption by increasing the diffusion rate as well as by decreasing the required diffusion length and facilitate higher mass transfer reactions. The smaller dimensions of nanoparticles can produce short diffusion paths to the materials' interiors and thus offer more favorable mass transfer reactions.
- Stronger, more ductile materials: The mechanical strength of nanomaterials is much higher than that of their bulk counterparts, due to the reduced number of defects. The physical stability of nanomaterial-based sorbents is adequate to prevent attrition losses in column operations and to provide good flow.
- Chemically very active materials: Because of the extremely small size of the grains (domains) of nanomaterials, a large fraction of the atoms in these materials are located at the grain boundaries. It is apparent that significant fractions of atoms are located on crystallite surfaces as well as on edges and corners which are usually chemically reactive and have polarizable surfaces, contributing to their high chemical potential. They can readily interact with chemical species that come in contact with them.
- They are adequately stable for almost all reasonable applications involving radioactivity. Nanoscale materials have radiation resistance due to the ideally unsaturable sink strength represented by free surfaces (Bai et al. 2010; Bringa et al. 2011). Radiation-induced point defects cannot accumulate in the presence of the high density of surfaces in these materials.
- Offer the possibility of tailoring the adsorption parameters independently without a change in chemical composition. The ability to manipulate the surface chemistry of nanomaterials allows a new level of selectivity control for adsorption of different metal ions on the surface.
- Nanomaterials provide almost unlimited combinations of various compositions, sizes, dimensions and shapes of materials, which can be tailored to adsorb different ionic species with desired attributes.

These distinctive properties of nanomaterials have sparked interest among researchers to explore their potential to yield substantial improvements in several conventional radionuclide generators in the near term and to enable the development of revolutionary products in future.

Nanostructured Metal Oxides

A variety of nanostructured metal oxides covering a wide range of compositions and tunable size/shape have been synthesized by various methods over the past decade (Cañas-Carrell et al. 2014; Chakravarty and Dash 2013b). Almost all the applications of nanomaterials in column chromatographic radionuclide generator have involved the use of nanostructured metal oxides due to the following practical advantages.

- They possess very high lattice energies and melting points due to their high ionicity. In theory very small crystals of material with high lattice energies should be stable and resistant toward melting, atom/ion migration and subsequent crystal growth, even at elevated temperatures.

- They offer high thermal and chemical stabilities combined with a well-developed porous structure and high surface areas (> 100 m^2/g) meeting the requirements for most column chromatographic separation applications.

- They can exist with numerous surface sites with enhanced surface reactivity, such as crystal corners, edges, or ion vacancies. Residual surface hydroxides and their chemistry is generally attributable to Lewis acid, Lewis base and Bronsted acid sites of varying coordination, that is, metal cations, oxide anions and surface –OH which can be isolated or lattice bound (Cao and Wang 2011; Chakravarty and Dash 2013b). Since the oxide can be chosen for the desired Lewis base/Lewis acid strengths and also since the oxides can be produced with thin layers of other chosen materials, surface chemical properties can be further modified as desired.

- In aqueous systems, metal oxide particles are hydrated and ≡M-OH groups cover their surface completely (Chakravarty and Dash 2013b). These amphoteric hydroxyl groups can undergo reaction with either H$^+$ or OH$^-$ and develop positive or negative charges on the surface depending on the pH, which can aid in selective adsorption and release of a wide variety of radiometal ions (Fig. 10.3).

- They can be easily prepared and there is a great flexibility in their design, giving the inspiration for designing new materials.

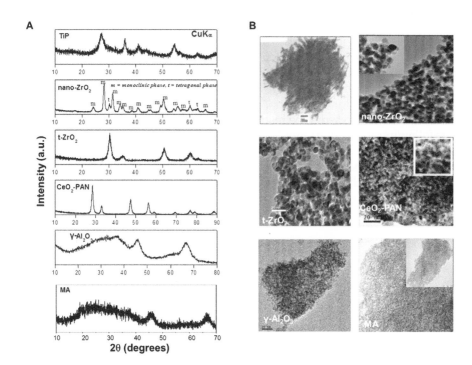

Fig. 10.3 Development of pH-dependent surface charge on nanosorbent which is responsible for sorption of parent radioisotope and elution of daughter radioisotope. (A) Surface acid-base properties of nanocrystalline metal oxides resulting in development of surface charge. (B) Variation in zeta potential of nanosorbents at different pH. Adapted from (Chakravarty et al. 2014) with permission from Taylor and Francis.

In the recent times, the development of a wide variety of column chromato-graphic radionuclide generators (such as ^{99}Mo/^{99m}Tc, ^{188}W/^{188}Re and ^{68}Ge/^{68}Ga generators) using nanostructured metal oxides such as polymer embedded nanocrystalline titania (TiP) (Chakravarty et al. 2009; Chakravarty et al. 2008), mixed phase nanocrystalline zirconia (nano-ZrO$_2$) (Chakravarty et al. 2010c), tetragonal nanozircona (t-ZrO$_2$) (Chakravarty et al. 2010a; Chakravarty et al. 2011a), nanocrystalline γ-alumina (γ-Al$_2$O$_3$) (Chakravarty et al. 2012b; Chakravarty et al. 2011b), mesoporous alumina (MA) (Chakravarty et al. 2013c) and nanoceria-polyacrylonitrile (CeO$_2$-PAN) composite (Chakravarty et al. 2010b) as sorbents has been reported. Synthesis procedures adopted and structural characteristics of these nanosorbents are summarized in Table 10.2. It is pertinent to point out that in case of all the nanosorbents used for preparation of radionuclide generators, the synthesis methods adopted were neither cumbersome nor used expensive precursors and were amenable for scale-up. All these nanomaterials exhibited good mechanical strength, gran-ular properties and were amenable for column operations. However, all these sorbents consist of agglomerated nanoparticles (Fig. 10.4). Agglomeration to a certain extent is essential for use of such materials as sorbent matrices in chromatographic columns (Chakravarty and Dash 2013a; Chakravarty et al.

2013c). Very fine particles without agglomeration are not suitable for column chromatographic application as such materials are impervious to the flow of liquid through the column bed. In view of the explicit need to use these sorbents for column chromatographic procedures, it is of utmost importance to establish their chemical stabilities to ensure that they resist chemical degradation during column operations. It is worth pointing out here that these materials were reported to be insoluble in common mineral acids and bases. The following subsections provide an overview of the radionuclide generators developed using nanomaterial-based sorbents. To date, ^{99}Mo/^{99m}Tc, ^{188}W/^{188}Re and ^{68}Ge/^{68}Ga generators have been developed using such sorbents, the characteristics of which are summarized in Table 10.3 and discussed below.

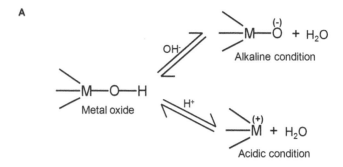

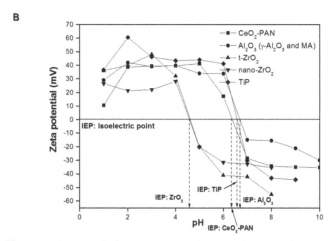

Fig. 10.4 Structural characterization of nanosorbents used in the preparation of column chromatographic radionuclide generators. (A) XRD patterns of nanosorbents. Adapted from (Chakravarty et al. 2013c; Chakravarty et al. 2012c; Chakravarty et al. 2011a; Chakravarty et al. 2011b; Chakravarty et al. 2010c) with permissions from Royal Society of Chemistry, American Chemical Society and Elsevier. (B) TEM micrographs of nanosorbents. Adapted from (Chakravarty et al. 2013c; Chakravarty et al. 2012c; Chakravarty et al. 2011a; Chakravarty et al. 2011b; Chakravarty et al. 2010c) with permissions from Royal Society of Chemistry, American Chemical Society and Elsevier.

Table 10.2 Synthesis and structural characterization of nanosorbents reported for the preparation of column chromatographic radionuclide generators.

Nanosorbent	Synthesis method	Structural characteristics	Radionuclide generators reported using this material	References
TiP	Controlled hydrolysis of $TiCl_4$ in isopropyl alcohol medium	Nanocrystalline, rutile phase, 5 nm crystallite size, surface area ~30 m^2/g	$^{99}Mo/^{99m}Tc$, $^{188}W/^{188}Re$	(Chakravarty et al. 2009; Chakravarty et al. 2008)
nano-ZrO_2	Controlled hydrolysis of $ZrOCl_2.8H_2O$ in isopropyl alcohol medium	Nanocrystalline, biphasic with monoclinic phase as major, 15 nm crystallite size, surface area ~45 m^2/g	$^{188}W/^{188}Re$	(Chakravarty et al. 2010c)
t-ZrO_2	Controlled hydrolysis of $ZrOCl_2.8H_2O$ in ammonical medium	Nanocrystalline, tetragonal phase, 7 nm crystallite size, surface area ~340 m^2/g	$^{99}Mo/^{99m}Tc$, $^{68}Ge/^{68}Ga$	(Chakravarty et al. 2010a; Chakravarty et al. 2011a)
γ-Al_2O_3	Mechanochemical reaction of aluminum nitrate with ammonium carbonate	Nanocrystalline, γ-phase, 2 nm crystallite size, surface area ~250 m^2/g	$^{99}Mo/^{99m}Tc$, $^{188}W/^{188}Re$	(Chakravarty et al. 2012b; Chakravarty et al. 2011b)
MA	Controlled hydrolysis of aluminum isopropoxide using glucose template	Nanocrystalline, γ-phase, 2-3 nm crystallite size, surface area ~230 m^2/g	$^{99}Mo/^{99m}Tc$	(Chakravarty et al. 2013c)
CeO_2-PAN	Decomposition of cerium oxalate precursor followed by incorporation in polyacrylonitrile matrix	Nanocrystalline, 10 nm crystallite size, surface area ~70 m^2/g	$^{68}Ge/^{68}Ga$	(Chakravarty et al. 2010b)

Table 10.3 Characteristics of column chromatographic radionuclide generators prepared using nanosorbents.

Generator	Sorbent	@Sorption capacity for parent radioisotope (mg/g)	Specific activity of parent radioisotope (GBq/g)	Parent activity loaded (GBq)	Elution yield (%)	Level of parent breakthrough (%)	Performance studied for time period (days)	References
$^{99}Mo/^{99m}Tc$	TiP	75	18.5	1.1	>80	$<10^{-3}$	7	(Chakravarty et al. 2008)
	t-ZrO$_2$	80	18.5	9.25	>80	$<10^{-4}$	10	(Chakravarty et al. 2010a)
	γ-Al$_2$O$_3$	148	17.6	13	>82	$<10^{-2}$	10	(Chakravarty et al. 2012b)
	MA	168	18.5	26	>80	$<10^{-2}$	10	(Chakravarty et al. 2013c)
$^{188}W/^{188}Re$	TiP	102	159	1.85	>80	$<10^{-3}$	120	(Chakravarty et al. 2009)
	nano-ZrO$_2$	120	159	1.85	>78	$<10^{-3}$	180	(Chakravarty et al. 2010c)
	γ-Al$_2$O$_3$	300	159	11.1	>80	$<10^{-4}$	180	(Chakravarty et al. 2011b)
$^{68}Ge/^{68}Ga$	CeO$_2$-PAN	20	NCA	0.74	>70	$<10^{-4}$	365	(Chakravarty et al. 2013b; Chakravarty et al. 2010b)
	t-ZrO$_2$	70	NCA	0.74	>80	$<10^{-4}$	365	(Chakravarty et al. 2011a)

NCA = no-carrier-added; @Sorption capacity is expressed as mg of parent radioisotope taken up by 1 g of nanosorbent under dynamic (column flow) conditions.

^{99}Mo/^{99m}Tc Generators

Over the last 5 decades, a variety of ^{99m}Tc/^{99}Mo generator systems have been thoroughly investigated all over the world due to the everlasting demand for ^{99m}Tc, which is the most commonly used medical radioisotope (Boyd 1982; Chakravarty and Dash 2013a; Dash et al. 2013; Eckelman 2009). This radioisotope is considered as the 'workhorse' of diagnostic nuclear medicine and is used for approximately 20-25 million procedures annually, comprising ~80% of all diagnostic nuclear medicine procedures (Eckelman 2009). The widespread interest in clinical utilization of this radioisotope is attributed to its attractive nuclear decay characteristics ($t_{1/2}$ = 6 h, emission of 140 keV γ-photon), convenient availability from ^{99}Mo/^{99m}Tc generator, versatile coordination chemistry of ^{99m}Tc for preparation of a wide variety of radiopharmaceuticals and commercial availability of lyophilized 'cold' kits which offer a cost-effective route for preparation of these radiopharmaceuticals suitable for human administration (Chakravarty and Dash 2013a). The column chromatographic ^{99m}Tc/^{99}Mo generator using a bed of acidic alumina has emerged as the most popular choice for availing ^{99m}Tc in nuclear medicine departments worldwide (Dash et al. 2013; Eckelman 2009; Knapp and Mirzadeh 1994). However, the capacity of bulk alumina for taking up molybdate ions is limited (2-20 mg Mo per gram of alumina), necessitating the use of NCA ^{99}Mo produced through fission route (Molinski 1982). Owing to the inherent complexities in production of fission ^{99}Mo and the vulnerability of irradiation services from 5 old research reactors which are currently in use for fission ^{99}Mo production, there is an increasing consensus to use low specific activity ^{99}Mo produced through the neutron activation route [(n, γ) ^{99}Mo] for preparation of clinically useful ^{99}Mo/^{99m}Tc generators (Pillai et al. 2013; Pillai and Knapp 2012; Ramamoorthy 2009). However, this would require the use of high capacity sorbents such as nanomaterials since the specific activity of (n, γ) ^{99}Mo is at least 1000-fold lesser than that of fission ^{99}Mo (Pillai and Knapp 2012).

The first utilization of a nanomaterial-based sorbent for preparation of ^{99}Mo/^{99m}Tc generator was reported in the year 2008 (Chakravarty et al. 2008). The authors synthesized TiP by controlled reaction of titanium tetrachloride in isopropyl alcohol medium, for use as a column matrix for the generator. A 1.1 MBq (30 mCi) ^{99}Mo/^{99m}Tc generator was prepared and its performance was evaluated for a period of 1 week. Technetium-99m could be regularly eluted from this generator with > 80% yield and the radionuclidic and radiochemical purity of the eluate were > 99.99% and > 95%, respectively. However, the utility of the sorbent in preparation of high-activity ^{99}Mo/^{99m}Tc generator suitable for clinical use was not demonstrated and the generator developed was more suitable for only preclinical studies in research laboratories. Nevertheless, this work set the stage for the use of nanomaterial-based sorbents in radionuclide generator technology. In a later development, the use of t-ZrO$_2$ for preparation of ^{99}Mo/^{99m}Tc generator was reported (Chakravarty et al. 2010a). t-ZrO$_2$ was synthesized by controlled hydrolysis of zirconyl chloride in ammonical medium. A 9.25 GBq (250 mCi) ^{99}Mo/^{99m}Tc generator was prepared and its performance evaluation was carried out by regular elution for 10 days. The elution yield of ^{99m}Tc was > 85% with acceptable radionuclidic, radiochemical and chemical purities for clinical applications. Further, the authors also demonstrated the compatibility of the ^{99m}Tc-eluate for preparation of several radiopharmaceuticals with high radiochemical purity and yield. Thus, it could amply be demonstrated that this medium activity generator was suitable for use in small nuclear medicine departments where there are a limited number of patients to undergo diagnosis using ^{99m}Tc-based radiopharmaceuticals.

Though TiP and t-ZrO$_2$ demonstrated desirable features for use as sorbents in the preparation of column chromatographic ^{99}Mo/^{99m}Tc generator, the use of nanostructured alumina would be the more preferred choice in view of potential clinical translation of this approach. Actually, bulk alumina-based ^{99}Mo/^{99m}Tc generator technology is well-established and globally accepted (Dash et al. 2013). Replacement of bulk alumina with nanostructured alumina while retaining the design characteristics of the well-established generators would find easier acceptance in the nuclear medicine community for future clinical use. With this objective, the preparation of 13 GBq (350 mCi) ^{99}Mo/^{99m}Tc generator using γ-Al$_2$O$_3$ sorbent synthesized by mechanochemical approach was reported (Chakravarty et al. 2012b). The overall yield of ^{99m}Tc obtained from this generator was > 80%, with radionuclidic purity > 99.99% and radiochemical purity > 99%. The radioactive concentration of ^{99m}Tc obtained from this generator was found adequate for the formulation of radiopharmaceuticals using commercially available lyophilized 'cold' kits. The performance of the generator remained consistent over an extended period of 10 days. A major advancement in column chromatographic ^{99}Mo/^{99m}Tc generator technology was the development of a high activity [26 GBq (700 mCi)] ^{99}Mo/^{99m}Tc generator using (n, γ) ^{99}Mo (specific activity ~18.5 GBq/g) (Chakravarty et al. 2013c). For this purpose, MA was used as the sorbent which was synthesized by hydrolysis of aluminium isopropoxide using glucose template. A novel 'tandem-column generator' concept was used for the preparation of the generator, in which two ^{99}Mo loaded columns were connected in series and ^{99m}Tc eluted in saline medium from the first column was fed to the second column to achieve high radioactive concentration as well as purity of ^{99m}Tc. The authors demonstrated that the double column approach reported in this study had several advantages over the conventional single column based generators, in terms of (a) higher elution yield of ^{99m}Tc and (b) sharper elution profile resulting in higher radioactive concentration of ^{99m}Tc eluate. The performance of this generator was found suitable for future clinical use. The 'tandem-column generator' concept can also be used for the preparation of higher activity generators (~185-370 GBq) by connecting multiple columns in series in centralized radiopharmacy setups to cater to the needs of several smaller hospitals in the region.

Recently, a comparative evaluation of the performance of the different nanosorbents reported was carried out to identify the best choice for preparation of ^{99}Mo/^{99m}Tc generators using (n, γ) ^{99}Mo (Chakravarty et al. 2014). Though, ^{99}Mo/^{99m}Tc generators prepared using any of the nanosorbents met the requirements for use in preparation of radiopharmaceuticals, MA and γ-Al$_2$O$_3$ were identified as the best choices in view of their higher sorption capacities which could be used for the preparation of clinical-scale ^{99}Mo/^{99m}Tc generator even while using (n, γ) ^{99}Mo produced in medium flux reactors (Table 10.3). The mechanism of adsorption of ^{99}Mo onto these nanosorbents and release of ^{99m}Tc from the same could be explained by the fact that the nanostructured metal oxide particles are hydrated and covered by amphoteric surface hydroxyl groups which can undergo reaction with either H$^+$ or OH$^-$ and develop positive or negative charges on the surface depending on the pH of the external solution (Fig. 10.3). At lower pH, these hydroxide groups become protonated and the nanosorbent surface develops a positive charge. In a weakly acidic solution, molybdate ions polymerize as follows (Spanos and Lycourghiotis 1994):

$$7 \, MoO_4^{2-} + 8 \, H^+ \rightarrow Mo_7O_{24}^{6-} + 4 \, H_2O$$

Based on this speciation, strong attraction between the positively charged nanosorbent surface and Mo$_7$O$_{24}^{6-}$ is anticipated, which accounts for strong affinity of the nanosorbents for ^{99}Mo. Subsequently, it may form a stable complex of the type [MMo$_6$O$_{24}$]$^{8-}$ (M = Ti, Zr or Al), similar to that reported with alumina (Steigman 1982). The decay of ^{99}Mo to ^{99m}Tc is not accompanied by any serious disruption of chemical bonds. As these polymolybdate ions start transforming into pertechnetate ion, which has only -1 charge, the binding would get weaker. Moreover, the saline solution used for elution of ^{99m}Tc has pH ~7 and in case of all the nanosorbents the surface charge is negative at that pH (Fig. 10.3). Owing to electrostatic repulsion, ^{99m}TcO$_4^-$ could easily be eluted with a normal saline (0.9% NaCl) solution.

^{188}W/^{188}Re Generator

The ^{188}W/^{188}Re generator is an excellent source for availing NCA grade ^{188}Re, which has immense potential for use in therapeutic nuclear medicine (Boschi et al. 2014; Callahan et al. 1989; Chakravarty et al. 2012a; Dash and Knapp 2015; Knapp Jr et al. 1994; Pillai et al. 2012). The pre-eminence of this radioisotope is primarily due its excellent nuclear decay characteristics [reasonable half-life (16.9 h), high-energy beta radiation (E$_{\beta max}$ = 2.118 MeV), 155 keV (15.8% abundance) suitable for scintigraphic imaging and dosimetry] and convenient on-site availability from ^{188}W/^{188}Re generators (Dash and

Knapp 2015; Pillai et al. 2012). The chemistry of Re is similar to that of Tc since they belong to the same group in the periodic table and this is an additional advantage towards preparing therapeutic analogues with molecules that have shown promising results in diagnosis as ^{99m}Tc-radiopharmaceuticals. Most of the separation methodologies which have been reported for ^{99}Mo/^{99m}Tc generators have also been exploited for the preparation of ^{188}W/^{188}Re generators (Dash and Knapp 2015; Pillai et al. 2012). Out of these procedures, the alumina-based column chromatographic approach wherein ^{188}W is absorbed on bulk alumina matrix and ^{188}Re is selectively eluted using saline solution at regular intervals, has been identified as the most reliable method for the preparation of ^{188}W/^{188}Re generator (Jeong and Knapp 2008). Owing to the limited sorption capacity of bulk alumina (~50 mg W/g), clinical-scale ^{188}W/^{188}Re generator can only be prepared using high specific activity (150-190 GBq/g) ^{188}W that can be produced in only very few high flux (~10^{15} n.cm^{-2}.s^{-1}) reactors available in the world (Jeong and Knapp 2008; Knapp Jr et al. 1994). Even while using high specific activity ^{188}W produced in these reactors, the ^{188}W/^{188}Re generators currently available yield low specific volume (activity/mL) of ^{188}Re and require post-elution concentration procedures prior to radiopharmaceutical preparation which is not always very convenient to perform in hospital radiopharmacy (Dash and Knapp 2015; Pillai et al. 2012). From this perspective, it is desirable to develop ^{188}W/^{188}Re generators where the concentration step can be avoided to simplify the operational procedure for their widespread clinical utility. The use of nanosorbents has been identified as an ideal choice in this direction.

The applicability of TiP which was previously used for the preparation of ^{99}Mo/^{99m}Tc generator was extended for the preparation of a 1.85 GBq (50 mCi) ^{188}W/^{188}Re generator (Chakravarty et al. 2009). The performance of this generator was studied for > 6 months. Rhenium-188 could be eluted from this generator as ^{188}ReO$_4^-$ using normal saline solution with > 80% yield. The level of ^{188}W present in the eluate was always below 10^{-3}%, which was within the acceptable limits as per the pharmacopeias (British Pharmacopoeia, 2008). The radiochemical purity of ^{188}ReO$_4^-$ was > 95% and it was found to be suitable for preparation of radiopharmaceuticals, without the need for post-elution concentration procedures. In another similar study, nano-ZrO$_2$ was used for preparation of a 1.85 GBq (50 mCi) ^{188}W/^{188}Re generator (Chakravarty et al. 2010c). The performance evaluation of this generator was carried out as done in the previous study and found to be satisfactory for use in preparation of radiopharmaceuticals. Subsequently, preparation of a clinical-scale [11.1 GBq (300 mCi) ^{188}W/^{188}Re generator was reported using γ-Al$_2$O$_3$ as sorbent (Chakravarty et al. 2011b). The sorption capacity of this material was found to be much higher than that of other nanosorbents reported earlier (Table 10.3). The generator was found to be amenable for clinical use and consistent performance was observed for a period of 6 months.

A comparative evaluation of the nanosorbents was carried out and γ-Al_2O_3 was identified as the best choice for the preparation of $^{188}W/^{188}Re$ generator since this material exhibited the highest sorption capacity for ^{188}W ions (Chakravarty and Dash 2012b). Using theoretical calculations, it was shown that while using this sorbent, even ^{188}W produced by long-term irradiation in medium flux (~10^{14} n.cm^{-2}.s^{-1}) reactors could be used for the preparation of clinical scale (> 11.1 GBq) $^{188}W/^{188}Re$ generators. Adoption of this sorbent for the preparation of $^{188}W/^{188}Re$ generator would reduce reliance on 3 high flux reactors in the world for ^{188}W-production. There are nearly 50 research reactors in the world having thermal neutron flux of ~1×10^{14} n cm^{-2} s^{-1}, many of which could be used for the production of ^{188}W for making $^{188}W/^{188}Re$ generators (Chakravarty and Dash 2012b). Therefore, adoption of this strategy would make ^{188}Re-radiopharmacy more practical, cost-effective and globally competitive. The mechanism of uptake of ^{188}W and elution of ^{188}Re from the nanosorbent based chromatographic columns could be explained on the basis of pH-dependent surface charge on nanosorbents (Fig. 10.3). There are various anionic species of tungsten, varying with pH of the medium. At pH 6, the predominant species is $W_{12}O_{41}^{10-}$ and between pH 2 and 4 the species $W_{12}O_{39}^{6-}$ predominates (Khalid et al. 2001). Owing to electrostatic attraction, high negatively charged tungsten anion is taken up on the positively charged surface of the nanosorbent under mildly acidic conditions (Fig. 10.3). Subsequently, it may form a stable complex of the type $[AlW_6O_{24}]^{8-}$, similar to that reported for bulk alumina (Steigman 1982). As these tungstate ions start transforming into perrhenate ion ($^{188}ReO_4^-$), which has only the mononegative charge, the binding gets weaker and an easy displacement of $^{188}ReO_4^-$ is expected from the nanosorbent.

$^{68}Ge/^{68}Ga$ Generator

The $^{68}Ge/^{68}Ga$ generator is an excellent source for availing ^{68}Ga ($t_{1/2}$ = 68 min), which is a positron emitter, with 89% positron branching accompanied by low photon emission (1,077 keV, 3.22%) (Roesch 2012; Roesch and Riss 2010; Rosch 2013). The cyclotron-independent availability of ^{68}Ga from this generator in an ionic form has led to the development of a wide variety of ^{68}Ga-based radiopharmaceuticals, which have opened new horizons for molecular diagnostics using PET (Banerjee and Pomper 2013; Chakravarty et al. 2013a; Kilian 2014; Mikuš et al. 2014; Morgat et al. 2013; Mueller et al. 2013; Rosch 2013; Smith et al. 2013). Despite the excellent attributes of ^{68}Ga-radiopharmacy, the low radioactive concentration, high acidity, unacceptable ^{68}Ge breakthrough and the presence of potential metal ion impurities in the generator eluate have emerged as the major deterrents towards the preparation of ^{68}Ga-based radiopharmaceuticals using ^{68}Ga eluted from majority of the $^{68}Ge/^{68}Ga$ generators available in the market (Roesch and Riss 2010). Also, most of these generator systems demonstrate deteriorating performance in terms of increased ^{68}Ge breakthrough and reduced ^{68}Ga elution yield on repeated elutions over a prolonged period of time (Roesch and Riss 2010). These limitations can be circumvented with the availability of 'state-of-the-art' automated modules for post-elution processing of ^{68}Ga eluate and subsequent radiopharmaceutical preparation (Boschi et al. 2013a; Boschi et al. 2013b; Le 2013). However, these automated modules are highly expensive and increase the production cost of ^{68}Ga-based radiopharmaceuticals. Therefore, it is desirable to develop ^{68}Ge/^{68}Ga generators which can directly be used in hospital radiopharmacies like the commercially available ^{99}Mo/^{99m}Tc generators.

The development of a ^{68}Ge/^{68}Ga generator which could directly be used for preparation of radiopharmaceuticals without the need for post-elution processing of ^{68}Ga was first reported in 2010 (Chakravarty et al. 2010b). CeO$_2$-PAN was used as the sorbent matrix in this generator, which was synthesized by decomposition of a cerium oxalate precursor to cerium oxide and its subsequent incorporation in PAN matrix. A 370 MBq (10 mCi) ^{68}Ge/^{68}Ga generator was prepared and its performance was evaluated for 7 months. Gallium-68 could be regularly eluted from this generator with > 70% elution yield with high radionuclidic purity (< 1×10^{-5}% of ^{68}Ge impurity), chemical purity (< 0.1 ppm of Ce, Fe and Mn ions) and was directly amenable for the preparation of ^{68}Ga-labeled radiopharmaceuticals (Chakraborty et al. 2013). The generator gave a consistent performance with respect to the elution yield and purity of ^{68}Ga throughout the period of investigation. Subsequently, the generator was scaled-up to 740 MBq (20 mCi) activity level and its performance was evaluated for 1 year (Chakravarty et al. 2013b). The ^{68}Ge/^{68}Ga generator was able to provide a ^{68}Ga activity with consistent yields (> 70%) and having acceptable radionuclidic, radiochemical and chemical purities over this period of time. In another similar study, a 740 MBq (20 mCi) ^{68}Ge/^{68}Ga generator was developed using t-ZrO$_2$ as the sorbent (Chakravarty et al. 2011a). The generator also gave a consistent performance with respect to the elution yield and purity of ^{68}Ga over a period of 1 year and ^{68}Ga availed from this generator was directly suitable for radiopharmaceutical preparation.

The mechanism of uptake of ^{68}Ge and elution of ^{68}Ga from the nanosorbent based chromatographic columns could be explained on the basis of pH dependent surface charge on nanosorbents (Fig. 10.3). In acid solutions (pH 1-3) the principal germanium species are [GeO(OH)$_3$]$^-$, [GeO$_2$(OH)$_2$]$^{2-}$ and [[Ge(OH)$_4$]$_8$(OH)$_3$]$^{3-}$, which are negatively charged (Neirinckx and Davis 1979). In this pH range, the surfaces of the nanosorbents develop positive charge. The strong affinity of ^{68}Ge for the nanosorbents in this pH range is attributed to the electrostatic interaction of negatively charged germanium species with the positively charged surface of the nanosorbent. In the same medium, ^{68}Ga formed from decay of ^{68}Ge, exists as Ga^{3+} ions and hence a nearly complete elution could be achieved because of electrostatic repulsion due to the same type of charges.

Hybrid Nanomaterials: The Next Generation Sorbents for Radionuclide Generators?

Although the term 'hybrid nanomaterials' has become incredibly popular for a wide variety of applications over the last few years (Alhassen et al. 2014; Fateixa et al. 2015; Gautier et al. 2013; Hadinoto et al. 2013; Qi et al. 2014), the utility of such materials in the preparation of column chromatographic radionuclide generators is yet to be explored. Hybrid nanoparticles are generally a combination of organic and inorganic matter at the nanoscopic scale. Such materials are able to combine the features of the integrated components to provide novel materials with improved characteristics. The final properties of the hybrid nanoparticles are very often not a simple addition of the properties of the independent components, but a unique result from synergetic effects. If hybrid nanoparticles are used as sorbents in column chromatographic radionuclide generators, they are expected to offer higher and better selectivity and sorption capacity for the parent radionuclide. The daughter activity can be separated with enhanced radioactive concentration and purity suitable for radiopharmaceutical preparation. However, due to the presence of the organic component which is generally more prone to damage in radiation environment, the radiation stability of hybrid nanoparticles might be questionable. Also, the consistency in radiochemical separation efficiency for such materials needs to be established before they can be routinely used in radionuclide generator technology.

Conclusions

In summary, an emerging paradigm in radionuclide generator technology has been described that lies at the intersection of nanotechnology and column chromatographic radiochemical separation technique. Taking advantage of the intrinsic properties of nanomaterials and the convenience of column chromatography based separation systems; captivating advances have been made towards the development of novel radionuclide generators suitable for clinical use. Recent advances in material science have paved the way for synthesis of a wide variety of nanosorbents through different routes to obtain tailored sizes, shapes and distributions. However, it must be ensured that the large-scale preparation of nanomaterial-based sorbents should be carried out in a sustainable and cost-effective way. The synthesis protocols adopted should be preferably based on the room-temperature aqueous solution route with negligible waste generation and should preclude the use of toxic precursors. While the utilization of any of these paradigm-changing new sorbents would be expected to reshape the present day radionuclide generator technology, it requires effective harnessing of emerging technology. Inspired vision from research scientists as well as industrial partners would be required to develop clinically useful radionuclide generator systems in appropriate volumes and at the right cost.

While the advances made so far are exciting and the efforts to develop a new generation of radionuclide generators using nanosorbents are evolving persistently, we still have a long way to go in terms of using this class of novel materials in clinical context. Completing the technology development as well as establishing the economics of this approach are the cornerstones for its survival and strength. Its success largely depends on the cost-effective availability of nanosorbents, the overall cost associated with the production of radionuclide generators while complying with current good manufacturing practices (cGMP) standards and the process of obtaining regulatory approval. In order to surmount the regulatory barrier, there is need for concerted efforts, disciplined adherence to basic regulatory rules and a joint effort by all professionals in the field including radiopharmacists, radiochemists, system designers, system engineers, automation engineers and nuclear medicine physicians. While the transition of the nanomaterial-based adsorbent from laboratory research into clinical generator prototype requires a sustained effort, the potential rewards at the end are substantial.

Acknowledgements

Research at the Bhabha Atomic Research Centre is part of the ongoing activities of the Department of Atomic Energy, India and is fully supported by government funding. The authors are grateful to Dr. K. L. Ramakumar, Director, Radiochemistry and Isotope Group, Bhabha Atomic Research Centre for his valuable support to the isotope program.

References

Alhassen, H., Antony, V., Ghanem, A., Yajadda, M.M., Han, Z.J. and Ostrikov, K.K. 2014. Organic/hybrid nanoparticles and single-walled carbon nanotubes: preparation methods and chiral applications. Chirality 26: 683-691.

Bai, X.M., Voter, A.F., Hoagland, R.G., Nastasi, M. and Uberuaga, B.P. 2010. Efficient annealing of radiation damage near grain boundaries via interstitial emission. Science 327: 1631-1634.

Banerjee, S.R. and Pomper, M.G. 2013. Clinical applications of Gallium-68. Appl Radiat Isot 76: 2-13.

Boschi, S., Lodi, F., Malizia, C., Cicoria, G. and Marengo, M. 2013a. Automation synthesis modules review. Appl Radiat Isot 76: 38-45.

Boschi, S., Malizia, C. and Lodi, F. 2013b. Overview and perspectives on automation strategies in ^{68}Ga radiopharmaceutical preparations. In Recent Results in Cancer Research, pp. 17-31.

Boschi, A., Uccelli, L., Pasquali, M., Duatti, A., Taibi, A., Pupillo, G., et al. 2014. ^{188}W/^{188}Re generator system and its therapeutic applications. J Chem 2014: 529406.

Boyd, R.E. 1982. Technetium-99m generators–the available options. Int J Appl Radiat Isot 33: 801-809.

Bringa, E.M., Monk, J.D., Caro, A., Misra, A., Zepeda-Ruiz, L., Duchaineau, M., et al. 2011. Are nanoporous materials radiation resistant? Nano Lett 12: 3351-3355.

British Pharmacopoeia Commission, British Pharmacopoeia, The Stationery Office, Norwich, UK 2008 (www.pharmacopoeiaorg.uk).

Bushra, R., Shahadat, M., Raeisssi, A.S. and Nabi, S.A. 2012. Development of nano-composite adsorbent for removal of heavy metals from industrial effluent and synthetic mixtures; its conducting behavior. Desalination 289: 1-11.

Callahan, A.P., Rice, D.E. and Knapp Jr, F.F. 1989. Rhenium-188 for therapeutic applications from an alumina-based tungsten-188/rhenium-188 radionuclide generator. A radionuclide generator system based on the adsorption of [188]W sodium or potassium tungstate on alumina gives good yields of [188]Re for radiolabelling of therapeutic agents. NUC Compact 20: 3-6.

Cañas-Carrell, J.E., Li, S., Parra, A.M. and Shrestha, B. 2014. Metal oxide nanomaterials: health and environmental effects. pp. 200-221. *In*: J. Njuguna, K. Pielichowski and H. Zhu (eds.). Health and Environmental Safety of Nanomaterials: Polymer Nanocomposites and Other Materials Containing Nanoparticles. Elsevier, United Kingdom.

Cao, G. and Wang, Y. 2011. Nanostructures and Nanomaterials: Synthesis, Properties and Applications. World Scientific, Singapore.

Chakraborty, S., Chakravarty, R., Sarma, H.D., Dash, A. and Pillai, M.R. 2013. The practicality of nanoceria-PAN-based [68]Ge/^{68}Ga generator toward preparation of ^{68}Ga-labeled cyclic RGD dimer as a potential PET radiotracer for tumor imaging. Cancer Biother Radiopharm 28: 77-83.

Chakravarty, R. and Dash, A. 2012a. Availability of yttrium-90 from strontium-90: a nuclear medicine perspective. Cancer Biother Radiopharm 27: 621-641.

Chakravarty, R. and Dash, A. 2012b. Nano structured metal oxides as potential sorbents for [188]W/[188]Re generator: a comparative study. Sep Sci Technol 48: 607-616.

Chakravarty, R. and Dash, A. 2013a. Development of Radionuclide Generators for Biomedical Applications. LAP Lambert Academic Publishing, Germany.

Chakravarty, R. and Dash, A. 2013b. Role of nanoporous materials in radiochemical separations for biomedical applications. J Nanosci Nanotechnol 13: 2431-2450.

Chakravarty, R. and Dash, A. 2014. Nanomaterial-based adsorbents: the prospect of developing new generation radionuclide generators to meet future research and clinical demands. J Radioanal Nucl Chem 299: 741-757.

Chakravarty, R., Shukla, R., Gandhi, S., Ram, R., Dash, A., Venkatesh, M., et al. 2008. Polymer embedded nanocrystalline titania sorbent for [99]Mo-^{99m}Tc generator. J Nanosci Nanotechnol 8: 4447-4452.

Chakravarty, R., Dash, A. and Venkatesh, M. 2009. Separation of clinical grade [188]Re from [188]W using polymer embedded nanocrystalline titania. Chromatographia 69: 1363-1372.

Chakravarty, R., Shukla, R., Ram, R., Tyagi, A., Dash, A. and Venkatesh, M. 2010a. Practicality of tetragonal nano-zirconia as a prospective sorbent in the preparation of [99]Mo/^{99m}Tc generator for biomedical applications. Chromatographia 72: 875-884.

Chakravarty, R., Shukla, R., Ram, R., Venkatesh, M., Dash, A. and Tyagi, A.K. 2010b. Nanoceria-PAN composite-based advanced sorbent material: a major step forward in the field of clinical-grade [68]Ge/^{68}Ga generator. ACS Appl Mater Interfaces 2: 2069-2075.

Chakravarty, R., Shukla, R., Tyagi, A.K., Dash, A. and Venkatesh, M. 2010c. Nanocrystalline zirconia: a novel sorbent for the preparation of ^{188}W/^{188}Re generator. Appl Radiat Isot 68: 229-238.

Chakravarty, R., Shukla, R., Ram, R., Tyagi, A.K., Dash, A. and Venkatesh, M. 2011a. Development of a nano-zirconia based ^{68}Ge/^{68}Ga generator for biomedical applications. Nucl Med Biol 38: 575-583.

Chakravarty, R., Shukla, R., Ram, R., Venkatesh, M., Tyagi, A.K. and Dash, A. 2011b. Exploitation of nano alumina for the chromatographic separation of clinical grade ^{188}Re from ^{188}W: a renaissance of the 188W/188Re generator technology. Anal Chem 83: 6342-6348.

Chakravarty, R., Dash, A. and Pillai, M.R.A. 2012a. Electrochemical separation is an attractive strategy for development of radionuclide generators for medical applications. Curr Radiopharm 5: 271-287.

Chakravarty, R., Ram, R., Dash, A. and Pillai, M.R. 2012b. Preparation of clinical-scale ^{99}Mo/^{99m}Tc column generator using neutron activated low specific activity ^{99}Mo and nanocrystalline gamma-Al_2O_3 as column matrix. Nucl Med Biol 39: 916-922.

Chakravarty, R., Shukla, R., Kumar Tyagi, A. and Dash, A. 2012c. Separation of medically useful radioisotopes: Role of nano-sorbents. pp. 259-301. *In*: K. Ariga (ed.). Manipulation of Nanoscale Materials: an Introduction to Nanoarchitectonics. The Royal Society of Chemistry, United Kingdom.

Chakravarty, R., Chakraborty, S., Dash, A. and Pillai, M.R.A. 2013a. Detailed evaluation on the effect of metal ion impurities on complexation of generator eluted ^{68}Ga with different bifunctional chelators. Nucl Med Biol 40: 197-205.

Chakravarty, R., Chakraborty, S., Ram, R., Dash, A. and Pillai, M.R. 2013b. Long-term evaluation of 'BARC ^{68}Ge/^{68}Ga generator' based on the nanoceria-polyacrylonitrile composite sorbent. Cancer Biother Radiopharm 28: 631-637.

Chakravarty, R., Ram, R., Mishra, R., Sen, D., Mazumder, S., Pillai, M.R.A., et al. 2013c. Mesoporous Alumina (MA) based double column approach for development of a clinical scale ^{99}Mo/^{99m}Tc generator using (n, γ) ^{99}Mo: an enticing application of nanomaterial. Ind Eng Chem Res 52: 11673-11684.

Chakravarty, R., Ram, R. and Dash, A. 2014. Comparative assessment of nano-structured metal oxides: a potential step forward to develop clinically useful ^{99}Mo/^{99m}Tc generators using (n, γ) ^{99}Mo. Sep Sci Technol 49: 1825-1837.

Chang, C., Wang, X., Bai, Y. and Liu, H. 2012. Applications of nanomaterials in enantioseparation and related techniques. TrAC Trends Anal Chem 39: 195-206.

Dash, A. and Chakravarty, R. 2014. Pivotal role of separation chemistry in the development of radionuclide generators to meet clinical demands. RSC Adv 4: 42779-42803.

Dash, A. and Knapp, F.F. 2015. An overview of radioisotope separation technologies for development of ^{188}W/^{188}Re radionuclide generators providing 188Re to meet future research and clinical demands. RSC Adv 5: 39012-39036.

Dash, A., Knapp, F.F., Jr. and Pillai, M.R. 2013. ^{99}Mo/^{99m}Tc separation: an assessment of technology options. Nucl Med Biol 40: 167-176.

Eckelman, W.C. 2009. Unparalleled contribution of technetium-99m to medicine over 5 decades. JACC Cardiovasc Imaging 2: 364-368.

Fateixa, S., Nogueira, H.I. and Trindade, T. 2015. Hybrid nanostructures for SERS: materials development and chemical detection. Phys Chem Chem Phys (In press; DOI: 10.1039/C5CP01032B).

Gautier, J., Allard-Vannier, E., Herve-Aubert, K., Souce, M. and Chourpa, I. 2013. Design strategies of hybrid metallic nanoparticles for theragnostic applications. Nanotechnology 24: 432002.

Hadinoto, K., Sundaresan, A. and Cheow, W.S. 2013. Lipid-polymer hybrid nanoparticles as a new generation therapeutic delivery platform: a review. Eur J Pharm Biopharm 85: 427-443.

Jeong, J.M. and Knapp, F.F., Jr. 2008. Use of the Oak Ridge National Laboratory tungsten-188/rhenium-188 generator for preparation of the rhenium-188 HDD/lipiodol complex for trans-arterial liver cancer therapy. Semin Nucl Med 38: S19-29.

Khalid, M., Mushtaq, A. and Iqbal, M.Z. 2001. Sorption of tungsten(VI) and rhenium(VII) on various ion-exchange materials. Sep Sci Technol 36: 283-294.

Kilian, K. 2014. ^{68}Ga-DOTA and analogs: current status and future perspectives. Rep Pract Oncol Radiother 19: S13-S21.

Knapp, F.F., Jr. and Mirzadeh, S. 1994. The continuing important role of radionuclide generator systems for nuclear medicine. Eur J Nucl Med 21: 1151-1165.

Knapp, F.F., Jr. and Baum, R.P. 2012. Radionuclide generators – a new renaissance in the development of technologies to provide diagnostic and therapeutic radioisotopes for clinical applications. Curr Radiopharm 5: 175-177.

Knapp Jr, F.F., Callahan, A.P., Beets, A.L., Mirzadeh, S. and Hsieh, B.T. 1994. Processing of reactor-produced ^{188}W for fabrication of clinical scale alumina-based ^{188}W/^{188}Re generators. Appl Radiat Isot 45: 1123-1128.

Le, V.S. 2013. ^{68}Ga generator integrated system: elution-purification- concentration integration. In Recent Results in Cancer Research, pp. 43-75.

Mikuš, P., Melník, M., Forgácsová, A., Krajčiová, D. and Havránek, E. 2014. Gallium compounds in nuclear medicine and oncology. Main Group Metal Chem 37: 53-65.

Mirzadeh, S. and Knapp Jr, F.F. 1996. Biomedical radioisotope generator systems. J Radioanal Nucl Chem 203: 471-488.

Molinski, V.J. 1982. A review of ^{99m}Tc generator technology. Int J Appl Radiat Isot. 33: 811-819.

Morgat, C., Hindie, E., Mishra, A.K., Allard, M. and Fernandez, P. 2013. Gallium-68: chemistry and radiolabeled peptides exploring different oncogenic pathways. Cancer Biother Radiopharm 28: 85-97.

Mueller, D., Klette, I. and Baum, R.P. 2013. Purification and labeling strategies for ^{68}Ga from ^{68}Ge/^{68}Ga generator eluate. In Recent Results in Cancer Research, pp. 77-87.

Mushtaq, A. 2004. Inorganic ion-exchangers: their role in chromatographic radionuclide generators for the decade 1993-2002. J Radioanal Nucl Chem 262: 797-810.

Neirinckx, R.D. and Davis, M.A. 1979. Potential column chromatography generators for ionic Ga-68. I. Inorganic substrates. J Nucl Med 20: 1075-1079.

Pillai, M.R. and Knapp, F.F., Jr. 2012. Molybdenum-99 production from reactor irradiation of molybdenum targets: a viable strategy for enhanced availability of technetium-99m. Q J Nucl Med Mol Imaging 56: 385-399.

Pillai, M.R., Dash, A. and Knapp, F.F., Jr. 2013. Sustained availability of ^{99m}Tc: possible paths forward. J Nucl Med 54: 313-323.

Pillai, M.R.A., Dash, A. and Knapp Jr, F.F. 2012. Rhenium-188: availability from the ^{188}W/^{188}Re generator and status of current applications. Curr Radiopharm 5: 228-243.

Qi, D., Cao, Z. and Ziener, U. 2014. Recent advances in the preparation of hybrid nanoparticles in miniemulsions. Adv Colloid Interface Sci 211: 47-62.

Ramamoorthy, N. 2009. Commentary: supplies of molybdenum-99–need for sustainable strategies and enhanced international cooperation. Nucl Med Commun 30: 899-905.

Roesch, F. 2012. Maturation of a key resource - the germanium-68/gallium-68 generator: development and new insights. Curr Radiopharm 5: 202-211.

Roesch, F. and Riss, P.J. 2010. The renaissance of the ^{68}Ge/^{68}Ga radionuclide generator initiates new developments in ^{68}Ga radiopharmaceutical chemistry. Curr Top Med Chem 10: 1633-1668.

Rosch, F. 2013. ^{68}Ge/^{68}Ga generators: past, present and future. Recent Results Cancer Res 194: 3-16.

Saha, G. 2010. Radionuclide generators. pp. 67-82. *In*: G.B. Saha (ed.). Fundamentals of Nuclear Pharmacy. Springer, New York.

Smith, D.L., Breeman, W.A.P. and Sims-Mourtada, J. 2013. The untapped potential of Gallium 68-PET: the next wave of 68Ga-agents. Appl Radiat Isot 76: 14-23.

Spanos, N. and Lycourghiotis, A. 1994. Molybdenum-oxo species deposited on alumina by adsorption: III. Advances in the mechanism of Mo(VI) deposition. J Catal 147: 57-71.

Steigman, J. 1982. Chemistry of the alumina column. Int J Appl Radiat Isot 33: 829-834.

11

Multifunctional Nanoparticles: Design and Application in Diagnosis of Cardiovascular Diseases

Guobing Liu,[1,2,3,a] *Hongcheng Shi* [1,2,3,b] and *Dengfeng Cheng*[1,2,3,*]

INTRODUCTION

With the rapid development of modern medicine, interdisciplinary cooperation has been more urgently needed in clinical practice. In corporating multiple fields including chemistry, physics, material science and medicine, nanomedicine provides a good example of interdisciplinary and multidisciplinary cooperation (Nune et al. 2009). The variety of nanomaterials available have provided new opportunities for the development of superior therapeutic and diagnostic agents (Wang et al. 2014). The recent years have witnessed the dramatic development of nanoparticles (NPs), which have become a major topic in nanomedicine and provide a promising surrogate for conventional diagnostic and therapeutic approaches (Lee et al. 2012). They are advantageous in that: 1) in comparison with conventional molecular imaging based on small molecules, NPs can incorporate various imaging agents, providing the potential for multifunctional platforms; 2) with large structures, NPs can load sufficient amount of imaging molecules or drugs; 3) appropriately sized and surface-modified NPs may increase the blood circulation time, mitigate opsonization and slow the elimination rate in reticuloendothelial system (RES); 4) with a large relative surface area, NPs allow chelation of multivalent targeting molecules, thus,

[1] Department of Nuclear Medicine, Zhongshan Hospital, Fudan University, Shanghai 200032, China.
[2] Institute of Nuclear Medicine, Fudan University, Shanghai 200032, China.
[3] Shanghai Institute of Medical Imaging, Shanghai 200032, China.
[a] E-mail: liuguobing0422@163.com
[b] E-mail: shi.hongcheng@zs-hospital.sh.cn
[*] Corresponding author: cheng.dengfeng@zs-hospital.sh.cn

when target-modified, enabling far more potent targeting as compared to conventional molecular imaging (Lee et al. 2012).

In recent years, NP technologies have gradually come into clinical use, with a number of NPs-based diagnostic and therapeutic agents being clinically approved and with even more under clinical or sub-clinical studies. One of the important directions is the design and application of multifunctional nanoparticles (MNPs), i.e., loading multiple imaging molecules or both imaging molecules and therapeutic agents onto the same NP to achieve multi-modal imaging or therapeutic/diagnostic benefit (Bao et al. 2013; Hrkach et al. 2012). MNPs allowing multi-modal imaging may provide important information for the diagnosis and prognosis of diseases from multiple perspectives (anatomy, metabolism, etc.) or at multiple stages, thus providing important guidance on disease management.

The design and application of MNPs rely on the conventional nanomaterial technologies as the support. To date a large variety of nanomaterials have been made available for modification of size, morphology and surface and physical modification for the assembly and construction of MNPs (Champion et al. 2007; Euliss et al. 2006; Jana et al. 2004; Xu et al. 2007). These include the inorganic nanomaterials such as metal ions, metal oxidants, semiconductors, rare earth minerals and silicon, etc., the organic nanomaterials such as liposome, micelles, high molecular polymers and dendrimers and some endogenic molecules such as HDL and LDL, etc. (Doshi and Mitragotri 2009; Euliss et al. 2006; Iqbal et al. 2012; Kaushik et al. 2015; Staicu et al. 2015). Some nanomaterials themselves provide good image contrast; for instance, gold NPs may enhance the contrast of optical image, photo-acoustic image and CT image, allowing for multi-modal imaging with only slight modification (Gobin et al. 2007; Popovtzer et al. 2008). In addition, a series of NP-related particle coating technologies, surface modification technologies, imaging molecule loading technologies and targeting molecule coupling technologies have also been under rapid development and application (Canelas et al. 2009; Wang and Uludag 2008; Yoo et al. 2011). These technologies play important roles in improving the biocompatibility and degradability of NPs, controlling the loading dose volume of contrast agent, reducing toxicity, increasing blood circulation time and enhancing targeting performance. All these have provided strong support for the design and application of MNPs (Bao et al. 2013; Lee et al. 2012; Nune et al. 2009).

Ideal MNPs used for multi-modal imaging should include: 1) a NP carrier platform for loading imaging molecules; 2) the optimal combination of multi-modal imaging modalities; 3) biological performance modification of MNPs for good biological behavior; 4) coupling of appropriate targeting molecules to increase imaging specificity. Figure 11.1 illustrates the composition of a typical MNP. Based on the research in recent years, this paper briefly introduces the selection, design and modification of MNPs and their application in cardiovascular diseases.

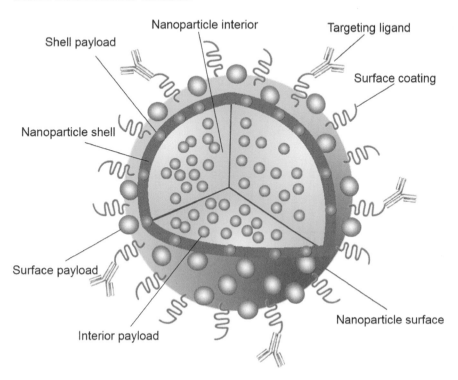

Fig. 11.1 Graphic Illustration of the Structure of a Typical Multifunctional Nanoparticle, which depicts the design and construction of a MNP including the loading routes of multiple imaging molecules or drugs, surface modification and targeting molecule coupling.

Common NPs Platform and Their Physico-Chemical Properties

NPs are generally defined as the nanoscopic particles with at least one dimension smaller than 100 nm. With significant scientific research value, NPs bridge the gap between bulk materials and atoms and molecules. While the physical properties of bulk materials are typically unrelated to their size, some of the physical properties of nanoscopic particles, e.g., surface plasmon resonance of metal NPs and superparamagnetism of magnetic materials, are closely related to the size. NPs can generally be classified into organic NPs such as liposome, micelles, high molecular polymers and dendrimers and inorganic NPs such as iron oxide, gold and silicon NPs. Table 11.1 summarizes the advantages and disadvantages of common NP materials, which will be described in the text below.

Table 11.1 Characteristics of Different Nanoparticles and Their Advantages and Disadvantages in Multifunctional Imaging.

NP	Intrinsic imaging modality	Advantages	Disadvantages	References
Organic				
Liposome	No	FDA-approved nano-carriers, ease to load hydrophilic and hydrophobic payloads	Poor *in vivo* stability	(Peer et al. 2007)
Micelles	No	Ease to load hydrophobic payloads, convenient synthesis and modification	Difficult to load hydrophilic payloads	(Koo et al. 2010; Min et al. 2010)
Dendrimers	No	Uniform and controllable size/chemical composition, high loading capacity of payloads	Limited chemical synthesis	(Torchilin 2007)
Biological NPs	No	Good biocompatibility and biodegradability, homogeneity, simple biosynthesis	Difficulty in further chemical modification	(Lin et al. 2011; Neumann et al. 2010)
Inorganic				
Iron NPs	MRI	Good biocompatibility, controllable size and morphology, clinically available contrast agents for MRI	Poor stability under aqueous condition	(Kim et al. 2009)
Gold NPs	CT, OI, PAI	Good biocompatibility, easy surface modification, controllable size and morphology, capability of optical quenching	Poor stability under aqueous condition	(Qian and Nie 2008; Sun et al. 2011)
QD	OI	Polychromatic fluorescence imaging, high fluorescence intensity, good biocompatibility	Potential toxicity	(Alivisatos et al. 2005; Lee et al. 2012)
Carbon NPs	PAI	Unique electrical characteristics, structural rigidity	Potential toxicity	(Dresselhaus and Araujo 2010; Ji et al. 2010)
Silicon NPs	No	Biocompatibility, high loading capacity	Mechanical stability	(Song et al. 1997; Tanaka et al. 2011)

Notes: FDA, food drug administration; MRI, magnetic resonance imaging; CT, computed tomography; PAI, photo-acoustic imaging; OI, optical imaging; SPECT, single photon emission tomography; PET: positron emission tomography; FMT, Fluorescent Molecular Tomography.

Organic NPs

Organic NPs can load the imaging agents including, e.g., radionuclide and near infrared fluorescent dyes, for the early diagnosis of diseases; they may also significantly increase blood circulation time and selectively concentrate in the lesions, which is known as the enhanced permeability and retention effect (Shi et al. 2009). The characteristics of the common organic NPs are described below:

Liposome

There have been extensive studies of liposome as the carrier for loading imaging molecules, probably due to the fact that it is composed of natural phospholipid molecules, which determines that it has good biocompatibility and biodegradability. Furthermore, liposome is suitable for loading various imaging molecules as it loads not only hydrophilic payloads through its aqueous interior but also hydrophobic payloads through its surface; with simple PEG modification, it is able to effectively increase the blood circulation time to facilitate sustained release of imaging molecules (Lee et al. 2012; Peer et al. 2007). Liposome has been used in the nano-carrier platforms of various imaging approaches including MRI (Mulder et al. 2004), CT (Kao et al. 2003), ultrasonography (Hamilton et al. 2002), PET (Marik et al. 2007), SPECT (Harrington et al. 2001) and optical imaging (Torchilin 2005).

Micelles

Micelles, also known as polymeric micelles, are the polymers consisting of the amphipathic chimera of heterobaric molecules, where the hydrophobic portion constitutes the core and the hydrophilic portion constitutes the shell; while the core can be used for loading hydrophobic imaging agent, the shell provides stability in the aqueous solution and the binding site for targeting molecules (Savic et al. 2003; Torchilin 2007). Thus micelles are typically used as the loading carriers for hydrophobic payloads, owing to their merits including narrow size distribution, ability to cross-link multiple functions in one entity, ability to increase water solubility of hydrophobic drugs, controlled release of loaded drug and reduced RES elimination, etc. (Savic et al. 2003). Stimuli-responsive micelles can release the loaded agents under appropriate environmental stimuli such as pH, temperature, light or ultrasound (Lee et al. 2012), which, combined with the controlled release feature of micelles, has been regarded as a common approach to targeting controlled release of biomedical imaging molecules in micelles (Min et al. 2010). Based on PEG (β amino ester) polychimera, Koo et al. (Koo et al. 2010) managed to encapsulate the hydrophobic photosensitizer protoporphyrin IX (PpIX), which is used for optical imaging, into the pH-responsive polymeric micelles. Since the pH of the extracellular fluid of tumor tissues (6.4-6.8) is normally lower than those in normal tissues (pH 7.4) and the hydrophobic amino ester is prone to receive protons in an acidic environment (pH < 6.5) leading to its reduction to a hydrophile, pH-responsive polymeric micelles are quickly demicellized in the tumor tissues leading to release of the loaded PpIX, thus allowing imaging of the tumor.

Dendritic Polymers

Dendritic polymer, named for its dendritic internal architecture, is a spherical dendritic giant molecule (Lee et al. 2012). The size, molecular weight and chemical structure of a dendritic polymer is adjustable by selecting appropriate monomer and controlling the extent of polymerization. Imaging molecules can be encapsulated into its interior or on its surface with high loading efficiency, resulting in high and controllable loading rate that makes polymeric micelles an ideal nano-platform for designing multifunctional nano-imaging agents. For instance, polyamidoamine (PAMAM) dendritic polymers have demonstrated superiority in disease diagnosis due to their 1) good biocompatibility; 2) quick kidney elimination; 3) small particle size (< 5 nm); 4) uniform morphology and size; and 5) extensive presence of amino groups to chelate multiple imaging molecules or targeting molecules (Torchilin 2007). Olson et al. coated fifth generation PAMAM dendritic polymers with six protease-activatable cell-penetrating peptides and then dually labeled it with Gd and Cy5 for MRI and optical imaging, respectively. Proteases in the tumor can cleave the peptide and activate cell penetrating peptides, which promote the internalization of the dually labeled dendritic polymer into the tumor cells and thereby contributing to successful visualization of tumor involvement in MR or optical imaging (Olson et al. 2010). Thus, these dually labeled dendrimers can be applied to aid in tumor resection, surgical planning, intraoperative guidance and postoperative evaluation after tumor resection.

Biological Polymers

Biological polymers originating from proteins, antibodies, enzymes, lipoproteins and virus, etc., may play an important role in disease diagnosis. With the advantages over other NPs, e.g., good biocompatibility and biodegradability, rapid and simple synthetic process and achievable homogeneity with the recipient, biological polymers provide superior carrier materials for the design of MNPs (Lee et al. 2012; Maham et al. 2009). For instance, with its intrinsic transportation capacity and good biocompatibility and degradability, plasma albumin can transport various molecules in the circulation system, making it a particularly feasible nano-carrier (Neumann et al. 2010). With the advantages including, e.g., good biodegradability, minimal inflammogenicity and simple modification, other biological polymers from natural sources, e.g., collagen protein, alginate, hyaluronic acid and chitosan, may also be of significant application value in loading and transporting imaging molecules (Lee et al. 2012). Ferritin, a human protein that has been extensively studied in recent years, can aggregate spontaneously to form a hollow cage structure, which can load various metallic substances or be multifunctionalized through chemical modification of surface. The aggregation or disaggregation of ferritin nanocage is related to the environmental pH: Disaggregation in an acidic environment is favorable for release of the molecules loaded in the interior of particles, while re-aggregation is allowed when the pH has risen to 7.4 or higher. Such a property gives ferritin an edge in the formulation of multifunctional imaging NPs (Lee et al. 2012). Recently, Lin et al. succeeded in designing a ferritin nanoprobe labeled with both ^{64}Cu and chelating Cy5.5 for PET and NIRF imaging (Lin et al. 2011).

Inorganic NPs
Magnetic NPs

Most magnetic NPs are composed of iron oxide and the minority is composed of iron or other magnetic elements. Two common magnetic NPs are magnetite (Fe_3O_4) and maghemite (γ-Fe_2O_3); others include the elements Cr, Fe and Co nanoparticles and Mn- and Zn-electroplated iron oxide particles, etc. (Jana et al. 2004; Jang et al. 2009; Lee et al. 2007; Sun et al. 2004). When the magnetic particle size is smaller than the width of the magnetic domain wall (typically 100 nm), the magnetic particle will produce a single domain, where the dipole moment generated by unpaired electrons is arranged in one direction, presenting superparamagnetism. The limit for the size of iron oxide particles to maintain superparamagnetism under room temperature is about 20-25 nm (Krishnan et al. 2006). In the absence of external magnetic field, there would be no magnetic interactions between the adjacent paramagnetic nanoparticles, which is essential for the nanoparticles to maintain the colloid stability.

With superparamagnetism, magnetic nanoparticles of iron oxide (MNPIO) provides unique contrast enhancement in enhanced MRI imaging, which, in addition to its good biosafety and surface modification potential, has already led to its practical application in biomedicine, where it has demonstrated tremendous application value (Sun et al. 2008; Urries et al. 2014). In comparison with Gd chelate, a conventional T1 contrast agent, MNPIO has exhibited better magnetic sensitivity and better MRI enhancement. While the lower limit of detection with Gd complex is 10^{-4} mol/L (Veiseh et al. 2010), the lower limit of detection with MNPIO reported so far can be as low as 10^{-12} mol/L (Gao and Hillebrenner 2008). In addition, MNPIO is less frequently associated with the adverse reactions including hepatic or pulmonary fibrosis. The MRI limit of detection with MNPIO is affected by, e.g., its size, crystallinity and surface modification (Veiseh et al. 2010). Presently, MNPIO can generally be classified, by the hydrodynamic size, into SPIO (small particle of iron oxide, > 40 nm) and USPIO (ultrasmall particle of iron oxide, < 40 nm), which have distinct behavior and metabolic pathways in the organism. By comparison, USPIO may be more advantageous in molecular targeting imaging, as that smaller particle size may make it more likely for USPIO to escape ingestion in the organism's immune system and reticuloendothelial system in particular (Moghimi et al. 2001; Qiao et al. 2009).

Gold NPs

The most common type of gold NP is nanosphere, which appears crimson in the water solution. Due to localized surface plasmon resonance (LSPR), gold NPs typically have complex optical properties, i.e., the valence electrons would oscillate synchronically with the incident light with specific wavelength (Boyer et al. 2002). A portion of the energy absorbed by gold NPs may be emitted in the form of scattered light, which serves as the basis for most gold NPs to be used in optical imaging (Jain et al. 2006). The absorption wavelength of gold NPs is mainly around 520 nm, at which the light would attenuate rapidly in the tissues. For *in vivo* imaging, the absorption peak of the imaging agent is preferable if it is within the optical window of human tissues (i.e., 650-1300 nm) so as to allow the incident light to penetrate the deep tissues (Tromberg et al. 2000). The absorption spectrum of gold NPs can be tuned through adjusting its geometry (Prodan et al. 2003). For instance, the LSPR of gold nanorods has two directions, one along the long axis and the other along the short axis; with increased long/short axis ratio, the frequency of oscillation along the long axis would redshift from the visible to the near-infrared region (Huang et al. 2006). Other gold nanostructures with adjustable LSPR frequency include nanoshells and nanocages, whose absorption spectrum would change with the overall size and thickness of their walls (Halas 2005; Skrabalak et al. 2007). As a result of coupled plasma resonance, aggregated gold nanosphere may also present near-infrared absorption, which is frequently used to construct the infrared photo-thermal property of gold nanostructures for *in vivo* diagnosis and treatment of diseases (Zharov et al. 2005).

As gold NPs produce strong scattered light within their LSPR frequency, they have a good prospect of extensive application in optical imaging (Jain et al. 2008). Furthermore, with the presence of LSPR, gold NPs may increase local electromagnetic field, leading to marked enhancement of fluoregenes or surface-enhanced Raman scattering (SERS) reporter molecules coupled to their surface (Qian et al. 2008). For *in vivo* application, gold NPs emitting near-infrared scattered light is most frequently used as favorable contrast agents in optical coherence tomography (Hirsch et al. 2003); while the photo-thermal attributes of gold NPs make them potent contrast agents in photo-acoustic (PA) imaging (Li and Chen 2015). In addition, taking advantage of its high material density and high atomic number, gold nanospheres and nanorods are also used as contrast agents in CT (Popovtzer et al. 2008; von Maltzahn et al. 2009).

Quantum Dots (QDs)

QDs, which by nature are semiconductor NPs emitting fluorescence, have attracted considerable attention in optical imaging (Alivisatos et al. 2005). Most QDs are semiconductor alloys of two metals, e.g., CdSe alloy, InP alloy. In order to increase the quantum yield, QDs are often coated with an isolated inorganic shell, e.g., ZnS shell; the optical attributes of QDs are the result of the quantum confinement of their valence electrons at nanometer scales (Staicu et al. 2015). The fluorescence wavelength emitted by QDs is dependent on the energy difference between the quantum energy levels and is affected by the particle size and composition of QDs, i.e., the larger the particle size is, the more redshifts the emission peak undergoes (Medintz et al. 2005). In comparison with organic fluorescent materials, QDs feature narrower emission peak and broader absorption spectrum, covering the range from ultraviolet to visible wavelengths. Therefore QDs with different emission wavelengths may all be excited simultaneously by ultraviolet light, facilitating polychromatic fluorescence imaging. In recent years, given their near infrared excitation and emission characteristics, QDs consisting of CdTe, PbS, InAs and InP have been commonly used in *in vivo* imaging (Hong et al. 2012). Mulder et al. (Mulder et al. 2006) coated QDs, as modified by trioctylphosphine oxide and hexadecylamine, with paramagnetic material to produce the fluorescence/MRI dual-modality NPs, which, with PEG and targeting molecule modification, displayed the expression of $\alpha_v\beta_3$ integrin in human umbilical venous cells, both under fluorescence microscope and in MRI.

Other Inorganic NPs

There are many other types of inorganic NPs, e.g., silicon NPs, calcium phosphate NPs, carbon nanotubes and hafnium oxide NPs (Epple et al. 2010; Maggiorella et al. 2012; Piao et al. 2008). Despite the absence of unique quantum mechanics, silicon and calcium phosphate NPs feature excellent biocompatibility, adjustable size, abundant chemical chelating agents and superior drug loading strategies, supporting their extensive application in loading drug molecules, genes and imaging molecules (Larson et al. 2008; Liong et al. 2008). Silicon NPs loaded with organic fluorochrome and radioactive iodine, known as Cornell dots or C dots, have been approved for clinical trial. Being able to enhance the efficiency of X-ray in radiation therapy, NPs formed by hafnium oxide have also been under phase I clinical studies (Bao et al. 2013). Perfluorocarbon NPs can be used in MRI and acoustic imaging (Myerson et al. 2011). For instance, Lanza et al. succeeded in designing the $\alpha_v\beta_3$- targeting perfluorocarbon NPs for MRI imaging of vascular injuries (Cyrus et al. 2008).

Optimal Strategies in Combing Multiple Imaging Modalities

Given these nanoparticles as payload platform for conveying imaging agents, what we need to consider next is how to combine the imaging modalities for complementary formulation of multi-modal NPs. The imaging modalities commonly used in biomedicine include ultrasonography, CT, MRI, SPECT, PET and optical imaging, whose characteristics are summarized in Table 11.2.

Table 11.2 Comparison of Various Imaging Modalities.

Imaging modality	Sensitivity (mol)	Resolution	Penetration depth	Imaging duration	Imaging basis	Dose	Imaging agent
Ultrasound	$10^{-6} \sim 10^{-9}$	$30 \sim 500\ \mu m$	$mm \sim cm$	$sec \sim min$	Anatomy, physiology	$\mu g \sim mg$	Microbubble
CT	Poor	$25 \sim 150\ \mu m$	Indefinite	$sec \sim min$	Anatomy, physiology	mg	Iodine
MRI	$10^{-3} \sim 10^{-5}$	$50 \sim 250\ \mu m$	Indefinite	$min \sim hr$	Anatomy, physiology, molecule	$\mu g \sim mg$	Paramagnetic: Gd, Mn, etc.; superparamagnetic: Fe_3O_4, γ-Fe_2O_3
SPECT	$10^{-10} \sim 10^{-11}$	$0.5 \sim 1\ mm$	Indefinite	min	Physiology, molecule	ng	^{99m}Tc, ^{111}In chelates
PET	$10^{-11} \sim 10^{-12}$	$1 \sim 2\ mm$	Indefinite	min	Physiology, molecule	ng	^{18}F, ^{11}C, ^{15}O
Optical imaging							
BI	$10^{-15} \sim 10^{-17}$	$3 \sim 5\ mm$	$1 \sim 2\ mm$	$sec \sim min$	Physiology, molecule	$ng \sim \mu g$	Fluorescein
FI	$10^{-9} \sim 10^{-12}$	$2 \sim 3\ mm$	$< 1\ mm$	$sec \sim min$	Physiology, molecule	$ng \sim \mu g$	Fluorochrome, QD, etc.
FMT	$10^{-6} \sim 10^{-12}$	$1 \sim 3\ mm$	$< 5\ cm$	min	Physiology, molecule	μg	Fluorochrome, QD, etc.

Notes: CT, computed tomography; MRI, magnetic resonance imaging; SPECT, single photon emission tomography; PET, positron emission tomography; BI, Bioluminescence imaging; FI, Fluorescence imaging; FMT, Fluorescent Molecular Tomography.

Despite the good safety and low cost, ultrasonic imaging has low spatial resolution as compared to CT and MRI. Being a classic anatomic imaging modality, CT allows clear visualization of organs and tissues such as the bronchovascular shadows of the lungs. MRI has even higher resolution for soft tissue pathologies and to some extent, conveys valuable information related to physiological process, but has lower diagnostic sensitivity than nuclear medicine imaging. Nuclear medicine imaging modalities (such as SPECT or PET) are sensitive tools for diagnosing diseases at an early stage, analyzing drug biodistribution and assessing the efficacy in living subjects, but the spatial resolutions of SPECT and PET are much behind the ultrasound, CT and MRI. Optical imaging technology including fluorescence and bioluminescence imaging are highly sensitive, but only accessible at limited depths of a few millimeters. It is difficult to provide accurate disease information or one can only provide the information on one aspect or at one stage of the disease simply depending on one imaging modality, as each modality is associated with the intrinsic defects such as high cost, long imaging duration or insufficient resolution or specificity or sensitivity, despite their respective advantages (Bao et al. 2013; Lee et al. 2012; Willmann et al. 2008). Furthermore, given the distinctive advantages and disadvantages of various imaging modalities, it is difficult to essentially address the problems by improvement of the imaging instruments alone. A feasible approach is the complementary combination of several modalities. For instance, PET/CT and PET/MRI produced markedly improved accuracy of disease diagnosis, attributed to the high diagnostic sensitivity of PET and the high resolution of CT and MRI to provide anatomic information. Hence, appropriate combination of multiple imaging modalities may not only achieve high diagnostic sensitivity but may also provide detailed anatomic and biological information on the disease.

In recent years, it has not been uncommon that NPs are designed to be used in multi-modal imaging including optical/MRI, optical/CT, optical/PET, PET/CT, PET/MRI, SPECT/CT, SPECT/MRI or even PET/NIRF/MRI (Jennings and Long 2009; Louie 2010). NPs can load multiple imaging agents and target transport them to the lesions, overcoming the shortcomings of any single imaging modality (Jennings and Long 2009; Louie 2010). Some inorganic NPs have an intrinsic capacity of contrast enhancement, e.g., iron oxide NPs for MRI enhancement, gold NPs for CT enhancement and QD NPs for enhanced optical imaging; organic NPs may also encapsulate into their interior or couple on their surface with various imaging agents to achieve multi-modal imaging. It therefore is possible to design multi-modal nanoprobes to incorporate several imaging modalities. For instance, Xie et al. (Xie et al. 2010) succeeded in chelating ^{64}Cu-DOTA and ^{111}In-DOTA into the optical/MRI dual-modality NPs to produce the triple-modal NPs in corporating PET/MRI/optical imaging.

Two problems should be paid attention to when designing multi-modal NPs for multi-modal imaging. First, overlap of advantages should be minimized while compensation for the weak points of each modality should be maximized to achieve the best additive benefit. For instance, CT/MRI may not be an ideal combination as they are both advantageous in presenting anatomic details; in contrast, SPECT, PET or optical imaging are often used in combination with MRI or CT to formulate multi-modal imaging modalities (Willmann et al. 2008). Second, the sensitivity difference between different imaging modalities also requires attention, as this may have an impact on the dose of imaging agent used. For instance, with high imaging sensitivity, PET and NIRF agents can be used in extremely low concentrations, while MRI and CT with low sensitivity require relatively high concentrations. Hence attention should be paid to the amount and ratio of the loaded imaging agents with different functions when designing MNPs.

Functionalization and Modification of MNPs

With the diversity of nano-carriers and optimal strategies of multi-modal combinations, considering that NPs may pass through cellular membrane and have complex reactions with it, some problems regarding NP assembly and modification technologies remain to be addressed, including surface opsonization (i.e., non-specific protein absorption) of NPs, ingestion and digestion of NPs in RES and long-term toxicity of NPs, etc. (Karagkiozaki et al. 2015; Naahidi et al. 2013). After being injected into the human body, NPs will first encounter dilution by about 5L of blood, followed by about 0.3 m/s of aortic blood flow determining the distribution and 0.6 ± 0.4 mm/s of capillary blood flow determining the absorption. From bolus injection to complete entry into blood circulation, NPs will pass through about 100,000 km of vessel and will be in contact with 8,000 m^2 of vascular surface area, during which it is inevitable that the NPs will be coated with a layer of protein from the plasma or interstitial fluid, leading to macrophage-mediated opsonization, RES phagocytosis and loss of targeting potential, etc.; The physiological factors including plasma dilution and blood circulation may affect the concentration of imaging molecules reaching the lesions (Karagkiozaki et al. 2015; Song et al. 1997). After a long period of circulation, while some particles may aggregate in the lesions, most of them would eventually be modified by plasma protein and then would be eliminated in the liver or kidney, thus preventing long-term toxicity retention. At this point, the physicochemical characteristics of NPs, e.g., hydrodynamic diameter (HD), hydrophilicity/hydrophobicity, flexibility and surface charge, may affect their elimination rate. For instance, hard spherical NPs with HD < 6 nm are primarily eliminated in the kidney, while NPs with HD > 8 nm are primarily phagocytized by the RES, metabolized in the liver and spleen and eventually eliminated in the biliary tract. With increased surface charge, non-specific plasma protein adhesion of NPs would be intensified, leading to increased HD; meanwhile the organism's opsonization on NPs is also enhanced, leading to increased elimination rate of NPs in the RES. Hence the design of NPs requires protective modification of their surface in order to allow them to escape elimination in the immune system, thus increasing the circulation time.

Some polymer molecules with isotropic, hydrophilic or neutral biophase, e.g., phospholipid (van Tilborg et al. 2006), dextran (Berry et al. 2003), polyvinylpyrrolidone (Liu et al. 2007), polyethylene glycol (PEG) (Longmire et al. 2011) and silicon (Selvan et al. 2005), are often used to modify the surface of NPs. On the one hand, when modified by these molecules, NPs are more likely to couple imaging molecules and targeting molecules to realize multiple functions; on the other hand, the hydrophilicity can be enhanced to prevent particle aggregation. In addition, appropriate molecular modification for heterogenic NPs may reduce opsonization and RES ingestion, thus helping NPs to escape immunological rejection of the organism. Another approach is to construct NPs using the endogenic materials such as HDL and viral particles, so that the NPs would not even be identified by the organism's defense system. Examples include chelating metal ions and fluorochromes onto the surface of HDL or viral particles or encapsulating some inorganic NPs into the interior of HDL or viral particles (Anderson et al. 2006; Cormode et al. 2008). In conclusion, ideal NPs must fulfill the following conditions: 1) ideal blood circulation time; 2) favorable biological behavior; 3) effective metabolic and elimination pathway; 4) good targeting efficiency; and 5) low long-term toxicity in the organism.

Special caution should be exercised when using NPs in the diagnosis of cardiovascular disease. Usually, these NPs may either react specifically with functionally disordered endothelial cells or pass through the endothelial barrier to reach the sub-endothelia region, e.g., site of atheromatous plaques. For the former, it should be considered whether the NPs have loaded enough targeting molecules and whether there are sufficient targeting receptors on the surface of the endothelial cells to secure the binding that is intense enough to resist the flush of blood flow. For the latter, the key is to control the NPs' size and surface charge to allow them to penetrate the endothelial cell space (< 10 nm) (Jayagopal et al. 2007; Karagkiozaki et al. 2015; Tassa et al. 2011). In addition, in order to ensure sufficient concentration and retention time of the NPs reaching the lesion site to release sufficient molecular probes, attention should also be paid to various physical attributes of the NPs including material, size, surface charge, curvature, morphology and surface functional groups, etc., so as to optimize the blood circulation time and to ensure their bioactivity while avoiding non-specific cell reactions (Jayagopal et al. 2007; Tassa et al. 2011).

Targeting Design of MNPs

In order to improve specificity for disease diagnosis, MNPs should also undergo targeting design, which can be either passive targeting or active targeting. In passive targeting or natural targeting, the drug-loading particles are ingested by the monocyte-macrophage system and are delivered to the organs including liver and spleen through the normal physiological process. Following intravenous injection, the distribution of NPs is primarily dependent on the particle size. Generally, when the particle size is 2.5~10 µm, most NPs accumulate in the macrophages; among them, those sized < 7 µm are generally ingested by liver and spleen macrophage; those sized 200 ~400 nm concentrate in the liver and are quickly eliminated by the liver; those sized < 10 nm accumulate slowly in the bone marrow; and those sized > 7 µm are generally retained in the form of mechanical filtration by the pulmonary capillary and are then ingested by leucocytes from pulmonary tissues or alveoli (Koo et al. 2011). In addition to particle size, the surface attributes (e.g., zeta potential) also have a major impact on the distribution of particles. Nevertheless, it is difficult to reach other target sites by passive targeting. Therefore, active targeting is more frequently used, i.e., the drug-loading particles undergo surface modification to avoid macrophage ingestion and to prevent entrapment by the liver. In the meanwhile, the specific targeting molecules, e.g., antibody or antibody segment, polysaccharide, polypeptide, polypeptide analogue, vitamin or some small molecular aptamers, are linked to the NPs, which then bind specifically to receptors of the target cell to shift the NPs from the site of natural distribution in the organism towards the specific target site. It is also acceptable that the drug is modified into a prodrug, which is activated in the targeted site to take effect (Mulder et al. 2008). Given these, accurate selection of target is also essential for improving specificity of MNP imaging. For cardiovascular diseases, a variety of multi-modal imaging targets have been identified. Taking atherosclerosis as an example, Table 11.3 lists the common imaging targets.

Table 11.3 Selection of Typical Targets for Detection of Atherosclerosis.

Pathological process	Category	Example targets
Inflammation	Macrophage surface receptors	Scavenger receptor (Ox-LDL receptor), dextran receptor, CD36, etc.
	Cellular adhesion molecules	VCAM-1, ICAM-1
	Selectins	E-selectin, P-selectin, etc.
	Proteases	MMP, myeloperoxidase, etc.
Vascularization	Endothelial cell	VCAM-1, $\alpha_v\beta_3$ receptor, E-selectin, VEGF receptor
Apoptosis	Cell membrane	PS
Thrombus		Phospholipid, fibrin, fibrinogen, IIb-IIa factors, platelet, etc.

Notes: Ox-LDL, oxidized low-density lipoprotein; VCAM-1, vascular cell adhesion molecule 1; ICAM-1, intercellular adhesion molecule 1; MMP, matrix metalloproteinase; VEGF: vascular endothelial growth factor; PS, phosphatidylserine.

Application of MNPs in The Diagnosis of Cardiovascular Disease

Cardiovascular disease (CVD) refers to a group of pathological conditions affecting the heart and vessels. In the USA, CVD has been the primary fatal disease over the past few decades, with approximately 235.5 out of 100,000 people being affected, contributing to a mortality of about 31.0%. When taking stroke into consideration, the number of people dying of CVD would account for 1/3 of the total deaths (Go et al. 2014). The availability of various clinical imaging technologies has made significant contribution to CVD diagnosis. However some problems, e.g., how to identify unstable plaques before rupture and how to monitor the therapeutic responses for patients with myocardial infarction, remain to be solved (Chen and Wu 2011; Shaw 2009). Multi-modal imaging based on MNPs promises to help solve these problems. In terms of example enumeration, some of its applications in CVD will be described below.

Atherosclerosis

Atherosclerosis is a latent, chronic, progressive disease mainly characterized by plaque formation in the arterial wall. It may remain dormant for many years, but it may, in the event of onset, lead to an acute cardiovascular event that can be life-threatening (Stone et al. 2011). In most cases, CVD is associated with atherosclerotic vascular diseases and their sequelae, e.g., myocardial infarction and stroke. Interestingly, among the factors determining the atherosclerosis-associated acute cardiovascular events, the composition of plaques is far more important than arterial stenosis, with plaque disruption and the secondary thromboembolism being the primary etiologies (Fleg et al. 2012). Despite the unavailability of exact diagnostic criteria for unstable plaque, the pathological characteristics including, e.g., fibrous cap thickness < 65 μm, area of lipid/necrosis core > 40% of the total area of the plaque, concentration of macrophages in fibrous cap > 25% or > 25 cells per 0.3 mm microscopic field in diameter and intra-plaque hemorrhage or angiogenesis are predictive of significantly increased probability of plaque disruption (Lafont 2003; Vancraeynest et al. 2011). Early diagnosis of unstable plaque is the current challenge facing the clinical physicians. In recent years, there have been extensive studies of multi-modal imaging based on MNPs in the diagnosis of atherosclerosis. Table 11.3 lists some of the common study targets. Devaraj et al. (Devaraj et al. 2009) designed an ^{18}F-CLIO (cross-linked iron oxide) MNP, which successfully displayed unstable plaque through PET-CT/MRI multi-modal imaging. Cheng et al. (Cheng et al. 2015) designed the Annexin V-targeting, ^{99}Tc-labeled iron oxide MNP, which visualized the macrophages with active apoptosis in unstable plaques through SPECT/MRI dual-modality imaging. Amirbekian et al. (Amirbekian et al. 2007) designed a magnetic/fluorescence micelle MNP, while Skajaa et al. (Skajaa et al. 2010) designed a magnetic/fluorescence/CT HDL nanocrystal, both of which, by targeted activation of antigen determinant on the macrophage surface, visualized the active macrophages in unstable plaques through MRI/optical dual-modality imaging and MRI/optical/CT tri-modal imaging, respectively. Li et al. (Li et al. 2010) designed an ^{111}In-labeled, Ga-coupled liposome MNP, which, using Anti-LOX-1 as the targeting molecule, visualized the lipid/necrosis core in the atheromatous plaques through MRI/SPECT-CT multi-modal imaging. Southworth et al. (Southworth et al. 2009) designed a ^{19}F-PFC (perfluorocarbon) MNP, which, by targeting VINP-28 (peptide ligand for VCAM-1) at VCAM-1, displayed high expression of VCAM-1 in the endothelial cells at the plaque through MRI/fluorescence imaging. Marsh et al. (Marsh et al. 2007) designed a collagen protein-coated PFC MNP which displayed target thrombus at the plaque using MRI/acoustic dual-modality imaging. The above text aimed only to list some of the many similar examples.

Vasculitis

Vasculitis, also known as angiitis, is characterized by infiltration inflammatory cell in the vascular wall or around the vessels and is associated with the vascular injuries involving a variety of pathological changes including, e.g., fibrin deposition, collagen fibrosis and endothelial cell or muscular cell necrosis. The key steps in its inflammatory cascade include dysfunction of vascular endothelial cells, activation of cellular adhesion molecules, monocytes recruitment and their differentiation toward macrophage, protein and extracellular matrix degradation, smooth muscular cell apoptosis and necrosis and neovascularity, making possible a series of targets for multimodal imaging (McAteer and Choudhury 2013). For instance, Tsourkas et al. (Tsourkas et al. 2005) visualized the high expression of VCAM-1 in the TNF-α-mediated mouse vascular injury model through MRI/fluorescence dual-modality imaging by using cross-linked iron oxide (CLIO) NP coupled with NIRF molecule Cy5.5 (i.e., CLIO/Cy5.5) and VCAM-1 antibody as the targeting molecule. Kelly et al. (Kelly et al. 2005) also visualized the high expression of VCAM-1 in the TNF-α-mediated mouse vascular injury model through MRI/fluorescence dual-modality imaging based on MRI/fluorescence dual-modality NP by substituting VCAM-1 antibody with VCAM-1 polypeptide originating from bacteriophage as the targeting molecule. Terashima et al. (Terashima et al. 2011) managed to detect the high macrophage activity under inflammatory vascular endothelium through MRI/optical dual-modality imaging by using human transferrin cage NP coupled with Cy5.5.

Myocardial Infarction

Myocardial cell apoptosis is one of the modes of cell death in myocardial ischemic process. It has been reported that during the first 4-6 hours of myocardial infarction, apoptosis is the predominant mode of death (Kajstura et al. 1996). Normally, phosphatidylserine (PS) is located on the internal surface of the myocardial cell membrane. In case of myocardial infarction, it is transferred to the external surface of cell membrane. Based on this, many scholars have designed MNPs for early diagnosis of myocardial infarction and evaluation of its extent. For instance, Sosnovik et al. (Sosnovik et al. 2005) and Schellenberger et al. (Schellenberger et al. 2004) designed the CLIO-Cy5.5 MNPs that were successfully used in the MRI/optical dual-modality imaging of acute myocardial infarction under Annexin V targeting. In addition, an inevitable pathological process accompanying myocardial apoptosis and necrosis is monocyte/macrophage infiltration, which is also an important target of acute myocardial infarction. For instance, Sosnovik et al. (Sosnovik et al. 2007) successfully used iron oxide NPs coupled with NIRF fluorochrome in MRI/FMT dual-modality imaging for visualizing the extent of myocardial infarction at the animal *in vivo* level.

Aneurysm

The incidence rate of aneurysm in elderly people is about 5%; among them about 50% of the large aneurysm rupture, leading to a mortality as high as 50% (Dillavou et al. 2006; Nahrendorf et al. 2011). While clinical treatment of aneurysm depends on imaging evaluation, the conventional anatomic imaging modalities (e.g., ultrasonography, CT, MRI) are only informative of the size and location, etc., of the aneurysm. Hence, the clinical treatment of aneurysm is restricted to therapeutic decision simply based on the size of the aneurysm, i.e., surgical or interventional therapy is indicated only for aneurysm with diameter > 5.5 cm or that has quickly enlarged recently, while aneurysms with smaller diameter may only be subject to periodical check in order to monitor the size changes, since surgical or interventional therapy itself may place the patient at a risk of aneurysm rupture (Chaikof et al. 2009). It was observed in a study that infiltration of monocytes/macrophage and the proteases such as elastase and matrix metalloproteinase that the macrophages excrete would increase significantly in the aneurysmal wall, which plays an important role in the destruction of intima-media of the aneurysmal wall and therefore, may serve as an important target for evaluating the risk of aneurysm rupture (Hellenthal et al. 2009). Nahrendorf et al. (Nahrendorf et al. 2011) designed an iron oxide NP that was coated with dextran labeled with F-18 and transformed it into a MNP by coupling the near infrared fluorescence molecule VT680 and used it in PET/CT/fluorescence multi-modal imaging, resulting in successful visualization of active macrophage metabolism in the aneurysmal wall in the ApoE$^{-/-}$ mouse model of aortic aneurysm induced by angiotensin II.

Valvular Heart Disease

Valvular heart disease is an important component of CVD, with aortic valvular heart disease being the most common (Rabkin-Aikawa et al. 2005).

The development and progression of valvular heart disease is a chronic, latent process, of which early diagnosis is vital for preventing secondary ventricular remodeling and cardiac dysfunction. Morphology-based ultrasonography, CT and MRI are only reflective of the morphological changes in valves, which however are already irreversible when identified. Studies have found many similarities in pathological changes, e.g., monocytes/macrophage infiltration, enzymolysis of matrix protein and calcium deposition, between aortic valvular heart disease and atherosclerosis (Allison et al. 2006; Drolet et al. 2006). Therefore, there may be targets for molecular imaging diagnosis of aortic valvular heart disease that are similar to those for atherosclerosis. For instance, Aikawa et al. (Aikawa et al. 2007) designed a magnetic/near infrared fluorescence MNP, which, through MRI/fluorescence dual-modality imaging, realized early diagnosis of aortic valvular heart disease induced by high-fat diet in ApoE$^{-/-}$ mice.

In addition, there have also been reports of nano-carriers being used for the molecular imaging diagnosis of diseases such as endocarditis, myocarditis and heart failure; however they were mostly used in single-modal imaging (Nahrendorf et al. 2009; Panizzi et al. 2014). It is positive that these studies at least provide feasible molecular imaging targets, e.g., endocardial vegetation for endocarditis and angiotensin II receptor 1 for heart failure, for the design of MNPs for the diagnosis of these diseases in the future (Nahrendorf et al. 2009; Panizzi et al. 2014).

Conclusion

In conclusion, in comparison with conventional small molecular imaging, MNPs have demonstrated marked advantages, especially when their application value in cardiovascular diseases has been preliminarily manifested. At present, a variety of nano-carriers and modification technologies are available for the design and application of MNPs. When designing MNPs for multi-modal imaging, the following technical issues are worth of attention: 1) Attention should always be paid to the biocompatibility of MNPs, for which organic NPs are superior over inorganic metal NPs. The designer should also ensure the degradability of the metabolites of NPs, for which the NPs from organic natural materials are advantageous by nature. 2) The ratio of various imaging agents used for MNPs should be carefully considered. For instance, the doses of imaging agents required for PET and NIRF imaging are far lower than those required for MRI and CT. 3) It is of vital importance to use targeting molecules, as they may not only improve specificity of MNPs in multi-modal imaging, but also reduce the side effects associated with non-targeting deposition.

Biomedicine practitioners have long been in search of new approaches for the diagnosis of various diseases and for the problems emerging in their treatment. As a return to such an effort, MNPs, with superior physicochemical properties and broad prospect of application, are expected to make greater contribution to disease diagnosis and treatment in the future.

Acknowledgements

This work was funded in part by the National Natural Science Foundation of China (No. 81271608 and 81201130), Shanghai Pujiang Program (No.13PJ1401400) and Shanghai Municipal Commission of Health and Family Planning Excellent Youth Program (XYQ2013106).

References

Aikawa, E., Nahrendorf, M., Sosnovik, D., Lok, V.M., Jaffer, F.A., Aikawa, M., et al. 2007. Multimodality molecular imaging identifies proteolytic and osteogenic activities in early aortic valve disease. Circulation 115: 377-86.

Alivisatos, A.P., Gu, W. and Larabell, C. 2005. Quantum dots as cellular probes. Annu Rev Biomed Eng 7: 55-76.

Allison, M.A., Cheung, P., Criqui, M.H., Langer, R.D. and Wright, C.M. 2006. Mitral and aortic annular calcification are highly associated with systemic calcified atherosclerosis. Circulation 113: 861-6.

Amirbekian, V., Lipinski, M.J., Briley-Saebo, K.C., Amirbekian, S., Aguinaldo, J.G., Weinreb, D.B., et al. 2007. Detecting and assessing macrophages in vivo to evaluate atherosclerosis noninvasively using molecular MRI. P Natl Acad Sci USA 104: 961-6.

Anderson, E.A., Isaacman, S., Peabody, D.S., Wang, E.Y., Canary, J.W. and Kirshenbaum, K. 2006. Viral nanoparticles donning a paramagnetic coat: conjugation of MRI contrast agents to the MS2 capsid. Nano Lett 6: 1160-4.

Bao, G., Mitragotri, S. and Tong, S. 2013. Multifunctional nanoparticles for drug delivery and molecular imaging. Annu Rev Biomed Eng 15: 253-282.

Berry, C.C., Wells, S., Charles, S. and Curtis, A.S. 2003. Dextran and albumin derivatised iron oxide nanoparticles: influence on fibroblasts in vitro. Biomaterials 24: 4551-7.

Boyer, D., Tamarat, P., Maali, A., Lounis, B. and Orrit, M. 2002. Photothermal imaging of nanometer-sized metal particles among scatterers. Science 297: 1160-3.

Canelas, D.A., Herlihy, K.P. and DeSimone, J.M. 2009. Top-down particle fabrication: control of size and shape for diagnostic imaging and drug delivery. Wiley Interdiscip Rev Nanomed Nanobiotechnol 1: 391-404.

Chaikof, E.L., Brewster, D.C., Dalman, R.L., Makaroun, M.S., Illig, K.A., Sicard, G.A., et al. 2009. The care of patients with an abdominal aortic aneurysm: the society for vascular surgery practice guidelines. J Vasc Surg 50: S2-49.

Champion, J.A., Katare, Y.K. and Mitragotri, S. 2007. Particle shape: a new design parameter for micro- and nanoscale drug delivery carriers. J Control Release 121: 3-9.

Chen, I.Y. and Wu, J.C. 2011. Cardiovascular molecular imaging: focus on clinical translation. Circulation 123: 425-43.

Cheng, D., Li, X., Zhang, C., Tan, H., Wang, C., Pang, L., et al. 2015. Detection of vulnerable atherosclerosis plaques with a dual-modal single-photon-emission computed tomography/magnetic resonance imaging probe targeting apoptotic macrophages. ACS Appl Mater Inter 7: 2847-55.

Cormode, D.P., Skajaa, T., van Schooneveld, M.M., Koole, R., Jarzyna, P., Lobatto, M.E., et al. 2008. Nanocrystal core high-density lipoproteins: a multimodal molecular imaging contrast agent platform. Nano Lett 8: 3715-3723.

Cyrus, T., Zhang, H., Allen, J.S., Williams, T.A., Hu, G., Caruthers, S.D., et al. 2008. Intramural delivery of rapamycin with alphavbeta3-targeted paramagnetic nanoparticles inhibits stenosis after balloon injury. Arterioscler Thromb Vasc Biol 28: 820-6.

Devaraj, N.K., Keliher, E.J., Thurber, G.M., Nahrendorf, M. and Weissleder, R. 2009. 18F labeled nanoparticles for in vivo PET-CT imaging. Bioconjug Chem 20: 397-401.

Dillavou, E.D., Muluk, S.C. and Makaroun, M.S. 2006. A decade of change in abdominal aortic aneurysm repair in the United States: have we improved outcomes equally between men and women? J Vasc Surg 43: 230-8; discussion 238.

Doshi, N. and Mitragotri, S. 2009. Designer biomaterials for nanomedicine. Adv Funct Mater 19: 3843-3854.

Dresselhaus, M.S. and Araujo, P.T. 2010. Perspectives on the 2010 nobel prize in physics for graphene. ACS Nano 4: 6297-302.

Drolet, M.C., Roussel, E., Deshaies, Y., Couet, J. and Arsenault, M. 2006. A high fat/high carbohydrate diet induces aortic valve disease in C57BL/6J mice. J Am Coll Cardiol 47: 850-5.

Epple, M., Ganesan, K., Heumann, R., Klesing, J., Kovtun, A. and Neumann, S. 2010. Application of calcium phosphate nanoparticles in biomedicine. J Mater Chem 20: 18-23.

Euliss, L.E., DuPont, J.A., Gratton, S. and DeSimone, J. 2006. Imparting size, shape and composition control of materials for nanomedicine. Chem Soc Rev 35: 1095-104.

Fleg, J.L., Stone, G.W., Fayad, Z.A., Granada, J.F., Hatsukami, T.S., Kolodgie, F.D., et al. 2012. Detection of high-risk atherosclerotic plaque: report of the NHLBI working group on current status and future directions. JACC Cardiovasc Imaging 5: 941-55.

Gao, J. and Hillebrenner, H.L. 2008. Nanotubular probes as ultrasensitive MR contrast agent. U.S. Patent #0124281.

Go, A.S., Mozaffarian, D., Roger, V.L., Benjamin, E.J., Berry, J.D., Blaha, M.J., et al. 2014. Heart disease and stroke statistics–2014 update: a report from the American Heart Association. Circulation 129: e28-e292.

Gobin, A.M., Lee, M.H., Halas, N.J., James, W.D., Drezek, R.A. and West, J.L. 2007. Near-infrared resonant nanoshells for combined optical imaging and photo-thermal cancer therapy. Nano Lett 7: 1929-34.

Halas, N. 2005. Playing with plasmons: tuning the optical resonant properties of metallic nanoshells. MRS Bulletin 30: 362-367.

Hamilton, A., Huang, S.L., Warnick, D., Stein, A., Rabbat, M., Madhav, T., et al. 2002. Left ventricular thrombus enhancement after intravenous injection of echo-genic immunoliposomes: studies in a new experimental model. Circulation 105: 2772-8.

Harrington, K.J., Mohammadtaghi, S., Uster, P.S., Glass, D., Peters, A.M., Vile, R.G., et al. 2001. Effective targeting of solid tumors in patients with locally advanced cancers by radiolabeled pegylated liposomes. Clin Cancer Res 7: 243-54.

Hellenthal, F.A., Buurman, W.A., Wodzig, W.K. and Schurink, G.W. 2009. Biomarkers of abdominal aortic aneurysm progression. Part 2: inflammation. Nat Rev Cardiol 6: 543-52.

Hirsch, L.R., Stafford, R.J., Bankson, J.A., Sershen, S.R., Rivera, B., Price, R.E., et al. 2003. Nanoshell-mediated near-infrared thermal therapy of tumors under magnetic resonance guidance. P Natl Acad Sci USA 100: 13549-54.

Hong, G., Robinson, J.T., Zhang, Y., Diao, S., Antaris, A.L., Wang, Q., et al. 2012. In vivo fluorescence imaging with Ag2S quantum dots in the second near-infrared region. Angew Chem Int Ed Engl 51: 9818-21.

Hrkach, J., Von Hoff, D., Mukkaram, A.M., Andrianova, E., Auer, J., Campbell, T., et al. 2012. Preclinical development and clinical translation of a PSMA-targeted docetaxel nanoparticle with a differentiated pharmacological profile. Sci Transl Med 4: 128ra39.

Huang, X., El-Sayed, I.H., Qian, W. and El-Sayed, M.A. 2006. Cancer cell imaging and photothermal therapy in the near-infrared region by using gold nanorods. J Am Chem Soc 128: 2115-20.

Iqbal, M.A., Md, S., Sahni, J.K., Baboota, S., Dang, S. and Ali, J. 2012. Nanostructured lipid carriers system: recent advances in drug delivery. J Drug Target 20: 813-30.

Jain, P.K., Lee, K.S., El-Sayed, I.H. and El-Sayed, M.A. 2006. Calculated absorption and scattering properties of gold nanoparticles of different size, shape and composition: applications in biological imaging and biomedicine. J Phys Chem B 110: 7238-48.

Jain, P.K., Huang, X., El-Sayed, I.H. and El-Sayed, M.A. 2008. Noble metals on the nanoscale: optical and photothermal properties and some applications in imaging, sensing, biology and medicine. Acc Chem Res 41: 1578-86.

Jana, N.R., Chen, Y. and Peng, X. 2004. Size- and shape-controlled magnetic (Cr, Mn, Fe, Co, Ni) oxide nanocrystals via a simple and general approach. Chem Mater 16: 3931-3935.

Jang, J.T., Nah, H., Lee, J.H., Moon, S.H., Kim, M.G. and Cheon, J. 2009. Critical enhancements of MRI contrast and hyperthermic effects by dopant-controlled magnetic nanoparticles. Angew Chem Int Ed Engl 48: 1234-38.

Jayagopal, A., Russ, P.K. and Haselton, F.R. 2007. Surface engineering of quantum dots for in vivo vascular imaging. Bioconjug Chem 18: 1424-33.

Jennings, L.E. and Long, N.J. 2009. 'Two is better than one'–probes for dual-modality molecular imaging. Chem Commun 28: 3511-24.

Ji, S.R., Liu, C., Zhang, B., Yang, F., Xu, J., Long, J., et al. 2010. Carbon nanotubes in cancer diagnosis and therapy. Biochim Biophys Acta 1806: 29-35.

Kajstura, J., Cheng, W., Reiss, K., Clark, W.A., Sonnenblick, E.H., Krajewski, S., et al. 1996. Apoptotic and necrotic myocyte cell deaths are independent contributing variables of infarct size in rats. Lab Invest 74: 86-107.

Kao, C.Y., Hoffman, E.A., Beck, K.C., Bellamkonda, R.V. and Annapragada, A.V. 2003. Long-residence-time nano-scale liposomal iohexol for X-ray-based blood pool imaging. Acad Radiol 10: 475-83.

Karagkiozaki, V., Logothetidis, S. and Pappa, A. 2015. Nanomedicine for atherosclerosis: molecular imaging and treatment. J Biomed Nanotechnol 11: 191-210.

Kaushik, A.Y., Tiwari, A.K. and Gaur, A. 2015. Role of excipients and polymeric advancements in preparation of floating drug delivery systems. Int J Pharma Investig 5: 1-12.

Kelly, K.A., Allport, J.R., Tsourkas, A., Shinde-Patil, V.R., Josephson, L. and Weissleder, R. 2005. Detection of vascular adhesion molecule-1 expression using a novel multimodal nanoparticle. Circ Res 96: 327-36.

Kim, J., Piao, Y. and Hyeon, T. 2009. Multifunctional nanostructured materials for multimodal imaging and simultaneous imaging and therapy. Chem Soc Rev 38: 372-90.

Koo, H., Lee, H., Lee, S., Min, K.H., Kim, M.S., Lee, D.S., et al. 2010. In vivo tumor diagnosis and photodynamic therapy via tumoral pH-responsive polymeric micelles. Chem Commun 46: 5668-70.

Koo, H., Huh, M.S., Sun, I.C., Yuk, S.H., Choi, K., Kim, K., et al. 2011. In vivo targeted delivery of nanoparticles for theranosis. Acc Chem Res 44: 1018-28.

Krishnan, K.M., Pakhomov, A.B., Bao, Y., Blomqvist, P., Chun, Y., Gonzales, M.ô.G.K., et al. 2006. Nanomagnetism and spin electronics: materials, microstructure and novel properties. J Mater Sci 41: 793-815.

Lafont, A. 2003. Basic aspects of plaque vulnerability. Heart 89: 1262-7.

Larson, D.R., Ow, H., Vishwasrao, H.D., Heikal, A.A., Wiesner, U. and Webb, W.W. 2008. Silica nanoparticle architecture determines radiative properties of encapsulated fluorophores. Chem Mater 20: 2677-2684.

Lee, D.E., Koo, H., Sun, I.C., Ryu, J.H., Kim, K. and Kwon, I.C. 2012. Multifunctional nanoparticles for multimodal imaging and theragnosis. Chem Soc Rev 41: 2656-72.

Lee, J.H., Huh, Y.M., Jun, Y.W., Seo, J.W., Jang, J.T., Song, H.T., et al. 2007. Artificially engineered magnetic nanoparticles for ultra-sensitive molecular imaging. Nat Med 13: 95-9.

Li, D., Patel, A.R., Klibanov, A.L., Kramer, C.M., Ruiz, M., Kang, B., et al. 2010. Molecular imaging of atherosclerotic plaques targeted to oxidized LDL receptor LOX-1 by SPECT/CT and magnetic resonance clinical perspective. Circulation 3: 464-472.

Li, W. and Chen, X. 2015. Gold nanoparticles for photoacoustic imaging. Nanomedicine 10: 299-320.

Lin, X., Xie, J., Niu, G., Zhang, F., Gao, H., Yang, M., et al. 2011. Chimeric ferritin nanocages for multiple function loading and multimodal imaging. Nano Lett 11: 814-9.

Liong, M., Lu, J., Kovochich, M., Xia, T., Ruehm, S.G., Nel, A.E., et al. 2008. Multifunctional inorganic nanoparticles for imaging, targeting and drug delivery. ACS Nano 2: 889-96.

Liu, H.L., Ko, S.P., Wu, J.H., Jung, M.H., Min, J.H., Lee, J.H., et al. 2007. One-pot polyol synthesis of monosize PVP-coated sub-5 nm Fe3O4 nanoparticles for biomedical applications. J Magn Magn Mater 310: e815-e817.

Longmire, M.R., Ogawa, M., Choyke, P.L. and Kobayashi, H. 2011. Biologically optimized nanosized molecules and particles: more than just size. Bioconjug Chem 22: 993-1000.

Louie, A. 2010. Multimodality imaging probes: design and challenges. Chem Rev 110: 3146-95.

Maggiorella, L., Barouch, G., Devaux, C., Pottier, A., Deutsch, E., Bourhis, J., et al. 2012. Nanoscale radiotherapy with hafnium oxide nanoparticles. Future Oncol 8: 1167-81.

Maham, A., Tang, Z., Wu, H., Wang, J. and Lin, Y. 2009. Protein-based nanomedicine platforms for drug delivery. Small 5: 1706-21.

Marik, J., Tartis, M.S., Zhang, H., Fung, J.Y., Kheirolomoom, A., Sutcliffe, J.L., et al. 2007. Long-circulating liposomes radiolabeled with [18F]fluorodipalmitin ([18F]FDP). Nucl Med Biol 34: 165-71.

Marsh, J.N., Senpan, A., Hu, G., Scott, M.J., Gaffney, P.J., Wickline, S.A., et al. 2007. Fibrin-targeted perfluorocarbon nanoparticles for targeted thrombolysis. Nanomedicine 2: 533-43.

McAteer, M.A. and Choudhury, R.P. 2013. Targeted molecular imaging of vascular inflammation in cardiovascular disease using nano- and micro-sized agents. Vascul Pharmacol 58: 31-8.

Medintz, I.L., Uyeda, H.T., Goldman, E.R. and Mattoussi, H. 2005. Quantum dot bio-conjugates for imaging, labelling and sensing. Nat Mater 4: 435-46.

Min, K.H., Kim, J.H., Bae, S.M., Shin, H., Kim, M.S., Park, S., et al. 2010. Tumoral acidic pH-responsive MPEG-poly (beta-amino ester) polymeric micelles for cancer targeting therapy. J Control Release 144: 259-66.

Moghimi, S.M., Hunter, A.C. and Murray, J.C. 2001. Long-circulating and target-specific nanoparticles: theory to practice. Pharmacol Rev 53: 283-318.

Mulder, W.J., Strijkers, G.J., Griffioen, A.W., van Bloois, L., Molema, G., Storm, G., et al. 2004. A liposomal system for contrast-enhanced magnetic resonance imaging of molecular targets. Bioconjug Chem 15: 799-806.

Mulder, W.J., Koole, R., Brandwijk, R.J., Storm, G., Chin, P.T., Strijkers, G.J., et al. 2006. Quantum dots with a paramagnetic coating as a bimodal molecular imaging probe. Nano Lett 6: 1-6.

Mulder, W.J., Cormode, D.P., Hak, S., Lobatto, M.E., Silvera, S. and Fayad, Z.A. 2008. Multimodality nanotracers for cardiovascular applications. Nat Clin Pract Cardiovasc Med 5 Suppl 2: S103-11.

Myerson, J., He, L., Lanza, G., Tollefsen, D. and Wickline, S. 2011. Thrombin-inhibiting perfluorocarbon nanoparticles provide a novel strategy for the treatment and magnetic resonance imaging of acute thrombosis. J Thromb Haemost 9: 1292-300.

Naahidi, S., Jafari, M., Edalat, F., Raymond, K., Khademhosseini, A. and Chen, P. 2013. Biocompatibility of engineered nanoparticles for drug delivery. J Control Release 166: 182-94.

Nahrendorf, M., Sosnovik, D.E., French, B.A., Swirski, F.K., Bengel, F., Sadeghi, M.M., et al. 2009. Multimodality cardiovascular molecular imaging, Part II. Circ Cardiovasc Imaging 2: 56-70.

Nahrendorf, M., Keliher, E., Marinelli, B., Leuschner, F., Robbins, C.S., Gerszten, R.E., et al. 2011. Detection of macrophages in aortic aneurysms by nanoparticle positron emission tomography-computed tomography. Arterioscler Thromb Vasc Biol 31: 750-7.

Neumann, E., Frei, E., Funk, D., Becker, M.D., Schrenk, H.H., Muller-Ladner, U., et al. 2010. Native albumin for targeted drug delivery. Expert Opin Drug Deliv 7: 915-25.

Nune, S.K., Gunda, P., Thallapally, P.K., Lin, Y., Forrest, M.L. and Berkland, C.J. 2009. Nanoparticles for biomedical imaging. Expert Opin Drug Del 6: 1175-1194.

Olson, E.S., Jiang, T., Aguilera, T.A., Nguyen, Q.T., Ellies, L.G., Scadeng, M., et al. 2010. Activatable cell penetrating peptides linked to nanoparticles as dual probes for in vivo fluorescence and MR imaging of proteases. P Natl Acad Sci USA 107: 4311-6.

Panizzi, P., Stone, J.R. and Nahrendorf, M. 2014. Endocarditis and molecular imaging. J Nucl Cardiol 21: 486-95.

Peer, D., Karp, J.M., Hong, S., Farokhzad, O.C., Margalit, R. and Langer, R. 2007. Nanocarriers as an emerging platform for cancer therapy. Nat Nanotechnol 2: 751-60.

Piao, Y., Burns, A., Kim, J., Wiesner, U. and Hyeon, T. 2008. Designed fabrication of silica-based nanostructured particle systems for nanomedicine applications. Adv Funct Mater 18: 3745-3758.

Popovtzer, R., Agrawal, A., Kotov, N.A., Popovtzer, A., Balter, J., Carey, T.E., et al. 2008. Targeted gold nanoparticles enable molecular CT imaging of cancer. Nano Lett 8: 4593-6.

Prodan, E., Radloff, C., Halas, N.J. and Nordlander, P. 2003. A hybridization model for the plasmon response of complex nanostructures. Science 302: 419-22.

Qian, X., Peng, X.H., Ansari, D.O., Yin-Goen, Q., Chen, G.Z., Shin, D.M., et al. 2008. In vivo tumor targeting and spectroscopic detection with surface-enhanced Raman nanoparticle tags. Nat Biotechnol 26: 83-90.

Qian, X.M. and Nie, S.M. 2008. Single-molecule and single-nanoparticle SERS: from fundamental mechanisms to biomedical applications. Chem Soc Rev 37: 912-20.

Qiao, R., Yang, C. and Gao, M. 2009. Superparamagnetic iron oxide nanoparticles: from preparations to in vivo MRI applications. J Mater Chem 19: 6274-6293.

Rabkin-Aikawa, E., Mayer, J.E.J. and Schoen, F.J. 2005. Heart valve regeneration. Adv Biochem Eng Biotechnol 94: 141-179.

Savic, R., Luo, L., Eisenberg, A. and Maysinger, D. 2003. Micellar nanocontainers distribute to defined cytoplasmic organelles. Science 300: 615-8.

Schellenberger, E.A., Sosnovik, D., Weissleder, R. and Josephson, L. 2004. Magneto/optical annexin V, a multimodal protein. Bioconjug Chem 15: 1062-7.

Selvan, S.T., Tan, T.T. and Ying, J.Y. 2005. Robust, non-cytotoxic, silica-coated CdSe quantum dots with efficient photoluminescence. Adv Mater 17: 1620-1625.

Shaw, S.Y. 2009. Molecular imaging in cardiovascular disease: targets and opportunities. Nat Rev Cardiol 6: 569-79.

Shi, M., Lu, J. and Shoichet, M.S. 2009. Organic nanoscale drug carriers coupled with ligands for targeted drug delivery in cancer. J Mater Chem 19: 5485-5498.

Skajaa, T., Cormode, D.P., Falk, E., Mulder, W.J., Fisher, E.A. and Fayad, Z.A. 2010. High-density lipoprotein-based contrast agents for multimodal imaging of atherosclerosis. Arterioscler Thromb Vasc Biol 30: 169-76.

Skrabalak, S.E., Au, L., Li, X. and Xia, Y. 2007. Facile synthesis of Ag nanocubes and Au nanocages. Nat Protoc 2: 2182-90.

Song, C.X., Labhasetwar, V., Murphy, H., Qu, X., Humphrey, W.R., Shebuski, R.J., et al. 1997. Formulation and characterization of biodegradable nanoparticles for intravascular local drug delivery. J Controlled Release 43: 197-212.

Sosnovik, D.E., Schellenberger, E.A., Nahrendorf, M., Novikov, M.S., Matsui, T., Dai, G., et al. 2005. Magnetic resonance imaging of cardiomyocyte apoptosis with a novel magneto-optical nanoparticle. Magn Reson Med 54: 718-24.

Sosnovik, D.E., Nahrendorf, M., Deliolanis, N., Novikov, M., Aikawa, E., Josephson, L., et al. 2007. Fluorescence tomography and magnetic resonance imaging of myocardial macrophage infiltration in infarcted myocardium in vivo. Circulation 115: 1384-91.

Southworth, R., Kaneda, M., Chen, J., Zhang, L., Zhang, H., Yang, X., et al. 2009. Renal vascular inflammation induced by western diet in ApoE-null mice quantified by (19)F NMR of VCAM-1 targeted nanobeacons. Nanomedicine 5: 359-67.

Staicu, L.C., Ackerson, C.J., Cornelis, P., Ye, L., Berendsen, R.L., Hunter, W.J., et al. 2015. Pseudomonas moraviensis subsp. stanleyae, a bacterial endophyte of hyperaccumulator stanleya pinnata, is capable of efficient selenite reduction to elemental selenium under aerobic conditions. J Appl Microbiol 22: 47-52.

Stone, G.W., Marso, S.P., Parise, H., Templin, B., White, R., Zhang, Z.ô.S.P., et al. 2011. A prospective natural-history study of coronary atherosclerosis. New Engl J Med 364: 226-235.

Sun, C., Lee, J.S.H. and Zhang, M. 2008. Magnetic nanoparticles in MR imaging and drug delivery. Adv Drug Deliver Rev 60: 1252-1265.

Sun, I.C., Eun, D.K., Koo, H., Ko, C.Y., Kim, H.S., Yi, D.K., et al. 2011. Tumor-targeting gold particles for dual computed tomography/optical cancer imaging. Angew Chem Int Ed Engl 50: 9348-51.

Sun, S., Zeng, H., Robinson, D.B., Raoux, S., Rice, P.M., Wang, S.X., et al. 2004. Monodisperse MFe2O4 (M = Fe, Co, Mn) nanoparticles. J Am Chem Soc 126: 273-9.

Tanaka, K., Kitamura, N. and Chujo, Y. 2011. Bimodal quantitative monitoring for enzymatic activity with Simultaneous signal increases in 19F NMR and fluorescence using silica nanoparticle-based molecular probes. Bioconjugate Chem 22:1484-1490.

Tassa, C., Shaw, S.Y. and Weissleder, R. 2011. Dextran-coated iron oxide nanoparticles: a versatile platform for targeted molecular imaging, molecular diagnostics and therapy. Acc Chem Res 44: 842-52.

Terashima, M., Uchida, M., Kosuge, H., Tsao, P.S., Young, M.J., Conolly, S.M., et al. 2011. Human ferritin cages for imaging vascular macrophages. Biomaterials 32: 1430-7.

Torchilin, V.P. 2005. Recent advances with liposomes as pharmaceutical carriers. Nat Rev Drug Discov 4: 145-160.

Torchilin, V.P. 2007. Micellar nanocarriers: pharmaceutical perspectives. Pharm Res 24: 1-16.

Tromberg, B.J., Shah, N., Lanning, R., Cerussi, A., Espinoza, J., Pham, T., et al. 2000. Non-invasive in vivo characterization of breast tumors using photon migration spectroscopy. Neoplasia 2: 26-40.

Tsourkas, A., Shinde-Patil, V.R., Kelly, K.A., Patel, P., Wolley, A., Allport, J.R., et al. 2005. In vivo imaging of activated endothelium using an anti-VCAM-1 magnetooptical probe. Bioconjug Chem 16: 576-81.

Urries, I., Munoz, C., Gomez, L., Marquina, C., Sebastian, V., Arruebo, M., et al. 2014. Magneto-plasmonic nanoparticles as theranostic platforms for magnetic resonance imaging, drug delivery and NIR hyperthermia applications. Nanoscale 6: 9230-40.

van Tilborg, G.A., Mulder, W.J., Deckers, N., Storm, G., Reutelingsperger, C.P., Strijkers, G.J., et al. 2006. Annexin A5-functionalized bimodal lipid-based contrast agents for the detection of apoptosis. Bioconjug Chem 17: 741-9.

Vancraeynest, D., Pasquet, A., Roelants, V., Gerber, B.L. and Vanoverschelde, J.J. 2011. Imaging the vulnerable plaque. J Am Coll Cardiol 57: 1961-1979.

Veiseh, O., Gunn, J.W. and Zhang, M. 2010. Design and fabrication of magnetic nanoparticles for targeted drug delivery and imaging. Adv Drug Deliv Rev 62: 284-304.

von Maltzahn, G., Park, J.H., Agrawal, A., Bandaru, N.K., Das, S.K., Sailor, M.J., et al. 2009. Computationally guided photothermal tumor therapy using long-circulating gold nanorod antennas. Cancer Res 69: 3892-900.

Wang, D., Lin, B. and Ai, H. 2014. Theranostic nanoparticles for cancer and cardiovascular applications. Pharm Res 31: 1390-406.

Wang, G. and Uludag, H. 2008. Recent developments in nanoparticle-based drug delivery and targeting systems with emphasis on protein-based nanoparticles. Expert Opin Drug Deliv 5: 499-515.

Willmann, J.K., van Bruggen, N., Dinkelborg, L.M. and Gambhir, S.S. 2008. Molecular imaging in drug development. Nat Rev Drug Discov 7: 591-607.

Xie, J., Chen, K., Huang, J., Lee, S., Wang, J., Gao, J., et al. 2010. PET/NIRF/MRI triple functional iron oxide nanoparticles. Biomaterials 31: 3016-22.

Xu, Z., Hou, Y. and Sun, S. 2007. Magnetic core/shell Fe3O4/Au and Fe3O4/Au/Ag nanoparticles with tunable plasmonic properties. J Am Chem Soc 129: 8698-9.

Yoo, J.W., Doshi, N. and Mitragotri, S. 2011. Adaptive micro and nanoparticles: temporal control over carrier properties to facilitate drug delivery. Adv Drug Deliv Rev 63: 1247-56.

Zharov, V.P., Galitovskaya, E.N., Johnson, C., Kelly, T. 2005. Synergistic enhancement of selective nanophotothermolysis with gold nanoclusters: potential for cancer therapy. Lasers Surg Med 37: 219-26.

12

Theranostic Nanoparticles for Photothermal Therapy of Cancer

Christopher G. England[1,*] and *Weibo Cai*[2,*]

INTRODUCTION

Cancer is a set of diseases that can be characterized by the overactive proliferation and differentiation of abnormal cells. Currently, it is estimated that 1 in 7 deaths are attributed to some type of cancer, making malignant neoplasms the second leading cause of death worldwide, claiming the lives of more than 8 million people each year (Torre et al. 2015). While lung cancer is the most common malignancy found in males worldwide, breast cancer remains the most diagnosed and lethal form of cancer for women (Torre et al. 2015). In developed countries, prostate cancer has risen to become the highest diagnosed malignancy for men. In general, solid tumors account for more than 85 percent of the malignancies diagnosed each year, with the remaining cancers consisting of hematological tumors affecting the blood, bone marrow and lymph nodes (Tang et al. 2007). Since the early 2000s, significant improvements in cancer treatment have been accomplished, including the development of a new class of cancer treatment, known as immunotherapy (Gravitz 2013). In addition, there are several platforms currently being investigated as potential cancer treatment modalities (Scott et al. 2012a). Despite significant improvements in treating some forms of cancer, there are several types of cancer that have maintained dismal survival rates with limited treatment options for patients.

Pancreatic cancer and lung cancer are two examples of malignancies that remain difficult to treat with overall five-year survival rates of 7 and 17 percent

[1] Postdoctoral Fellow, Department of Medical Physics, University of Wisconsin – Madison, Room 7148, 1111 Highland Ave, Madison, WI 53705-2275, USA.
[2] Associate Professor, Departments of Radiology and Medical Physics, University of Wisconsin – Madison, Room 7137, 1111 Highland Ave, Madison, WI 53705-2275, USA.
* Corresponding authors: cengland2@wisc.edu and wcai@uwhealth.org

in the United States, respectively (ACS 2015). The treatment of some forms of cancer, including breast, melanoma, prostate, testis, thyroid and uterine, has seen five-year survival rates improve to greater than 80%. The dismal outcomes associated with highly lethal cancers have been attributed to several factors, including the limited number of effective chemotherapeutics available for treatment, the lack of biomarkers for screening high-risk patients, the inability to overcome resistance mechanisms displayed by malignant cells and ineffective screening modalities for monitoring treatment response in patients. In addition, the microenvironment of solid tumors is vastly different from that of normal tissue (Tredan et al. 2007). Studies conducted in the mid-1900s revealed that tumor masses undergo sporadic changes in vasculature, termed angiogenesis, which results in the formation of hypoxic and necrotic regions of tumor. Treatment of highly hypoxic tumors remains difficult, as chemotherapeutics have limited access to intratumoral regions (Khawar et al. 2015; Tredan et al. 2007). In addition, increased intratumoral pressure and the highly acidic microenvironment of tumors limit the effectiveness of most available treatments. Lastly, radiation therapy is dependent upon intratumoral oxygen, thus hypoxic regions of tissue may be resistant to the anticancer effects of radiotherapy (Brown 2007). For these reasons, there is a dire need for the creation of new and improved modalities for the clinical treatment of cancer.

Malignant tissues display several unique characteristics different from normal tissue that promote their growth and survival. For example, the harsh microenvironment of solid tumors effectively limits the delivery of many small molecules and other therapeutic compounds to the diseased tissue (Sriraman et al. 2014). Since the treatment of cancer relies upon the selective elimination of all malignant cells residing in a tumor, ineffective delivery of therapeutic agents to the entire tumor results in suboptimal therapeutic response or complete treatment failure. In addition, highly toxic chemotherapeutics must be delivered specifically to cancerous cells, while avoiding off-target accumulation that would result in unnecessary damage to the healthy surrounding tissue. While most chemotherapeutics are effective in treating cancer by slowing the disease progression, most of these highly toxic compounds result in severe systemic side effects (Malhotra and Perry 2003). While surgical intervention offers the highest cure rates for cancer, patients with advanced disease are often not candidates for surgical resection as the disease may have metastasized to other organs or tissues. In addition, malignant cells have gained several evolutionary tools for displaying resistance to various chemotherapeutics and other therapies, including mechanisms that allow cells to pump chemotherapeutics out of the cell back into the extracellular space (Holohan et al. 2013).

Most cancers are diagnosed according to a scale determined by the World Health Organization (WHO) that describes the tumor grade from G1 to G4, ranging from low grade (G1) to intermediate grade (G2) to high-grade (G3/

G4) disease. Staging of solid tumors is essential for determining the optimal treatment plan for patients. For patients with low-grade disease with well-differentiated tumors, surgical resection is often the best option, offering the highest potential cure rate. For those patients with advanced stage disease (G3/G4), surgical intervention may not be an option, thus physicians will rely upon chemotherapeutics, radiation therapy and other treatment modalities for improving survival and the patient's quality of life. Currently, there are several chemotherapeutic agents available that have been divided into classes based upon their mechanism of action (Mihlon et al. 2010). Common chemotherapeutic classes include alkylating antineoplastic agents that directly disrupt DNA synthesis, antimetabolites that disrupt both DNA and RNA synthesis, topoisomerase inhibitors that cause destructive changes in DNA structure and certain antibiotics with anti-cancer properties that may cause DNA intercalation, alter ion transport or result in free radical formation.

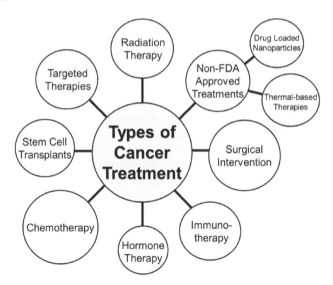

Fig. 12.1 Current options for cancer treatment include surgical intervention, chemotherapy, radiotherapy, targeted therapies, immunotherapy, hormone therapy, stem cell transplants and some non-FDA approved treatments. Non-FDA approved treatments consist of drug-loaded nanoparticle systems and thermal-based therapies, including photothermal therapy (PTT), photodynamic therapy (PDT) and photoimmunotherapy (PIT).

In addition, there are several relatively new treatments available for cancer therapy, including immunotherapy, targeted therapies, hormone therapies and stem cell transplants (Fig. 12.1). During an initial cancer evaluation, physicians will determine a treatment plan specific for the patient that is dependent upon several factors. Most treatment plans will rely upon the concurrent or sequential use of two or more therapeutic modalities. Immunotherapy uses the immune system of the patient to elicit an anti-tumor response. Immunotherapy includes the use of monoclonal antibodies and cytokines and has been shown to be highly effective in the treatment of several cancers (Kyi and Postow 2014). At this time, there are over 25 monoclonal antibodies approved for the treatment of cancer in the United States, with several others in the preclinical pipeline (Scott et al. 2012b). Targeted therapies are small-molecule drugs or monoclonal antibodies that target specific receptors or target proteins. Hormone therapy and stem cell transplants are relatively uncommon cancer treatments. In addition, there are several treatment modalities currently being investigated as potential cancer treatments. One of these treatments includes drug-loaded nanoparticles for increasing the delivery of drugs specifically to tumors, while avoiding off-target accumulation in healthy tissues (Davis et al. 2008). Photothermal therapy (PTT) is another new cancer treatment currently being investigated in clinical trials that will be thoroughly discussed in this chapter (Day et al. 2011). While there are several treatments for cancer available, surgical intervention, chemotherapy and radiotherapy remain the most commonly employed treatment modalities.

Diagnostic and Therapeutic Applications of Nanoparticles in Oncology

Since chemotherapeutics undergo systemic distribution *in vivo*, these toxic compounds often accumulate in both cancerous and non-cancerous tissues. The accumulation of chemotherapeutics in non-cancerous tissues may result in severe systemic toxicity. For example, cisplatin is an anti-cancer drug that commonly causes widespread toxicities, including nephrotoxicity, ototoxicity and neutropenia (Adams 2000). In addition, some chemotherapeutics suffer from low water solubility and poor pharmacokinetic profiles, thus there is an increased risk of dose-dependent hepatotoxicity and nephrotoxicity in patients needing high drug dosages to elicit a therapeutic response (Narvekar et al. 2014). Several strategies have been developed to overcome the current limitations of cancer treatments, yet many of these novel treatment platforms have shown modest clinical benefits. However, nanoparticle-based treatment platforms have become increasingly popular during the last decade.

Nanotherapeutics have been investigated as a means to improve upon the current limitations of cancer treatment by delivering drugs using particles ranging in size from 1-200 nm, selectively targeted to cancerous lesions using various targeting ligands. Some common entities added to the surface of nanoparticles for selective targeting of cancer include antibodies, small proteins, peptides and others (Ruoslahti et al. 2010). While this chapter discusses the use of nanomaterials as theranostic agents for cancer treatment and imaging, nanoparticles have been extensively investigated for treatment of numerous diseases, ranging from cardiovascular disease to viral infections (Godin et al. 2010).

As an alternative treatment approach, nanoparticles have several advantages over current chemotherapeutics. First, the small size of nanoparticles allows them to passively accumulate in tumor tissue through the enhanced permeation and retention (EPR) effect (Bertrand et al. 2014). The formation of new blood vessels from preexistent vasculature (angiogenesis), in combination with inefficient lymphatic drainage, results in leaky blood vasculature and increased intratumoral pressure. These two factors are the leading contributors to the EPR effect in solid tumors. Secondly, the high surface area of nanoparticles allows the loading of large drug payloads. In addition, multiple therapeutic compounds can be loaded in or onto the surface of nanoparticles. For example, gold nanoparticles and silica-gold nanoshells were layered with bilayer lipid coating, which allowed the concurrent loading of hydrophobic (paclitaxel) and hydrophilic anti-cancer compounds (cisplatin) (England et al. 2015; England et al. 2013). In addition, several chemotherapeutics display poor pharmacokinetic profiles as they are limited by their physicochemical properties (e.g., low water solubility), which can be overcome by loading drugs in the nanoparticle. These drug-loaded nanoparticles display the pharmacokinetic profile of the nanoparticle, virtually masking the physicochemical properties of the drug. Lastly, active targeting of nanoparticles through surface modifications leads to specific accumulation of nanoparticles in the tissues of interest. For example, nanoparticles are often targeted to cell surface receptors that are upregulated in malignant cells (e.g., epidermal growth factor receptor (EGFR)) (Wang and Thanou 2010). In some instances, researchers have designed nanoparticles specifically targeting tumor vasculature, such as CD105 (Chen et al. 2015b). Many cancer cell-targeted nanoplatforms display limited tumor accumulation as they must extravasate from the vasculature to reach the malignant cells, yet vasculature extravasation is not required for angiogenesis-targeted nanoparticles.

While nanoparticles have shown great promise as potential candidates for improving cancer treatment, there have also been several investigations into the use of nanoparticles for cancer imaging (Fukumori and Ichikawa 2006). Certain nanomaterials absorb light at specific wavelengths optimal for their visualization in tissue and living cells through optical imaging. Nanoparticles capable of absorbing light in the near infrared region (NIR) have been widely exploited for visualizing and treating cancer and other diseases. The NIR region of light is the wavelength range between 650-1200 nm. In this spectral region, the autofluorescence that arises from chromophores that commonly absorb light between 250-650 nm (e.g., hemoglobin, melanin and water) can be diminished (Weissleder 2001). In addition, tissue penetration is highest in the NIR, which allows researchers to investigate biological activities occurring up to 4 cm below the skin (Escobedo et al. 2010). In return, it is possible to image orthotopic tumors or visualize treatment response in tumors deeper from the anatomical surface. Overall, these optical properties of nanoparticles have been exploited for several reasons, including cancer detection through various imaging modalities (e.g., optical and photoacoustic imaging) and cancer treatment through light-activated therapies.

Applications of Thermal-Based Therapies in Cancer Treatment

The human body is constantly maintaining a state of homeostasis, in which the body temperature is kept constant at 37°C. Since early times, people saw a connection between increased body temperature and illness. Later, it was discovered that temperatures above 37°C could cause irreversible damage to cellular structures, resulting in various bodily responses from cellular death to organ failure (Van Drie 2011). In the 19th century, it was first noted that elevated body temperature could cure some diseases, such as cancer (Hobohm 2005). While not understood at that time, physicians would induce fevers in the patients through the use of bacteria, which would slow the progression of the cancer (Habash et al. 2006). In the 21st century, scientists discovered the biochemical actions that occur during localized temperature increases. For example, proteins begin to lose their conformational structure and start to denature at temperatures above 41°C, with irreversible cellular damage occurring after 60 min of exposure to this temperature (Van Drie 2011). When the temperature rises above 48°C for short time periods, irreversible damage to the cellular machinery occurs rapidly. While systemic temperature changes are not commonly utilized in disease treatment (e.g., injection of pathogen to elicit a fever); localized temperature changes have found their place in the treatment of several diseases. For example, there are numerous investigations into the effects of temperature and cancer treatment, as chemotherapy and radiotherapy are more successful when given after the patients undergo a cycle of hyperthermia (Kaur et al. 2011).

While the concept of using light for cancer therapy may seem unique, it is a common phenomenon in the natural world. There are many light-dependent biological processes that occur in the human body, including the synthesis of vitamin D, the metabolism of bilirubin and light therapy has been shown to be effective in the treatment of seasonal depression. For the treatment of cancer (and some other diseases), there are three light-based therapies that utilize photons of particular wavelengths to elicit a therapeutic effect. The three thermal-based therapies are photothermal therapy (PTT), photodynamic therapy (PDT) and photoimmuno therapy (PIT) (Xie et al. 2012; Yi et al. 2014). While PDT and PIT are dependent upon the release of oxidative radicals by activated photosensitizing agents, PTT does not rely upon oxygen to interact with the target cells (Fig. 12.2). The advantage of using PTT, in comparison to PDT, is that many tumors experience varying levels of hypoxia, or low oxygen concentration, thus PDT and PIT lack effectiveness in certain solid tumors. Figure 12.2 demonstrates the mechanisms by which PTT and PDT light-based therapies function. The applications of PTT have been widely explored in the last two decades as a means of treating malignancies, while effectively limiting the possible damage to healthy tissues (Hirsch et al. 2003). While PDT and PIT have been studied for cancer therapy, this chapter focuses on the application of theranostic nanoparticles for specifically PTT.

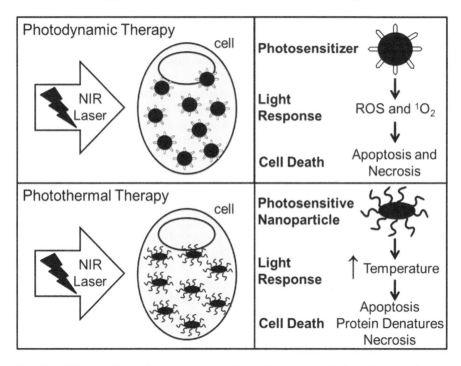

Fig. 12.2 Therapeutic mechanisms of photothermal therapy and photodynamic therapy. While both light-based therapies are similar, PDT is oxygen-dependent, while PTT does not require oxygen.

Exposure of NIR-absorbing nanoparticles to NIR irradiation results in the excitement of electrons to a higher energy state. When the electrons return to ground state, they release energy in the form of heat or electromagnetic radiation. This heat released during this transition results in localized hyperthermia of the tissue, which can cause the temperature in the tissue to rise from 37°C to between 41–47°C (Huang and El-Sayed 2011). While normal tissues can function in hyperthermia conditions for short amounts of time, cancerous tissues are more prone to cellular damage in these circumstances.

The biological effects that arise from the localized hyperthermia (i.e., protein denaturation, cellular apoptosis and cellular necrosis) are contributed to the intratumoral hypoxia and heat dissipation found in tumor tissue (Skitzki et al. 2009).

Theranostic Nanoparticles for Photothermal Therapy of Cancer

Theranostic agents are nanoparticle platforms that combine both treatment and imaging capabilities into a single nanoparticle-based construct. For more information regarding theranostic nanoparticles, please see the in-depth review written by Janib et al. that was published in *Advanced Drug Delivery Reviews* (Janib et al. 2010). Some nanoparticles are excellent theranostic agents due to their unique physicochemical properties, including their chemical composition, shape and size. As briefly mentioned earlier, nanoparticles have a high surface area to volume ratio that may be readily functionalized with passive and active targeting ligands. In addition, this high surface area allows the loading of large drug payloads. Besides drug delivery, many nanoplatforms display unique chemical properties that allow their usage for molecular imaging. For example, quantum dots are effective contrast agents for optical and photoacoustic imaging of tumors (Popescu and Toms 2006), while some magnetic nanoparticles are highly effective magnetic resonance imaging (MRI) contrast agents (Hadjipanayis et al. 2008). In addition, the radiolabeling of almost any nanoparticle-based platform allows for the tracking of the nanoparticle *in vivo* using positron emission tomography (PET) or single-photon emission computer tomography (SPECT) imaging modalities (Sun et al. 2015a).

The nanoplatforms utilized for PTT often have similar characteristics, as they must have an absorbance in the NIR spectral region. Besides displaying a strong absorbance in the NIR region, successful nanoparticle-based PTT agents should be highly efficient at converting solar radiation into heat (i.e., photothermal conversion) and should display good thermal conductivity (Sailor and Park 2012). In addition, highly biocompatible or biodegradable nanoparticles have more clinical potential than many other nanoparticle platforms (e.g., metallic nanoparticles) that may result in systemic toxicity (Kempen et al. 2015).

In the past, PTT was considered an ineffective treatment modality as several proteins in human tissues display strong extinction coefficients in the visible range. It was expected that localized heating would result in severe tissue damage to healthy tissues. With the development of highly selective nanoparticles capable of photothermal conversion, it became possible to specifically target tumor tissues for PTT. In addition, advancements in PTT have shown that unnecessary damage to healthy tissue can be limited by using nanoparticles that absorb light in two specific spectral regions, known as the first and second biological windows (Fig. 12.3) (Smith et al. 2009). The first biological window is found between 700 nm to 980 nm, while the second window extends from 1000 nm to 1350 nm. In these biological windows, optical absorption and scattering is minimized, which allows for imaging tissue at increased depths from the surface. Several studies have shown the feasibility of achieving optical penetration depths of up to 4 cm in small animals (Ntziachristos 2010). While this method remains limited for the treatment of orthotopic diseases, PTT remains a suitable treatment modality for diseases close to the skin layer and for diseased tissues accessible by endoscopic techniques.

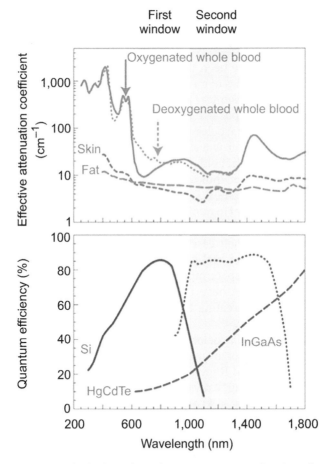

Fig. 12.3 The biological windows are two spectral regions that allow for optimal tissue penetration and limited scattering effects. The biological windows are between 700 nm to 980 nm and 1000 nm to 1350 nm. Adapted with permission from (Smith et al. 2009).

This chapter examines the recent advances in PTT using theranostic nanoparticles. Since there are several nanoplatforms that have been success-fully employed as PTT agents, the sections of this chapter have been arranged by nanoparticle type. The first section covers metallic nanoparticles or those nanoparticles that consist of a single metallic core (e.g., gold or silver), for their PTT properties. Next, nanoparticles synthesized using core-shell processes (i.e., nanoshells) are examined before discussing the applications of carbon nanotubes for PTT. Lastly, several nanoparticle-based platforms consisting of multiple nanoparticles in a single construct are discussed. As this chapter focuses on the development of theranostic agents for both imaging and treatment of cancer, we discuss the applications of nanoparticles for concurrent molecular imaging and PTT in each section. Literature searches have been limited to the past five years in most cases.

Metallic Nanoparticles for Photothermal Therapy of Cancer

Metallic nanoparticles have been the most studied nanoparticle-based PTT agents. The optical properties of metallic nanoparticles are readily modifiable by changing the physicochemical properties of the nanoparticle, including chemical composition, shape, size and agglomeration state. For example, as the size of spherical nanoparticles increases, the extinction peak shifts from lower wavelengths to higher wavelengths (Kelly et al. 2003; Khlebtsov et al. 2005). In the case of spherical gold nanoparticles, extinction peaks are limited to approximately 570-610 nm, thus not reaching the biological window needed for successful PTT. To overcome this limitation, the shape of gold nanoparticles can be altered from spherical to more complex structures, including nanorods, nanocages and nanoprisms. Gold nanorods have gained significant impact in the last decade due to their ease of synthesis and high reproducibility. In addition, gold nanorods are highly tunable as their extinction peak can be adjusted by altering the aspect ratios (Park et al. 2014). Overall, there have been several groups which have utilized the unique optical properties of metallic nanoparticles in creating theranostic nanoplatforms for concurrent imaging and PTT of cancer.

Many theranostic nanoparticles have been designed for multimodality imaging and treatment. For example, Liu et al. designed a gold nanostar probe for surface-enhanced Raman scattering (SERS) detection, two-photon luminescence (TPL) imaging, x-ray computed tomography (CT) and PTT (Liu et al. 2015b). A comparison of 30 and 60 nm gold nanostars was performed to determine which size allowed for optimal tumor uptake. The 30 nm gold nanostars were shown to accumulate better than the 60 nm gold nanostars. In addition, the accumulation was dose-dependent, thus an increased bolus dose of nanoparticles resulted in higher tumor accumulation. In reference to the PTT properties of gold nanostars, the star-like shape made the nanoparticles highly effective photothermal conversion agents due to the tip-enhanced plasmonic effect (Zhang et al. 2013). This resulted in the heating effect being highly concentrated in tumor tissue with no damage to healthy tissues. The authors concluded this study by noting that gold nanostars could be useful for intraoperative detection of malignancies.

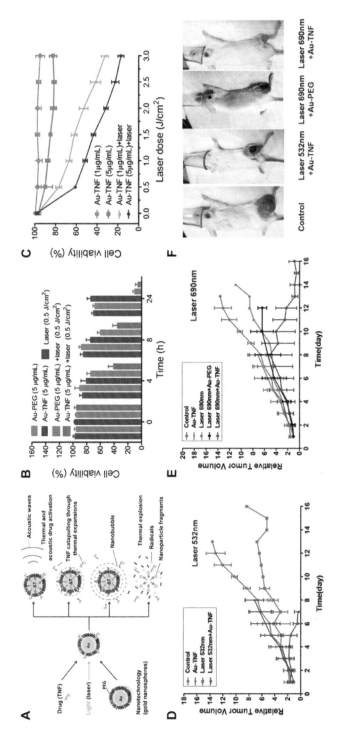

Fig. 12.4 Photothermal properties of TNF-gold nanospheres. (A) Laser excitation of TNF-conjugated gold nanospheres resulted in the formation of acoustic waves, thermal expansion, acoustic nanobubbles and thermal explosion into nanoparticle fragments. (B) Cell viability was measured after exposure to PEGylated nanoparticles or TNF-nanoparticles with and without laser treatment. (C) The photothermal effects were dose-dependent as the higher dose of TNF-nanoparticles resulted in significant cell death. Relative tumor size was monitored after select treatments with a (D) 532 nm laser and (E) 690 nm laser. (F) Tumor images revealed minimal tissue damage for 690 nm laser with TNF-nanoparticles, while other groups showed significant damage to surrounding tissues. Adapted with permission from (Shao et al. 2013).

Recently, Shao et al. examined the synergistic biochemical and physical effects of photothermally-activated gold ellipsoidal nanospheres loaded with tumor necrosis factor alpha (TNF-α), a thermosensitizing cytokine, to result in a multiple dynamic phenomena, which included nanoparticle temperature increases, thermal expansion of tissue and production of acoustic waves, microbubbles and nanobubbles (Fig. 12.4A) (Shao et al. 2013). It was hypothesized that the anti-cancer properties of the TNF-gold nanoparticles could be enhanced through the mechanisms shown in Fig. 12.4A. PEGylated gold nanoparticles were used as the control for all studies. The cytotoxicity of TNF-coated gold nanospheres was investigated by incubating cancer cells with TNF-coated gold nanospheres or PEGylated gold nanospheres (Fig. 12.4B). Cells incubated with TNF-coated gold nanoparticles and exposed to 0.5 J/cm^2 of laser displayed the most significant cell death, which became noticeable 8 h post laser ablation. In addition, two concentrations of TNF-coated nanoparticles were examined with or without laser ablation (Fig. 12.4C). Cell death was shown to be concentration dependent with 5 µg/mL TNF-coated gold nanoparticles with laser ablation displaying significantly more cytotoxicity than 1 µg/mL TNF-coated gold nanoparticles with laser ablation. The authors noted that the gold nanoparticles would accumulate in tumor tissue as clusters. This clustering effect caused a shift in the plasmon resonance toward higher wavelengths with a broader peak, which made the nanoparticles optimal for PTT. For this reason, the efficacy of PTT *in vivo* was compared using two lasers at different wavelengths (532 nm and 690 nm). While TNF-coated nanoparticles, in combination with PTT at 532 nm, resulted in significant tumor inhibition (Fig. 12.4D), all mice experienced severe tissue damage and had to be euthanized before the conclusion of the study (Fig. 12.4F). For mice receiving TNF-coated gold nanoparticles with the 690 nm laser ablation, tumor size was significantly inhibited through 16 d (Fig. 12.4E); with no visible signs of tissue damage (Fig. 12.4F).

In a similar study, Zelasko-Leon and colleagues synthesized albumin-stabilized gold nanorods actively targeted to the upregulated receptor mucin 1 (MUC1) (Zelasko-Leon et al. 2015). The surface of the nanoplatform was shown to be highly stable and bioinert through two-photon luminescence confocal and darkfield imaging techniques. In addition, the MUC1-targeted nanoparticles were assessed for their PTT properties by incubating the nanoparticles with cells before NIR laser exposure. In a similar study, Chen et al. developed gold nanoparticles specifically targeted to Mucin 7 (MUC7) (Chen et al. 2015a). They showed that non-conjugated nanoparticles required significantly higher energy levels of 30 W/cm^3 to reach similar anti-tumor efficacy, in comparison to targeted nanoparticles, which required only 10 W/cm^3. In another study, Kim et al. developed silica-coated gold nanorods for photoacoustic imaging and PTT (Kim et al. 2014). Using a xenograft mouse model, they evaluated the delivery and spatial distribution of nanoparticles in tumor tissue. After the nanoparticles were allowed to accumulate in the tumor, PTT was performed using an 808 nm continuous wave laser. The authors captured the PTT process using photoacoustic imaging to monitor temperature changes within the tumor and healthy surrounding tissues. Tumors were exposed to laser ablation for 5 min and the maximum temperature achieved in the tumor was 53°C.

Since the optical properties of nanoparticles are shape-dependent, researchers have designed several uniquely shaped nanoplatforms. For example, Huang et al. designed gold bellflowers with multiple-branched petals for photoacoustic image-guided PTT (Huang et al. 2014b). Nanoparticles were constructed through a liquid-liquid-gas triphase interface system and the resulting nanoparticles displayed a significantly high photothermal conversion efficiency of 74 percent. Besides PTT, this nanoplatform could be used for multimodality molecular imaging and treatment, including SERS, infrared (IR) sensors and NIR-triggered drug delivery. In general, gold nanorods are the most widely investigated platform for nanoparticle-based PTT (Dickerson et al. 2008; Huang et al. 2006). A recent study compared the photothermal efficiency of three gold nanorods, ranging in length from 17 to 38 nm. Gold nanoparticles of the middle size, with a length of 28 nm, produced the most photothermal heat (Mackey et al. 2014). Furthermore, the 28 nm gold nanorods were evaluated as PTT agents for ablation of human oral squamous cell carcinomas.

Several groups have examined the use of gold nanocages for PTT, both *in vitro* and *in vivo* (Menon et al. 2013). Piao et al. designed a natural stealth coating to improve the blood circulation lifetime of gold nanoparticles through a red blood cell (RBC) membrane coating (Piao et al. 2014). The RBC-coated nanocages showed high tumor accumulation and all mice that received PTT survived over 45 d after PTT, Huang et al. synthesized ferritin nanocages loaded with an NIR dye for PTT, fluorescence imaging and photoacoustic imaging (Huang et al. 2014a). Excitation of the nanoparticle at 550 nm allowed for high quantum yield fluorescence imaging, while excitation of the dye at 808 nm made PTT and photoacoustic imaging possible. In addition to imaging and treatment, several nanoplatforms have been designed for multimodality treatment. For example, You et al. prepared a liposomal nanoparticle containing hollow gold nanospheres and doxorubicin (You et al. 2014). Doxorubicin was released in a controlled manner by irradiating the liposomes with an NIR laser. The combination of PTT and chemotherapeutic delivery, with laser activation, resulted in significant decreases in tumor size, as compared to PTT or doxorubicin alone.

Magnetic nanoparticles have been widely explored for their excellent biocompatibility and magnetic properties. Since iron is an endogenous nutrient, it is metabolized in the body via the transferrin pathway, thus iron nanoparticles are considered highly biocompatible. In addition, iron-based nanoparticles readily pass through the plasma membrane and their magnetic properties allow for magnetic hyperthermia. While current studies require a large dose of superparamagnetic iron oxide nanoparticles (SPIONs) for PTT, many researchers are developing novel magnetic nanoparticles better suited for PTT. Shen and colleagues were the first group to report the use of magnetic nanoparticle clusters for photothermal ablation of tumors *in vivo* (Shen et al. 2015). Tumors were irradiated with a laser at 808 nm after receiving an injection of clustered SPIONs. In comparison to single nanoparticles, nanoparticle clusters displayed superior photothermal conversion efficiency.

Nanoshells for Photothermal Therapy of Cancer

Nanoshells have a dielectric core covered with a thin metallic shell. These nanoparticles are optically tunable and optimizable for PTT and NIR-based molecular imaging. By altering the ratio of core radius to the thickness of the metallic shell, the peak extinction can be shifted to a specific wavelength (Lal et al. 2008). For example, nanoshells with 60 nm cores and 20 nm shells display a maximum extinction peak around 700 nm, while a reduction in the shell thickness to 5 nm will shift the peak to approximately 1050 nm (Kado et al. 2014). While the other nanoparticles described in this chapter are still in preclinical investigations, nanoshell-mediated PTT is in clinical trials for neck tumors and prostate cancer. While PEGylated nanoshells have shown great promise as PTT agents, early results from a Phase 1 trial have shown that 150 nm PEGylated nanoshells are slow to clear from the body (Gad et al. 2012). In addition, Stern et al. evaluated the safety of gold nanoshells for PTT in patients with human prostate cancer (Stern et al. 2015). The study consisted of 15 patients that underwent single-fiber laser ablation prior to a radical prostatectomy. Patients were monitored for six months after the procedure and only two patients displayed adverse effects linked to the nanoparticle injection, including an allergic reaction resolved with antihistamines and a transient burning sensation. Overall, gold nanoshells displayed an excellent safety profile.

Some researchers have developed complex nanoplatforms that include multiple layers. For example, Jiang et al. synthesized doxorubicin-loaded poly (lactic acid) (PLA) cores that were covered with a gold shell (Jing et al. 2014). In addition, the nanoparticles received a surface modification containing Mn-porphyrin, which provided a greater relaxivity for MRI. The Mn-porphyrin-conjugated gold nanoshells were evaluated as potential theranostic agents for concurrent MRI-guided PTT and doxorubicin delivery for the treatment of colorectal cancer *in vivo*. The release of doxorubicin from the nanoparticles was dependent upon the laser, in which PTT and drug release occurred simultaneously. By combining two treatment modalities into a single construct, the authors noted a synergistic improvement in therapeutic outcome.

Recently, a novel nanoplatform, known as nanomatryoshkas, was developed for PTT that consisted of a three-layer system, in which a gold core is encapsulated by a silica layer and finally coated with a gold layer (Ayala-Orozco et al. 2014a). The optical properties of nanomatryoshkas arise from the strong interactions between the solid gold core and gold layer, which effectively allows for NIR tuning (Mukherjee et al. 2010). Using these three layered nanoparticles, Ayala-Orozco and colleagues compared the photothermal efficacy of 90 nm gold nanomatryoshkas and 150 nm gold nanoshells in breast cancer tumor-bearing mice (Ayala-Orozco et al. 2014b). The nanomatryoshkas were shown to absorb light five times more effectively than gold nanoshells. In addition, the mice that underwent PTT treatment with nanomatryoshkas displayed significantly higher survival rates with 83% of the mice remaining healthy and tumor free at 60 d post treatment. For comparison, only 33% of the mice that underwent PTT with nanoshells survived 60 d post treatment.

While the majority of nanoshells discussed in this section have been gold-based, it is important to consider nanoshells constructed from other metals. For example, Fu and colleagues designed a nanoshell by growing Prussian blue nanoshells around SPIO nanocores (Wang et al. 2014a). For MRI, the nanoshell platform displayed significant T2-weighted contrast enhancements. In addition, the nanoparticle increased PTT efficacy, resulting in more than 80% cell death. In mice, an injection of nanoparticles in combination with PTT resulted in an 87.2% inhibition of tumor growth. Another type of nanoshell was constructed by Ji et al., who developed a three layer platform for photoacoustic and MRI-guided PTT (Ji et al. 2007). The nanoshell contained an SPIO core, covered by a silica interface, followed by a gold shell. The nanoparticles displayed a strong NIR spectral band, which allowed for highly efficient photothermal conversion. In addition, the nanoparticles exhibited high transverse relaxivity and could undergo T2-weighted MRI.

Silver is another metal that may be utilized for developing novel theranostic systems. For example, a silver core was functionalized with small SPIOs for MRI and PTT by Lin et al. (Lin et al. 2014). The nanoparticles were less than 100 nm, which promoted tumor accumulation and cell internalization. In addition, the nanoparticles displayed a strong NIR absorption peak optimal for PTT. The PTT efficiency was examined in a melanoma murine model and showed that tumor size was significantly decreased when mice received PTT after an intravenous injection of Fe_3O_4-coated silver nanoshells. In addition, the overall survival of the mice was extended by approximately 5 d in the treatment group. It was noted by the authors that most studies utilize a single large SPION for MRI, yet this study employed smaller SPIOs for improved magnetic properties. In addition, they were able to keep the overall size of the nanoplatform small, while maintaining its NIR peak. In another study employing silver nanoshells, the PTT efficiency was shown to be largely dependent upon the spatial distribution of nanoparticles in the tumor. Thompson et al. showed that a concentration of 5-250 µg/mL of silver nanoparticles was needed to produce enough heat to induce cell death in three different breast cancer cell lines (Thompson et al. 2014). Cell uptake of silver nanoparticles was significantly different between each cell line, with MCF7 cells displaying high cytoplasm uptake, MDA-MB-231 cells displaying high nuclei localization and MCF-10A cells have equal cytoplasm and nuclei localization.

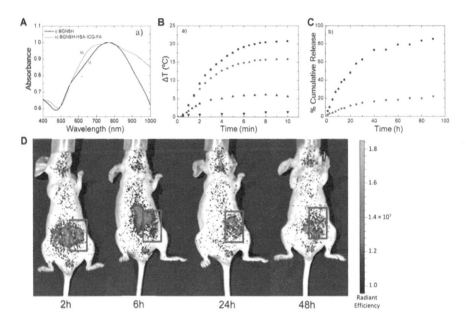

Fig. 12.5 Gold-branched PLGA/doxorubicin (BGNSH) modified with human serum albumin (HSA), indocyanine green (ICG) and targeted with folic acid (FA) for PTT. (A) Biostability was examined by incubating nanoparticles in PBS with 10% FBS. The peak of bare BGNSH (ii) slightly shifted and broadened, while there was no shift or change for the peak of BGNSH-HSA-ICG-FA (i). (B) Nanoparticles were exposed to an 808 nm laser at 2 W/cm². BGNSH-HSA-ICG-FA (squares) showed the highest temperature change, followed by bare BGNSH (circles), free ICG (upright triangles) and PBS (downright triangles). (C) Doxorubicin released from nanoparticles was shown to be laser-dependent (square), with circles depicting no laser exposure. (D) Fluorescence imaging confirmed that nanoparticles actively reached the tumor within 48 h post-injection. Adapted with permission from (Topete et al. 2014a).

Attempting to design a nanoplatform for chemotherapy, phototherapy and molecular imaging, Topete et al. synthesized poly (lactic-co-glycolicacid) (PLGA) cores loaded with doxorubicin and modified with a porous gold-branched shell (Topete et al. 2014a) (Fig. 12.5). In addition, the nanoplatform was functionalized with human serum albumin (HSA) to increase the circulation time of the nanoparticle *in vivo* and indocyanine green (ICG) for PDT and fluorescence imaging capabilities. Lastly, folic acid (FA) was used to target the nanoparticles to tumor tissue. The nanoplatform displayed a shape similar to that of a virus, with an NIR absorbance peak near 780 nm. To determine the biostability of this novel nanoplatform, bare nanoparticles (BGNSH) and surface modified nanoparticles (BGNSH-HSA-ICG-FA) were incubated in PBS with 10% FBS. BGNSH-HSA-ICG-FA was highly stable with no shift in absorbance (Fig. 12.5A). The photothermal efficiency was examined in solution by exposing the nanoparticles to an 808 nm laser at 2 W/cm². As depicted in Fig. 12.5B, BGNSH-HSA-ICG-FA (squares) showed the highest temperature change followed by bare BGNSH (circles). Free ICG (upright triangles) and PBS (downright triangles) experienced minimal temperature changes. The release of doxorubicin from BGNSH-HSA-ICG-FA nanoparticles was measured in the presence (squares) and absence (circle) of NIR irradiation (Fig. 12.5C). Since drug release was much higher in the presence of laser irradiation, this platform would be less likely to elicit chemotherapeutic toxicity in healthy tissues. Lastly, the nanoparticles were examined *in vivo* using fluorescence imaging with the tumor being outlined in a square (Fig. 12.5D).

Carbon Nanotubes for Photothermal Therapy of Cancer

Carbon nanotubes display unique characteristics that make them optimal PTT agents. They display a broad electromagnetic absorbance spectrum and can be easily modified with surface modifications. In addition, their high energy transduction efficiency makes them a leading contender against other photosensitive agents (Sailor and Park 2012). While carbon nanoparticles have been extensively evaluated as drug delivery vehicles, only recently were they investigated for their photothermal properties. Carbon nanotubes are a member of the fullerene structural family and are constructed from graphite sheets that are rolled into the shape of a cylinder, with varying dimensions. Since the sheets can be rolled into different geometries, there are many types of carbon nanotubes in existence with unique physical and optical characteristics. For the purpose of PTT, single-walled carbon nanotubes (SWCNT) are commonly employed, yet multi-walled carbon nanotubes have also been used (MWCNT). There are several signature extinction peaks for SWCNTs, corresponding to different Van Hove transitions, including pronounced peaks at 550 nm and 1000 nm (Weisman 2008).

There have been several studies examining the use of carbon nanotubes for concurrent PTT and drug delivery. For example, Zhang et al. designed poly-L-lactic acid (PLA) nanofibers containing MWCNTs and doxorubicin (Zhang et al. 2015b). The efficacy of this dual-treatment nanoparticle was significantly enhanced both *in vitro* and *in vivo*. Similar studies have proposed novel methodologies for treating cancer with drug-loaded carbon nanotubes (Qin et al. 2015; Zhang et al. 2014; Zhu et al. 2015). Currently, the use of unmodified carbon nanotubes is limited by severe aggregation (Chen et al. 2004). To overcome this barrier, it was recently shown that the addition of Evans Blue can enhance the functionalization of SWCNTs, which prevents aggregation by allowing the nanoparticles to be dispersed in solution (Zhang et al. 2015a). In addition, Zhang et al. showed that SWCNTs, dispersed with Evans Blue, could be further modified with albumin and paclitaxel for dual chemo- and thermo-based treatment of xenograft tumors (Zhang et al. 2015a).

In another investigation, the optical properties of SWCNTs were utilized for imaging the movement of carbon nanotubes from primary tumors to sentinel lymph nodes in mice (Liang et al. 2014). A comparison of two treatment groups, either PTT of the primary tumor alone or PTT of the primary tumor and corresponding lymph node, allowed researchers to evaluate the efficacy of multi-locational PTT. The study showed that survival was significantly higher for mice receiving both primary tumor and sentinel lymph node PTT. In addition, optical and MRI demonstrated the potential use of this nanoplatform for image-guided PTT.

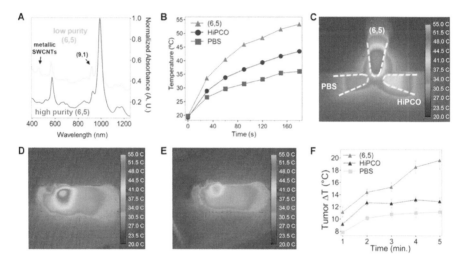

Fig. 12.6 Carbon nanotubes were purified for enhanced PTT properties. (A) In comparison to the low purity sample, purified carbon nanotubes displayed a significant peak near 1000 nm, with less background noise. (B) The purified carbon nanotubes showed improved photothermal heating efficiency as compared to the non-purified nanotubes, which was also shown through thermal imaging. (C) The improved photothermal heating efficiency of purified carbon nanotubes was compared to HiPCO and the control (PBS) before animal injection. (D) Mice were injected with purified (E) or non-purified carbon nanotube samples and thermal imaging revealed the temperature changes. (F) The overall temperature change was highest for purified carbon nanotubes (5, 6) with a 20°C temperature increase, in comparison to the 12°C increase of non-purified carbon nanotubes (HiPCO). Adapted with permission from (Antaris et al. 2013).

A similar investigation by Antaris et al. examined the intrinsic fluorescence and NIR absorption properties of SWCNTs (Antaris et al. 2013). The authors noted a significant weakness of the previous studies using SWCNTs for PTT. Researchers were using as-synthesized SWCNTs. There was no sorting method to eliminate the carbon nanotubes that were off resonance. Removal of the off resonance SWCNTs would effectively reduce the dose needed to elicit a therapeutic response. They were able to purify the nanotubes to obtain a high purity sample with a maximum absorbance of 1000 nm (Fig. 12.6A). The photothermal conversion efficiency of purified (6, 5) carbon nanotubes was shown to be higher than that of non-purified nanotubes, as depicted in Fig. 12.6B. Thermal images were obtained after 3 min of 980 nm laser-induced heating of the nanoparticle solutions with a power density of 0.6 W/cm^2 (Fig. 12.6C). Thermal images were used to compare the PTT efficiency of purified and non-purified SWCNTs (Fig. 12.6D, E). A significantly higher temperature change was seen with (6, 5) SWCNTs (Fig. 12.6E), while a smaller difference in temperature was seen in the group of mice that received non-purified SWCNTs with laser ablation (Fig. 12.6E, F).

Recent studies have raised several questions regarding the immune response to the use of SWCNTs for PTT. Wang et al. showed that SWCNTs triggered an adaptive immune response *in vivo*, which did not occur if the tumor had been surgically resected (Wang et al. 2014b). They combined SWCNTs with anti-CTLA-4 antibody therapy for the prevention of metastatic cancer in mice. The combination therapy was effective at inhibiting the growth of malignant cells that survived PTT. Furthermore, Yang and colleagues found that subcutaneously injected MWCNTs induced phagocytosis of macrophages and recruited macrophages from other surrounding tissues, which resulted in decreased vessel density around the tumor (Yang et al. 2012). In return, the immune response elicited by the injected MWCNTs effectively inhibited tumor progression.

Combination Nanoplatforms for Photothermal Therapy of Cancer

Many researchers are designing novel nanoplatforms that combine multiple nanoparticles into a single construct. This is especially true of theranostics, as researchers attempt to harness the properties of several nanoparticles to develop highly complex nanoparticle systems capable of multimodality imaging and treatment of cancer. Recently, Bai et al. developed a novel theranostic nanoplatform by combining the MRI properties of SPIOs with the optical properties of hollow gold nanospheres (HGNs) (Bai et al. 2015). The HGN-SPIO nanoplatform was water soluble, highly biocompatible and displayed a significant NIR absorbance peak. Overall, the nanoplatform displayed optimal photothermal efficacy and enhanced T2-weighted MR contrast.

In a recent study, Wang et al. designed a complex theranostic platform consisting of a magnetic core modified with carbon dots, ultra-small gold nanoparticles and a carbon shell (Wang et al. 2015). The nanoparticles were developed for multimodality imaging and treatment, as they could be used for multicolor cell imaging and MRI-guided PTT and drug delivery. The nanoparticles could carry a large payload of drug to tumor tissues and were readily internalized by mouse melanoma cells. In a similar study, SPIONs were functionalized with ultra-small gold nanoseeds and doxorubicin (Topete et al. 2014b). The MRI, chemotherapy and PTT properties of this novel nanoplatform were examined in a human cervical cancer cell line. The nanoparticles displayed a high NIR absorbance that promoted optimal PTT and allowed for steady heat-dependent release of doxorubicin. In addition, active targeting of the nanoparticles with folic acid (FA) significantly enhanced overall tumor accumulation and cell internalization of the theranostic nanoparticles. The authors noted that this nanoplatform might be used in combination with an external magnet for enhancing the overall accumulation of nanoparticles in tumor tissue.

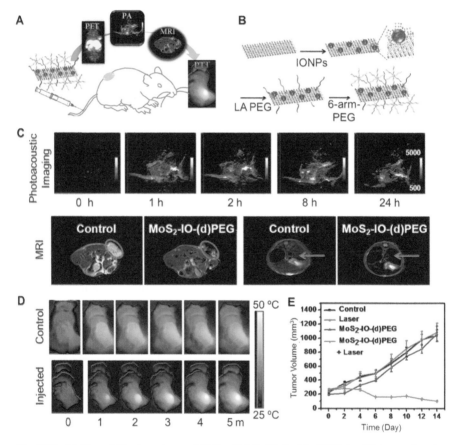

Fig. 12.7 Multimodality imaging and PTT of MoS$_2$-IO-double (PEG) in 4T1 tumor-bearing mice. (A) Mice were injected with a single nanoparticle construct capable of PET, photoacoustic imaging, MRI and PTT. (B) IONPs were double PEGylated to increase circulation time. (C) Photoacoustic imaging and MRI allowed for clear delineation of tumor boundaries, as circled in MRI. (D) Temperature change was higher for mice injected with nanoparticles with (E) MoS$_2$-IO-double (PEG) with laser irradiation resulting in significant decreases in tumor volume through 14 d post injection. Adapted with permission from (Liu et al. 2015a).

Recently, Liu et al. synthesized IONPs functionalized with MoS$_2$ nanosheets for multimodality (PET, MRI and photoacoustic) image-guided PTT in 4T1 tumor-bearing mice (Fig. 12.7A) (Liu et al. 2015a). For enhancement of circulation and biostability, modified IONPs on MoS$_2$ sheets underwent double PEGylation, as shown in Fig. 12.7B. Besides PET imaging, both MRI and photoacoustic imaging were utilized for visualizing the biodistribution and tumor accumulation of nanoparticles at various time points after mice were injected with 6.85 mg/kg of MoS$_2$-IO-double (PEG) (Fig. 12.7C). Photoacoustic imaging allowed for delineation of tumor boundaries and T2-weighted MRI clearly showed the tumor (red circles) and differences in tumor contrast (blue arrow) (Fig. 12.7C). For PET imaging, it was found that this novel nanoplatform could undergo chelator-free radiolabeling, as ^{64}Cu was strongly absorbed onto the surface of MoS$_2$. Lastly, the authors investigated the PTT efficiency of MoS$_2$-IO-double (PEG) in 4T1 tumor-bearing mice. An 808 nm laser irradiation of 0.78 W/cm^2 was used for treatment with varying time intervals. Mice underwent PTT 8 h post nanoparticle injection and temperature changes were monitored using infrared imaging (Fig. 12.7D). Mice that received the nanoparticle injection experienced a temperature increase that was 15°C higher than the control group, which allowed for efficient inhibition of tumor growth through 14 d post PTT (Fig. 12.7E). This highly versatile theranostic agent was successfully utilized with several imaging modalities, including PET, MRI, photoacoustic imaging and optical imaging.

The surface of mesoporous silica coated core-shell up-conversion nanoparticles (UCNPs) was modified with photosensitive gold clusters, a pH sensitive polymer and doxorubicin to create an image-guided nanoparticle for PDT, PTT and pH-sensitive drug release (Lv et al. 2015). The nanoparticles were highly effective in the treatment of cancer *in vitro* and *in vivo*. Several other groups have developed combination nanoplatforms. For example, Liu and colleagues designed carbon nanotubes filled with ferrite nanoparticles for MRI-guided PTT (Liu et al. 2014). Similarly, Wang et al. developed water soluble and highly biocompatible MWCNTs loaded with RGD-conjugated silica-coated gold nanorods for photoacoustic image-guided PTT of gastric cancer in mice (Wang et al. 2014a). In another instance, Nair et al. explored the combination of SWCNTs and quantum dots for cellular imaging and PTT (Nair et al. 2014).

Conclusions and Future Perspectives

The development of nanoparticles has made it easier to examine the anti-cancer properties of PTT *in vivo*. While effective in preclinical studies, more investigations need to be conducted to ensure its safety and reproducibility in living organisms. As PTT is dependent upon nanoparticles, the development of nanoparticles for PTT should be standardized and highly reproducible between batches. This is a common problem at this time, as batches of the same nanoparticle may display slightly different physicochemical properties. By setting up standard protocols, these slight alterations can be limited. In addition, methods for ensuring that nanoparticles are biocompatible or nontoxic should be developed.

Several factors should be considered during the synthesis and design of nanoparticles optimized for image-guided PTT. The most important factor is the biocompatibility and biodegradability of the nanoparticle *in vivo*. Nanoparticles that display higher clearance rates or biodegradability will be more likely to obtain FDA-approval in the future as they pose limited toxicity risks. Metallic nanoparticles and other larger nanoparticles are prone to liver and spleen accumulation, which may result in hepatotoxicity. In addition, there have been no studies investigating the long-term toxicities associated with exposure to nanoparticles. Biodegradable silica-based nanoparticles have been effectively utilized in drug delivery and PTT. In addition, liposomal-based nanoparticles are also biodegradable and nanoparticles less than 50 nm are known to undergo renal filtration.

Many researchers have begun investigating other nanoparticle systems as potential PTT agents. For example, quantum dots (QDs) are primarily utilized for fluorescence imaging, as they are highly efficient at producing light when excited in the NIR spectrum. In addition, QDs display unique photophysical properties that make them superior fluorescence probes in comparison to traditional organic proteins and dyes. Cu-based QDs display extinction spectral profiles with an NIR band near 980 nm, making them optimal nanoparticles for PTT. Most particles less than 50 nm can be effectively cleared through the liver and kidney, thus QDs would be expected to cause limited toxicity. In the past, QDs have been exploited for their photosensitizer properties in PDT for the production of reactive oxygen species, yet QDs are far less efficient than common photosensitizers.

Recently, the PTT properties of CdTe and CdSe QDs were examined by Chu et al. in mice bearing melanoma tumors using a 671 nm laser for irradiation (Chu et al. 2012). The physicochemical properties of the CdTe QDs displayed photoluminescence emission spectra that ranged from 530 to 730 nm. In addition, the heating effects of each QD were examined in an aqueous solution, with CdTe (710) possessing the most efficient heating capabilities. In another study, Sun et al. examined the PTT properties of ultra-small black phosphorus QDs, as phosphorus is a common element found in the body (Sun et al. 2015b). The black phosphorus QDs exhibited an NIR absorbance peak near 808 nm and were found to be 2.6 nm in size, with a thickness of 1.5 nm. The photothermal conversion efficiency was found to be 28.4%, which is lower than that of other nanoparticle platforms. The QDs displayed enhanced serum stability with no signs of cellular toxicity *in vitro*. In addition, these ultra-small nanoparticles undergo renal clearance and testing of the PTT effects *in vitro* revealed significant cell death.

Currently, another challenge in the development of effective PTT agents involves active targeting concerns. PTT requires that a large percent of the injected dose of nanoparticles accumulate in the tissue undergoing laser ablation. Suboptimal accumulation of nanoparticle in the tumor will result in diminished efficacy or treatment failure. To overcome this issue, nanoplatforms are actively targeted to tumors using surface modifications. While active targeting can significantly enhance tumor accumulation of certain nanoparticles, only a small portion of the injected dose will localize in the tumor (up to 20 percent). There is a need for improved methods for actively targeting nanoparticles to tumor tissue for PTT. This was previously accomplished using external magnetics for localizing magnetic nanoparticles in tumors. Several other strategies have been developed including dual-targeting strategies with heterodimeric antibodies. As PTT is dependent upon the nanoparticle localization for effectiveness, this is a critical concern that must be addressed to ensure optimal PTT outcomes in the future.

While the future of PTT remains unclear at this time, there is strong evidence supporting its use in cancer therapy. While more studies are needed to develop specific treatment protocols, engineers will continue to construct novel nanoplatforms for cancer treatment and imaging. In addition, the future of personalized medicine is dependent upon the development of theranostic agents. As the field of image-guided cancer therapy expands in the future, we predict that PTT will become a vital tool in the clinical treatment of cancer worldwide.

Acknowledgements

This work is supported, in part, by the University of Wisconsin–Madison, the National Institutes of Health (NIBIB/NCI 1R01CA169365, P30CA014520 and T32CA009206), the Department of Defense (W81XWH-11-1-0644) and the American Cancer Society (125246-RSG-13-099-01-CCE).

References

ACS 2015. Cancer Facts & Figures 2015. American Cancer Society, Atlanta.

Adams, V.R. 2000. Adverse events associated with chemotherapy for common cancers. Pharmacotherapy 20: 96S-103S.

Antaris, A.L., Robinson, J.T., Yaghi, O.K., Hong, G., Diao, S., Luong, R., et al. 2013. Ultra-low doses of chirality sorted (6, 5) carbon nanotubes for simultaneous tumor imaging and photothermal therapy. ACS Nano 7: 3644-3652.

Ayala-Orozco, C., Urban, C., Bishnoi, S., Urban, A., Charron, H., Mitchell, T., et al. 2014a. Sub-100 nm gold nanomatryoshkas improve photo-thermal therapy efficacy in large and highly aggressive triple negative breast tumors. J Control Release 191: 90-97.

Ayala-Orozco, C., Urban, C., Knight, M.W., Urban, A.S., Neumann, O., Bishnoi, S.W., et al. 2014b. Au nanomatryoshkas as efficient near-infrared photothermal transducers for cancer treatment: benchmarking against nanoshells. ACS Nano 8: 6372-6381.

Bai, L.Y., Yang, X.Q., An, J., Zhang, L., Zhao, K., Qin, M.Y., et al. 2015. Multifunctional magnetic-hollow gold nanospheres for bimodal cancer cell imaging and photo-thermal therapy. Nanotechnology 26: 315701.

Bertrand, N., Wu, J., Xu, X., Kamaly, N. and Farokhzad, O.C. 2014. Cancer nanotechnology: the impact of passive and active targeting in the era of modern cancer biology. Adv Drug Deliv Rev 66: 2-25.

Brown, J.M. 2007. Tumor hypoxia in cancer therapy. Methods Enzymol 435: 297-321.

Chen, C.H., Wu, Y.J. and Chen, J.J. 2015a. Gold nanotheranostics: photothermal therapy and imaging of mucin 7 conjugated antibody nanoparticles for urothelial cancer. Biomed Res Int 2015: 813632.

Chen, F., Hong, H., Goel, S., Graves, S.A., Orbay, H., Ehlerding, E.B., et al. 2015b. In vivo tumor vasculature targeting of CuS@MSN based theranostic nanomedicine. ACS Nano 9: 3926-3934.

Chen, Q., Saltiel, C., Manickavasagam, S., Schadler, L.S., Siegel, R.W. and Yang, H. 2004. Aggregation behavior of single-walled carbon nanotubes in dilute aqueous suspension. J Colloid Interface Sci 280: 91-97.

Chu, M., Pan, X., Zhang, D., Wu, Q., Peng, J. and Hai, W. 2012. The therapeutic efficacy of CdTe and CdSe quantum dots for photothermal cancer therapy. Biomaterials 33: 7071-7083.

Davis, M.E., Chen, Z.G. and Shin, D.M. 2008. Nanoparticle therapeutics: an emerging treatment modality for cancer. Nat Rev Drug Discov 7: 771-782.

Day, E.S., Thompson, P.A., Zhang, L., Lewinski, N.A., Ahmed, N., Drezek, R.A., et al. 2011. Nanoshell-mediated photothermal therapy improves survival in a murine glioma model. J Neurooncol 104: 55-63.

Dickerson, E.B., Dreaden, E.C., Huang, X., El-Sayed, I.H., Chu, H., Pushpanketh, S., et al. 2008. Gold nanorod assisted near-infrared plasmonic photothermal therapy (PPTT) of squamous cell carcinoma in mice. Cancer Lett 269: 57-66.

England, C.G., Priest, T., Zhang, G., Sun, X., Patel, D.N., McNally, L.R., et al. 2013. Enhanced penetration into 3D cell culture using two and three layered gold nanoparticles. Int J Nanomedicine 8: 3603-3617.

England, C.G., Miller, M.C., Kuttan, A., Trent, J.O. and Frieboes, H.B. 2015. Release kinetics of paclitaxel and cisplatin from two and three layered gold nanoparticles. Eur J Pharm Biopharm 92: 120-129.

Escobedo, J.O., Rusin, O., Lim, S. and Strongin, R.M. 2010. NIR dyes for bioimaging applications. Curr Opin Chem Biol 14: 64-70.

Fukumori, Y. and Ichikawa, H. 2006. Nanoparticles for cancer therapy and diagnosis. Adv Powder Technol 17: 1-28.

Gad, S.C., Sharp, K.L., Montgomery, C., Payne, J.D. and Goodrich, G.P. 2012. Evaluation of the toxicity of intravenous delivery of auroshell particles (gold-silica nanoshells). Int J Toxicol 31: 584-594.

Godin, B., Sakamoto, J.H., Serda, R.E., Grattoni, A., Bouamrani, A. and Ferrari, M. 2010. Emerging applications of nanomedicine for the diagnosis and treatment of cardiovascular diseases. Trends Pharmacol Sci 31: 199-205.

Gravitz, L. 2013. Cancer immunotherapy. Nature 504: S1.

Habash, R.W., Bansal, R., Krewski, D. and Alhafid, H.T. 2006. Thermal therapy, part 2: hyperthermia techniques. Crit Rev Biomed Eng 34: 491-542.

Hadjipanayis, C.G., Bonder, M.J., Balakrishnan, S., Wang, X., Mao, H. and Hadjipanayis, G.C. 2008. Metallic iron nanoparticles for MRI contrast enhancement and local hyperthermia. Small 4: 1925-1929.

Hirsch, L.R., Stafford, R.J., Bankson, J.A., Sershen, S.R., Rivera, B., Price, R.E., et al. 2003. Nanoshell-mediated near-infrared thermal therapy of tumors under magnetic resonance guidance. Proc Natl Acad Sci USA 100: 13549-13554.

Hobohm, U. 2005. Fever therapy revisited. Br J Cancer 92: 421-425.

Holohan, C., Van Schaeybroeck, S., Longley, D.B. and Johnston, P.G. 2013. Cancer drug resistance: an evolving paradigm. Nat Rev Cancer 13: 714-726.

Huang, P., Rong, P., Jin, A., Yan, X., Zhang, M.G., Lin, J., et al. 2014a. Dye-loaded ferritin nanocages for multimodal imaging and photothermal therapy. Adv Mater 26: 6401-6408.

Huang, P., Rong, P., Lin, J., Li, W., Yan, X., Zhang, M.G., et al. 2014b. Triphase interface synthesis of plasmonic gold bellflowers as near-infrared light mediated acoustic and thermal theranostics. J Am Chem Soc 136: 8307-8313.

Huang, X. and El-Sayed, M.A. 2011. Plasmonic photo-thermal therapy (PPTT). Bull Alexandria Fac 47: 1-9.

Huang, X., El-Sayed, I.H., Qian, W. and El-Sayed, M.A. 2006. Cancer cell imaging and photothermal therapy in the near-infrared region by using gold nanorods. J Am Chem Soc 128: 2115-2120.

Janib, S.M., Moses, A.S. and MacKay, J.A. 2010. Imaging and drug delivery using theranostic nanoparticles. Adv Drug Deliv Rev 62: 1052-1063.

Ji, X., Shao, R., Elliott, A.M., Stafford, R.J., Esparza-Coss, E., Bankson, J.A., et al. 2007. Bifunctional gold nanoshells with a superparamagnetic iron oxide-silica core suitable for both MR imaging and photothermal therapy. J Phys Chem C Nanomater Interfaces 111: 6245-6251.

Jing, L., Liang, X., Li, X., Lin, L., Yang, Y., Yue, X., et al. 2014. Mn-porphyrin conjugated Au nanoshells encapsulating doxorubicin for potential magnetic resonance imaging and light triggered synergistic therapy of cancer. Theranostics 4: 858-871.

Kado, S., Yokomine, S. and Kimura, K. 2014. One-pot synthesis of silver nanoshells with near-infrared extinction by a thiocyanate-assisted approach. Rsc Advances 4: 10830-10833.

Kaur, P., Hurwitz, M.D., Krishnan, S. and Asea, A. 2011. Combined hyperthermia and radiotherapy for the treatment of cancer. Cancers (Basel) 3: 3799-3823.

Kelly, K.L., Coronado, E., Zhao, L.L. and Schatz, G.C. 2003. The optical properties of metal nanoparticles: the influence of size, shape and dielectric environment. J Phys Chem B 107: 668-677.

Kempen, P.J., Greasley, S., Parker, K.A., Campbell, J.L., Chang, H.Y., Jones, J.R., et al. 2015. Theranostic mesoporous silica nanoparticles biodegrade after pro-survival drug delivery and ultrasound/magnetic resonance imaging of stem cells. Theranostics 5: 631-642.

Khawar, I.A., Kim, J.H. and Kuh, H.J. 2015. Improving drug delivery to solid tumors: priming the tumor microenvironment. J Control Release 201: 78-89.

Khlebtsov, N.G., Trachuk, L.A. and Mel'nikov, A.G. 2005. The effect of the size, shape and structure of metal nanoparticles on the dependence of their optical properties on the refractive index of a disperse medium. Opt Spectrosc 98: 77-83.

Kim, S., Chen, Y.S., Luke, G.P. and Emelianov, S.Y. 2014. In-vivo ultrasound and photoacoustic image-guided photothermal cancer therapy using silica-coated gold nanorods. IEEE Trans Sonics Ultrason 61: 891-897.

Kyi, C. and Postow, M.A. 2014. Checkpoint blocking antibodies in cancer immunotherapy. FEBS Lett 588: 368-376.

Lal, S., Clare, S.E. and Halas, N.J. 2008. Nanoshell-enabled photothermal cancer therapy: impending clinical impact. Acc Chem Res 41: 1842-1851.

Liang, C., Diao, S., Wang, C., Gong, H., Liu, T., Hong, G., et al. 2014. Tumor metastasis inhibition by imaging-guided photothermal therapy with single-walled carbon nanotubes. Adv Mater 26: 5646-5652.

Lin, A.Y., Young, J.K., Nixon, A.V. and Drezek, R.A. 2014. Encapsulated Fe3O4/Ag complexed cores in hollow gold nanoshells for enhanced theranostic magnetic resonance imaging and photothermal therapy. Small 10: 3246-3251.

Liu, T., Shi, S., Liang, C., Shen, S., Cheng, L., Wang, C., et al. 2015a. Iron oxide decorated MoS2 nanosheets with double PEGylation for chelator-free radiolabeling and multimodal imaging guided photothermal therapy. ACS Nano 9: 950-960.

Liu, X., Marangon, I., Melinte, G., Wilhelm, C., Menard-Moyon, C., Pichon, B.P., et al. 2014. Design of covalently functionalized carbon nanotubes filled with metal oxide nanoparticles for imaging, therapy and magnetic manipulation. ACS Nano 8: 11290-11304.

Liu, Y., Ashton, J.R., Moding, E.J., Yuan, H., Register, J.K., Fales, A.M., et al. 2015b. A plasmonic gold nanostar theranostic probe for in vivo tumor imaging and photothermal therapy. Theranostics 5: 946-960.

Lv, R., Yang, P., He, F., Gai, S., Yang, G., Dai, Y., et al. 2015. An imaging-guided platform for synergistic photodynamic/photothermal/chemo-therapy with pH/temperature-responsive drug release. Biomaterials 63: 115-127.

Mackey, M.A., Ali, M.R., Austin, L.A., Near, R.D. and El-Sayed, M.A. 2014. The most effective gold nanorod size for plasmonic photothermal therapy: theory and in vitro experiments. J Phys Chem B 118: 1319-1326.

Malhotra, V. and Perry, M.C. 2003. Classical chemotherapy: mechanisms, toxicities and the therapeutic window. Cancer Biol Ther 2: S2-4.

Menon, J.U., Jadeja, P., Tambe, P., Vu, K., Yuan, B. and Nguyen, K.T. 2013. Nanomaterials for photo-based diagnostic and therapeutic applications. Theranostics 3: 152-166.

Mihlon, F.t., Ray, C.E., Jr. and Messersmith, W. 2010. Chemotherapy agents: a primer for the interventional radiologist. Semin Intervent Radiol 27: 384-390.

Mukherjee, S., Sobhani, H., Lassiter, J.B., Bardhan, R., Nordlander, P. and Halas, N.J. 2010. Fanoshells: nanoparticles with built-in Fano resonances. Nano Lett 10: 2694-2701.

Nair, L.V., Nagaoka, Y., Maekawa, T., Sakthikumar, D. and Jayasree, R.S. 2014. Quantum dot tailored to single wall carbon nanotubes: a multifunctional hybrid nanoconstruct for cellular imaging and targeted photothermal therapy. Small 10: 2771-2775, 2740.

Narvekar, M., Xue, H.Y., Eoh, J.Y. and Wong, H.L. 2014. Nanocarrier for poorly water-soluble anticancer drugs–barriers of translation and solutions. AAPS Pharm Sci Tech 15: 822-833.

Ntziachristos, V. 2010. Going deeper than microscopy: the optical imaging frontier in biology. Nat Methods 7: 603-614.

Park, K., Biswas, S., Kanel, S., Nepal, D. and Vaia, R.A. 2014. Engineering the optical properties of gold nanorods: independent tuning of surface plasmon energy, extinction coefficient and scattering cross section. J Phys Chem C 118: 5918-5926.

Piao, J.G., Wang, L., Gao, F., You, Y.Z., Xiong, Y. and Yang, L. 2014. Erythrocyte membrane is an alternative coating to polyethylene glycol for prolonging the circulation lifetime of gold nanocages for photothermal therapy. ACS Nano 8: 10414-10425.

Popescu, M.A. and Toms, S.A. 2006. In vivo optical imaging using quantum dots for the management of brain tumors. Expert Rev Mol Diagn 6: 879-890.

Qin, Y., Chen, J., Bi, Y., Xu, X., Zhou, H., Gao, J., et al. 2015. Near-infrared light remote-controlled intracellular anti-cancer drug delivery using thermo/pH sensitive nanovehicle. Acta Biomater 17: 201-209.

Ruoslahti, E., Bhatia, S.N. and Sailor, M.J. 2010. Targeting of drugs and nanoparticles to tumors. J Cell Biol 188: 759-768.

Sailor, M.J. and Park, J.H. 2012. Hybrid nanoparticles for detection and treatment of cancer. Adv Mater 24: 3779-3802.

Scott, A.M., Wolchok, J.D. and Old, L.J. 2012a. Antibody therapy of cancer. Nat Rev Cancer 12: 278-287.

Scott, A.M., Wolchok, J.D. and Old, L.J. 2012b. Antibody therapy of cancer. Nat Rev Cancer 12: 278-287.

Shao, J., Griffin, R.J., Galanzha, E.I., Kim, J.W., Koonce, N., Webber, J., et al. 2013. Photothermal nanodrugs: potential of TNF-gold nanospheres for cancer theranostics. Sci Rep 3: 1293.

Shen, S., Wang, S., Zheng, R., Zhu, X., Jiang, X., Fu, D., et al. 2015. Magnetic nanoparticle clusters for photothermal therapy with near-infrared irradiation. Biomaterials 39: 67-74.

Skitzki, J.J., Repasky, E.A. and Evans, S.S. 2009. Hyperthermia as an immunotherapy strategy for cancer. Curr Opin Invest Drugs 10: 550-558.

Smith, A.M., Mancini, M.C. and Nie, S. 2009. Bioimaging: second window for in vivo imaging. Nat Nanotechnol 4: 710-711.

Sriraman, S.K., Aryasomayajula, B. and Torchilin, V.P. 2014. Barriers to drug delivery in solid tumors. Tissue Barriers 2: e29528.

Stern, J.M., Solomonov, V.V., Sazykina, E., Schwartz, J.A., Gad, S.C. and Goodrich, G.P. 2015. Initial evaluation of the safety of nanoshell-directed photothermal therapy in the treatment of prostate disease. Int J Toxicol 35: 38-46.

Sun, X., Cai, W. and Chen, X. 2015a. Positron emission tomography imaging using radiolabeled inorganic nanomaterials. Acc Chem Res 48: 286-294.

Sun, Z., Xie, H., Tang, S., Yu, X.F., Guo, Z., Shao, J., et al. 2015b. Ultrasmall black phosphorus quantum dots: synthesis and use as photothermal agents. Angew Chem Int Ed Engl 54: 11526-11530.

Tang, N., Du, G., Wang, N., Liu, C., Hang, H. and Liang, W. 2007. Improving penetration in tumors with nanoassemblies of phospholipids and doxorubicin. J Natl Cancer Inst 99: 1004-1015.

Thompson, E.A., Graham, E., MacNeill, C.M., Young, M., Donati, G., Wailes, E.M., et al. 2014. Differential response of MCF7, MDA-MB-231 and MCF 10A cells to hyperthermia, silver nanoparticles and silver nanoparticle-induced photothermal therapy. Int J Hyperthermia 30: 312-323.

Topete, A., Alatorre-Meda, M., Iglesias, P., Villar-Alvarez, E.M., Barbosa, S., Costoya, J.A., et al. 2014a. Fluorescent drug-loaded, polymeric-based, branched gold nanoshells for localized multimodal therapy and imaging of tumoral cells. ACS Nano 8: 2725-2738.

Topete, A., Alatorre-Meda, M., Villar-Alvarez, E.M., Carregal-Romero, S., Barbosa, S., Parak, W.J., et al. 2014b. Polymeric-gold nanohybrids for combined imaging and cancer therapy. Adv Healthc Mater 3: 1309-1325.

Torre, L.A., Bray, F., Siegel, R.L., Ferlay, J., Lortet-Tieulent, J. and Jemal, A. 2015. Global cancer statistics, 2012. CA Cancer J Clin 65: 87-108.

Tredan, O., Galmarini, C.M., Patel, K. and Tannock, I.F. 2007. Drug resistance and the solid tumor microenvironment. J Natl Cancer Inst 99: 1441-1454.

Van Drie, J.H. 2011. Protein folding, protein homeostasis and cancer. Chin J Cancer 30: 124-137.

Wang, C., Bao, C., Liang, S., Fu, H., Wang, K., Deng, M., et al. 2014a. RGD-conjugated silica-coated gold nanorods on the surface of carbon nanotubes for targeted photoacoustic imaging of gastric cancer. Nanoscale Res Lett 9: 264.

Wang, C., Xu, L., Liang, C., Xiang, J., Peng, R. and Liu, Z. 2014b. Immunological responses triggered by photothermal therapy with carbon nanotubes in combination with anti-CTLA-4 therapy to inhibit cancer metastasis. Adv Mater 26: 8154-8162.

Wang, H., Cao, G., Gai, Z., Hong, K., Banerjee, P. and Zhou, S. 2015. Magnetic/NIR-responsive drug carrier, multicolor cell imaging and enhanced photothermal therapy of gold capped magnetite-fluorescent carbon hybrid nanoparticles. Nanoscale 7: 7885-7895.

Wang, M. and Thanou, M. 2010. Targeting nanoparticles to cancer. Pharmacol Res 62: 90-99.

Weisman, R.B. 2008. Optical spectroscopy of single-walled carbon nanotubes. pp. 109-133. *In:* Susumo Saito and Alex Zettl (eds.). Carbon Nanotubes: Quantum Cylinders of Graphene. Elsevier, New York.

Weissleder, R. 2001. A clearer vision for in vivo imaging. Nat Biotechnol 19: 316-317.

Xie, B.W., Mol, I.M., Keereweer, S., van Beek, E.R., Que, I., Snoeks, T.J., et al. 2012. Dual-wavelength imaging of tumor progression by activatable and targeting near-infrared fluorescent probes in a bioluminescent breast cancer model. PLoS One 7: e31875.

Yang, M., Meng, J., Cheng, X., Lei, J., Guo, H., Zhang, W., et al. 2012. Multiwalled carbon nanotubes interact with macrophages and influence tumor progression and metastasis. Theranostics 2: 258-270.

Yi, X., Wang, F., Qin, W., Yang, X. and Yuan, J. 2014. Near-infrared fluorescent probes in cancer imaging and therapy: an emerging field. Int J Nanomedicine 9: 1347-1365.

You, J., Zhang, P., Hu, F., Du, Y., Yuan, H., Zhu, J., et al. 2014. Near-infrared light-sensitive liposomes for the enhanced photothermal tumor treatment by the combination with chemotherapy. Pharm Res 31: 554-565.

Zelasko-Leon, D.C., Fuentes, C.M. and Messersmith, P.B. 2015. MUC1-targeted cancer cell photothermal ablation using bioinspired gold nanorods. PLoS One 10: e0128756.

Zhang, H., Hou, L., Jiao, X., Yandan, J., Zhu, X., Hongji, L., et al. 2014. In vitro and in vivo evaluation of antitumor drug-loaded aptamer targeted single-walled carbon nanotubes system. Curr Pharm Biotechnol 14: 1105-1117.

Zhang, L., Rong, P., Chen, M., Gao, S. and Zhu, L. 2015a. A novel single walled carbon nanotube (SWCNT) functionalization agent facilitating in vivo combined chemo/thermo therapy. Nanoscale 7: 16204-16213.

Zhang, M., Wang, J. and Tian, Q. 2013. Tip-enhanced Raman spectroscopy based on plasmonic lens excitation and experimental detection. Opt Express 21: 9414-9421.

Zhang, Z., Liu, S., Xiong, H., Jing, X., Xie, Z., Chen, X., et al. 2015b. Electrospun PLA/MWCNTs composite nanofibers for combined chemo- and photothermal therapy. Acta Biomater 26: 115-123.

Zhu, X., Huang, H., Zhang, Y., Xie, Y., Hou, L., Zhang, H., et al. 2015. Investigation on single-walled carbon nanotubes-liposomes conjugate to treatment tumor with dual-mechanism. Curr Pharm Biotechnol 16: 927-936.

13

Biomedical Applications of Nanoscale Metal-Organic Frameworks

Tran Hoang[a] and *Shengqian Ma**

INTRODUCTION

Metal-organic frameworks (MOFs) are one of the latest classes of hybrid materials. Since their discovery for the first time in 1989 by Robson (Hoskins and Robson 1989), MOFs have been well known for their outstanding performance in gas storage and separation, catalysis and great potential in magnetism, nonlinear optics, light harvesting, etc. Composed of metal centers linked together with organic ligands, the number of possible MOF structures is almost infinite. Recently, MOFs have been scaled down to nanometer sizes, called nanoscale-MOFs (NMOFs), which have the exact same structures as MOFs but fall in the nanoregime in terms of crystal size, exhibiting several potential advantages over conventional hybrid nanomaterials. One of the biggest advantages is the tunability of the pore size and shape, depending on the choice of the organic linkers and the metal centers. Furthermore, the interaction between the guest molecules and the host framework can be enhanced through functionalization of the organic linkers. This property has made NMOFs a very attractive and promising candidate in biomedical applications such as drug/therapeutic molecule delivery, biosensor and imaging. The ability to tune the framework via functionalization can improve administration, biodistribution, excretion profile and selectivity of the drug/ gasotransmitter delivery process as well as enhance signals, sensitivity and selectivity of biosensors and imaging applications.

Department of Chemistry, University of South Florida, CHE 205A, 4202 E. Fowler Avenue, Tampa FL 33620.

[a] E-mail: tranhoang@mail.usf.edu

* Corresponding author: sqma@usf.edu

Preparation of Nanoscale Metal-Organic Frameworks for Biomedical Applications

Choice of Metals and Organic Ligands

The choice of metal centers and organic linkers is crucial in designing functional NMOFs for biomedical application. Many metals build stable and robust frameworks but are highly toxic and therefore cannot be used for biomedical application (Horcajada et al. 2012). So far, iron has proven to be the most suitable metal center for biomedical NMOFs with great potential in drug delivery. Since the types of organic ligands are limitless, the variety of biomedical NMOFs likely depends on the choice of the organic ligands. Generally, there are two types of organic ligands: exogenous and endogenous. Exogenous is the more common type of ligand and is synthesized or obtained from nature. Once inside a biological system, these ligands will not intervene with its activity; therefore, the main concerns for using exogenous ligands are absorption, biodistribution and excretion. In contrast, endogenous ligands are those that can naturally be found in biological systems. This type of ligand, of course, is less toxic and results in less undesired side effects (Rabone et al. 2010). However, endogenous ligand-based MOFs tend to be less stable and less porous as compared to those built from exogenous ligands, possibly due to the tendency of these ligands to dissolve in aqueous solutions, leading to breakdown of framework.

Synthesis Method

Four general methods that have been used to synthesize NMOFs are hydro-solvothermal, surfactant-templated solvothermal reactions, sonochemical and microwave assisted hydro/solvothermal synthesis. Hydrosolvothermal method, the most commonly used by far, depends on the reaction time, temperature, stoichiometry, pressure, etc. to control the size of crystals (Li et al. 1999; Yaghi et al. 2003). Surfactant-templated solvothermal reactions use surfactants to control the size and shape of the pores. Cationic cetyltrimethylammonium bromide, a very popular surfactant, has been used to synthesize many NMOFs with great potential for bio-applications (Rieter et al. 2006; Taylor et al. 2008a). For a better control of the size of NMOFs, sonochemical and microwave assisted methods are preferred over the two described above. Sonochemical synthesis utilizes ultrasound to create acoustic cavitation which eventually leads to the formation of high energy microreactors for rapid crystallization of MOFs. This method has been reported mainly for the synthesis of micro and NMOFs with uniform sizes and great potential for biomedical applications (Qiu et al. 2008; Li et al. 2009; Haque et al. 2010). Microwave assisted methods, on the other hand, use polar solvents with high dielectric constants to create superheated local spots, which increase the nucleation rate and crystal formation of NMOFs. Due to the consistency of microwave, this method delivers a homogenous environment that has both better control of size and faster rate of crystal nucleation and growth (Ni and Masel 2006; Jhung et al. 2007; Taylor-Pashow et al. 2009).

Post-Synthetic Modification and Formulation

Post-synthetic modification is the most common method for the tuning and functionalization of NMOFs. Most of the modifications are done on the NMOF surface to improve selectivity, adsorption, distribution, metabolism, excretion profiles or to introduce visualization groups for imaging purposes or tracking the delivery of therapeutic agents. To delay the degradation of the NMOF particles, multifunctional polymer chains such as polyethylene glycol (Rowe et al. 2009a) and isopropylacrylamide (Rowe et al. 2009b) can be used. Surface coating using a silica shell is also common (Rieter et al. 2007; Zou et al. 2008). Peptide, RNA or DNA fragments specific to receptors of target cells can be conjugated to the NMOF surface for selectivity (Morris et al. 2014; Wu et al. 2015).

Depending on the purpose, NMOFs will need to be formulated into liquids, tablets, capsules, creams, or suspensions. Since physical and chemical processing can affect the integrity of the NMOF framework, the stability of NMOFs must be considered when choosing the formulation method. Binders or other additives that are biologically friendly are typically used to improve the stability of the framework. Thanks to the advanced development of MOFs in other areas (gas storage, separation and catalysis etc.), several MOFs such as MIL-53, MOF-5 and HKUST-1 have been commercially produced and sold in tablet or pellet forms. In addition, many groups are focusing on developing thin film NMOFs or using sol-gel method for NMOF synthesis. Thin-film NMOFs are greatly beneficial for topical application of encapsulated therapeutic gas such as NO for antibacterial purposes. Sol-gel synthesis avoids the use of binders or pelletizing processes, maintaining the integrity of the NMOFs structure.

Stability and Toxicity

For biomedical applications, high water stability is crucial; the framework must remain intact until the material has served its purpose, for example, delivering drug molecules or imaging. However, due to the bonding properties between metal and organic ligands in NMOFs, the NMOF structures are prone to solvent exchange. In addition, biological aqueous solutions consist of many substances and are typically slightly basic. Some NMOFs are stable in pure water solution, but degrade quickly within a few hr, few d or a few wk soaking in common biological buffers such as phosphate buffer (Horcajada et al. 2010; Miller et al. 2010).

There have not been many studies focused on the toxicity of NMOFs in biological systems. The effects of topology, surface charges, composition, particle size and biological stability on toxicity are normally examined in toxicity studies. Zn-based NMOFs were considered the most toxic, followed by Zr-based, with Fe-based NMOFs being the least toxic (Tamames-Tabar et al. 2014). The toxicity of Zn is attributed to its competition with Fe^{2+} and Ca^{2+} cations through ion channels, leading to cellular damage (Wahab et al. 2011).

NMOF toxicity also depends on the organic linkers. The organic ligands of several iron-based NMOFs from the MIL family were evaluated based on their IC-50. In general, the organic linkers in the study were non-toxic with IC-50 values ranging from 0.60 to 0.80 g/mL. Other ligands such as terephthalate or benzenedicarboxylic acid, 2-amino terephthalates and 3.3′.5.5′-azobenzene tetracarboxylate, with IC-50 values greater than 1.00 mg/mL, are considered to be more toxic (Tamames-Tabar et al. 2014). The polarity and hydrophobic-hydrophilic balance (log P) of the organic linkers also affect the materials' cytotoxicity. Interestingly, the toxicities of the organic ligands differ between cell lines, indicating that the specialization of the cells could lead to different tolerance towards NMOFs.

Unfortunately, most of the toxicity studies of MOFs in biological systems have only been done *in vitro*. The only *in vivo* study reported was done for MIL-100, MIL-88B_$4CH_3$) and MIL-88A (Baati et al. 2013). High doses (~220 mg/kg) of these MOFs were administered intravenously into rats and toxicity was evaluated by observing the behavior, body and organ weights, histology and enzymatic activity. Concentration of iron in kidneys, heart and brain of the organism showed no significant changes, except a slight increase in the brain's iron concentration. Only a small amount of ligand MIL-88B_$4CH_3$ ligand was detected in the brain. The iron metals and the ligands were almost completely eliminated from the body after seven days via urine and feces with no sign of ligand degradation.

Biomedical Applications of NMOFs

NMOFs for Drug Delivery

NMOFs are seen as optimal drug delivery materials thanks to the tunability of the framework' spore sizes and functional groups, enhancing framework-drug molecule interaction and enabling the attachment of fluorescence/luminescence molecules for tracking purposes. The first reported study of NMOFs for drug delivery was in 2008 with the physical encapsulation of ibuprofen by Horcajada and co-workers (Horcajada et al. 2008) using MIL-53 (Cr and Fe). The group later demonstrated the controlled release of challenging antitumoral and antiviral drugs against cancer and AIDS using low-toxic iron-based MIL series (MIL-53, MIL-88A, MIL-88Bt, MIL-89, MIL-100 and MIL-101_NH_2) (Horcajada et al. 2010). The tested NMOFs showed a significant improvement in drug uptake and controlled releasing by MIL-100 and MIL-101 as compared to others in the MIL series and other nanocarriers. This improvement was credited to the cage-like structures of both MOFs and their high pore volume (Fig. 13.1). However, MIL-100 outperformed MIL-101 for all the tested drugs in term of drug uptake and delivery profiles, indicating higher volume does not necessarily mean higher uptake since too large volume and the lack of guest-host interactions can lead to leaching.

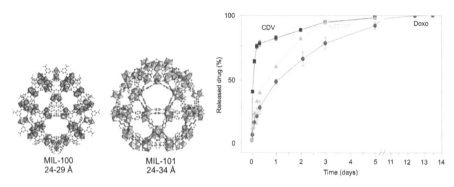

Fig. 13.1 (Left and middle) Cage structures of mesoporous MIL-100 and MIL 101 nanoparticles. (Right) Delivery profiles of Cidofovir (CDV), Doxorubicin (Doxo) and Azidothimidine triphosphate (AZT-TP) from MIL-100 under simulated physiological conditions (PBS, 37°C). Reprinted by permission from Macmillan Publisher Ltd: Nature Materials (Horcajada et al. 2010), copyright 2010.

Direct coordination of drug molecules to the metal centers of the framework (Miller et al. 2010) or into the framework structure as organic linkers (Burrows et al. 2013) showed high drug uptake; however, the framework stability might be a concern. "Ship-in-a-bottle" strategy enhanced uptake and prevented leaching of the hydrophilic drugs (di Nunzio et al. 2014). Utilizing the large pores and open metal sites, Lin and co-workers co-delivered cisplatin and pooled small interfering RNAs (siRNAs) to overcome drug resistance in ovarian cancer cells (He et al. 2014a). The large pores of NMOFs were loaded with chemotherapeutics cisplatin, while the metal ions on the NMOF surfaces were used to bind siRNAs. The NMOF protected siRNAs from nuclease degradation and enhanced siRNA cellular uptake. Surface modification has also been applied to enhance cellular uptake, selectivity (Morris et al. 2014) and *in vivo* drug tracking (Rieter et al. 2008). Short DNA segments were conjugated onto the surface of UiO-66-N_3, increasing the stability and cellular uptake as compared to the un-functionalized UiO-66-N_3 (Morris et al. 2014).

One of the main reasons for NMOFs being the ideal materials for drug delivery is the ability of highly controlled release of drug molecules. Many materials suffer from a "burst release" effect in which the maximal delivery occurs within hours of administration. This "burst release" effect can be caused by the rapid degradation of the framework in contact with physiological buffer solutions such phosphate buffer saline (PBS). To avoid this problem, Zhang and co-workers rationalized that the use of an interpenetrated network could enhance the stability of the framework, resulting in a pH-dependent controlled release of the drug molecules (Duan et al. 2015). In fact, pH-dependent controlled release is the most commonly used strategy for NMOFs (Zhu et al. 2014; Zhuang et al. 2014; Tan et al. 2015). Controlled release can also be achieved by optimizing the interaction between the NMOF framework and drug molecules. Zhong and co-workers used a Zn-based penetrated 3D network enriched with hydrogen bonds to achieve high ibuprofen uptake and a controlled release profile of four days (Zhong et al. 2014). Another strategy for prolonged drug delivery for ionic drug molecules is to utilize the NMOF surface or pore's charges to deliver the charged drug molecules. Anionic bioMOF-1 was used to encapsulate positively charged procainamide HCl with significant increase in uptake and complete drug release after 72 hr (An et al. 2009). Complete drug release can also be achieved through the degradation of MOFs in a biological solution. As already mentioned in section 2.4, the degradation in PBS of some MOF materials occurs from hours to weeks after incubation (Miller et al. 2010). Therefore, by knowing the degradation profile of the NMOFs particles, one can determine the releasing profile of the encapsulated drug molecules.

MOFs for Gasotransmitter Storage and Delivery

Gasotransmitters are endogenously produced and critical signaling molecules in biological systems, which can easily diffuse through membranes without the help of a membrane receptor. Due to their extremely short lifetime in blood, only a few sec, these gasotransmitters would only cause local effects upon being released, reducing the possibility of unwanted side effects (Horcajada et al. 2012). There are several advantages of using MOFs to deliver gasotransmitters over other materials, such as the absence of significant by-products from the storage and delivery process, possibility of highly controlled release and feasible formulation procedure to produce desired form of administration (cream, thin films, pellets, et. al) (Horcajada et al. 2012).

Nitric Oxide (NO)

Nitric oxide (NO) is one of the three important gasotransmitters of biological systems, commonly used as a vasodilator whose effect on the body depends on its concentration. Too high of a concentration can lead to hypotension, excessive bleeding and high level of inflammation, whereas too low concentration leads to hypertension and decrease in immune response to an infection. The current methods used to deliver nitric oxide to patients are through drugs such as sodium nitroprusside, glyceryl trinitrate, or gas cylinders, which exhibit many undesired side-effects and low practicality.

The existence of open metal sites (OMSs) in MOFs yields a big advantage when it comes to gas adsorption. Many studies have shown that MOFs with OMSs have a higher gas adsorption uptake and longer shelf-life of storage than those without (Rowsell and Yaghi 2005; Wu et al. 2009). Nitric oxide can bind to the metal ions and is released in contact with water. OMSs with their high surface area and large porosity increase the amount of adsorption and the affinity of nitric oxide molecules to the framework. The importance of OMSs for NO storage has been extensively studied in several structures (Fig. 13.2) such as HKUST-1 (Xiao et al. 2007) CPO-27 (McKinlay et al. 2008), MIL-88 (Ma et al. 2013a; McKinlay et al. 2013), Fe-MIL-100 (Horcajada et al. 2007) and bioMIL-3 (Miller et al. 2013). To increase the number of OMSs in the framework, ion-exchange can be employed (Pinto et al. 2014). The loading of NO into ETS-4 significantly increased after the cation-exchange to replace Na^+ with Cu^{2+} or Co^{2+} due to the higher affinity of Cu^{2+} and Co^{2+} for NO molecules.

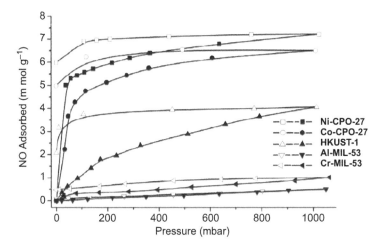

Fig. 13.2 The NO adsorption/desorption isotherms of three MOFs at room temperature. The NO uptakes are higher where coordinative unsaturated sites are available. Reprinted with permission from (Hinks et al. 2010). Copyright 2009 Elsevier Inc.

Post-synthetic modification has also been used to functionalize the frameworks to enhance the interaction between the framework and NO molecules. Secondary amines have been engineered inside the pores of MOF HKUST-1 using open metal sites to form NONOate compound with NO (Ingleson et al. 2009). Modifying the organic linkers instead of the open metal sites to form NONOate compounds is an alternative to improve host-guest interaction (Nguyen et al. 2010). A new strategy that has been introduced by Jacqueline and co-workers for NO adsorption was the use of a bioactive MOF catalyst to generate NO from endogenous sources (Harding et al.

2014). Cu (III) 1, 3, 5-Benzene-tris-triazole (CuBTTri), a very water stable MOF was used to catalyze NO generation from S-nitrosothiols (RSNOs), which are storage and transport vehicles for NO in the blood stream. By carefully tuning the amount of catalyst CuBTTri, the amount of NO generated was fixed within the biological therapeutic range and the catalytic activity remained consistent even after immersion in biological fluids such as whole blood.

NMOFs with high selectivity for NO have also been introduced based on the strong interaction between NO and atoms of the MOF framework. After the activation process, MOF Cu-SIP-3 entered a non-porous phase transition which was reversed only in the presence of NO due to strong interaction between NO and the framework (Xiao et al. 2009), resulting in ultra-selectivity for NO. Electron transfer reactions between organic linkers and NO molecules were also used to increase NO selectivity of the framework (Shimomura et al. 2010; Kosaka et al. 2013).

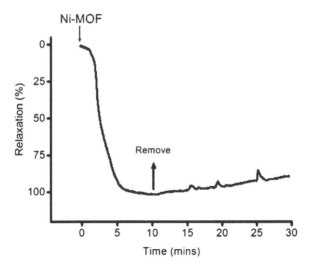

Fig. 13.3 Vasodilatory effect of Ni-MOF in a pre-contracted porcine coronary artery. A pellet (~5 mg) of Ni-MOF was placed in the organ bath (10 mL) and removed as indicated. Within 4 min of the addition of the pellet, 100 percent relaxation was achieved and slowly recovered upon pellet removal. Without removal of the pellet, no recovery was seen 1 hr after pellet addition. (Reprinted with permission from (McKinlay et al. 2008). Copyright 2008 American Chemical Society).

Strategies to achieve controlled delivery of NO are based on the manipulation of changes in temperature, pressure, the use of a chemical or photolytic trigger. However, the most practical strategy for *in vivo* delivery of NO is to use water to displace NO molecules adsorbed in the host framework. The key to the success of this strategy is the framework stability in water, which undoubtedly is also a critical property of MOFs for biomedical applications. Differences in water stability results in a significant difference in the amount of NO delivered (Dietzel et al. 2005; Dietzel et al. 2006). In addition, different metals result in different delivery potentials as shown in the M-CPO-27 family. Co and Ni structures showed almost perfect delivery while Mg (Dietzel et al. 2008a), Zn (Dietzel et al. 2008b) and Mn (Zhou et al. 2008) delivered only 10 percent of their capacities. ETS-4 MOF with unsaturated Ti^{4+} and Cu^{2+} or Co^{2+} released 85 percent of its NO with slow release kinetics of 0.023 nM/min, which was preferred for controlled *in vivo* releasing of NO (Pinto et al. 2014).

To date, the biological activity of NO-loaded MOFs has been tested for three purposes: antiplatelet activation, blood vessel relaxation (or vasodilatory) and antibacterial properties. Ni-MOF loaded with NO has been demonstrated to have great vessel relaxation effect on pre-contracted pig coronary arteries *in vitro* (McKinlay et al. 2008). The placement of Ni-CPO pellets at a distance of 2 mm from the vessel resulted in a rapid 10 percent relaxation of the vessel. The contraction was gradually recovered after the removal of the pellets, proving the reversible vasodilatory property of NO-loaded MOFs (Fig. 13.3).

Hydrogen Sulfide (H_2S)

In contrast to its commonly known toxicity and highly unpleasant odor, H_2S has been reported, at low concentration, to have vasodilatation (Shibuya et al. 2009), neuroprotection, asthma anti-inflammation (Chen et al. 2009) and interestingly to create a beneficial hypothermia/hypometabolic state that could improve organ preservation or conditions such as ischemia/reperfusion injury or trauma (Blackstone et al. 2005). The most current method of H_2S delivery is the use of H_2S donor, NaHS, which produces H_2S *in situ* via chemical reaction (Papapetropoulos et al. 2009). This method lacks specificity and produces side products, which in turn could lead to unwanted side effects (Wheatley et al. 2006).

Due to its aggressive properties and undesired presence in fuels or gas mixture, research in H_2S adsorption using MOFs has been very active even before the discovery of its therapeutic properties. One character of H_2S which makes its storage more challenging than NO'sis its strong interaction with the host frameworks which sometimes leads to some degree of permanent adsorption or even destruction of the framework. Copper-based MOFs are unsuitable for H_2S storage since H_2S molecules bind to the copper centers and gradually react to form copper sulfide (Petit et al. 2010). Flexible MIL-53 and CPO-27 families are considered to be a more suitable host for H_2S for there is no evidence of damage done to the framework after H_2S adsorption (Hamon et al. 2009; Hamon et al. 2011). Adsorption uptake of H_2S onto M-CPO-27 (M=Ni or Zn) was ~10 mmol H_2S per gram of MOF (Allan et al. 2012).

Similar to NO delivery mechanism, the release of H_2S from MOF hosts is currently based on solvent exchange with water molecules in solution or from moisture. Only 30 percent of H_2S adsorbed into Ni- or Zn-CPO-27 was released (1.8 mmol/g and 0.5 mmol/g, respectively) in contact with moisture, possibly due to the strong interaction between H_2S and the framework (Allan et al. 2012). The release from both materials was completed after 1 hr. The biological effect of H_2S upon releasing from MOFs was examined following a similar experiment for NO gas. A pellet of H_2S-loaded Zn-CPO-27 was placed 2 mm from pre-contracted coronary artery, resulting in a reversible 39 percent relaxation after 15 min. Experiments on the releasing of H_2S in MOFs are still limited in number (Allan et al. 2012). However, current results indicate a promising future for MOFs as H_2S delivery agents.

Carbon Monoxide (CO)

Carbon monoxide, often known as "the silent killer", is responsible for many household deaths. Its deadly effect on human bodies is due to its much stronger binding affinity to the iron metal center of hemoglobin as compared to that of oxygen molecules, which interrupts the delivery of oxygen to organs via blood and eventually results in tissue hypoxia and poisoning (Gorman et al. 2003). Research on materials for CO delivery has been done with many materials with carbon-monoxide-releasing molecules (CORMs) being the most intensively studied, especially the photoactive CORMs (Rimmer et al. 2012, Gonzales and Mascharak 2014). Others such as silica (Gonzales et al. 2014) and peptide-based materials (Matson et al. 2012) have also been experimented for CO delivery purpose. However, as promising as MOFs are for gas storage and delivery application, to the extent of our knowledge, there is still no study on CO biological delivery using MOFs. Studies have been done on the interaction and binding affinity of CO gas molecules with MOF frameworks (Supronowicz et al. 2013). This will provide basics for CO delivery research using MOFs in the future.

MOFs as Biosensors

A commonly used strategy to design biosensor MOFs is to incorporate luminescence or fluorescence quencher/producer into the framework. The exposure of the reactive sites of the framework to the targeted compound will trigger the significant enhancement or reduction in luminescence or fluorescence signal. For example, Duan's group used Cu-based MOF CU-TCA (TCA=tricarboxytriphenyl amine) to quench the luminescence of incorporated triphenylamine for the detection of NO in living cells (Wu et al.

2012). Upon addition of NO, the luminescence increased 700 times with high selectivity for NO apart from the other reactive species present in biological systems such as H_2O_2, NO_3^-, NO_2^-, $ONOO^-$, ClO^-. Similarly, several other MOFs for detection of gasotransmitters such as NO (Wu et al. 2012) and H_2S (Li et al. 2014; Ma et al. 2014) were also reported with high selectivity and low toxicity. Exposure of malonitrile-functionalized MN-ZIF-90 to H_2S significantly enhanced its fluorescence intensity (Li et al. 2014). The material also showed high selectivity toward biothiols such as cysteine over other amino acids.

Biomolecule detection such as DNA and microRNA (miRNA) is also a focus for biosensor development since they are integral parts in genetic, pathological and forensic science. MiRNAs are small non-protein-coding RNAs which are key biomolecules in regulating important biological processes. Changes in miRNA expressions have been linked to various diseases (Wu et al. 2015). Existing and common methods such as PCR offer highly accurate results, but suffer from high cost, high risk of contamination and sometimes false-negative results, while DNA detection using NMOFs provides a simpler, faster and cost-effective alternative. The first demonstration of MOF for sensing DNA was done by Guonan's group using a Cu-based MOF (Zhu et al. 2013). A year later, a MOF with the ability to distinguish complementary and mismatched target sequences with high sensitivity and selectivity was introduced by functionalizing UiO-66 with amine (UiO-66-NH_2) (Zhang et al. 2014). Wu and co-workers conjugated fluorophores-labeled peptide nucleic acid (PNA) with NMOF vehicles to monitor miRNAs in living cancer cells, where the NMOF works as a fluorophorescence quencher of the labeled PNA that is firmly bound to the metal center. In the presence of a target miRNA, PNA is hybridized and released from the NMOF leading to the recovery of fluorescence, allowing the detection of multiplexed miRNA in living cells and monitoring of the spatiotemporal changes of miRNA expression (Wu et al. 2015).

Fast-response biosensors to determine the concentration of glucose are made by immobilizing enzymes such as peroxidases and glucose oxidase (GO) to achieve a colorimetric response (Ma et al. 2013b; Qin et al. 2013). This is often described as a cascade reaction in which the GO first catalyzes the oxidation of glucose to produce hydrogen peroxide (H_2O_2), which in turn oxidizes a chosen compound that can produce or change color once oxidized, which could be measured using a spectrophotometer.

In 2014, Wenbin's group reported the first real-time intracellular pH sensing of NMOFs by covalently attaching fluorescein isothiocyanate (FITC) with UiO NMOF (F-UiO) for excellent live cell imaging of the endocytosis and intracellular trafficking process (He et al. 2014a). FITC acts as a pH sensor by providing pH-specific ratios of fluorescence intensities at 520 nm when excited at 488 (for F-UiO) and 435 nm (for free fluorescence) (Fig. 13.4).

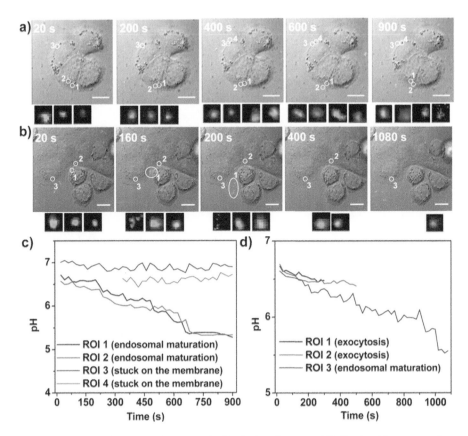

Fig. 13.4 Time relapse pH changes in individual endosomes. (A) Confocal laser scanning microscopy (CLSM) images shown are overlays of differential interference contrast (DIC) microscopy, green and red. The color changes according to pH level from pH 4 (red) to pH 8 (green). (B) Each region of interest (ROI)was labeled with a white circle and number. Small images showing under each CLSM image are zoomed-in views of overlays of green and red representing ROI 1 to ROI 4 or ROI 1 to ROI 3 from left to right. Bar: 10 μm. (C, D) The pH evolution curve of each ROI, indicating the drop in pH caused by endosomal maturation of cell-internalized ROI 1 and ROI 2, whereas ROI 3 and 4 remained at constant pH since they were stuck on the membrane. (Reprinted with permission from (He et al. 2014a). Copyright 2014 American Chemical Society).

MOFs as Imaging Agents

Magnetic Resonance Imaging (MRI)

NMOFs are extensively studied as MRI contrast agents. Most of the currently used contrast agents are nonspecific and less efficient, thus, require relatively large amounts to be administered (Andre et al. 2013). Good contrast agents are those with high relaxivity R1, meaning that the agents can reduce the tissue water proton relaxation time at a large magnitude, resulting in higher contrast (Sim and Parker 2015). Complexes of gadolinium Gd^{3+} are the current commonly used contrast agents.

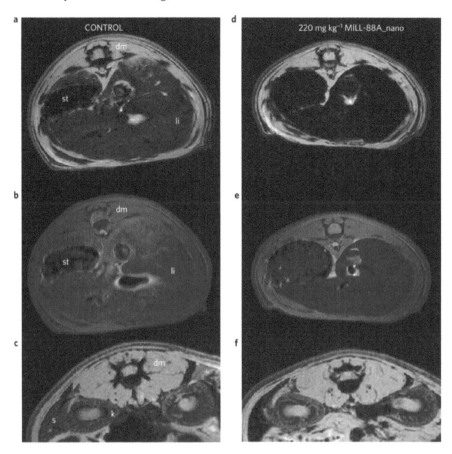

Fig. 13.5 Magnetic resonance images of control rats (left: A-C) and female rats after being injected with 220 mg/kg MIL-88A nanoparticles (right; D-F), in liver (A, B, D, E) and spleen (C, F) regions. The images were acquired with gradient echo (A, C, D, F) or spin echo (B, E) sequence. Thirty min after injection, product effect is observable on the liver and spleen. (dm, dorsal muscle; k, kidney; li, liver; s, spleen; st, stomach). Reprinted by permission from Macmillan Publisher Ltd: Nature Materials (Horcajada et al. 2010), Copyright 2010.

The first reported study of Gd-based MOF as a contrast agent was done by Lin's group with high relaxivities with several orders of magnitude higher than those of current Gd-based MRI contrast agents (Rieter et al. 2006). This Gd-based MOF was later coated with a thin shell of silica for controlled release of Gd^{3+} (Rieter et al. 2007). However, due to *in vivo* toxicity of Gd^{3+} ions, Mn-based NMOFs were proposed as an alternative (Taylor et al. 2008b). Even though the relaxivities obtained were lower than that of Gd-based MOFs, the improvement in selectivity was significant and was observed using confocal microscopy, *in vitro* MRI and inductively coupled plasma mass spectrometry (ICP-MS). Coated with a surface-functionalized silica shell, these Mn-based NMOFs showed improved controlled release and selectivity toward tumor cells (Taylor et al. 2008b). Attempts of using other metal-based MOFs such as lanthanides (Pereira et al. 2010), Pt (Rieter et al. 2006) have been reported; however, though the toxicity profiles were favorable, the relaxivities were relatively low as compared to Gd- and Mn-based MOFs.

An *in vivo* study demonstrated by Horcajada and co-workers showed the use of MIL-88A and MIL-100 as MRI contrast agents in female rats (Fig. 13.5) (Horcajada et al. 2010). After 30 min of the injection, darker images appeared on the treated liver and spleen organs, indicating the promising application of iron-based MOFs as contrast agents. In addition, three months after the injection, the liver and spleen returned to the original appearance, showing the reversibility of the MOF contrast agents. Fluorophore can also be implanted to Fe-based NMOF framework, MIL-101, the toxicity and uptake profile of which have been studied quite thoroughly (Horcajada et al. 2006, Horcajada et al. 2008), in order to track the delivery of encapsulated drug molecules within the body (Taylor-Pashow et al. 2009a).

Computed Tomography (CT)

In addition to MRI contrast agents, MOFs have also showed promising results for use in computed tomography (CT) imaging. For CT imaging, iodinated aromatic molecules are the typical contrast agents used. Studies have used Cu (II) and Zn (II) MOFs as carriers for iodinated aromatic molecules (Dekrafft et al. 2009). More recently, Lin and co-workers prepared Zr-BDC and Hf-BDC MOFs of UiO-66 structure. The Hf-MOF was determined to be twice as efficient as CT contrast agents such as Iodixanol. An *in vivo* CT imaging study conducted for Hf-BDC MOF coated with silica and PEG as the contrast agent showed enhanced attenuation in the liver and spleen 15 min after injection (Dekrafft et al. 2009).

Optical Imaging (OI)

Optical Imaging is a powerful tool in cancer diagnosis thanks to its noninvasive nature and higher sensitivity as compared to MRI and CT. Efforts have been made using Zn^{2+} and Zr^{4+} with $Ru(bpy)_3^{2+}$ as bridging ligands to make nanoparticles with high dye loadings (78.7 percent and 57.4 percent, respectively), in order to overcome the self-quenching and photo-bleaching problems that typical CT contrast agents have (Liu et al. 2011). Kimizuka and co-workers demonstrated a different method for capturing and delivering functional molecules by using nucleotide and lanthanide ion building blocks to synthesize a material that can encapsulate fluorescent dyes into the material (Nishiyabu et al. 2009). The field of OI itself is still relatively new; therefore, studies using NMOFs for OI are still at the infant stage. However, the fast development of NMOFs for applications in other fields indicates a promising future of NMOFs development in OI research.

Conclusions

Due to their tunable structure, pore sizes and pore volume along with easy functionalization, metal-organic frameworks have many advantages for biomedical applications. The numbers of metals that are biocompatible with the human body are somewhat limited but the choices of organic linkers are almost infinite. So far, metal-organic frameworks surpass other existing inorganic porous solids (zeolites and mesoporous silica) and organic polymers in terms of drug uptake and prolonged release. However, to make MOFs practical for medical applications, more *in vivo* studies on the delivery kinetics, the degradation mechanism and toxicity of MOFs are needed. However, despite all the existing problems, MOFs for biomedical applications are still of great interest with promising results and will attract more and more enthusiastic scientists to this field of study.

Acknowledgment

The authors acknowledge the National Science Foundation (DMR-1352065) and the University of South Florida for financial support of this work.

References

Allan, P.K., Wheatley, P.S., Aldous, D., Mohideen, M.I., Tang, C., Hriljac, J.A., et al. 2012. Metal-organic frameworks for the storage and delivery of biologically active hydrogen sulfide. Dalton Trans 41(14): 4060-4066.

An, J.Y., Geib, S.J. and Rosi, N.L., 2009. Cation-triggered drug release from a porous zinc-adeninate metal-organic framework. J Am Chem Soc 131(24): 8376-8377.

Andre S., Merbach, Helm, L. and Toth, E. 2013. General principle of MRI. pp. 1-23. *In:* A.S. Merbach, L. Helm and E. Toth [eds.]. The Chemistry of Contrast Agents in Medical Magnetic Resonance Imaging. Wiley Chichester, U.K.: 45.

Baati, T., Njim, L., Neffati, F., Kerkeni, A., Bouttemi, M., Gref, R., et al. 2013. In depth analysis of the in vivo toxicity of nanoparticles of porous iron (III) metal-organic frameworks. Chem Sci 4(4): 1597-1607.

Blackstone, E., Morrison, M. and Roth, M.B. 2005. H2S induces a suspended animation-like state in mice. Science 308(5721): 518-518.

Burrows, A.D., Jurcic, M., Keenan, L.L., Lane, R.A., Mahon, M.F., Warren, M.R., et al. 2013. Incorporation by coordination and release of the iron chelator drug deferiprone from zinc-based metal-organic frameworks. Chem Comm 49(96): 11260-11262.

Chen, Y.H., Wu, R., Geng, B., Qi, Y.F., Wang, P.P., Yao, W.Z., et al. 2009. Endogenous hydrogen sulfide reduces airway inflammation and remodeling in a rat model of asthma. Cytokine 45(2): 117-123.

Dekrafft, K.E., Xie, Z.G., Cao, G.H., Tran, S., Ma, L.Q., Zhou, O.Z., et al. 2009. Iodinated nanoscale coordination polymers as potential contrast agents for computed tomography. Angew Chem Int Ed 48(52): 9901-9904.

di Nunzio, M.R., Agostoni, V., Cohen, B., Gref, R. and Douhal, A. 2014. A "Ship in a bottle" strategy to load a hydrophilic anticancer drug in porous metal organic framework nanoparticles: efficient encapsulation, matrix stabilization and photodelivery. J Med Chem 57(2): 411-420.

Dietzel, P.D.C., Morita, Y., Blom, R. and Fjellvag, H. 2005. An in situ high-temperature single-crystal investigation of a dehydrated metal-organic framework compound and field-induced magnetization of one-dimensional metaloxygen chains. Angew Chem Int Ed 44(39): 6354-6358.

Dietzel, P.D.C., Panella, B., Hirscher, M., Blom, R. and Fjellvag, H. 2006. Hydrogen adsorption in a nickel based coordination polymer with open metal sites in the cylindrical cavities of the desolvated framework. Chem Comm (9): 959-961.

Dietzel, P.D.C., Blom, R. and Fjellvag, H. 2008a. Base-induced formation of two magnesium metal-organic framework compounds with a bifunctional tetratopic ligand. Eur J Inorg Chem (23): 3624-3632.

Dietzel, P.D.C., Johnsen, R.E., Blom, R. and Fjellvag, H. 2008b. Structural changes and coordinatively unsaturated metal atoms on dehydration of honeycomb analogous microporous metal-organic frameworks. Chem Eur J 14(8): 2389-2397.

Duan, L.N., Dang, Q.Q., Han, C.Y. and Zhang, X.M. 2015. An interpenetrated bioactive nonlinear optical MOF containing a coordinated quinolone-like drug and Zn(II) for pH-responsive release. Dalton Trans 44(4): 1800-1804.

Gonzales, M.A. and Mascharak, P.K. 2014. Photoactive metal carbonyl complexes as potential agents for targeted CO delivery. J Inorg Biochem 133: 127-135.

Gonzales, M.A., Han, H., Moyes, A., Radinos, A., Hobbs, A.J., Coombs, N., et al. 2014. Light-triggered carbon monoxide delivery with Al-MCM-41-based nanoparticles bearing a designed manganese carbonyl complex. J Mater Chem B2(15): 2107-2113.

Gorman, D., Drewry, A., Huang, Y.L. and Sames, C. 2003. The clinical toxicology of carbon monoxide. Toxicology 187(1): 25-38.

Hamon, L., Serre, C., Devic, T., Loiseau, T., Millange, F., Ferey, G., et al. 2009. Comparative study of hydrogen sulfide adsorption in the MIL-53(Al, Cr, Fe), MIL-47(V), MIL-100(Cr) and MIL-101(Cr) metal-organic frameworks at room temperature. J Amer Chem Soc 131(25): 8775-8777.

Hamon, L., Leclerc, H., Ghoufi, A., Oliviero, L., Travert, A., Lavalley, J.C., et al. 2011. Molecular insight into the adsorption of H2S in the flexible MIL-53(Cr) and rigid MIL-47(V) MOFs: infrared spectroscopy combined to molecular simulations. J Phys Chem C115(5): 2047-2056.

Haque, E., Khan, N.A., Park, J.H. and Jhung, S.H. 2010. Synthesis of a metal–organic framework material, iron terephthalate, by ultrasound, microwave and conventional electric heating: a kinetic study. Chem Eur J 16(3): 1046-1052.

Harding, J.L., Metz, J.M. and Reynolds, M.M. 2014. A tunable, stable and bioactive MOF catalyst for generating a localized therapeutic from endogenous sources. Adv Funct Mater 24(47): 7503-7509.

He, C., Lu, K., Liu, D. and Lin, W. 2014a. Nanoscale metal–organic frameworks for the co-delivery of cisplatin and pooled siRNAs to enhance therapeutic efficacy in drug-resistant ovarian cancer cells. J Amer Chem Soc 136(14): 5181-5184.

He, C.B., Lu, K.D. and Lin, W.B. 2014b. Nanoscale metal-organic frameworks for real-time intracellular pH sensing in live cells. J Am Chem Soc 136(35): 12253-12256.

Hinks, N.J., McKinlay, A.C., Xiao, B., Wheatley, P.S. and Morris, R.E. 2010. Metal organic frameworks as NO delivery materials for biological applications. Microporous and Mesoporous Mater 129(3): 330-334.

Horcajada, P., Serre, C., Vallet-Regi, M., Sebban, M., Taulelle, F. and Ferey, G. 2006. Metal-organic frameworks as efficient materials for drug delivery. Angew Chem Int Ed 45(36): 5974-5978.

Horcajada, P., Surble, S., Serre, C., Hong, D.Y., Seo, Y.K., Chang, J.S., et al. 2007. Synthesis and catalytic properties of MIL-100 (Fe), an iron (III) carboxylate with large pores. Chem Comm (27): 2820-2822.

Horcajada, P., Serre, C., Maurin, G., Ramsahye, N.A., Balas, F., Vallet-Regi, M., et al. 2008. Flexible porous metal-organic frameworks for a controlled drug delivery. J Amer Chem Soc 130(21): 6774-6780.

Horcajada, P., Chalati, T., Serre, C., Gillet, B., Sebrie, C., Baati, T., et al. 2010. Porous metal-organic-framework nanoscale carriers as a potential platform for drug delivery and imaging. Nat Mater 9(2): 172-178.

Horcajada, P., Gref, R., Baati, T., Allan, P.K., Maurin, G., Couvreur, P., et al. 2012. Metal-organic frameworks in biomedicine. Chem Rev 112(2): 1232-1268.

Hoskins, B.F. and Robson, R. 1989. Infinite polymeric frameworks consisting of 3 dimensionally linked rod-like segments. J Am Chem Soc 111(15): 5962-5964.

Ingleson, M.J., Heck, R., Gould, J.A. and Rosseinsky, M.J. 2009. Nitric oxide chemisorption in a postsynthetically modified metal-organic framework. Inorg Chem 48(21): 9986-9988.

Jhung, S.H., Lee, J.H., Yoon, J.W., Serre, C., Férey, G. and Chang, J.S. 2007. Microwave synthesis of chromium terephthalate MIL-101 and its benzene sorption ability. Adv Mater 19(1): 121-124.

Kosaka, W., Yamagishi, K., Hori, A., Sato, H., Matsuda, R., Kitagawa, S., et al. 2013. Selective NO trapping in the pores of chain-type complex assemblies based on electronically activated paddlewheel-type [Ru2(II,II)]/[Rh2(II,II)] dimers. J Am Chem Soc 135(49): 18469-18480.

Li, H., Eddaoudi, M., O'Keeffe, M. and Yaghi, O.M. 1999. Design and synthesis of an exceptionally stable and highly porous metal-organic framework." Nature 402(6759): 276-279.

Li, H., Feng, X., Guo, Y., Chen, D., Li, R., Ren, X., et al. 2014. A malonitrile-functionalized metal-organic framework for hydrogen sulfide detection and selective amino acid molecular recognition. Sci Rep 4: 4366.

Li, Z.-Q., Qiu, L.-G., Xu, T., Wu, Y., Wang, W., Wu, Z.-Y., et al. 2009. Ultrasonic synthesis of the microporous metal–organic framework Cu3(BTC)2 at ambient temperature and pressure: an efficient and environmentally friendly method. Mater Lett 63(1): 78-80.

Liu, D.M., Huxford, R.C. and Lin, W.B. 2011. Phosphorescent nanoscale coordination polymers as contrast agents for optical imaging. Angew Chem Int Ed 50(16): 3696-3700.

Ma, M., Noei, H., Mienert, B., Niesel, J., Bill, E., Muhler, M., et al. 2013a. Iron metalorganic frameworks MIL-88B and NH2-MIL-88B for the loading and delivery of the gasotransmitter carbon monoxide. Chem Eur J 19(21): 6785-6790.

Ma, W., Jiang, Q., Yu, P., Yang, L. and Mao, L. 2013b. Zeolitic imidazolate framework-based electrochemical biosensor for in vivo electrochemical measurements. Anal Chem 85(15): 7550-7557.

Ma, Y., Su, H., Kuang, X., Li, X.Y., Zhang, T.T. and Tang, B. 2014. Heterogeneous nano metal-organic framework fluorescence probe for highly selective and sensitive detection of hydrogen sulfide in living Cells. Anal Chem 86(22): 11459-11463.

Matson, J.B., Webber, M.J., Tamboli, V.K., Weber, B. and Stupp, S.I. 2012. A peptide-based material for therapeutic carbon monoxide delivery. Soft Matter 8(25): 6689-6692.

McKinlay, A.C., Xiao, B., Wragg, D.S., Wheatley, P.S., Megson, I.L. and Morris, R.E. 2008. Exceptional behavior over the whole adsorption-storage-delivery cycle for NO in porous metal organic frameworks. J Am Chem Soc 130(31): 10440-10444.

McKinlay, A.C., Eubank, J.F., Wuttke, S., Xiao, B., Wheadey, P.S., Bazin, P., et al. 2013. Nitric oxide adsorption and delivery in flexible MIL-88(Fe) metal-organic frameworks. Chem Mater 25(9): 1592-1599.

Miller, S.R., Heurtaux, D., Baati, T., Horcajada, P., Greneche, J.M. and Serre C. 2010. Biodegradable therapeutic MOFs for the delivery of bioactive molecules. Chem Commun 46(25): 4526-4528.

Miller, S.R., Alvarez, E., Fradcourt, L., Devic, T., Wuttke, S., Wheatley, P.S., et al. 2013. A rare example of a porous Ca-MOF for the controlled release of biologically active NO. Chem Commun 49(71): 7773-7775.

Morris, W., Briley, W.E., Auyeung, E., Cabezas, M.D. and Mirkin, C.A. 2014. Nucleic acid–metal organic framework (MOF) nanoparticle conjugates. J Am Chem Soc 136(20): 7261-7264.

Nguyen, J.G., Tanabe, K.K. and Cohen, S.M. 2010. Postsynthetic diazeniumdiolate formation and NO release from MOFs. Cryst Eng Comm 12(8): 2335-2338.

Ni, Z. and Masel, R.I. 2006. Rapid production of metal–organic frameworks via microwave-assisted solvothermal synthesis. J Am Chem Soc 128(38): 12394-12395.

Nishiyabu, R., Hashimoto, N., Cho, T., Watanabe, K., Yasunaga, T., Endo, A., et al. 2009. Nanoparticles of adaptive supramolecular networks self-assembled from nucleotides and lanthanide ions. J Am Chem Soc 131(6): 2151-2158.

Papapetropoulos, A., Pyriochou, A., Altaany, Z., Yang, G.D., Marazioti, A., Zhou, Z.M., et al. 2009. Hydrogen sulfide is an endogenous stimulator of angiogenesis. Proc Natl Acad Sci USA 106(51): 21972-21977.

Pereira, G.A., Peters, J.A., Paz, F.A.A., Rocha, J. and Geraldes, C.F.G.C. 2010. Evaluation of [Ln(H(2)cmp)(H2O)] metal organic framework materials for potential application as magnetic resonance imaging contrast agents. Inorg Chem 49(6): 2969-2974.

Petit, C., Mendoza, B. and Bandosz, T.J. 2010. Hydrogen sulfide adsorption on MOFs and MOF/graphite oxide composites. Chemphyschem 11(17): 3678-3684.

Pinto, M.L., Fernandes, A.C., Rocha, J., Ferreira, A., Antunes, F. and J. Pires. 2014. Microporous titanosilicates Cu2+- and Co2+-ETS-4 for storage and slow release of therapeutic nitric oxide. J Mater Chem B 2(2): 224-230.

Qin, F.-X., Jia, S.-Y., Wang, F.-F., Wu, S.-H., Song, J. and Liu, Y. 2013. Hemin@metal-organic framework with peroxidase-like activity and its application to glucose detection. Catal Sci Technol 3(10): 2761-2768.

Qiu, L.-G., Li, Z.-Q., Wu, Y., Wang, W., Xu, T. and Jiang, X. 2008. Facile synthesis of nanocrystals of a microporous metal-organic framework by an ultrasonic method and selective sensing of organoamines. Chem Commun (31): 3642-3644.

Rabone, J., Yue, Y.-F., Chong, S.Y., Stylianou, K.C., Bacsa, J., Bradshaw, D., et al. 2010. An adaptable peptide-based porous material. Science 329(5995): 1053-1057.

Rieter, W.J., Taylor, K.M.L., An, H.Y., Lin, W.L. and Lin, W.B. 2006. Nanoscale metal-organic frameworks as potential multimodal contrast enhancing agents. J Am Chem Soc 128(28): 9024-9025.

Rieter, W.J., Taylor, K.M.L. and Lin, W.B. 2007. Surface modification and functionalization of nanoscale metal-organic frameworks for controlled release and luminescence sensing. J Am Chem Soc 129(32): 9852-9853.

Rieter, W.J., Pott, K.M., Taylor, K.M.L. and Lin, W.B. 2008. Nanoscale coordination polymers for platinum-based anticancer drug delivery. J Am Chem Soc 130(35): 11584-11585.

Rimmer, R.D., Pierri, A.E. and Ford, P.C. 2012. Photochemically activated carbon monoxide release for biological targets. Toward developing air-stable photo-CORMs labilized by visible light. Coord Chem Rev 256(15-16): 1509-1519.

Rowe, M.D., Chang, C.-C., Thamm, D.H., Kraft, S.L., Harmon, J.F., Vogt, A.P., et al. 2009a. Tuning the magnetic resonance imaging properties of positive contrast agent nanoparticles by surface modification with RAFT polymers. Langmuir 25(16): 9487-9499.

Rowe, M.D., Thamm, D.H., Kraft, S.L. and Boyes, S.G. 2009b. Polymer-modified gadolinium metal-organic framework nanoparticles used as multifunctional nanomedicines for the targeted imaging and treatment of cancer. Biomacromolecules 10(4): 983-993.

Rowsell, J.L.C. and Yaghi, O.M. 2005. Strategies for hydrogen storage in metal–organic frameworks. Angew Chem Int Ed 44(30): 4670-4679.

Shibuya, N., Mikami, Y., Kimura, Y., Nagahara, N. and Kimura, H. 2009. Vascular endothelium expresses 3-mercaptopyruvate sulfurtransferase and produces hydrogen sulfide. J Biochem 146(5): 623-626.

Shimomura, S., Higuchi, M., Matsuda, R., Yoneda, K., Hijikata, Y., Kubota, Y., et al. 2010. Selective sorption of oxygen and nitric oxide by an electron-donating flexible porous coordination polymer. Nat Chem 2(8): 633-637.

Sim, N. and Parker, D. 2015. Critical design issues in the targeted molecular imaging of cell surface receptors. Chem Soc Rev 44(8): 2122-2134.

Supronowicz, B., Mavrandonakis, A. and Heine, T. 2013. Interaction of small gases with the unsaturated metal centers of the HKUST-1 Metal Organic Framework. J Phys Chem C 117(28): 14570-14578.

Tamames-Tabar, C., Cunha, D., Imbuluzqueta, E., Ragon, F., Serre, C., Blanco-Prieto, M.J., et al. 2014. Cytotoxicity of nanoscaled metal-organic frameworks. J Mater Chem B 2(3): 262-271.

Tan, L.-L., Li, H., Qiu, Y.-C., Chen, D.-X., Wang, X., et al. 2015. Stimuli-responsive metal-organic frameworks gated by pillar[5]arene supramolecular switches. Chem Sci 6(3): 1640-1644.

Taylor, K.M.L., Jin, A. and Lin, W.B. 2008a. Surfactant-assisted synthesis of nanoscale gadolinium metal-organic frameworks for potential multimodal imaging. Angew Chem Int Ed 47(40): 7722-7725.

Taylor, K.M.L., Rieter, W.J. and Lin, W.B. 2008b. Manganese-based nanoscale metal-organic frameworks for magnetic resonance imaging. J Am Chem Soc 130(44): 14358-14359.

Taylor-Pashow, K.M.L., Rocca, J.D., Xie, Z., Tran, S. and Lin, W. 2009. Postsynthetic modifications of Iron-carboxylate nanoscale metal–organic frameworks for imaging and drug delivery. J Am Chem Soc 131(40): 14261-14263.

Wahab, R., Kaushik, N.K., Verma, A.K., Mishra, A. Hwang, I.H., Yang, Y.B., et al. 2011. Fabrication and growth mechanism of ZnO nanostructures and their cytotoxic effect on human brain tumor U87, cervical cancer HeLa and normal HEK cells. J Biol Inorg Chem 16(3): 431-442.

Wheatley, P.S., Butler, A.R., Crane, M.S., Fox, S., Xiao, B., Rossi, A.G., et al. 2006. NO-releasing zeolites and their antithrombotic properties. J Am Chem Soc 128(2): 502-509.

Wu, H., Zhou, W. and Yildirim, T. 2009. High-capacity methane storage in metal–organic frameworks M2(dhtp): the important role of open metal sites. J Am Chem Soc 131(13): 4995-5000.

Wu, P.Y., Wang, J., He, C., Zhang, X.L., Wang, Y.T., Liu, T., et al. 2012. Luminescent metal-organic frameworks for selectively sensing nitric oxide in an aqueous solution and in living cells. Adv Funct Mater 22(8): 1698-1703.

Wu, Y., Han, J., Xue, P., Xu, R. and Kang, Y. 2015. Nano metal-organic framework (NMOF)-based strategies for multiplexed microRNA detection in solution and living cancer cells. Nanoscale 7(5): 1753-1759.

Xiao, B., Wheatley, P.S., Zhao, X., Fletcher, A.J., Fox, S., Rossi, A.G., et al. 2007. High-capacity hydrogen and nitric oxide adsorption and storage in a metal-organic framework. J Am Chem Soc 129(5): 1203-1209.

Xiao, B., Byrne, P.J., Wheatley, P.S., Wragg, D.S., Zhao, X., Fletcher, A.J., et al. 2009. Chemically blockable transformation and ultraselective low-pressure gas adsorption in a non-porous metal organic framework. Nat Chem 1(4): 289-294.

Yaghi, O.M., O'Keeffe, M., Ockwig, N.W., Chae, H.K., Eddaoudi, M. and Kim, J. 2003. Reticular synthesis and the design of new materials. Nature 423(6941): 705-714.

Zhang, H.-T., Zhang, J.-W., Huang, G., Du, Z.-Y. and Jiang, H.-L. 2014. An amine-functionalized metal-organic framework as a sensing platform for DNA detection. Chem Commun 50(81): 12069-12072.

Zhong, D.-C., Liao, L.-Q., Deng, J.-H., Chen, Q., Lian, P. and Luo, X.-Z. 2014. A rare (3, 4, 5)-connected metal-organic framework featuring an unprecedented 1D + 2D [rightward arrow] 3D self-interpenetrated array and an O-atom lined pore surface: structure and controlled drug release. Chem Commun 50(99): 15807-15810.

Zhou, W., Wu, H. and Yildirim, T. 2008. Enhanced H-2 adsorption in isostructural metal-organic frameworks with open metal sites: strong dependence of the binding strength on metal ions. J Am Chem Soc 130(46): 15268-15269.

Zhu, X., Zheng, H., Wei, X., Lin, Z., Guo, L., Qiu, B., et al. 2013. Metal-organic framework (MOF): a novel sensing platform for biomolecules. Chem Commun 49(13): 1276-1278.

Zhu, X., Gu, J., Wang, Y., Li, B., Li, Y., Zhao, W., et al. 2014. Inherent anchorages in UiO-66 nanoparticles for efficient capture of alendronate and its mediated release. Chem Commun 50(63): 8779-8782.

Zhuang, J., Kuo, C.-H., Chou, L.-Y., Liu, D.-Y., Weerapana, E. and Tsung, C.-K. 2014. Optimized metal–organic-framework nanospheres for drug delivery: evaluation of small-molecule encapsulation. ACS Nano 8(3): 2812-2819.

Zou, H., Wu, S. and Shen, J. 2008. Polymer/silica nanocomposites: preparation, characterization, properties and applications. Chem Rev 108(9): 3893-3957.

14

Silica/Gold Hybrid Nanoparticles for Imaging and Therapy

Shengqin Su[a] *and Jesse V. Jokerst**

INTRODUCTION

Nanoparticles (NPs) have been used as medical tools for many decades and have been made from a variety of materials and with a variety of designs. NPs are particularly useful in imaging and therapy because of their unique physical properties including: 1) intense and longitudinally stable imaging signal; 2) passive or molecular targeting to their target; 3) high avidity (large association constant due to multiple ligands per NP); 4) theranostic (both therapeutic and diagnostic) capabilities; 5) capacity for multimodal signal, e.g., fluorescence and magnetic resonance imaging; 6) capacity for multiplexed imaging of different targets in the same sample; 7) concurrent imaging of different targets by coating a single NP with two or more homing ligands and 8) circulation times longer than the rapid washout of small molecules but shorter than proteins. In selecting a material type, the researcher should first focus on the task at hand. That is, what medical problem needs to be solved? In selecting a nanoparticle type, the researcher should ask himself or herself many questions including:

- What do I want the nanoparticle to do?
- Where is the target site?
- Are there target cells or proteins?
- What is the route of delivery?
- How long should the nanoparticles remain at the target site?
- What happens to the nanoparticle when the task is completed?

Department of NanoEngineering, University of California, San Diego, USA.

[a] E-mail: shs071@eng.ucsd.edu

* Corresponding author: jjokerst@ucsd.edu

After answering these questions, the researcher can then start to design a nanoparticle that matches these requirements. Many types of materials are used to make nanoparticles including carbon nanotubes, precious metals (Boote et al. 2010; Hirsch et al. 2003; Huang et al. 2006; Jain et al. 2008; Park et al. 2008; Thakor et al. 2011; von Maltzahn et al. 2009a; von Maltzahn et al. 2009b; Wuelfing et al. 1998), semiconductors (Delehanty et al. 2009; Gao et al. 2004; Peng et al. 2000), dendrimers (Ku et al. 2011; Kukowska-Latallo et al. 1996), silica (Ashley et al. 2011; Benezra et al. 2011; Liberman et al. 2012; Liu et al. 2009; Ow et al. 2005; Patel et al. 2008; Santra et al. 2001; Zavaleta et al. 2009), superparamagnetic iron oxide (Arbab et al. 2003; Boddington et al. 2011; de Vries et al. 2005; Murase et al. 2011; Vogl et al. 2003; Winter et al. 2003), polymers (Borden et al. 2006; Nilsson 2004; Perry et al. 2011; Shi et al. 2011) and hybrid materials (Ananta et al. 2010; Boyer et al. 2010; Rabin et al. 2006). Hybrid materials have become especially promising in recent decades because they combine the attributes of multiple material types into a single design and the most common hybrids seen in nanomedicine involve silica and gold.

In this Chapter, we will review the use of gold/silica hybrid nanoparticles (HNPs) in medicine. We offer a brief history of these materials and explain the benefits they offer to both researchers and physicians. We then explain the particular advantages they offer when they are combined together— gold/silica hybrids. Next, we discuss shape effects and explain how nanoparticle morphology can affect *in vivo* biodistribution and imaging/ delivery performance. Although there are an exhaustive number of literature reports describing the synthesis of gold, silica and gold/silica HNPs, we next offer a brief overview of some of the most common methods to create these contrast agents. The next two sections are the most extensive in which we describe the imaging and therapeutic features of gold/silica HNPs. We offer many examples, discuss the advantages of these approaches and address the limitations in the field. We close with some examples of gold/silica HNPs that offer both therapeutic and diagnostic ("theranostic") features and close with ideas on how the field will advance in the coming years.

A Brief History of Silica and Gold

Silica and gold are some of the oldest materials known to man. Gold has been used for medicinal purposes for centuries even going back to the age of the ancient Chinese and Egyptians (Antonovych 1981; Lewis and Walz 1982). Multiple gold elixirs were used in the Middle Ages and 17th-19th century practitioners used gold to treat fevers and syphilis (Fricker 1996). The use of gold was first reported in modern medicine in 1890 by the German bacteriologist Robert Koch who identified the toxic effects that gold cyanide has on *tubercle bacillus* (Antonovych 1981), which led to the use of gold as a treatment for tuberculosis in the early 1900s. This was extended to rheumatoid arthritis in 1927, which was considered to be a form of tuberculosis at the time (Merchant 1998). Although no benefits were ever shown from gold therapy in tuberculosis patients, multiple reports showed utility in rheumatism and arthritic patients (Clark et al. 2000; Fricker 1996). Subsequently, gold therapy has been expanded to include juvenile arthritis, psoriatic arthritis (Ujfalussy et al. 2003) and discoid lupus erythematosus (Champion et al. 1990; Dalziel et al. 1986). While still approved and available under trade names such as Aurolate, Auranofin and Aurolate, gold-based therapy has largely been replaced by other drugs. Other uses of gold in medicine include dentistry (Belies 1994) as well as ophthalmology (Bair et al. 1995) and gold-coated coronary stents (Kastrati et al. 2000) and renal stents (Nolan et al. 2005).

Silica (SiO_2) was historically known as quartz and is familiar to most people as sand and glass, which are the most common forms of silica. Indeed, oxygen and silica are the two most common elements on earth. Silica does not have as much of a history in human health, but is used in a variety of foods as an anti-caking agent due to its hygroscopic properties. Silica gel is also used as a desiccant. Silica nanoparticles were first reported by Stöber in 1968 (Stöber et al. 1968). In the 1970s and 1980s, the morphology of these nanoparticles was changed to make more porous structures—the so-called mesoporous silica nanoparticles (MSNs). MSNs have been used as transfection reagents or to deliver drugs because of their very large surface area.

Features of Gold and Silica

Materials scientists like to work with gold nanoparticles (GNPs) for multiple reasons. First, the shape of the particles is easily tuned via surfactants and other additives during bottom-up synthesis (Fig. 14.1A). Second, gold and other metal nanoparticles have a large surface plasmon resonance that can harness high amounts of light energy. Third, this resonance is tunable based on the gold morphology (Fig. 14.1B). Fourth, gold is very corrosion resistant and stable in biological environments. Silica is advantageous because it has *in vivo* stability superior to many polymeric nanoparticles, but less than that of gold. Porous silica is especially attractive because it can be loaded with a variety of cargo for imaging and/or therapy. The size of these pores are quite tunable from ~2-30 nm.

Silica nanoparticles can also be synthesized in multiple and different sizes, shapes and porosities (Fig. 14.2). The advantages of silica include its very high surface area (~1000 m^2/g of material) and its chemical stability. This is useful in drug delivery applications. Silica nanoparticles can also easily be capped and interfaced with other materials.

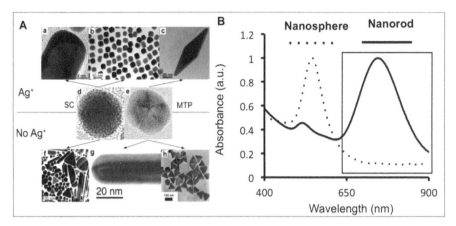

Fig. 14.1 Gold nanoparticle properties. (A) GNPs can be created in multiple shapes from single crystals (d) or (e) seeds. The resulting product can be tuned by adding (a-c) or excluding (f-h) Ag(NO₃). (B) The shape of the GNPs influences their interaction with visible light. Spherical particles absorb near 540 nm, while rod-shaped gold nanoparticles can be tuned to have a resonance deeper into the infrared region. The box in B represents the optical window used in imaging (see below). Panel A reproduced courtesy of the Royal Society of Chemistry (Grzelczak et al. 2008).

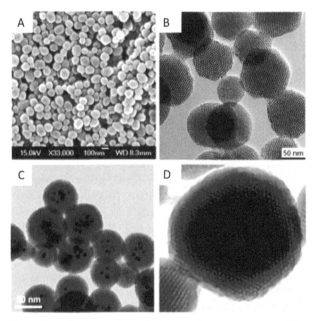

Fig. 14.2 Various formulations of mesoporous silica nanoparticles. Panel (A) is low magnification scanning electron microscopy and (B)-(D) are transmission electron images of (B) MCM-41 nanoparticles, (C) mixed iron oxide and silver encapsulated nanoparticles and (D) Zn-doped iron oxide encapsulated MSNs. Reproduced courtesy of the Royal Society of Chemistry (Li et al. 2012).

Advantages of HNPs

HNPs are structures with a maximal dimension below 1000 nm and composed of one or more materials. HNPs include organic-organic, organic-inorganic and inorganic-inorganic designs and examples include lipid-polymer nanoparticles (Zhang et al. 2008), small molecule/GNPs (Su et al. 2005) and semiconductor/silica nanoparticles (Yi et al. 2005), respectively. People use HNPs because they combine the advantages of two different materials into a single platform. For example, lipid-polymer nanoparticles combine the following features: a hydrophobic polymeric core that encapsulates drugs with poor water solubility, a hydrophilic shell to increase solubility and circulation time and a lipid monolayer at the border of the core and shell to stabilize drug encapsulation (Zhang et al. 2008). Many excellent reviews have been published about the general features of hybrid materials elsewhere (Kim et al. 2007b; Tran and Nguyen 2013).

Gold/silica HNPs have many advantages that are useful for imaging and therapy—many of which complement the general features of nanoparticles described above. First, they have a small size that can facilitate passive targeting via the enhanced permeation and retention effect (EPR). Second, they also have exceptional stability. This inertness can be useful for long term imaging experiments or long term delivery protocols. Third, these materials are ideally suited for theranostic applications because the gold and silica components can act in parallel or in series to create a diagnostic or therapeutic event. Fourth, it is possible to use multiple types of imaging signals (multi-modal imaging) with a single nanoparticle type. Fifth, the size and shape of gold/silica nanoparticles are particularly easy to control. Thus, the cellular uptake and/or tissue extravasation can be altered by changing the shape of the nanoparticles. Sixth, these materials are easily made from affordable materials via simple—but careful—chemical approaches that are accessible to a broad variety of researchers. Finally, these materials have been extensively characterized for toxicity with minimal negative effects.

Design and Synthesis of Gold/Silica HNPs

We focus here on the experimental principles with additional details in the literature. We next separate the reports into "silica-coated GNPs", "gold-coated silica nanoparticles" and "gold-capped MSNs." Thus, we discuss how the design strategy affects the targeting and signaling properties and how the particles are synthesized.

In discussing the design and synthesis of these hybrids, it is necessary to first introduce the surface plasmon resonance (SPR), which is a key property of GNPs (Eustis and El-Sayed 2006). Surface electrons of GNPs can oscillate with light wave. As the frequency of light equals the resonance frequency of surface electrons, the oscillation is significantly enhanced. This can result in nearly 100% absorption of incoming light. The absorbed energy can be scattered, absorbed and converted into heat or be transferred to the molecules near the gold surface—this allows a broad range of potential applications within a single platform.

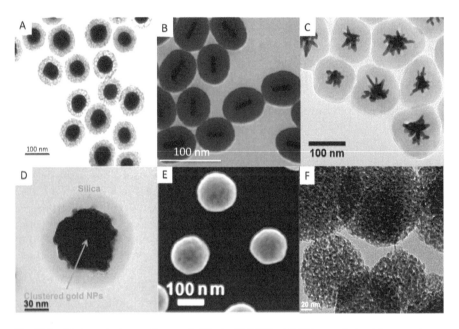

Fig. 14.3   Various design patterns of silica/gold HNPs. (A) Silica-coated gold nanospheres. (B) Silica-coated GBRs. (C) Silica-coated gold nanostars. (D) Silica-coated AuNCs. (E) Gold coated silica nanospheres. (F) Gold-capped MSNs. Reproduced courtesy of the American Chemical Society and the Royal Society of Chemistry (Ayala-Orozco et al. 2014b; Chen et al. 2011; Fales et al. 2011; Hayashi et al. 2013a; Karabeber et al. 2014; Yang et al. 2013).

Silica-Coated GNPs

The combination of a gold core with a silica shell is the most common gold/silica HNP. The next paragraph will focus on the synthesis of gold core, silica shell and the approaches to functionalize the nanoparticles.

Core Synthesis

Many different gold cores including spheres, stars, nanorods, clusters, etc. have been reported. Of these, the most common fundamental mechanism is the reduction of chloroauric acid ($HAuCl_4$) into solid gold (Enustun and Turkevich 1963). In the section below, we introduce the synthesis strategies for these materials.

Gold Sphere

There are currently two main routes to synthesize gold spheres. For ~15 nm spheres, $HAuCl_4$ is first mixed with sodium citrate at 100°C with stirring. The generation of GNPs is indicated by the presence of a wine-red color (Mine et al. 2003). To create larger spheres (Fig. 14.3A), a seed-mediated method is commonly used (Liu et al. 2006). The seeds are mixed with sodium citrate at ~70°C. The $HAuCl_4$ is then slowly added to the solution during which the gold atoms are reduced on the gold seed surface. After 1 hour, the solution is cooled down to room temperature. Alternatively, large spheres can be generated via seed aggregation, where the gold core has a larger surface/volume ratio that generates a stronger SPR over the former method. Hayashi (Hayashi et al. 2013a) has linked tetrakis (4-carboxyphenyl) porphyrin (TCPP) to 3-aminopro-pyltriethoxysilane (APTES) to get TCPP-APTES. This was then mixed with gold seeds to trigger their aggregation and cross-linking. After gold aggregation, a thin silica layer is coated on gold, which allows further thick silica growth and ensures that the cluster remains stable.

Gold Nanorod (GNR)

Gold nanorods are quite common because their interaction with the infrared region of the spectrum is useful for *in vivo* applications (Fig. 14.3B). GNRs have two SPR peaks—one is longitudinal and the other is transverse. As compared to gold spheres, GNRs have an adjustable aspect ratio and a relatively large surface area (Eustis and El-Sayed 2006). As the aspect ratio increases, the absorption peak of the longitudinal SPR undergoes a red shift to the near infrared region, while the transverse SPR peak remains the same. The most common GNR synthesis scheme uses a seed-mediated method. The first step is the preparation of the gold seed followed by seed growth. To prepare the gold seed, CTAB (cetyltrimethylammonium bromide) is mixed with $HAuCl_4$, which is reduced with $NaBH_4$ or sodium citrate to form gold seeds with a diameter of less than 20 nm (Gole and Murphy 2004; Liu et al. 2006). The next step is seed growth in a solution of CTAB, $AgNO_3$, $HAuCl_4$ and ascorbic acid. CTAB facilitates longitudinal growth (Perezjuste et al. 2005) and the concentration of the silver ions controls the aspect ratio of the GNRs.

Gold Nanostar

Gold nanostars have a larger surface/volume ratio than gold spheres (Fig. 14.3C). As the size increases, its SPR absorption peak is red-shifted (Khoury and Vo-Dinh 2008). Briefly, there are two methods for synthesizing gold nanostars. The first approach is the rapid mixing of $HAuCl_4$ and ascorbic acid at ~4°C (Harmsen et al. 2015b). The second method is seed-growth method (Fales et al. 2011). Gold seed is prepared the same as discussed in gold sphere, but the AgNO$_3$ and ascorbic acid are added simultaneously to the solution containing the gold seeds, HCl and HAuCl$_4$ under stirring.

Gold Nanocluster (AuNC)

Gold nanoclusters are an accumulation of several gold atoms (Fig. 14.3D) (Jadzinsky et al. 2007; Zheng et al. 2004). Unlike larger GNPs (> 2 nm), AuNCs are too small to support SPR but do show fluorescence. As the number of atoms per cluster increases, its emission peak turns from ultraviolet to infrared. One strategy in AuNC synthesis is a protein-directed synthesis method (Xie et al. 2009). The HAuCl$_4$ solution is first added to the bovine serum albumin (BSA). After 2 min of stirring, NaOH is added to the mixture to trigger reduction. After 12 h of stirring, AuNCs with ~25 Au atoms are further stabilized by BSA.

A recently reported approach to synthesis AuNC is through the reduction of chloro (triphenylphosphine), which is a gold (I) salt versus HAuCl$_4$, which is a gold (III) salt. Molly et al. (Hembury et al. 2015) synthesized the AuNCs in a hollow mesoporous silica shell. Next, chloro (triphenylphosphine) and HSs were dispersed in chloroform overnight at room temperature. The next day, 1-octanethiol was added to the mixture and stirred for 20 min. Borane *tert*-butylamine in chloroform was then added to reduce the gold.

Other Structures

The above-mentioned gold designs are the most common, but many others have also been reported including the gold nanothorn (Khajehpour et al. 2012), gold nanocube (Deng et al. 2013), gold nanodot (Pak and Yoo 2013) and gold nanoplate (Luke et al. 2013). However, these designs are less frequently utilized as hybrid materials in imaging and therapy and we do not discuss them further here.

Silica Coating

Silica coating can affect the gold core properties and generate new functionalities. The SPR peak of the gold core is red-shifted ~5 nm due to silica coating by ~10 nm (Chen et al. 2011; Wang et al. 2006). Silica shells can stabilize the gold core from deforming upon heating while offering additional sites (inside or on surface) for functionalization.

Procedures for silica coating are much simpler than those of core formation. In addition to the gold shape, the silica thickness also affects the SPR absorption of GNPs. Pratsinis proved (Sotiriou et al. 2014) that the NIR absorption becomes weaker as the thickness of the silica shell between the two gold spheres becomes larger. The fundamental principle of silica coating is the tetraethyl orthosilicate (TEOS) hydrolysis reaction, i.e., the

Stöber method $(Si(OC_2H_5)_4 + 2H_2O \rightarrow SiO_2 + 4C_2H_5OH)$ (Stöber et al. 1968). Both silica and mesoporous silica (MSN) shells have been used to coat gold cores. During the silica coating process, TEOS, ethanol and $NH_3 \cdot H_2O$ are added to the gold cores with stirring. The thickness of the silica increases with reaction time. For MSN coating, the gold solution is first mixed with NaOH (Gorelikov and Matsuura 2008). Then TEOS in methanol is added to the mixture every 30 min upon stirring. The MSN-coated gold is generated after 12-48 hours.

Functionalization of Silica-Coated GNPs

To achieve multiple imaging and therapy tasks, functional molecules or nanoparticles are incorporated onto the gold surface, inside the silica shell, or on the silica surface. Here we discuss these three cases.

Functionalization of Gold Surface

The attachment of Raman-active molecules is a common gold functionalization approach. Here, the Raman molecule creates much more signal when it is placed on the gold surface (see Imaging section below). The gold cores are incubated with the Raman dye or co-incubated with the Raman dye and some surfactant, e.g., CTAB or PEG as the stabilizer and 3, 3-diethylthiatricarbocyanine iodide (DTTC) or IR792 as the Raman dyes (Fales et al. 2011; von Maltzahn et al. 2009b). Finally, the nanoparticle is coated with a silica shell. The advantage of the silica is that it protects the Raman layer from passive diffusion (maintains high signal) and also allows for targeting ligands to be attached. Alternatively, the silica and Raman molecules can be co-coated (Harmsen et al. 2015b).

Functionalization of Silica Shell

There are two types of silica shells. Amorphous silica is a non-porous shell commonly produced in the Stober synthesis. Mesoporous silica has templated pores that can be used to hold cargo. Amorphous silica can be impregnated with fluorescence molecules or QDs for imaging purposes (Burns et al. 2009). The most common way to embed functional molecules in a silica shell is to add these molecules into the silica coating solution (co-coating). In order to encapsulate QDs into a silica shell, a previously-formed silica shell can be sonicated in hydroxypropyl cellulose (HPC) first. Then the QDs and silica shell are added into a silica coating (Xia et al. 2014). For mesoporous silica coatings, the drugs or payloads can be loaded in MSNs for therapy purpose simply by mixing them together. It usually takes more than one day for drugs such as DOX to be absorbed into MSNs (Li et al. 2014b; Zhang et al. 2012c).

Functionalization of Silica Surface

Two steps are involved in surface functionalization: a) activation of the silica surface with functional groups; and 2) linking functional molecules to the functional group. The silica surface has to be activated before conjugating a functional molecule. For example, by mixing with 3-aminopropyl-trimethoxysilane (APTMS), the silica surface can be modified with $-NH_2$ (Huang et al. 2013b; Xia et al. 2014) or mercaptopropyltrimethoxysilane (MPTMS) for thiols (Kircher et al. 2012); octadecanol treatment can make silica surface hydrophobic (van Schooneveld et al. 2010).

After surface activation, many molecules can be bonded to those functional groups. Specific tumor targeting can be achieved by RGD (Shen et al. 2013; Wang et al. 2014a) or folic acid (Xia et al. 2014) conjugation. Bonding of fluorescence dyes/QDs or Gd chelate enables the hybrid particle to be used in fluorescence or magnetic resonance imaging, respectively (Hayashi et al. 2013b; Kircher et al. 2012; Liu et al. 2006; van Schooneveld et al. 2010). We can also control drug release from MSNs by linking them with switches that can turn on the MSN pores under stimulation. These switches include poly (N-isopropylacrylamide) (PNIPAM) (Yang et al. 2013), sulfonatocalix[4]arene (SC[4]A) (Li et al. 2014a), rotaxane (Li et al. 2014b), etc.

Synthesis of Gold Coated Silica Nanoparticles

Unlike the HNPs discussed above, gold-coated silica nanoparticles have also been applied in imaging and therapy, although they have received less attention (Fig. 14.3E). Most of these examples focus on MRI and photothermal therapy.

Silica Synthesis

Silica synthesis is based on the Stöber method discussed above. Generally, there are two categories of silica fabrication—silica spheres and silica shells. First, the silica sphere can be directly formed by mixing TEOS, $NH_3 \cdot H_2O$, ethanol and water (Kim et al. 2006). The second approach forms a silica layer on a magnetic core for MRI. In Ji's report (Ji et al. 2007), superparamagnetic iron oxide (SPIO) is diluted in water and ethanol. Then TEOS and $NH_3 \cdot H_2O$ are added to the solution under stirring to coat a silica layer on the SPIO.

Gold Coating

Gold/silica hybrids are also used to make gold shells, which also have size-tunable SPR resonance—this is useful for photothermal therapy (Ayala-Orozco et al. 2014b). As mentioned before, gold coating reduces the $HAuCl_4$. Silica is first treated with APTMS to obtain $-NH_2$ groups on its surface. Next, the activated silica core absorbs gold colloids by mixing them together (Hu et al. 2009). These gold colloids serve as growth sites for gold shell formation. This is done by adding formaldehyde, $HAuCl_4$ and K_2CO_3 to the solution.

Synthesis of Gold Capped MSN

Gold-capped MSNs are not core-shell materials and people use the linkage between gold and silica for controlled drug release instead of focusing on the SPR property (Vivero-Escoto et al. 2009). The basic structure of these nanoparticles is gold-linkage-MSN, where gold spheres are capped on MSN pores by chemical bonding. The key procedure is to find a linkage that decomposes with external stimulation to "unclog" the MSN pores and release the cargo. In a pH-responsive drug delivery example (Liu et al. 2010), carboxylic acid groups are added to the MSN surface by reacting with succinic anhydride in triethylamine solution (Park et al. 2007). Then, the MSN-COOH NPs are linked with 3, 9-bis (3-aminopropyl) –2, 4, 8, 10-tetraoxaspiro [5.5] undecane with acetal-NH_2. At the same time, a gold sphere is synthesized by mixing $NaBH_4$, $HAuCl_4$ and mercaptosuccinic acid that is then modified with –COOH to bind –NH_2 on the silica surface by dehydration (Yao et al. 2001). This gold sphere then effectively blocks the pores of the silica particle. However, low pH can catalyze the acetal hydrolysis to detach the Au caps and release drug from the MSNs (Fig. 14.3F).

Imaging Using Silica/Gold HNPs

Contrast agents specifically amplify the signal in target regions. Radiologists use contrast agents to study diseases such as cancer and cardiovascular disease by analyzing the signal unique to the contrast agent (and thus disease). Recently, gold/silica HNPs have become common contrast agents for many imaging strategies. We next discuss the applications of gold/silica HNPs in photoacoustic imaging, surface enhanced Raman scattering imaging, X-ray computed tomography, magnetic resonance imaging and fluorescence imaging.

Photoacoustic Imaging (PAI)

PAI shows deep tissue penetration (up to 3 cm) and high resolution (100 s of μm) (Wang and Hu 2012; Xu and Wang 2006). In PAI, ultrasound is generated due to a rapid pressure difference because of the temperature increase caused by contrast absorption of incident pulsed laser (Beard 2011). PAI combines the contrast and spectral performance of optical imaging with the deep tissue penetration and spatial/temporal resolution of ultrasound.

PAI can use endogenous (Yao et al. 2010) or exogenous (Galanzha et al. 2009) contrast agents. Endogenous molecules such as melanin and hemoglobin offer label-free PAI, but have a relatively weak signal and peaks that are often blue-shifted. Exogenous labels such as GNPs or CuS offer larger absorption cross sections and more red-shifted peaks, which reduces attenuation and produces more signal at the transducer (Ku et al. 2012; Luke et al. 2012; Wang and Hu 2012; Zhang et al. 2012a).

GNPs are a common material to generate photoacoustic signal because their SPR property allows them to absorb NIR and generate large amounts of heat. However, under nanosecond NIR pulses, gold nanorods can deform and become spherical (Link et al. 2000). The silica shell coating of the GNP significantly enhances photoacoustic signal for the following two reasons: 1) The silica stabilizes the gold core and increases its resistance to morphological changes due to heating after optical excitation (Chen et al. 2010); and 2) The silica shell facilitates heat transfer to generate a PAI signal. A 20 nm silica shell thickness can increase the photoacoustic signal by 3-fold under NIR radiation (Fig. 14.4) (Chen et al. 2011).

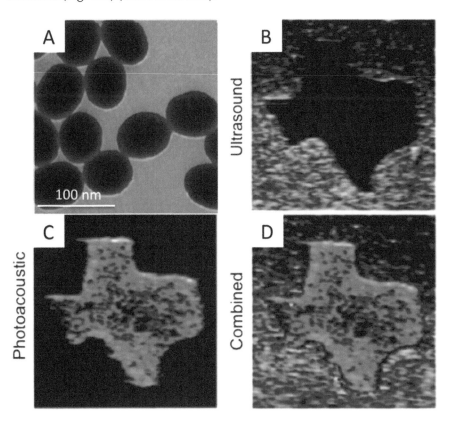

Fig. 14.4   (A) TEM images of 75 nm silica-coated gold nanorods with an aspect ratio of 3.9 under TEM. B-D) Silica-coated gold nanorods are loaded into gelatin on poly (vinyl alcohol) (PVA) tissue-mimicking phantoms. Different imaging modes are performed: (B) ultrasound imaging, (C) photoacoustic imaging and (D) combined ultrasound and photoacoustic imaging. PAI generates a high contrast against surroundings. Reproduced courtesy of the American Chemical Society (Chen et al. 2011).

Silica/gold HNPs can identify cancer cells. Bayer and coworkers (Bayer et al. 2011) synthesized two different silica-coated GNRs with absorption peaks at 780 and 830 nm, respectively. They targeted MCF7 and A431 cells using these two particles and detected different PAI signals. Yang et al. (Wang et al. 2014a) conjugated silica-coated gold nanorods with RGD (Arg-Gly-Asp) for PAI of targeted gastric cancer cells. Furthermore, silica/gold HNPs have promise for *in vivo* deep tissue imaging (Razansky et al. 2009). Emelianov et al. fabricated silica-coated gold nanoplates to map sentinel lymph node (SLN) (Luke et al. 2013). They also applied silica-coated GNRs to monitor photothermal therapy of epithelial tumor (Kim et al. 2014), during which 3-D pictures of the tumor were created with real-time PAI at 800 nm. Silica/gold HNPs also facilitate stem cell therapy. Jokerst and coworkers (Jokerst et al. 2012) synthesized silica-coated gold nanorods and labeled them with mesenchymal stem cells (MSCs). These cells were injected into muscles under PAI for musculature regeneration.

Surface Enhanced Raman Scattering (SERS) Imaging

Raman scattering is an inelastic scatter that results in a frequency shift between incident and scattered light (Kneipp 2007). Noble metals enhance Raman scattering through two mechanisms: 1) Electromagnetic (EM) enhancement, which is proportional to the fourth power of the enhancement of the near-surface EM field (Moskovits 2005). For GNPs, the SERS enhancement from SPR can be up to 10^{10}. 2) Chemical enhancement due to metal-molecule electron interaction can go up to 10^3 or higher (Jensen et al. 2008). The SPR wavelength of gold is in the NIR region (700-900 nm) (Gabudean et al. 2012), which provides deeper tissue penetration (~3 mm) than visible light (Stolik et al. 2000).

Although it was originally proposed that single gold NPs were suitable for surface-enhanced Raman, recent evidence has suggested that < 1 nm "hotspots" between NPs produce the majority of SERS signal (Mulvaney et al. 2003b; Xie et al. 2014). Silica coating on the gold surface has two advantages: 1) The silica coating is able to protect the Raman dyes from desorption (Doering and Nie 2003); and 2) The silica outer surface can be modified with probes or antibodies, which makes it a good candidate for specific binding (Ko et al. 2013). The ligand-target interaction satisfies the purpose of molecular imaging—to measure biological features at the molecular and cellular level (Jokerst et al. 2013).

Although many previous studies have demonstrated that non-silica coated GNPs could be used for SERS imaging (Freeman et al. 1995; Mulvaney et al. 2003a; Qian et al. 2008; von Maltzahn et al. 2009a), most people now take advantage of silica-coated GNPs (Fales et al. 2011) such as silica-coated gold nanospheres (Doering and Nie 2003), silica-coated gold nanostars (McNay et al. 2011) and gold nanothorns coated with macroporous silica (Khajehpour et al. 2012). In one example, affibodies were conjugated to a silica surface to target epidermal growth factor receptor (EGFR)-positive A431 tumor cells.

The tumor cells exhibited 7-fold more SERS intensity than adjacent tissues (Jokerst et al. 2011).

One advantage of SERS is high sensitivity (Fig. 14.5). Kircher et al. (Harmsen et al. 2015b) fabricated silica-coated gold nanostars for high-precision cancer imaging. The 1.5 fM detection limit makes premalignant tumor detection and resection possible (Fig. 14.1b). They also synthesized novel 2-thienyl-substituted chalcogenopyrylium dyes absorbed on the interface between silica and gold (Harmsen et al. 2015a). The increased sensitivity due to SERRS (surface enhanced Raman resonance scattering, where the incident laser matches the resonance frequency of both GNPs and Raman molecules (Kneipp et al. 1995)) has a detection limit of 0.1 fM. The other advantage of SERS is multiplexed imaging (Zavaleta et al. 2009). Zavaleta et al. (Zavaleta et al. 2013) modified silica surface with various tumor markers for real-time endoscopic imaging. This group used several different Raman absorbing dyes on the core/shell NP and could deconvolute the mixed spectrum. They also developed a novel fiber-optic based SERS device that can detect Raman signal within a distance of 10 mm. Similarly, Liu's group (Wang et al. 2014b) reported a fiber-optic SERS detector with ~0.1 s integration time, which can detect multiple cell surface markers with a functionalized silica surface.

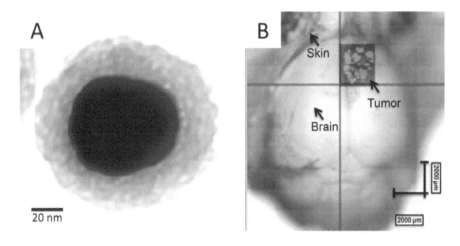

Fig. 14.5   (A) Silica-coated gold sphere with 110 nm diameters. (B) SERS signal is detected in brain tumor. Reproduced courtesy of the American Chemical Society (Karabeber et al. 2014).

X-Ray Computed Tomography (X-Ray CT)

X-ray absorption increases with increasing electron density. In X-ray CT, the body is tomographically scanned by X-ray and the X-ray absorption of different tissues are computed to create contrast (Liu et al. 2012c). Versus conventional iodine contrast agents (electron density 4.9 g/cm³), GNPs have a much higher electron density ($19.32g/cm^3$) and less toxicity (Kim et al. 2007a). These have also been applied in X-ray CT for experimental studies since 2006 (Rabin et al. 2006). Silica coating of GNPs generates new surfaces for modification and new bulk for encapsulation, enabling the silica/gold NPs to accomplish new tasks. However, there is no significant change of X-ray attenuation before and after silica coating (Kim et al. 2007a; Rabin et al. 2006; Xie et al. 2014).

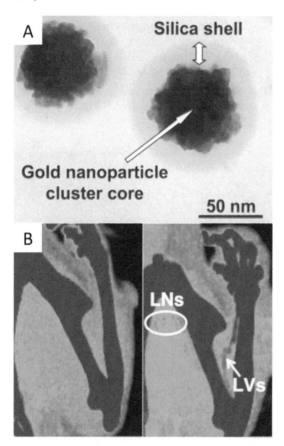

Fig. 14.6 (A) Silica-coated AuNCs with ~100 nm diameter. (B) X-ray CT image of lymph vessels (LVs). The images were taken 18 h after injection. Reproduced courtesy of the Wiley Online Library (Hayashi et al. 2013a).

Current examples of X-ray CT imaging with HNPs all leverage multimodality studies (Fig. 14.6). The first example was conducted by Mulder for CT/MRI dual model mice abdominal imaging (van Schooneveld et al. 2010). They fabricated PEGylated silica/gold HNPs linked with a paramagnetic lipid as the MRI contrast. After that, three groups reported intra-tumoral injection of silica/gold HNPs for CT/NIR fluorescence dual mode imaging. ICG (indo-cyanine green) (Luo et al. 2011), FITC (fluorescein isothiocyanate) (Feng et al. 2014) and porphyrins (Hayashi et al. 2013b) have all been embedded in the silica layer for fluorescence.

Magnetic Resonance Imaging (MRI)

In MRI, the hydrogen protons in a water molecule are aligned along the external magnetic field (B_0). Given an extra magnetic field pulse (B_1) that is perpendicular to B_0, the protons are excited to B_1 direction. After the removal of B_1, the protons relax to the B_0 direction and release energy, which is detected and transformed into an image (Cheon and Lee 2008). Magnetic nanoparticles have an extra magnetic moment that can modulate the relaxation of their surrounding protons generating a much higher MRI contrast.

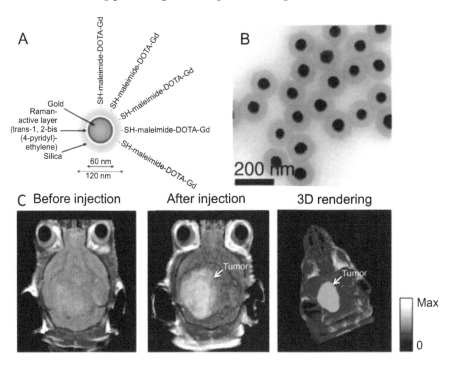

Fig. 14.7 (A-B) Schematic of MRI-positive silica-coated gold nanospheres (~100 nm) and corresponding TEM image. Gd ions are conjugated on the silica surface. (C) MRI image of mouse brain tumor and tumor 3D structure shows enhanced T1-weighted signal in the tumor region. Reproduced courtesy of the Nature Publishing Group (Kircher et al. 2012).

Silica/gold HNPs must be functionalized for MRI because neither gold nor silica has a significant magnetic moment (Hirsch et al. 2003). The most common approach is to take magnetic particles as the core including Fe_3O_4 (Ma et al. 2012), $MnFe_2O_4$ (Lee et al. 2008), SPIO (superparamagnetic iron oxide) (Ji et al. 2007), $Gd_2O(CO_3)_2 \cdot H_2O$ (Hu et al. 2009) and lanthanide-doped upconverting (UC) nanophosphors (Liu et al. 2013). This magnetic core can be coated with a silica shell and a subsequent gold coat. The second approach is to encapsulate magnetic molecules on a silica/gold interface (Kim et al. 2006). Finally, the outer silica surface can be tagged with a gadolinium chelate such as Gd-DTPA-DSA [gadolinium diethylenetriaminepentaacetic acid di (stearylamide)] (van Schooneveld et al. 2010) or maleimide-DOTA (1, 4, 7, 10-tetraazacyclododecane-1, 4, 7, 10-tetraacetic acid)-Gd (Kircher et al. 2012) (Fig. 14.7).

Fluorescence Imaging

Current fluorescence imaging strategies using silica/gold HNPs include visible light fluorescence imaging and near infrared (NIR) fluorescence imaging (Frangioni 2003). In visible light fluorescence imaging fluorophores must be located at a proper distance from the gold surface because the SPR can cause both plasmon-enhanced fluorescence and quenching (Burns et al. 2006). This distance where fluorescence intensities are increased varies but is almost always 5-30 nm (Komarala et al. 2006; Tovmachenko et al. 2006), This is because gold induces quenching when the fluorophores are absorbed on the gold surface so the silica shell is needed as a graft to prevent the fluorophores from quenching (Dulkeith et al. 2002). The fluorophores can be embedded in silica (to decrease photobleaching (Burns et al. 2009) shell (Priyam et al. 2012; Xia et al. 2014), in the sphere (Cao et al. 2011) and can be conjugated on the outer surface of silica (Chen et al. 2013a; Deng et al. 2013; Li et al. 2010; Wang et al. 2011). Due to the poor penetration distance of visible light, however, most studies have focused on imaging cell surface proteins *in vitro* such as cancer cell markers. One *in vivo* study was conducted by Zhao et al. (Li et al. 2014b) who encapsulated doxorubicin (DOX, an anti-cancer drug with fluorescence signal) into mesoporous silica shell. Under NIR irradiation, DOX is released and its distribution is detected by confocal laser microscopy (CLSM, Fig. 14.8). Surgical resection has also been performed with silica embedded dyes (Bradbury et al. 2013).

NIR fluorescence imaging has a deeper tissue penetration because NIR has less absorption versus visible light. GNPs can generate NIR fluorescence signal because the SPR peak of the GNP is in the NIR region and can absorb and emit NIR, which is converted into an image (Le Guével et al. 2011). The silica coating stabilizes the shape of the GNPs because they sometimes deform due to heat generation from NIR (Le Guével et al. 2011; Prevo et al. 2008). These particles can be used for *in vivo* tumor imaging and photothermal therapy, which will be discussed later.

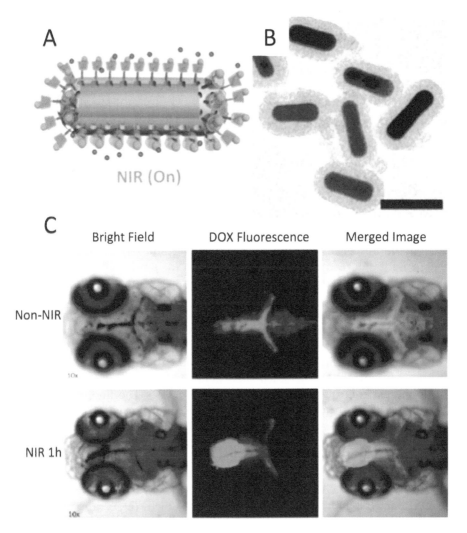

Fig. 14.8 (A) After NIR radiation, DOX (red dots) is released from MSN. (B) TEM image of MSN coated gold nanorod. Scale bar = 100 nm. (C) DOX (red) is released from MSN in zebrafish head. 1 h after radiation, DOX has spread from center of head to surroundings. Reproduced courtesy of the Royal Society of Chemistry (Li et al. 2014b).

The Therapeutic Applications of Gold/Silica HNPs

Nanoparticles have a long history of therapy, especially in drug delivery. Their main contributions have been to increase the circulation time of small molecules and modulate the hydrophobicity of certain chemotherapeutics. These clinical applications primarily use "soft" polymeric nanoparticles rather than "hard" nanoparticles like gold and silica. Nevertheless, there are many exciting therapeutic applications of gold/silica HNPs currently under development. Here we describe four therapeutic applications of the gold/silica hybrids as applied in cancer and stem cell therapy. These include photothermal therapy, photodynamic therapy, small molecule drug delivery and surgical resection. In each, we describe some of the materials used for the therapy, main results and present imaging data that validate the claims.

Photothermal Therapy (PTT)

Current strategies to minimize tumor burden include chemotherapy, surgical resection and radiation therapy. Photothermal therapy (PTT) is a novel approach that uses heat created by incident photons to ablate the tumor and minimize the tumor bulk (Fig. 14.9). A key requirement is that PTT uses a sensitizer of contrast agent to absorb light and convert it into heat—tumor tissue alone does not absorb sufficient light to produce cancer-killing temperatures (usually > 42°C). Electrons in photoabsorbing agents absorb light and convert it into heat (Huang et al. 2008). This leads to a hyperthermia (41-65°C) in the surrounding tissue, which destroys the cells within minutes (Svaasand et al. 1990). The IR radiation is commonly used because of its deep tissue penetration.

For PTT tumor treatment, NIR-absorbing dyes often suffer from photobleaching and low efficiency (Jori et al. 1996). Indeed, indocyanine green (Chen et al. 1996) can only last for 3-5 min in tumor during PTT and posed a slight positive effect. Therefore, GNPs have soon replaced NIR dyes because GNPs have a stronger absorption in the NIR region and are more photostable (O'Neal et al. 2004). Both GNRs (Bagley et al. 2013; Lin et al. 2010; Lo et al. 2013; Park et al. 2010) and gold nanoshells (Cole et al. 2009; Lal et al. 2008) have shown significant promise in the ongoing clinical validation studies. Studies have shown that the use of silica/gold has four advantages over simple colloidal gold: 1) The silica core provides a positively charged surface to efficiently grow a gold nanostructure (Oldenburg et al. 1998); 2) The mesoporous silica shell can facilitate chemotherapy because the increased surface area is ideal for drug encapsulation (Liu et al. 2011); 3) The silica coating prevents gold accumulation in cells (Zhang et al. 2012c); and 4) The new silica surface enables further functionalization, i.e., for tumor targeting binding proteins or for surface coatings to modulate the charge and circulation time (Chen et al. 2013b).

The team led by Naomi Halas and Jennifer West have published extensively on a 50-100 nm silica core and 3-8 nm gold shell that has been shown to be very effective for PTT in cancer therapy (Ayala-Orozco et al. 2014a; Ayala-Orozco et al. 2014b; Khlebtsov et al. 2006). Various gold-coated silica nanoparticles have been used for cancer treatment, such as subcutaneous tumors (O'Neal et al. 2004), mouse bladder cancer (MBT-2) cells (Hu et al. 2009), breast cancer cells (Carpin et al. 2011), BE (2)-C neuroblastoma cells (Qian et al. 2011), MCF-7 (human breast adenocarcinoma cell) cells (Liu et al. 2012a)/tumors (Liu et al. 2012b) and Hela cells (Byeon and Kim 2014). Furthermore, because MSNs are capable of encapsulating drugs, chemotherapeutics can also be loaded and released during PTT. Liu (Liu et al. 2011) fabricated gold coated MSNs loaded with docetaxel (DOC) for H22 tumor destruction. Ming (Ma et al. 2012) and Shen (Shen et al. 2013) coated DOX-embedded MSNs on the surface of gold nanorods to destroy MCF-7 cells and A549 cells, respectively. Stevens (Hembury et al. 2015) embedded DOX into MSN-coated gold spheres to treat LS174T-luc tumor *in vivo* and Hela cells *in vitro*.

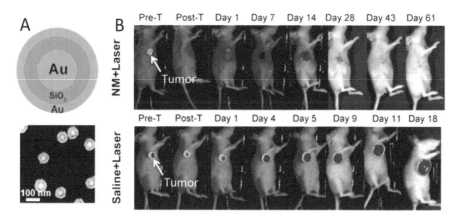

Fig. 14.9 (A) Gold/SiO₂/gold nanomatryoshkas (NM) with diameter of 90 nm. (B) Mice bearing highly aggressive triple-negative breast cancer (TNBC) tumors are treated with NM+Laser and Saline+Laser respectively. Bioluminescence signal is obtained by measuring luciferase activity. Tumor signal in NM treated mice disappeared post-treatment while tumors in contrast group still exist. Reproduced courtesy of the American Chemical Society (Ayala-Orozco et al. 2014b).

Photodynamic Therapy (PDT)

Instead of transferring light to heat, the photosensitizer (PS) used in PDT transfers photon energy to its surrounding oxygen molecules resulting in singlet oxygen generation to kill cancer cells (Dougherty et al. 1998). There are currently many small molecules under study as PDT sensitizers including porphyrins, chlorophylls, aminolevulinic acid and cyanine-based dyes (Abdel-Kader 2014). The utility of these agents can be increased by using them with a GNP because the amount of singlet oxygen produced by the gold/PS conjugate is higher than that produced by the PS alone (Stuchinskaya et al. 2011). The gold can also be used to bind targeting ligands or can be used for SERS imaging or NIR fluorescence imaging. The PS can be encapsulated in silica shell or covalently incorporated on the silica surface. It usually has an absorption peak in the NIR region that allows deep penetration (Cheng et al. 2008; Huang et al. 2008). Versus free PS, the use of silica to absorb PS is more favorable because it prevents PS from accumulating in the blood pool and can increase tumor accumulation due to the enhanced permeation and retention effect (Qian et al. 2009).

There are three examples of using silica-coated gold cores for PDT. Fales coated gold nanostars with MSNs that contain methylene blue (MB, Fig. 14.10) (Fales et al. 2011). These NPs were applied in SERS imaging and *in vitro* PDT for BT549 breast cancer cells. Chen fabricated gold QDs embedded silica sphere and conjugated chlorin e6 (Ce6) on the silica surface. Under NIR radiation, they performed NIR fluorescence imaging and PDT to kill MDA-MB-435 cells *in vitro*. Finally, an *in vivo* experiment was reported by Zhang (Zhang et al. 2013) in which this group fabricated silica-coated gold nanorods for SERS imaging and PDT of Hela cells tumor in mice. After injection through the tail, these NPs accumulated in the tumor and destroyed more than 40% of cancer cells with 15 min of radiation.

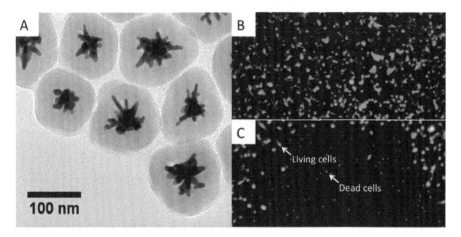

Fig. 14.10 Silica-coated gold nanostars for PDT. (A) Silica-coated gold nanostars are fabricated with a diameter of 100 nm. (B-C) After putting nanoparticles in cell culture dishes, NIR is illuminated on these nanoparticles for 1 h. The living cells are stained green by hydrolyzed fluorescein diacetate. Dead cells are stained red by DNA-intercalated propidium iodide that only crosses dead cell membranes. Reproduced courtesy of the American Chemical Society (Fales et al. 2011).

Drug Delivery

Various drug delivery systems have been reported using silica/gold HNPs. The main goal of these systems is to increase tumor accumulation by increasing the circulation time and modulating hydrophobicity. Mesoporous silica is the most common approach to absorb drugs (Slowing et al. 2008), while GNPs are usually related to the release of drugs.

The first set of examples deals with controlled release, where the drugs are loaded into MSNs and are then released by a switch. There are two mechanisms to release the drugs. First, the gold spheres are covalently bonded with silica to block their pores. In order to release the drugs inside, several switches have been used to trigger the uncapping of gold spheres including UV radiation (Vivero-Escoto et al. 2009) and pH (Liu et al. 2010; Ma et al. 2012; Radhakumary and Sreenivasan 2011). The other switch is NIR. The energy of NIR is converted into heat by GNPs, which heats the drugs and accelerates their release into the surroundings (Fig. 14.11) (Yang et al. 2013).

Successful examples include DOC (Liu et al. 2011), DOX (Shen et al. 2013; Yang et al. 2013; Zhang et al. 2012c) and ibuprofen (IBU)/poly-l-lysine (PLL) (Byeon and Kim 2014). Furthermore, in order to prevent drug leaking before NIR radiation, people have used thermal-sensitive caps to block MSNs. Li (Li et al. 2014a) synthesized sulfo-natocalix[4]arene (SC[4]A) to block the MSN pores. After NIR radiation, SC[4]A was dissociated from MSNs and rhodamine B (RhB) was released. Another example of cap is rotaxane, which switches to *trans* conformation under NIR and opens the channels.

Macrophages can also be used with HNPs to act as a drug delivery system (Dou et al. 2006). The key advantage is that macrophages can penetrate blood–brain barrier so they can make PTT in brain possible. Madsen reported that gold-coated silica core can be phagocytosed by macrophages for brain delivery in a Trojan Horse-like approach (Madsen et al. 2012). The growth of glioma spheroids was suppressed after PTT treatment.

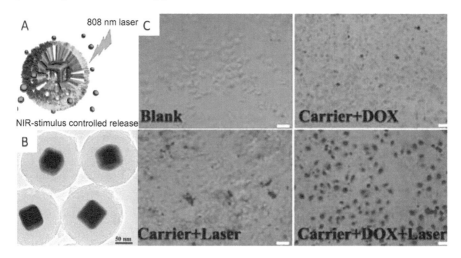

Fig. 14.11   NIR-triggered DOX release from MSN-coated gold nanocages. (A) The incident NIR laser is converted to heat by gold nanocages. (B) TEM image shows that nanoparticle has a diameter of ~200 nm. (C) Laser-triggered DOX release kills Hela cells. Nanocarrier + DOX + cancer triggered Hela cell death, which is stained by Trypan-blue (black in this figure). Living cells are left blank because Trypan-blue can only stain dead cells. Reproduced courtesy of the American Chemical Society (Yang et al. 2013).

Surgery

Conventional tumor surgery suffers from recurrences mainly because of failed identification of residual cancer cells by standard imaging strategies (Petrecca et al. 2013). However, residual tumor cells in adjacent tissues can be detected by molecular imaging methods including SERS imaging due to its high sensitivity (Rutka et al. 2014). Silica-coated GNPs have shown great promise in complete tumor resection under SERS imaging.

Silica-coated gold spheres can guide brain tumor resection. Kircher (Kircher et al. 2012) injected these particles via tail vein into mice, which underwent craniotomy and tumor resection under SERS imaging. After remove of the entire tumor, there were several small foci of Raman signals remaining, which indicated residual tumors (Fig. 14.12). After surgery, all SERS-positive tissues were removed. Furthermore, Kircher tested a hand-held SERS scanner company that is flexible during surgery produced by B&W Tek (Karabeber et al. 2014). During brain tumor surgery, the flexible SERS detector was able to resect lateral tumors by angulated scan. This same group has also developed high sensitive SERS using silica-coated gold nanostars that can direct the resection of tumors in various stages, including primary tumors, precancerous lesions and tumor infiltration (Harmsen et al. 2015b). Liu (Kang et al. 2015) developed a SERS imaging platform for *ex vivo* detection of resected tumors with silica-coated gold sphere.

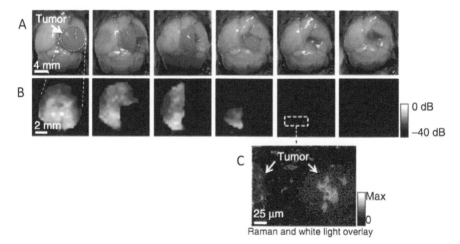

Fig. 14.12 SERS guided brain tumor resection. (A) Stepwise mouse brain tumor resection. (B) SERS images of every photograph in A. (C) After four rounds of resection the surgical field was clear by naked eye, but closer study with Raman imaging showed that residual tumor was left. This was confirmed with microscopy and completely removed in a final round of surgery. Reproduced courtesy of the Nature Publishing Group (Kircher et al. 2012).

The other surgical applications of silica/gold HNPs is in stem cell therapy. Jokerst (Jokerst et al. 2012) labeled mesenchymal stem cells (MSCs) with photoacoustic responsive silica-coated gold nanorods. Under real time PAI, stem cells were accurately injected to damaged muscles. This method achieved a good detection limit of MSCs (90,000 *in vivo*).

Conclusion and Outlook

Although silica/gold HNPs have achieved great success in imaging and therapy, there are still multiple challenges and opportunities in this field. In this closing section we discuss the challenges to their clinical translation as well as the potential solutions. A key limitation of many nanoparticles—and especially of gold and silica nanoparticles—is poor biodegradation and clearance *in vivo*. While many nanoparticle types approved by the FDA degrade and clear through the kidneys or liver in a matter of days, gold and silica both experience long-term accumulation by the reticuloendothelial system (RES). The RES consists of the bone marrow, liver and spleen including kuppfer cells and other specialized macrophages that remove foreign particles from the blood. Nanoparticles are easily recognized as foreign bodies and are removed from circulation.

Toxicity

Although GNPs have achieved great success, this technique suffers from poor biodegradation time. As discussed above, colloidal gold has been used in humans for many years and has little toxicity. Small GNPs (< 10 nm) can be cleared by the kidney through the urine but large GNPs (> 40 nm) are not (Choi et al. 2007; Zhang et al. 2012b). Therefore, there have been reports where people utilize biodegradable AuNCs to form a gold sphere rather than directly synthesize it. After the imaging or therapy task is complete, the cluster can degrade and the small nanoparticles can be cleared through the kidneys. Sokolov et al. (Tam et al. 2010) synthesized biodegradable polymer (PLA(2K)-b-PEG(10K)-b-PLA(2K)) to aggregate ~5 nm gold nanoclusters (AuNCs) into a ~100 nm sphere. The AuNCs dispersed into primary ~5 nm gold particles under low pH. Chen et al. (Huang et al. 2013a) synthesized ~200 nm AuNCs by holding 26 nm gold particles together via PEG-b-PCL biodegradable polymer package. Recently Stevens's group (Hembury et al. 2015) embedded < 2 nm gold quantum dots (Au QDs) into mesoporous silica shell for *in vivo* PAI of LS174T-luc tumor.

Silica nanoparticles can also cause toxicity to the plasma membrane (Yu et al. 2011). It has been shown that silica nanoparticles can cause inflammation in human lungs (McCarthy et al. 2012). However, utilizing small nanoparticles can solve these problems. Sailor (Park et al. 2009) has synthesized biodegradable porous silicon nanoparticles by etching in HF solution. The particles are degraded into orthosilicic acid $(Si(OH)_4)$ *in vivo*, which is excreted out of the body through urine. Furthermore, the only organic nanoparticle to be used in humans is silica—a group led by Bradbury and Wiesner have moved silica nanoparticles to human use for combined PET (positron emission tomography) and fluorescence imaging (Phillips et al. 2014). They synthesized ~6 nm silica-coated fluorescence dyes that can be extracted through urine after 60 h.

Theranostics

Theranostics combines diagnosis and therapy together (Funkhouser 2002). With only one injection of nanoparticles, physicians can diagnose where the tumor is by multiple imaging techniques and perform treatment after that. Because of this, theranostics decreases the number of interventions that the patient receives—this also decreases patient cost. However theranostic nanoparticle fabrication is a complicated process because it needs integration of both diagnostic and therapeutic properties. This always leads to difficulties to effectively characterize theranostic nanoparticles and fabricate these nanoparticles reproducibly. So far, there have been no theranostic nanoparticles with FDA approval.

Multiple Imaging Techniques

We can easily achieve multimodal imaging with one silica/gold HNP. The SPR property of GNPs enables both CT and PAI. Furthermore, by integrating fluorescent and Raman-active molecules into the HNP, these nanoparticles can be used for optical imaging. Therefore, by taking advantage of multiple imaging techniques, people can image both deep and surface tissue images with only a single nanoparticle. For example, PAI is used to locate tumors in deep tissue and SERS can show tumor boundaries during surgical resection (Kircher et al. 2012).

In Vivo Experiments

Early studies were mostly focused on *in vitro* applications. Recently, many experiments have been performed *in vivo* such as fish, mammalians and even humans. There have been many examples performed in mice, such as PDT, CT, MRI and PAI. Further reductions in toxicity and biocompatibility are needed to move nanoparticles to clinical trials. This can be done by modulating the HNPs' size or through PEGylation to prevent macrophage uptake of silica (Pilloni et al. 2010) and gold (Santos-Martinez et al. 2014). Another approach is erythrocyte membrane (Hu et al. 2011) or mimicking the erythrocyte membrane coating (Roggers et al. 2014) of gold or silica nanoparticles, which produces more macrophage resistance than PEGylation.

Final Thoughts

In summary, we believe that there will be continued innovation in gold/silica HNPs because of the advantages these materials offer. Clinically, these materials will likely only have niche applications, but due to high and stable signal as well as the unique homing abilities of gold/silica hybrids, they will likely remain robust pre-clinical and *in vitro* tools. We emphasize that there is also a large body of literature that uses gold/silica nanoparticles for catalysis, electronics, solar panels, etc. and that cross-disciplinary advances will likely benefit all fields in the coming decades.

Acknowledgements

We thank the University of California and the NIH for funding including R00 HL117048.

Abbreviations

NPs: Nanoparticles.
SPR: Surface plasmon resonance.
HNPs: Hybrid nanoparticles.
MSNs: Mesoporous silica nanoparticles.
GNPs: Gold nanoparticles.
GNRs: Gold nanorods.
AuNCs: Gold nanoclusters.
NIR: Near infrared.
PAI: Photoacoustic imaging.
MRI: Magnetic resonance imaging.
CT: Computed tomography.
PTT: Photothermal therapy.
PDT: Photodynamic therapy.

References

Abdel-Kader, M.H. 2014. Photodynamic Therapy: From Theory to Application. Springer Science & Business Media, Heidelberg, Germany.
Ananta, J.S., Godin, B., Sethi, R., Moriggi, L., Liu, X., Serda, R.E., et al. 2010. Geometrical confinement of gadolinium-based contrast agents in nanoporous particles enhances T1 contrast. Nat Nanotechnol 5: 815-821.
Antonovych, T.T. 1981. Gold nephropathy. Ann Clin Lab Sci 11: 386-391.
Arbab, A.S., Bashaw, L.A., Miller, B.R., Jordan, E.K., Lewis, B.K., Kalish, H., et al. 2003. Characterization of biophysical and metabolic properties of cells labeled with superparamagnetic iron oxide nanoparticles and transfection agent for cellular MR Imaging. Radiology 229: 838.

Ashley, C.E., Carnes, E.C., Phillips, G.K., Padilla, D., Durfee, P.N., Brown, P.A., et al. 2011. The targeted delivery of multicomponent cargos to cancer cells by nanoporous particle-supported lipid bilayers. Nat Mater 10: 389-397.

Ayala-Orozco, C., Urban, C., Bishnoi, S., Urban, A., Charron, H., Mitchell, T., et al. 2014a. Sub-100 nm gold nanomatryoshkas improve photo-thermal therapy efficacy in large and highly aggressive triple negative breast tumors. J Control Release 191: 90-97.

Ayala-Orozco, C., Urban, C., Knight, M.W., Urban, A.S., Neumann, O., Bishnoi, S.W., et al. 2014b. Au nanomatryoshkas as efficient near-infrared photothermal transducers for cancer treatment: benchmarking against nanoshells. ACS Nano 8: 6372-6381.

Bagley, A.F., Hill, S., Rogers, G.S. and Bhatia, S.N. 2013. Plasmonic photothermal heating of intraperitoneal tumors through the use of an implanted near-infrared source. ACS Nano 7: 8089-8097.

Bair, R.L., Harris, G.J., Lyon, D.B. and Komorowski, R.A. 1995. Noninfectious inflammatory response to gold weight eyelid implants. Ophthal Plast Reconstr Surg 11: 209-214.

Bayer, C.L., Chen, Y.S., Kim, S., Mallidi, S., Sokolov, K. and Emelianov, S. 2011. Multiplex photoacoustic molecular imaging using targeted silica-coated gold nanorods. Biomed Opt Express 2: 1828-1835.

Beard, P. 2011. Biomedical photoacoustic imaging. Interface focus 1: 602-631.

Belies, R.P., Ed. 1994. Patty's Industrial Hygeine and Toxicology, 4th edn. John Wiley & Sons Inc, New York.

Benezra, M., Penate-Medina, O., Zanzonico, P.B., Schaer, D., Ow, H., Burns, A., et al. 2011. Multimodal silica nanoparticles are effective cancer-targeted probes in a model of human melanoma. J Clin Invest 121: 2768.

Boddington, S.E., Sutton, E.J., Henning, T.D., Nedopil, A.J., Sennino, B., Kim, A., et al. 2011. Labeling human mesenchymal stem cells with fluorescent contrast agents: the biological impact. Mol Imag Biol 13: 3-9.

Boote, E., Fent, G., Kattumuri, V., Casteel, S., Katti, K., Chanda, N., et al. 2010. Gold nanoparticle contrast in a phantom and juvenile swine: models for molecular imaging of human organs using X-ray computed tomography. Acad Radiol 17: 410-417.

Borden, M.A., Martinez, G.V., Ricker, J., Tsvetkova, N., Longo, M., Gillies, R.J., et al. 2006. Lateral phase separation in lipid-coated microbubbles. Langmuir 22: 4291-4297.

Boyer, J.C., Manseau, M.P., Murray, J.I. and van Veggel, F.C. 2010. Surface modification of upconverting NaYF4 nanoparticles with PEG-phosphate ligands for NIR (800 nm) biolabeling within the biological window. Langmuir 26: 1157-1164.

Bradbury, M.S., Phillips, E., Montero, P.H., Cheal, S.M., Stambuk, H., Durack, J.C., et al. 2013. Clinically-translated silica nanoparticles as dual-modality cancer-targeted probes for image-guided surgery and interventions. Integr Biol 5: 74-86.

Burns, A., Ow, H. and Wiesner, U. 2006. Fluorescent core–shell silica nanoparticles: towards "Lab on a Particle" architectures for nanobiotechnology. Chem Soc Rev 35: 1028-1042.

Burns, A.A., Vider, J., Ow, H., Herz, E., Penate-Medina, O., Baumgart, M., et al. 2009. Fluorescent silica nanoparticles with efficient urinary excretion for nanomedicine. Nano Lett 9: 442-448.

Byeon, J.H. and Kim, Y.W. 2014. Aero-self-assembly of ultrafine gold incorporated silica nanobunches for NIR-induced chemo-thermal therapy. Small 10: 2331-2335.

Cao, F., Deng, R., Liu, D., Song, S., Wang, S., Su, S., et al. 2011. Fabrication of fluorescent silica-Au hybrid nanostructures for targeted imaging of tumor cells. Dalton Trans 40: 4800-4802.

Carpin, L.B., Bickford, L.R., Agollah, G., Yu, T.K., Schiff, R., Li, Y., et al. 2011. Immunoconjugated gold nanoshell-mediated photothermal ablation of trastuzumab-resistant breast cancer cells. Breast Cancer Res Treat 125: 27-34.

Champion, G.D., Graham, G.G. and Ziegler, J.B. 1990. The gold complexes. Baillieres Clin Rheumatol 4: 491-534.

Chen, T., Hu, Y., Cen, Y., Chu, X. and Lu, Y. 2013a. A dual-emission fluorescent nanocomplex of gold-cluster-decorated silica particles for live cell imaging of highly reactive oxygen species. J Am Chem Soc 135: 11595-11602.

Chen, W.R., Adams, R.L., Higgins, A.K., Bartels, K.E. and Nordquist, R.E. 1996. Photothermal effects on murine mammary tumors using indocyanine green and an 808-nm diode laser: an in vivo efficacy study. Cancer Lett 98: 169-173.

Chen, Y., Chen, H. and Shi, J. 2013b. In vivo bio-safety evaluations and diagnostic/therapeutic applications of chemically designed mesoporous silica nanoparticles. Adv Mater 25: 3144-3176.

Chen, Y.S., Frey, W., Kim, S., Homan, K., Kruizinga, P., Sokolov, K., et al. 2010. Enhanced thermal stability of silica-coated gold nanorods for photoacoustic imaging and image-guided therapy. Opt Express 18: 8867-8878.

Chen, Y.S., Frey, W., Kim, S., Kruizinga, P., Homan, K. and Emelianov, S. 2011. Silica-coated gold nanorods as photoacoustic signal nanoamplifiers. Nano Lett 11: 348-354.

Cheng, Y., C. Samia, A., Meyers, J.D., Panagopoulos, I., Fei, B. and Burda, C. 2008. Highly efficient drug delivery with gold nanoparticle vectors for in vivo photodynamic therapy of cancer. JACS 130: 10643-10647.

Cheon, J. and Lee, J.H. 2008. Synergistically integrated nanoparticles as multimodal probes for nanobiotechnology. Acc Chem Res 41: 1630-1640.

Choi, H.S., Liu, W., Misra, P., Tanaka, E., Zimmer, J.P., Itty Ipe, B., et al. 2007. Renal clearance of quantum dots. Nat Biotechnol 25: 1165-1170.

Clark, P., Tugwell, P., Bennet, K., Bombardier, C., Shea, B., Wells, G., et al. 2000. Injectable gold for rheumatoid arthritis. Cochrane Database Syst Rev. CD000520.

Cole, J.R., Mirin, N.A., Knight, M.W., Goodrich, G.P. and Halas, N.J. 2009. Photothermal efficiencies of nanoshells and nanorods for clinical therapeutic applications. J Phys Chem C 113: 12090-12094.

Dalziel, K., Going, G., Cartwright, P.H., Marks, R., Beveridge, G.W. and Rowell, N.R. 1986. Treatment of chronic discoid lupus erythematosus with an oral gold compound (auranofin). Br J Dermatol 115: 211-216.

de Vries, I.J., Lesterhuis, W.J., Barentsz, J.O., Verdijk, P., van Krieken, J.H., Boerman, O.C., et al. 2005. Magnetic resonance tracking of dendritic cells in melanoma patients for monitoring of cellular therapy. Nat Biotechnol 23: 1407-1413.

Delehanty, J.B., Mattoussi, H. and Medintz, I.L. 2009. Delivering quantum dots into cells: strategies, progress and remaining issues. Anal Bioanal Chem 393: 1091-1105.

Deng, L., Liu, L., Zhu, C., Li, D. and Dong, S. 2013. Hybrid gold nanocube@silica@ graphene-quantum-dot superstructures: synthesis and specific cell surface protein imaging applications. Chem Commun (Camb) 49: 2503-2505.

Doering, W.E. and Nie, S. 2003. Spectroscopic tags using dye-embedded nanoparticles and surface-enhanced Raman scattering. Anal Chem 75: 6171-6176.

Dou, H., Destache, C.J., Morehead, J.R., Mosley, R.L., Boska, M.D., Kingsley, J., et al. 2006. Development of a macrophage-based nanoparticle platform for antiretroviral drug delivery. Blood 108: 2827-2835.

Dougherty, T.J., Gomer, C.J., Henderson, B.W., Jori, G., Kessel, D., Korbelik, M., et al. 1998. Photodynamic therapy. J Natl Cancer Inst 90: 889-905.

Dulkeith, E., Morteani, A.C., Niedereichholz, T., Klar, T.A., Feldmann, J., Levi, S.A., et al. 2002. Fluorescence quenching of dye molecules near gold nanoparticles: radiative and nonradiative effects. Phys Rev Lett 89: 203002.

Enustun, B.V. and Turkevich, J. 1963. Coagulation of colloidal gold. J Am Chem Soc 85: 3317-3328.

Eustis, S. and El-Sayed, M.A. 2006. Why gold nanoparticles are more precious than pretty gold: noble metal surface plasmon resonance and its enhancement of the radiative and nonradiative properties of nanocrystals of different shapes. Chem Soc Rev 35: 209-217.

Fales, A.M., Yuan, H. and Vo-Dinh, T. 2011. Silica-coated gold nanostars for combined surface-enhanced Raman scattering (SERS) detection and singlet-oxygen generation: a potential nanoplatform for theranostics. Langmuir 27: 12186-12190.

Feng, J., Chang, D., Wang, Z.F., Shen, B., Yang, J.J., Jiang, Y.Y., et al. 2014. A FITC-doped silica coated gold nanocomposite for both in vivo X-ray CT and fluorescence dual modal imaging. RSC Adv 4: 51950-51959.

Frangioni, J.V. 2003. In vivo near-infrared fluorescence imaging. Curr Opin Chem Biol 7: 626-634.

Freeman, R.G., Grabar, K.C., Allison, K.J., Bright, R.M., Davis, J.A., Guthrie, A.P., et al. 1995. Self-assembled metal colloid monolayers: an approach to SERS substrates. Science 267: 1629-1632.

Fricker, S.P. 1996. Medical uses of gold compounds: past, present and future. Gold Bulletin 29.

Funkhouser, J. 2002. Reinventing pharma: the theranostic revolution. Curr Drug Discov 2: 17-19.

Gabudean, A.M., Biro, D. and Astilean, S. 2012. Hybrid plasmonic platforms based on silica-encapsulated gold nanorods as effective spectroscopic enhancers for Raman and fluorescence spectroscopy. Nanotechnology 23: 485706.

Galanzha, E.I., Shashkov, E.V., Kelly, T., Kim, J.W., Yang, L. and Zharov, V.P. 2009. In vivo magnetic enrichment and multiplex photoacoustic detection of circulating tumour cells. Nat Nanotechnol 4: 855-860.

Gao, X., Cui, Y., Levenson, R.M., Chung, L.W.K. and Nie, S. 2004. In vivo cancer targeting and imaging with semiconductor quantum dots. Nature Biotechnology 22: 969-976.

Gole, A. and Murphy, C.J. 2004. Seed-mediated synthesis of gold nanorods: role of the size and nature of the seed. Chem Mater 16: 3633-3640.

Gorelikov, I. and Matsuura, N. 2008. Single-step coating of mesoporous silica on cetyltrimethyl ammonium bromide-capped nanoparticles. Nano Lett 8: 369-373.

Grzelczak, M., Perez-Juste, J., Mulvaney, P. and Liz-Marzan, L.M. 2008. Shape control in gold nanoparticle synthesis. Chem Soc Rev 37: 1783-1791.

Harmsen, S., Bedics, M.A., Wall, M.A., Huang, R., Detty, M.R. and Kircher, M.F. 2015a. Rational design of a chalcogenopyrylium-based surface-enhanced resonance Raman scattering nanoprobe with attomolar sensitivity. Nat Commun 6: 6570.

Harmsen, S., Huang, R., Wall, M.A., Karabeber, H., Samii, J.M., Spaliviero, M., et al. 2015b. Surface-enhanced resonance Raman scattering nanostars for high-precision cancer imaging. Sci Transl Med 7: 271ra277.

Hayashi, K., Nakamura, M. and Ishimura, K. 2013a. Near-infrared fluorescent silica-coated gold nanoparticle clusters for X-ray computed tomography/optical dual modal imaging of the lymphatic system. Adv Healthc Mater 2: 756-763.

Hayashi, K., Nakamura, M., Miki, H., Ozaki, S., Abe, M., Matsumoto, T., et al. 2013b. Gold nanoparticle cluster-plasmon-enhanced fluorescent silica core-shell nanoparticles for X-ray computed tomography-fluorescence dual-mode imaging of tumors. Chem Commun (Camb) 49: 5334-5336.

Hembury, M., Chiappini, C., Bertazzo, S., Kalber, T.L., Drisko, G.L., Ogunlade, O., et al. 2015. Gold-silica quantum rattles for multimodal imaging and therapy. Proc Natl Acad Sci USA 112: 1959-1964.

Hirsch, L.R., Stafford, R., Bankson, J., Sershen, S., Rivera, B., Price, R., et al. 2003. Nanoshell-mediated near-infrared thermal therapy of tumors under magnetic resonance guidance. Proc Natl Acad Sci USA 100: 13549-13554.

Hu, C.M., Zhang, L., Aryal, S., Cheung, C., Fang, R.H. and Zhang, L. 2011. Erythrocyte membrane-camouflaged polymeric nanoparticles as a biomimetic delivery platform. Proc Natl Acad Sci USA 108: 10980-10985.

Hu, K.-W., Jhang, F.-Y., Su, C.-H. and Yeh, C.-S. 2009. Fabrication of $Gd_2O(CO_3)_2 \cdot H_2O$/silica/gold hybrid particles as a bifunctional agent for MR imaging and photothermal destruction of cancer cells. J Mater Chem 19: 2147.

Huang, P., Lin, J., Li, W., Rong, P., Wang, Z., Wang, S., et al. 2013a. Biodegradable gold nanovesicles with an ultrastrong plasmonic coupling effect for photoacoustic imaging and photothermal therapy. Angew Chem Int Ed Engl 52: 13958-13964.

Huang, P., Lin, J., Wang, S., Zhou, Z., Li, Z., Wang, Z., et al. 2013b. Photosensitizer-conjugated silica-coated gold nanoclusters for fluorescence imaging-guided photodynamic therapy. Biomaterials 34: 4643-4654.

Huang, X., El-Sayed, I.H., Qian, W. and El-Sayed, M.A. 2006. Cancer cell imaging and photothermal therapy in the near-infrared region by using gold nanorods. J Am Chem Soc 128: 2115-2120.

Huang, X., Jain, P.K., El-Sayed, I.H. and El-Sayed, M.A. 2008. Plasmonic photothermal therapy (PPTT) using gold nanoparticles. Lasers Med Sci 23: 217-228.

Jadzinsky, P.D., Calero, G., Ackerson, C.J., Bushnell, D.A. and Kornberg, R.D. 2007. Structure of a thiol monolayer-protected gold nanoparticle at 1.1 A resolution. Science 318: 430-433.

Jain, P.K., Huang, X., El-Sayed, I.H. and El-Sayed, M.A. 2008. Noble metals on the nanoscale: optical and photothermal properties and some applications in imaging, sensing, biology and medicine. Acc Chem Res 41: 1578-1586.

Jensen, L., Aikens, C.M. and Schatz, G.C. 2008. Electronic structure methods for studying surface-enhanced Raman scattering. Chem Soc Rev 37: 1061-1073.

Ji, X., Shao, R., Elliott, A.M., Stafford, R.J., Esparza-Coss, E., Bankson, J.A., et al. 2007. Bifunctional gold nanoshells with a superparamagnetic iron oxide-silica core suitable for both MR imaging and photothermal therapy. J P Chem C 111: 6245-6251.

Jokerst, J.V., Miao, Z., Zavaleta, C., Cheng, Z. and Gambhir, S.S. 2011. Affibody-functionalized gold-silica nanoparticles for Raman molecular imaging of the epidermal growth factor receptor. Small 7: 625-633.

Jokerst, J.V., Thangaraj, M., Kempen, P.J., Sinclair, R. and Gambhir, S.S. 2012. Photoacoustic imaging of mesenchymal stem cells in living mice via silica-coated gold nanorods. ACS Nano 6: 5920-5930.

Jokerst, J.V., Pohling, C. and Gambhir, S.S. 2013. Molecular imaging with surface-enhanced Raman spectroscopy nanoparticle reporters. MRS Bull 38: 625-630.

Jori, G., Schindl, L., Schindl, A. and Polo, L. 1996. Novel approaches towards a detailed control of the mechanism and efficiency of photosensitized processes in vivo. J Photochem Photobiol A Chem 102: 101-107.

Kang, S., Wang, Y.W., Khan, A., Leigh, S.Y. and Liu, J.T. 2015. Molecular imaging of topically applied SERS nanoparticles for guiding tumor resection. Paper presented at: optical Molecular Probes, Imaging and Drug Delivery (Optical Society of America) https://www.osapublishing.org/abstract.cfm?uri=OMP-2015-OW1D.7.

Karabeber, H., Huang, R., Iacono, P., Samii, J.M., Pitter, K., Holland, E.C., et al. 2014. Guiding brain tumor resection using surface-enhanced Raman scattering nanoparticles and a hand-held Raman scanner. ACS Nano 8: 9755-9766.

Kastrati, A., Schomig, A., Dirschinger, J., Mehilli, J., von Welser, N., Pache, J., et al. 2000. Increased risk of restenosis after placement of gold-coated stents: results of a randomized trial comparing gold-coated with uncoated steel stents in patients with coronary artery disease. Circulation 101: 2478-2483.

Khajehpour, K.J., Williams, T., Bourgeois, L. and Adeloju, S. 2012. Gold nanothorns-macroporous silicon hybrid structure: a simple and ultrasensitive platform for SERS. Chem Commun (Camb) 48: 5349-5351.

Khlebtsov, B., Zharov, V., Melnikov, A., Tuchin, V. and Khlebtsov, N. 2006. Optical amplification of photothermal therapy with gold nanoparticles and nanoclusters. Nanotechnology 17: 5167-5179.

Khoury, C.G. and Vo-Dinh, T. 2008. Gold nanostars for surface-enhanced Raman scattering: synthesis, characterization and optimization. J Phys Chem C Nanomater Interfaces 2008: 18849-18859.

Kim, D., Park, S., Lee, J.H., Jeong, Y.Y. and Jon, S. 2007a. Antibiofouling polymer-coated gold nanoparticles as a contrast agent for in vivo X-ray computed tomography imaging. J Am Chem Soc 129: 7661-7665.

Kim, J., Park, S., Lee, J.E., Jin, S.M., Lee, J.H., Lee, I.S., et al. 2006. Designed fabrication of multifunctional magnetic gold nanoshells and their application to magnetic resonance imaging and photothermal therapy. Angew Chem Int Ed Engl 45: 7754-7758.

Kim, J.S., Rieter, W.J., Taylor, K.M., An, H., Lin, W. and Lin, W. 2007b. Self-assembled hybrid nanoparticles for cancer-specific multimodal imaging. JACS 129: 8962-8963.

Kim, S., Chen, Y.S., Luke, G.P. and Emelianov, S.Y. 2014. In-vivo ultrasound and photoacoustic image-guided photothermal cancer therapy using silica-coated gold nanorods. IEEE Trans Ultrason Ferroelectr Freq Control 61: 891-897.

Kircher, M.F., de la Zerda, A., Jokerst, J.V., Zavaleta, C.L., Kempen, P.J., Mittra, E., et al. 2012. A brain tumor molecular imaging strategy using a new triple-modality MRI-photoacoustic-Raman nanoparticle. Nat Med 18: 829-834.

Kneipp, K. 2007. Surface-enhanced Raman scattering. Phys Today 60: 40-46.

Kneipp, K., Wang, Y., Dasari, R.R. and Feld, M.S. 1995. Approach to single-molecule detection using surface-enhanced resonance Raman-scattering (SERRS) – a study using Rhodamine 6g on colloidal silver. Appl Spectrosc 49: 780-784.

Ko, J., Lee, S., Lee, E.K., Chang, S.I., Chen, L., Yoon, S.Y., et al. 2013. SERS-based immunoassay of tumor marker VEGF using DNA aptamers and silica-encapsulated hollow gold nanospheres. Phys Chem Chem Phys 15: 5379-5385.

Komarala, V.K., Rakovich, Y.P., Bradley, A.L., Byrne, S.J., Gun'ko, Y.K., Gaponik, N., et al. 2006. Off-resonance surface plasmon enhanced spontaneous emission from CdTe quantum dots. Appl Phys Lett 89: 253118.

Ku, G., Zhou, M., Song, S., Huang, Q., Hazle, J. and Li, C. 2012. Copper sulfide nanoparticles as a new class of photoacoustic contrast agent for deep tissue imaging at 1064 nm. ACS Nano 6: 7489-7496.

Ku, T.H., Chien, M.P., Thompson, M.P., Sinkovits, R.S., Olson, N.H., Baker, T.S., et al. 2011. Controlling and switching the morphology of micellar nanoparticles with enzymes. J Am Chem Soc 133(22): 8392–8395.

Kukowska-Latallo, J.F., Bielinska, A.U., Johnson, J., Spindler, R., Tomalia, D.A. and Baker, J.R. 1996. Efficient transfer of genetic material into mammalian cells using Starburst polyamidoamine dendrimers. Proc Natl Acad Sci USA 93: 4897-4902.

Lal, S., Clare, S.E. and Halas, N.J. 2008. Nanoshell-enabled photothermal cancer therapy: impending clinical impact. Acc Chem Res 41: 1842-1851.

Le Guével, X., Hötzer, B., Jung, G. and Schneider, M. 2011. NIR-emitting fluorescent gold nanoclusters doped in silica nanoparticles. J Mater Chem 21: 2974-2981.

Lee, J., Yang, J., Ko, H., Oh, S.J., Kang, J., Son, J.H., et al. 2008. Multifunctional magnetic gold nanocomposites: human epithelial cancer detection via magnetic resonance imaging and localized synchronous therapy. Adv Funct Mater 18: 258-264.

Lewis, A.J. and Walz, D.T. 1982. Immunopharmacology of gold. Prog Med Chem 19: 1-58.

Li, H., Tan, L.L., Jia, P., Li, Q.L., Sun, Y.L., Zhang, J., et al. 2014a. Near-infrared light-responsive supramolecular nanovalve based on mesoporous silica-coated gold nanorods. Chem Sci 5: 2804-2808.

Li, M., Yan, H., Teh, C., Korzh, V. and Zhao, Y. 2014b. NIR-triggered drug release from switchable rotaxane-functionalized silica-covered Au nanorods. Chem Commun (Camb) 50: 9745-9748.

Li, X., Kao, F.J., Chuang, C.C. and He, S. 2010. Enhancing fluorescence of quantum dots by silica-coated gold nanorods under one- and two-photon excitation. Opt Express 18: 11335-11346.

Li, Z., Barnes, J.C., Bosoy, A., Stoddart, J.F. and Zink, J.I. 2012. Mesoporous silica nanoparticles in biomedical applications. Chem Soc Rev 41: 2590-2605.

Liberman, A., Martinez, H.P., Ta, C.N., Barback, C.V., Mattrey, R.F., Kono, Y., et al. 2012. Hollow silica and silica-boron nano/microparticles for contrast-enhanced ultrasound to detect small tumors. Biomaterials 33: 5124-5129.

Lin, K.Y., Bagley, A.F., Zhang, A.Y., Karl, D.L., Yoon, S.S. and Bhatia, S.N. 2010. Gold nanorod photothermal therapy in a genetically engineered mouse model of soft tissue sarcoma. Nano Life 01: 277-287.

Link, S., Burda, C., Nikoobakht, B. and El-Sayed, M.A. 2000. Laser-induced shape changes of colloidal gold nanorods using femtosecond and nanosecond laser pulses. J Phys Chem B 104: 6152-6163.

Liu, H., Chen, D., Li, L., Liu, T., Tan, L., Wu, X., et al. 2011. Multifunctional gold nanoshells on silica nanorattles: a platform for the combination of photothermal therapy and chemotherapy with low systemic toxicity. Angew Chem Int Ed Engl 50: 891-895.

Liu, H., Liu, T., Li, L., Hao, N., Tan, L., Meng, X., et al. 2012a. Size dependent cellular uptake, in vivo fate and light-heat conversion efficiency of gold nanoshells on silica nanorattles. Nanoscale 4: 3523-3529.

Liu, H., Liu, T., Wu, X., Li, L., Tan, L., Chen, D., et al. 2012b. Targeting gold nanoshells on silica nanorattles: a drug cocktail to fight breast tumors via a single irradiation with near-infrared laser light. Adv Mater 24: 755-761.

Liu, J., Stace-Naughton, A., Jiang, X. and Brinker, C.J. 2009. Porous nanoparticle supported lipid bilayers (protocells) as delivery vehicles. J Am Chem Soc 131: 1354-1355.

Liu, N., Prall, B.S. and Klimov, V.I. 2006. Hybrid gold/silica/nanocrystal-quantum-dot superstructures: synthesis and analysis of semiconductor-metal interactions. J Am Chem Soc 128: 15362-15363.

Liu, R., Zhang, Y., Zhao, X., Agarwal, A., Mueller, L.J. and Feng, P.Y. 2010. pH-responsive nanogated ensemble based on gold-capped mesoporous silica through an acid-labile acetal linker. J Am Chem Soc 132(5): 1500-1501.

Liu, S., Chen, G., Ohulchanskyy, T.Y., Swihart, M.T. and Prasad, P.N. 2013. Facile synthesis and potential bioimaging applications of hybrid upconverting and plasmonic NaGdF4: Yb3+, Er3+/silica/gold nanoparticles. Theranostics 3: 275-281.

Liu, Y., Ai, K. and Lu, L. 2012c. Nanoparticulate X-ray computed tomography contrast agents: from design validation to in vivo applications. Acc Chem Res 45: 1817-1827.

Lo, J.H., von Maltzahn, G., Douglass, J., Park, J.H., Sailor, M.J., Ruoslahti, E., et al. 2013. Nanoparticle amplification photothermal unveiling of cryptic collagen binding sites. J Mater Chem B Mater Biol Med 1: 5235-5240.

Luke, G.P., Yeager, D. and Emelianov, S.Y. 2012. Biomedical applications of photoacoustic imaging with exogenous contrast agents. Ann Biomed Eng 40: 422-437.

Luke, G.P., Bashyam, A., Homan, K.A., Makhija, S., Chen, Y.S. and Emelianov, S.Y. 2013. Silica-coated gold nanoplates as stable photoacoustic contrast agents for sentinel lymph node imaging. Nanotechnology 24: 455101.

Luo, T., Huang, P., Gao, G., Shen, G., Fu, S., Cui, D., et al. 2011. Mesoporous silica-coated gold nanorods with embedded indocyanine green for dual mode X-ray CT and NIR fluorescence imaging. Opt Express 19: 17030-17039.

Ma, M., Chen, H., Chen, Y., Wang, X., Chen, F., Cui, X., et al. 2012. Au capped magnetic core/mesoporous silica shell nanoparticles for combined photothermo-/chemo-therapy and multimodal imaging. Biomaterials 33: 989-998.

Madsen, S.J., Baek, S.K., Makkouk, A.R., Krasieva, T. and Hirschberg, H. 2012. Macrophages as cell-based delivery systems for nanoshells in photothermal therapy. Ann Biomed Eng 40: 507-515.

McCarthy, J., Inkielewicz-Stepniak, I., Corbalan, J.J. and Radomski, M.W. 2012. Mechanisms of toxicity of amorphous silica nanoparticles on human lung submucosal cells in vitro: protective effects of fisetin. Chem Res Toxicol 25: 2227-2235.

McNay, G., Eustace, D., Smith, W.E., Faulds, K. and Graham, D. 2011. Surface-enhanced Raman scattering (SERS) and surface-enhanced resonance Raman scattering (SERRS): a review of applications. Appl Spectrosc 65: 825-837.

Merchant, B. 1998. Gold, the noble metal and the paradoxes of its toxicology. Biologicals 26: 49-59.

Mine, E., Yamada, A., Kobayashi, Y., Konno, M. and Liz-Marzan, L.M. 2003. Direct coating of gold nanoparticles with silica by a seeded polymerization technique. J Colloid Interface Sci 264: 385-390.

Moskovits, M. 2005. Surface-enhanced Raman spectroscopy: a brief retrospective. J Raman Spectrosc 36: 485-496.

Mulvaney, S.P., He, L., Natan, M.J. and Keating, C.D. 2003a. Three-layer substrates for surface-enhanced Raman scattering: preparation and preliminary evaluation. J Raman Spectrosc 34: 163-171.

Mulvaney, S.P., Musick, M.D., Keating, C.D. and Natan, M.J. 2003b. Glass-coated, analyte-tagged nanoparticles: a new tagging system based on detection with surface-enhanced Raman scattering. Langmuir 19: 4784-4790.

Murase, K., Oonoki, J., Takata, H., Song, R., Angraini, A., Ausanai, P., et al. 2011. Simulation and experimental studies on magnetic hyperthermia with use of superparamagnetic iron oxide nanoparticles. Radiol Phys Technol 4: 194-202.

Nilsson, A. 2004. Contrast-enhanced ultrasound of the kidneys. European Radiology 14: 104-109.

Nolan, B.W., Schermerhorn, M.L., Powell, R.J., Rowell, E., Fillinger, M.F., Rzucidlo, E.M., et al. 2005. Restenosis in gold-coated renal artery stents. J Vasc Surg 42: 40-46.

O'Neal, D.P., Hirsch, L.R., Halas, N.J., Payne, J.D. and West, J.L. 2004. Photo-thermal tumor ablation in mice using near infrared-absorbing nanoparticles. Cancer Lett 209: 171-176.

Oldenburg, S.J., Averitt, R.D., Westcott, S.L. and Halas, N.J. 1998. Nanoengineering of optical resonances. Chem Phys Lett 288: 243-247.

Ow, H., Larson, D.R., Srivastava, M., Baird, B.A., Webb, W.W. and Wiesner, U. 2005. Bright and stable core-shell fluorescent silica nanoparticles. Nano Letters 5: 113-117.

Pak, J. and Yoo, H. 2013. Facile synthesis of spherical nanoparticles with a silica shell and multiple Au nanodots as the core. J Mater Chem A 1: 5408-5413.

Park, C., Oh, K., Lee, S.C. and Kim, C. 2007. Controlled release of guest molecules from mesoporous silica particles based on a pH-responsive polypseudorotaxane motif. Angew Chem Int Ed Engl 46: 1455-1457.

Park, J.H., Gu, L., von Maltzahn, G., Ruoslahti, E., Bhatia, S.N. and Sailor, M.J. 2009. Biodegradable luminescent porous silicon nanoparticles for in vivo applications. Nat Mater 8: 331-336.

Park, J.H., von Maltzahn, G., Ong, L.L., Centrone, A., Hatton, T.A., Ruoslahti, E., et al. 2010. Cooperative nanoparticles for tumor detection and photothermally triggered drug delivery. Adv Mater 22: 880-885.

Park, S.Y., Lytton-Jean, A.K., Lee, B., Weigand, S., Schatz, G.C. and Mirkin, C.A. 2008. DNA-programmable nanoparticle crystallization. Nature 451: 553-556.

Patel, K., Angelos, S., Dichtel, W.R., Coskun, A., Yang, Y.W., Zink, J.I., et al. 2008. Enzyme-responsive snap-top covered silica nanocontainers. J Am Chem Soc 130: 2382-2383.

Peng, X., Manna, L., Yang, W., Wickham, J., Scher, E., Kadavanich, A., et al. 2000. Shape control of CdSe nanocrystals. Nature 404: 59-61.

Perezjuste, J., Pastorizasantos, I., Lizmarzan, L. and Mulvaney, P. 2005. Gold nanorods: synthesis, characterization and applications. Coord Chem Rev 249: 1870-1901.

Perry, J.L., Herlihy, K.P., Napier, M.E. and DeSimone, J.M. 2011. PRINT: a novel platform toward shape and size specific nanoparticle theranostics. Acc Chem Res 44: 990-998.

Petrecca, K., Guiot, M.C., Panet-Raymond, V. and Souhami, L. 2013. Failure pattern following complete resection plus radiotherapy and temozolomide is at the resection margin in patients with glioblastoma. J Neurooncol 111: 19-23.

Phillips, E., Penate-Medina, O., Zanzonico, P.B., Carvajal, R.D., Mohan, P., Ye, Y., et al. 2014. Clinical translation of an ultrasmall inorganic optical-PET imaging nanoparticle probe. Sci Transl Med 6: 260ra149.

Pilloni, M., Nicolas, J., Marsaud, V., Bouchemal, K., Frongia, F., Scano, A., et al. 2010. PEGylation and preliminary biocompatibility evaluation of magnetite-silica nanocomposites obtained by high energy ball milling. Int J Pharm 401: 103-112.

Prevo, B.G., Esakoff, S.A., Mikhailovsky, A. and Zasadzinski, J.A. 2008. Scalable routes to gold nanoshells with tunable sizes and response to near-infrared pulsed-laser irradiation. Small 4: 1183-1195.

Priyam, A., Idris, N.M. and Zhang, Y. 2012. Gold nanoshell coated NaYF4 nanoparticles for simultaneously enhanced upconversion fluorescence and darkfield imaging. J Mater Chem 22: 960-965.

Qian, H.S., Guo, H.C., Ho, P.C., Mahendran, R. and Zhang, Y. 2009. Mesoporous-silica-coated up-conversion fluorescent nanoparticles for photodynamic therapy. Small 5: 2285-2290.

Qian, L.P., Zhou, L.H., Too, H.P. and Chow, G.M. 2011. Gold decorated NaYF4: Yb, Er/NaYF4/silica (core/shell/shell) upconversion nanoparticles for photothermal destruction of BE(2)-C neuroblastoma cells. J Nanopart Res 13: 499-510.

Qian, X., Peng, X.H., Ansari, D.O., Yin-Goen, Q., Chen, G.Z., Shin, D.M., et al. 2008. In vivo tumor targeting and spectroscopic detection with surface-enhanced Raman nanoparticle tags. Nat Biotechnol 26: 83-90.

Rabin, O., Manuel Perez, J., Grimm, J., Wojtkiewicz, G. and Weissleder, R. 2006. An X-ray computed tomography imaging agent based on long-circulating bismuth sulphide nanoparticles. Nat Mater 5: 118-122.

Radhakumary, C. and Sreenivasan, K. 2011. Synthesis and evaluation of pH responsive silica-polymer hybrid nano capsules. Soft Mater 9: 347-358.

Razansky, D., Distel, M., Vinegoni, C., Ma, R., Perrimon, N., Koster, R.W., et al. 2009. Multispectral opto-acoustic tomography of deep-seated fluorescent proteins in vivo. Nat Photon 3: 412-417.

Roggers, R.A., Joglekar, M., Valenstein, J.S. and Trewyn, B.G. 2014. Mimicking red blood cell lipid membrane to enhance the hemocompatibility of large-pore mesoporous silica. ACS Appl Mater Interfaces 6: 1675-1681.

Rutka, J.T., Kim, B., Etame, A. and Diaz, R.J. 2014. Nanosurgical resection of malignant brain tumors: beyond the cutting edge. ACS Nano 8: 9716-9722.

Santos-Martinez, M.J., Rahme, K., Corbalan, J.J., Faulkner, C., Holmes, J.D., Tajber, L., et al. 2014. Pegylation increases platelet biocompatibility of gold nanoparticles. J Biomed Nanotechnol 10: 1004-1015.

Santra, S., Zhang, P., Wang, K., Tapec, R. and Tan, W. 2001. Conjugation of biomolecules with luminophore-doped silica nanoparticles for photostable biomarkers. Analytical Chemistry 73: 4988-4993.

Shen, S., Tang, H., Zhang, X., Ren, J., Pang, Z., Wang, D., et al. 2013. Targeting mesoporous silica-encapsulated gold nanorods for chemo-photothermal therapy with near-infrared radiation. Biomaterials 34: 3150-3158.

Shi, J., Xiao, Z., Kamaly, N. and Farokhzad, O.C. 2011. Self-assembled targeted nanoparticles: evolution of technologies and bench to bedside translation. Acc Chem Res 44: 1123-1134.

Slowing, I.I., Vivero-Escoto, J.L., Wu, C.W. and Lin, V.S.Y. 2008. Mesoporous silica nanoparticles as controlled release drug delivery and gene transfection carriers. Adv Drug Deliv Rev 60: 1278-1288.

Sotiriou, G.A., Starsich, F., Dasargyri, A., Wurnig, M.C., Krumeich, F., Boss, A., et al. 2014. Photothermal killing of cancer cells by the controlled plasmonic coupling of silica-coated Au/Fe_2O_3 Nanoaggregates. Adv Funct Mater 24: 2818-2827.

Stöber, W., Fink, A. and Bohn, E. 1968. Controlled growth of monodisperse silica spheres in the micron size range. J Colloid Interface Sci 26: 62-69.

Stolik, S., Delgado, J., Perez, A. and Anasagasti, L. 2000. Measurement of the penetration depths of red and near infrared light in human "ex vivo" tissues. J Photochem Photobiol B 57: 90-93.

Stuchinskaya, T., Moreno, M., Cook, M.J., Edwards, D.R. and Russell, D.A. 2011. Targeted photodynamic therapy of breast cancer cells using antibody-phthalocyanine-gold nanoparticle conjugates. Photochem Photobiol Sci 10: 822-831.

Su, X., Zhang, J., Sun, L., Koo, T.-W., Chan, S., Sundararajan, N., et al. 2005. Composite organic-inorganic nanoparticles (COINs) with chemically encoded optical signatures. Nano letters 5: 49-54.

Svaasand, L.O., Gomer, C.J. and Morinelli, E. 1990. On the physical rationale of laser induced hyperthermia. Lasers Med Sci 5: 121-128.

Tam, J.M., Tam, J.O., Murthy, A., Ingram, D.R., Ma, L.L., Travis, K., et al. 2010. Controlled assembly of biodegradable plasmonic nanoclusters for near-infrared imaging and therapeutic applications. ACS Nano 4: 2178-2184.

Thakor, A.S., Jokerst, J.V., Zavaleta, C.L., Massoud, T.F. and Gambhir, S.S. 2011. Gold Nanoparticles: a revival in precious metal administration to patients. Nano Letters 11: 4029-4036.

Tovmachenko, O.G., Graf, C., van den Heuvel, D.J., van Blaaderen, A. and Gerritsen, H.C. 2006. Fluorescence enhancement by metal-core/silica-shell nanoparticles. Adv Mater 18: 91-95.

Tran, T.H. and Nguyen, T.D. 2013. Functional Inorganic Nanohybrids for Biomedical Diagnosis. INTECH Open Access Publisher.

Ujfalussy, I., Koo, E., Sesztak, M. and Gergely, P. 2003. Termination of disease-modifying antirheumatic drugs in rheumatoid arthritis and in psoriatic arthritis. A comparative study of 270 cases. Z Rheumatol 62: 155-160.

van Schooneveld, M.M., Cormode, D.P., Koole, R., van Wijngaarden, J.T., Calcagno, C., Skajaa, T., et al. 2010. A fluorescent, paramagnetic and PEGylated gold/ silica nanoparticle for MRI, CT and fluorescence imaging. Contrast Media Mol Imaging 5: 231-236.

Vivero-Escoto, J.L., Slowing, II, Wu, C.W. and Lin, V.S. 2009. Photoinduced intracellular controlled release drug delivery in human cells by gold-capped mesoporous silica nanosphere. J Am Chem Soc 131: 3462-3463.

Vogl, T.J., Schwarz, W., Blume, S., Pietsch, M., Shamsi, K., Franz, M., et al. 2003. Preoperative evaluation of malignant liver tumors: comparison of unenhanced and SPIO (Resovist)-enhanced MR imaging with biphasic CTAP and intraoperative US. Eur Radiol 13: 262-272.

von Maltzahn, G., Centrone, A., Park, J.H., Ramanathan, R., Sailor, M.J., Hatton, T.A., et al. 2009a. SERS-coded gold nanorods as a multifunctional platform for densely multiplexed near-infrared imaging and photothermal heating. Adv Mater 21: 3175-3180.

von Maltzahn, G., Park, J.H., Agrawal, A., Bandaru, N.K., Das, S.K., Sailor, M.J., et al. 2009b. Computationally guided photothermal tumor therapy using long-circulating gold nanorod antennas. Cancer Research 69: 3892-3900.

Wang, C., Ma, Z., Wang, T. and Su, Z. 2006. Synthesis, assembly and biofunctionalization of silica-coated gold nanorods for colorimetric biosensing. Adv Funct Mater 16: 1673-1678.

Wang, C., Bao, C., Liang, S., Fu, H., Wang, K., Deng, M., et al. 2014a. RGD-conjugated silica-coated gold nanorods on the surface of carbon nanotubes for targeted photoacoustic imaging of gastric cancer. Nanoscale Res Lett 9: 264.

Wang, L.V. and Hu, S. 2012. Photoacoustic tomography: in vivo imaging from organelles to organs. Science 335: 1458-1462.

Wang, Y.W., Khan, A., Som, M., Wang, D., Chen, Y., Leigh, S.Y., et al. 2014b. Rapid ratiometric biomarker detection with topically applied SERS nanoparticles. Technology (Singap World Sci) 2: 118-132.

Wang, Z., Zong, S., Yang, J., Li, J. and Cui, Y. 2011. Dual-mode probe based on mesoporous silica coated gold nanorods for targeting cancer cells. Biosens Bioelectron 26: 2883-2889.

Winter, P., Morawski, A., Caruthers, S., Fuhrhop, R., Zhang, H., Williams, T., et al. 2003. Molecular imaging of angiogenesis in early-stage atherosclerosis with {alpha} v {beta} 3-integrin-targeted nanoparticles. Circulation 108: 2270.

Wuelfing, W., Gross, S., Miles, D. and Murray, R. 1998. Nanometer gold clusters protected by surface-bound monolayers of thiolated poly (ethylene glycol) polymer electrolyte. J Am Chem Soc 120: 12696-12697.

Xia, H.X., Yang, X.Q., Song, J.T., Chen, J., Zhang, M.Z., Yan, D.M., et al. 2014. Folic acid-conjugated silica-coated gold nanorods and quantum dots for dual-modality CT and fluorescence imaging and photothermal therapy. J Mater Chem B 2: 1945-1953.

Xie, J., Zheng, Y. and Ying, J.Y. 2009. Protein-directed synthesis of highly fluorescent gold nanoclusters. J Am Chem Soc 131: 888-889.

Xie, W., Zhang, Y., Walkenfort, B., Yoon, J.H. and Schluecker, S. 2014. Gold and silver nanoparticle monomers are not SERS-active: a negative experimental study with silica-encapsulated, Raman-reporter-coated metal colloids. Phys Chem Chem Phys

Xu, M.H. and Wang, L.H.V. 2006. Photoacoustic imaging in biomedicine. Rev Sci Instrum 77: 041101.

Yang, J., Shen, D., Zhou, L., Li, W., Li, X., Yao, C., et al. 2013. Spatially confined fabrication of core–shell gold nanocages@ mesoporous silica for near-infrared controlled photothermal drug release. Chem Mater 25: 3030-3037.

Yao, D.K., Maslov, K., Shung, K.K., Zhou, Q. and Wang, L.V. 2010. In vivo label-free photoacoustic microscopy of cell nuclei by excitation of DNA and RNA. Opt Lett 35: 4139-4141.

Yao, H., Momozawa, O., Hamatani, T. and Kimura, K. 2001. Stepwise size-selective extraction of carboxylate-modified gold nanoparticles from an aqueous suspension into toluene with tetraoctylammonium cations. Chem Mater 13: 4692-4697.

Yi, D.K., Selvan, S.T., Lee, S.S., Papaefthymiou, G.C., Kundaliya, D. and Ying, J.Y. 2005. Silica-coated nanocomposites of magnetic nanoparticles and quantum dots. J Am Chem Soc 127: 4990-4991.

Yu, T., Malugin, A. and Ghandehari, H. 2011. Impact of silica nanoparticle design on cellular toxicity and hemolytic activity. ACS Nano 5: 5717-5728.

Zavaleta, C.L., Smith, B.R., Walton, I., Doering, W., Davis, G., Shojaei, B., et al. 2009. Multiplexed imaging of surface enhanced Raman scattering nanotags in living mice using noninvasive Raman spectroscopy. Proc Natl Acad Sci USA 106: 13511-13516.

Zavaleta, C.L., Garai, E., Liu, J.T., Sensarn, S., Mandella, M.J., Van de Sompel, D., et al. 2013. A Raman-based endoscopic strategy for multiplexed molecular imaging. Proc Natl Acad Sci USA 110: e2288-2297.

Zhang, C., Maslov, K., Yao, J. and Wang, L.V. 2012a. In vivo photoacoustic microscopy with 7.6-μm axial resolution using a commercial 125-MHz ultrasonic transducer. J Biomed Opt 17: 116016-116016.

Zhang, L., Chan, J.M., Gu, F.X., Rhee, J.-W., Wang, A.Z., Radovic-Moreno, A.F., et al. 2008. Self-assembled lipid-polymer hybrid nanoparticles: a robust drug delivery platform. ACS Nano 2: 1696-1702.

Zhang, X.D., Wu, D., Shen, X., Liu, P.X., Fan, F.Y. and Fan, S.J. 2012b. In vivo renal clearance, biodistribution, toxicity of gold nanoclusters. Biomaterials 33: 4628-4638.

Zhang, Y., Qian, J., Wang, D., Wang, Y. and He, S. 2013. Multifunctional gold nanorods with ultrahigh stability and tunability for in vivo fluorescence imaging, SERS detection and photodynamic therapy. Angew Chem Int Ed Engl 52: 1148-1151.

Zhang, Z., Wang, L., Wang, J., Jiang, X., Li, X., Hu, Z., et al. 2012c. Mesoporous silica-coated gold nanorods as a light-mediated multifunctional theranostic platform for cancer treatment. Adv Mater 24: 1418-1423.

Zheng, J., Zhang, C. and Dickson, R.M. 2004. Highly fluorescent, water-soluble, size-tunable gold quantum dots. Phys Rev Lett 93: 077402.

15

Mesoporous Silica-Based Hybrid Nanosystems for Magnetic Resonance Imaging-Based Cancer Theranostics

Yu Chen[1,2,a] and *Shengjian Zhang*[3,*]

INTRODUCTION

Cancer has become one of the most serious diseases all over the world. How to build a promising life bridge between the medicine and cancer death remains a great challenge to scientists. Recently developed nano-biotechnology provides an alternative and highly efficient approach for the early diagnosis and efficient treatment of cancer (Allen and Cullis 2004; Kostarelos et al. 2009; LaVan et al. 2003; Torchilin 2005; Wagner et al. 2006). It is generally accepted that nano-biotechnology depends on the elaborately designed and fabricated nanomaterials with desirable composition, nanostructure, performance and biocompatibility (Arvizo et al. 2012; Boisselier and Astruc 2009; Chen et al. 2014a; Lammers et al. 2011; Na et al. 2009; Park et al. 2009). These nanomaterials can act as the contrast agents (CAs) for diagnostic imaging, as the drug delivery systems (DDSs) for chemotherapy and as the scaffold materials for tissue engineering. Abundant nanomaterials have been synthesized for diverse biomedical applications, including organic, inorganic and organic-inorganic hybrid nanosystems. For instance, magnetic Fe_3O_4 nanoparticles showed excellent contrast-enhanced imaging performance in

[1] State Key Laboratory of High Performance Ceramic and Superfine Nanostructures, Shanghai Institute of Ceramics, Chinese Academy of Sciences, Shanghai, 200050, P. R. China.

[2] School of Chemical Engineering, The University of Queensland, Brisbane, 4072, Australia.

[a] E-mail: chenyu@mail.sic.ac.cn

[3] Department of Radiology, Cancer Hospital/Institute & Department of Oncology, Shanghai Medical College, Fudan University, Shanghai, 200032, P. R. China.

[*] Corresponding author: zhangshengjian@yeah.net

T_2-weighted magnetic resonance imaging (MRI) and magnetic hyperthermia of cancer tissues (Chen et al. 2014c; Chen et al. 2015; Chen et al. 2014e; Laurent et al. 2008; Piao et al. 2008; Sun et al. 2008). Gold nanorods/nanocages (Daniel and Astruc 2004; Lal et al. 2008; Patra et al. 2010; Skrabalak et al. 2008), CuS_x nanoparticles (Tian et al. 2011b; Zhou et al. 2010) and even two-dimensional graphene/graphene analogues (e.g., MoS_2, WS and TiS_2) can be utilized as photothermal agents for light-induced photothermal ablation of cancer cells. (Chen et al. 2014g; Chung et al. 2013; Tian et al. 2011a) Several organic nano-systems, including liposomes, emulsions or other synthetic biomacromolecules, have entered the clinical stage for anticancer drug delivery with high chemotherapeutic efficiency and mitigated side-effects (Allen and Cullis 2004; Farokhzad and Langer 2009; Riehemann et al. 2009; Torchilin 2005). In addition, various nanoparticles can be embedded into the implanted scaffolds for tissue engineering or cancer therapy (Lalwani et al. 2013; Li et al. 2008; Shi et al. 2013; Vallet-Regi 2006; Wu et al. 2009; Zhu et al. 2011).

Among the various organic or inorganic nanosystems, inorganic mesoporous silica nanoparticles (MSNs) have attracted tremendous attention especially in various biomedical fields (Ambrogio et al. 2011; Chen et al. 2014b; Lee et al. 2011; Tarn et al. 2013). MSNs are featured with a well-defined mesoporous structure, large surface area, high pore volume, abundant surface chemistry and high biocompatibility/biosafety (Chen et al. 2010b; Chen et al. 2010c; Gao et al. 2011). Anticancer drug molecules can be encapsulated into the mesopores with high cargo-loading amount, which exhibit sustained releasing behavior (Chen et al. 2013b; Chen et al. 2012c; Li et al. 2012b; Mamaeva et al. 2013; Slowing et al. 2008; Wang et al. 2012; Yang et al. 2012). Various designed nanovalves can also be capped onto the surface of mesopores for stimuli-responsive intelligent drug releasing. In addition, large pore-sized MSNs have been demonstrated as carriers for siRNA or DNA molecules in gene delivery and transfection (Hartono et al. 2012; Wu et al. 2015). Besides the satisfactory property of MSNs in drug delivery, the hybridization and multifunctionalization of MSNs can endow the carrier with concurrent diagnostic-imaging features. The typical hybridization strategy includes surface modification (Pan et al. 2012; Xia et al. 2009; Zhu et al. 2005), mesopore functionalization (Chen et al. 2011a) and framework hybridization. (Chen et al. 2014d; Chen et al. 2014f; Chen et al. 2013c) These multifunctional MSNs have been demonstrated as high-performance CAs in MR imaging (Liong et al. 2008), fluorescent imaging (Lee et al. 2009), ultrasound imaging (Hu et al. 2011; Martinez et al. 2010; Wang et al. 2013), computed tomography (Luo et al. 2011), photoacoustic imaging (Jokerst et al. 2012) and nuclear imaging (Chen et al. 2013a; Huang et al. 2012).

This chapter focuses on the functionalization and hybridization of MSNs for MR imaging-based cancer theranostics (concurrent diagnosis and therapy). These MSNs-based nanosystems can precisely provide the MRI signal of tumor tissues, which is also potentially to be used for targeted drug delivery, monitoring the therapeutic process and evaluating the therapeutic efficiency. The pre-loaded therapeutic agents are expected to be *in situ* released within the tumor to kill the cancer cells with enhanced efficiency and reduced side-effects. The MR imaging-guided cancer drug delivery can be realized using these multifunctional MSNs-based composite nanosystems as the CAs. On this ground, it is highly expected that these composite mesoporous nanosystems can exhibit high performance in MRI-guided cancer theranostics.

Synthetic Strategies of MSNs-Based Composite Nanosystems for Enhanced MR Imaging

Several strategies have been developed to construct MSNs-based composite nanosystems for MR imaging, typically based on sol-gel chemistry. It is well-known that mesoporous silica can be easily coated onto the surface of various nanocrystals, either hydrophilic or hydrophobic functional nanoparticles (Wang and Gu 2015). Based on this process, magnetic Fe_3O_4 nanocrystals can be elaborately coated by a layer of mesoporous silica to form the core/shell structure. Such a new type of magnetic core/shell nanoparticle is regarded as an excellent candidate for T_2-weighted MR imaging. For hydrophilic magnetic Fe_3O_4 nanocrystals, mesoporous silica layer is directly deposited onto the nanocrystals' surface by the hydrolysis of the silica precursor under alkaline condition (ammonia solution), which is assisted by structural-directing agents such as surfactants or saline coupling agents. For hydrophobic Fe_3O_4 nanocrystals, amphiphilic surfactants should be initially employed to transfer hydrophobic Fe_3O_4 nanocrystals from organic medium into aqueous solution via hydrophobic-hydrophobic interaction. The subsequent co-assembly of hydrolyzed silica precursor and surfactants can form a mesoporous silica layer on the surface of magnetic Fe_3O_4 nanocrystals. In this regard, the surfactants perform two roles. One is the transferring agent while the other is the structural-directing agent. Surface-modified magnetic Fe_3O_4 nanocrystals can also be anchored onto the surface of MSNs by a covalent bond to form a special MSNs@Fe_3O_4 composite nanoparticle. Based on the abundant saline chemistry, Gd-based chelates are easily anchored onto the mesopore surface of MSNs for enhanced T_1-weighted MR imaging because of the enhanced interaction between water molecules and paramagnetic centers. Therefore, the general multifunctional strategies to endow MSNs with MR imaging performance include either chemical coating procedures or organic modification. In addition, large hollow cavities are generally introduced between the Fe_3O_4 core and mesoporous silica shell of mesoporous composite nanoparticles to enhance the drug-loading capacity of these theranostic agents (Chen et al. 2010c; Zhu et al. 2010; Zhu et al. 2009).

Mesoporous Silica-Based Hybrid Materials for T_1-Weighted MR Imaging

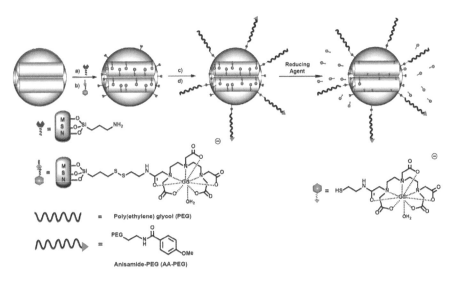

Fig. 15.1 Schematic illustration of the synthetic procedure for PEG-Gd-MSNs and AA-Gd-MSNs by several chemical reactions (Vivero-Escoto et al. 2011).

It is well-known that paramagnetic ions with a large number of unpaired electrons, such as Gd^{3+}, Mn^{2+} and Fe^{3+} are highly suitable for contrast-enhanced T_1-weighted MR imaging, which is due to the specific interactions between the protons of water molecules and the electron spins of paramagnetic ions (Schwert et al. 2002; Terreno et al. 2010; Viswanathan et al. 2010). The large surface area of mesopores within MSNs facilitates the high dispersity of various paramagnetic centers (e.g., Gd, Mn and Fe ions). In addition, the water molecules can freely diffuse into the mesopores to interact with the paramagnetic centers, resulting in substantially enhanced T_1-weighted MRI performance of mesoporous silica-based hybrid materials. To demonstrate this assumption, Lin et al. grafted Gd chelates (Gd-Si-DTTA) onto the mesopores of MSNs by a typical siloxane linkage (Taylor et al. 2008). They demonstrated that the accessibility of Gd paramagnetic centers to water molecules was maximized due to the high dispersity of Gd-Si-DTTA within the well-defined mesopores. The longitudinal relaxivity r_1 could reach as high as 28.8 $mM^{-1}s^{-1}$ at 3 T and 10.2 $mM^{-1}s^{-1}$ at 9.4 T. The transverse relaxivity r_2 was shown to be 65.5 $mM^{-1}s^{-1}$ at 3 T and 110.8 $mM^{-1}s^{-1}$ at 9.4 T. After the injection of Gd-grafted MSNs *via* tail vein into the mice (dose: 2.1 µmol/kg), the T_1-weighted contrast enhancement could be clearly presented in the aorta of mice after 15 min post-injection. This *in vivo* result indicated that the achieved Gd-grafted MSNs could act as the intravascular CAs for

T_1-weighted MR imaging. The modification of MSNs with Gd phosphonate can also enhance the T_1-weighted MRI performance. After the grafting of Gd phosphonate into MSNs with varied mesoporous parameters, the relaxivity of the mesoporous composites could be tuned from 2 to 10 $mM^{-1}s^{-1}$, much larger than commercial Gd-based agents with relaxivities in the range of 3.0-3.5 $mM^{-1}s^{-1}$(Duncan et al. 2012).

Similarly, Huang et al. grafted Gd-DTPA (DTPA: diethylenetriaminepenta-acetic acid) into the mesopores of MSNs for contrast-enhanced T_1-weighted MR imaging of human mesenchymal stem cells (hMSCs) (Hsiao et al. 2008). The viability, proliferation and differentiation capacities of hMSCs were not affected after the labeling with Gd-DTPA-grafted MSNs, demonstrating the high biocompatibility of Gd-DTPA-functionalized MSNs. More importantly, the labeled hMSCs clearly presented detectable MRI signals after long-term (14 day) *in vivo* growth and differentiation. Gd-labeled microparticles were also administrated by intraperitoneal injection into mice for potential real-time MRI tracking (Steinbacher et al. 2010). The urine excretion of these Gd-labeled microparticles was observed in T_1-weighted MR imaging. In addition to the endowed high T_1-weighted MRI performance, it is noted that the grafting of Gd^{3+} chelates within the mesopores could substantially enhance the transverse relaxivity of magnetic Fe_3O_4@mSiO_2 core/shell nanoparticles (from initial 97 $mM^{-1}s^{-1}$ to subsequent 681 $mM^{-1}s^{-1}$ with r_2/r_1 = 486), illustrating another advantage of grafting Gd chelates within the mesopores of MSNs (Huang et al. 2011).

The high accumulation of nanoparticulate CAs and slow clearance within the liver limit their clinical application in diagnostic MR imaging of tumor. To solve this critical issue, Lin et al. modified Gd chelates onto the surface of mesopores *via* cleavable disulfide bond (Fig. 15.1) (Vivero-Escoto et al. 2011). The redox-responsive disulfide bond could be broken up by the intracellular redox microenvironment or the reducing agents in the blood pool, facilitating their excretion from the body *via* urine and maintaining the high MRI performance of nanoparticles. Further surface functionalization with polyethylene glycol (PEG) and targeted anisamide (AA) improved the colloidal stability, biocompatibility and targeting specificity of Gd-MSNs.

In addition to Gd-based chelates, ultrasmall Gd-based oxide nanoparticles were also successfully encapsulated into the mesopores of MSNs for enhanced MR imaging. For example, the molecular assembly of Gd_2O_3 clusters within the mesopores of MSNs (designated as Gd_2O_3@MCM-41) could not only avoid the dissociation of free toxic Gd^{3+} from MSNs matrix, but also presented strong water-adsorptive ability to achieve large longitudinal relaxivity and enhanced T_1-weighted MRI performance (Li et al. 2011). The achieved Gd_2O_3@MCM-41 could be cleared *via* the hepatobiliary transport mechanism. More importantly, such Gd_2O_3@MCM-41 MRI CAs presented improved MRI performance as compared to commercial gadolinium diethylenetriaminepentaacetate (Shao et al. 2012).

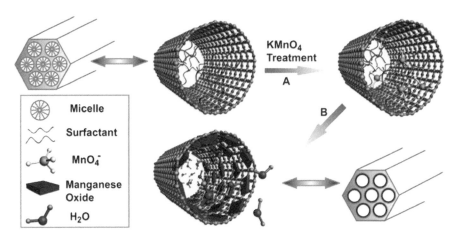

Fig. 15.2 Schematic illustration of dispersing ultrasmall MnO_x nanoparticles into the mesopores of MSNs by a redox reaction between the surfactants and $KMnO_4$.

Although Gd-based CAs have found the extensive applications in clinical diagnostic imaging, the U.S. Food and Drug Administration (FDA) has warned that clinical Gd-based MRI CAs are associated with nephrogenic systemic fibrosis (NSF), hypersensitivity reaction and nephrogenic fibrosing dermopathy (NFD) (Penfield and Reilly 2007; Perez-Rodriguez et al. 2009; Tromsdorf et al. 2009). It is of high significance to update the MRI CAs with lower toxicity and higher biocompatibility. As the alternative candidate, manganese (Mn)-based MRI CAs have attracted great attention recently because Mn element is the essential element for daily physical metabolism. In addition, Mn (II) ions have five unpaired electrons with long electronic relaxation time, facilitating their application in MR imaging (Huang et al. 2010; Shin et al. 2009). Two typical strategies have been developed to integrate MnO_x nanoparticles with MSNs for potential T_1-weighted MRI. On one hand, MnO_x nanoparticles were coated by a mesoporous silica layer to form the core/shell nanostructure (Peng et al. 2011). On the other hand, MnO_x nanoparticles were dispersed into the mesopores of MSNs for T_1-weighted MR imaging, similar to aforementioned Gd_2O_3@MCM-41 (Chen et al. 2012b). Hyeon et al. coated a mesoporous silica layer onto the surface of hollow manganese oxide (HMnO@mSiO$_2$) for application as T_1-weighted MRI CAs (Kim et al. 2011b). The longitudinal relaxation r_1 of the achieved CAs was calculated to be 0.99 mM^{-1}s^{-1} at 11.7 T, which showed enhanced/long-term positive labeling and MRI tracking of adipose-derived mesenchymal stem cells. To further improve the MRI performance of Mn-based CAs, Chen et al. recently dispersed ultrasmall MnO_x nanoparticles into the mesopores of MSNs to improve the accessibility of Mn paramagnetic centers to water molecules (Fig. 15.2), which was realized by the redox reaction between strongly oxidative $KMnO_4$ and reducing surfactants (Chen et al. 2012a).

The longitudinal relaxation r_1 could reach as high as 2.28 mM^{-1}s^{-1} at 3.0 T. The enhanced T_1-weighted MR imaging of tumor was also preliminarily demonstrated after intravenous administration. It is noted that MnO$_x$-based T_1 CAs exhibited the unique break-up nature for pH-responsive MR imaging, which is highly desirable for tumor imaging because of the mildly acidic microenvironment of tumor tissues (Chen et al. 2014e; Chen et al. 2014g; Chen et al. 2012d). Further encapsulation of anticancer drug doxorubicin could reverse the multidrug resistance of cancer cells due to the enhanced intracellular concentration of chemotherapeutic drug molecules, showing their concurrent diagnostic and therapeutic functions (Chen et al. 2012d).

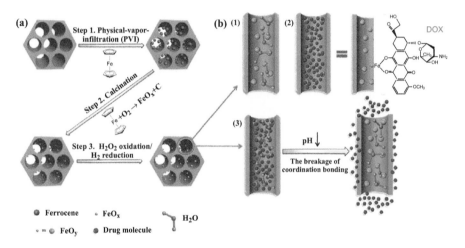

Fig. 15.3 (a) Schematic illustration of the fabrication procedure for ultrasmall FeOy-dispersed MSNs composite via physical-vapor-infiltration (PVI) strategy and (b) their encapsulation/ sustained releasing behavior for anticancer drugs (Wu et al. 2014).

Mn-based CAs for T_1-weighted MRI suffer from potential toxicity in diagnostic imaging. As compared to Gd- and Mn-based MRI CAs, Fe-based CAs show much higher biocompatibility. Typically, Fe-based CAs were used for T_2-weighted MR imaging because of their unique superparamagnetic feature in the form of magnetic Fe$_3$O$_4$ (Chen et al. 2011c; Chen et al. 2014e). When the particle size of Fe$_3$O$_4$ nanoparticles was reduced into ultrasmall range, it was reported that these ultrasmall Fe$_3$O$_4$ nanoparticles showed high performance in T_2-weighted MR imaging (Kim et al. 2011a; Li et al. 2012a; Ninjbadgar and Brougham 2011; Taboada et al. 2007; Tromsdorf et al. 2009; Zeng et al. 2012). Wu et al. recently developed a physical-vapor-infiltration (PVI) method to introduce ferrocene into the mesopores of MSNs (Fig. 15.3) (Wu et al. 2014). Through further thermal treatment, the confined ferrocene *in situ* decomposed to form ultrasmall FeOy nanoparticles within the mesopores. The valance of Fe in FeOy could be controlled by the subsequent thermal-reduction process. Specifically, the r_1 value reached 1.50 mM^{-1}s^{-1}, which was increased by precisely controlling the valance of Fe species. *In vivo* contrast-enhanced T_1-weighted MRI was preliminarily demonstrated by intratumor injection of FeOy-MSNs nanocomposites. This report provides the first evidence that Fe-based nanoparticles can be integrated with MSNs for high-performance T1-weighted MRI-based diagnostic imaging.

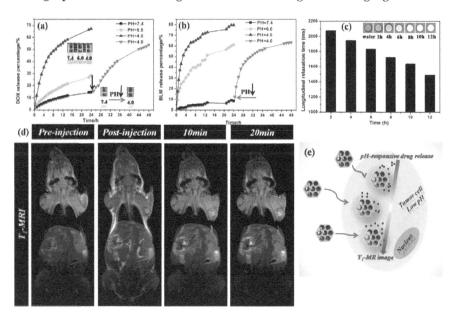

Fig. 15.4 PH-responsive (a) DOX- and (b) BLM-releasing patterns from the mesopores. (c) The longitudinal relaxation time of the drug-releasing media after different releasing durations. (d) *In vivo* T_1-weighted MR imaging of a tumor-bearing mouse after the intratumor injection of DOX-loaded FeOy-MSNs for different durations. (e) Schematic illustration of FeOy-MSNs for pH-responsive drug releasing assisted by T_1-weighted MR imaging (Wu et al. 2014).

The dispersed FeOy nanoparticles within the mesopores facilitated the encapsulation and controlled release of anticancer drug molecules. The Fe species could form the specific metal ion-ligand coordination bond with the drug molecules. Wu et al. demonstrated that the amino, carbonyl and hydroxyl groups of doxorubicin (DOX) and bleomycin (BLM) could act as the binding sites to metal ions to form the coordination bond (Wu et al. 2014). Such a coordination bond was responsive to pH changes where the acidic microenvironment could fast trigger the release of anticancer drugs from the mesopores, which was demonstrated by investigating the releasing patterns of DOX (Fig. 15.4a) and BLM (Fig. 15.4b) from FeOy-MSNs at different pHs. More importantly, the release of drug molecules made more paramagnetic centers available to the water molecules, thus the T_1-weighted MRI performance recovered upon drug release. It was found that the

MRI-positive signal was enhanced after the drug release, both *in vitro* (Fig. 15.4c) and *in vivo* (Fig. 15.4d). Therefore, the FeOy-integrated MSNs could not only act as the positive CAs for MRI, but also as the drug delivery system for MRI-visible and pH-responsive drug release.

Mesoporous Silica-Based Composite Materials for T_2-Weighted MR Imaging

It has been extensively examined that superparamagnetic Fe_3O_4 nanoparticles are the desirable CAs for T_2-weighted MR imaging (Sun et al. 2008). Therefore, the integration of Fe_3O_4 nanoparticles into MSNs matrix can endow MSNs with the specific functions of concurrent T_2-weighted MR imaging and drug delivery. Shi's group for the first time successfully fabricated uniform magnetic composite spheres by directly coating a mesoporous silica layer onto the surface of iron oxide nanoparticles (Zhao et al. 2005). These magnetic composite mesoporous materials exhibited sustained drug-releasing behavior. As early as 2008, Mou et al. integrated $Fe_3O_4@SiO_2$ nanoparticles with MSNs for intracellular labeling and *in vivo* MRI evaluations (Wu et al. 2008). The time-dependent contrast-enhanced T_2-weighted MR imaging was preliminarily demonstrated in liver and kidney after the administration of nanocomposites through the eye vein. The significant breakthrough was made by Hyeon et al. in reducing the particle size of $Fe_3O_4@mSiO_2$ into the range of less than 100 nm, which was highly desirable for *in vivo* tumor imaging (Kim et al. 2008; Kim et al. 2006). They employed amphiphilic surfactants to transfer as-synthesized hydrophobic Fe_3O_4 nanoparticles from organic medium into aqueous solution. Further introduction of silica source could form a uniform mesoporous silica layer on the surface of magnetic Fe_3O_4 nanoparticles. Enhanced tumor imaging by using these small $Fe_3O_4@mSiO_2$ nanoparticles as the T_2-weighted MRI CAs was preliminarily demonstrated.

The outer surface of MSNs provides the anchoring points for superparamagnetic nanoparticles. To demonstrate this assumption, Hyeon et al. firstly grafted amine groups onto the surface of MSNs by 3-aminopropytriethoxysilane (Lee et al. 2010). The as-synthesized Fe_3O_4 nanoparticles were initially reacted with 2-bromo-2-methylpropionic acid, which was further assembled onto the surface of amine groups-modified MSNs by direct nucleophilic substitution reaction (Fig. 15.5). After the PEGylation process, Fe_3O_4-MSNs composites exhibited average particle size of 93 nm in aqueous solution. Such a small particle size guaranteed the high accumulation of nanocomposites at the tumor site. It was found that the intravenous administration of Fe_3O_4-MSNs could enhance the T_2-weighted MR imaging of tumor. The presence of mesopores of Fe_3O_4-MSNs exhibited delivery performance for doxorubicin into the tumor (Lee et al. 2010).

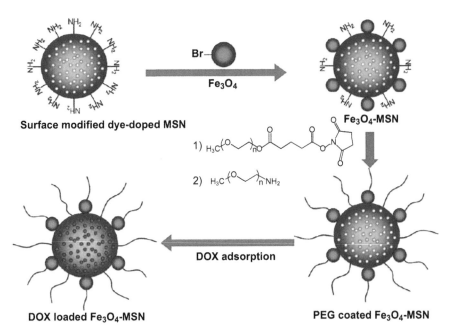

Fig. 15.5　Schematic illustration of assembling Fe_3O_4 nanoparticles onto the surface of MSNs and their capability for drug loading (Lee et al. 2010).

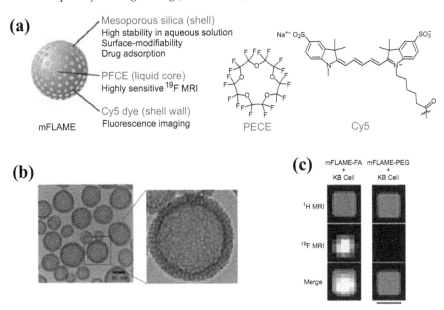

Fig. 15.6　(a) Schematic illustration of mFLAME (mesoporous Fluorine Accumulated silica nanoparticles for MRI enhancement) structure, including liquid PFC as the core, mesoporous silica as the shell and Cy5 dye incorporated within the core. (b) TEM image of mFLAME at different magnifications. (c) $^1H/^{19}F$ MR images of KB cells after the co-incubation with FA-modified mFLAME and PEGylated mFLAME (Nakamura et al. 2015).

Mesoporous Silica Nanoparticles for ^{19}F MR Imaging

Gd-, Mn- and Fe-based MRI-T_1 CAs and magnetic Fe_3O_4-based MRI-T_2 CAs are being extensively explored for the diagnosis of diseases such as cancer. However, the traditional 1H MRI is difficult to provide the high contrast-images of some specific functions, such as molecular imaging or other disease-related biological events, which is mainly due to the high background signal within the *in vivo* biological systems. Because of the high sensitivity and low background signals, ^{19}F MRI CAs show the promising applications in diagnostic imaging (Boehm-Sturm et al. 2011; Chen et al. 2010a; Ebner et al. 2010; Srinivas et al. 2013). Kikuchi et al. recently encapsulated perfluorocarbon (PFC: perfluoro-15-crown-5-ether) into the interior of MSNs as the CAs for ^{19}F MR imaging (Fig. 15.6a) (Nakamura et al. 2015). In addition, NIR dye Cy5 was incorporated into the mesoporous silica shell for fluorescent imaging. The mesopores were further used for drug delivery. The core/shell and mesoporous structure of mFLAME was clearly distinguished in TEM image (Fig. 15.6b). The PFC core for ^{19}F MRI was demonstrated by the MR imaging of KB cells after co-incubation with mFLAME (Fig. 15.6c). The surface folic acid-targeting modification further enhanced the uptake efficiency of KB cells for mFLAME, as demonstrated by the MRI signal variations between the samples of FA-modified mFLAME and PEGylated mFLAME. The incorporated Cy5 dye was shown to be effective in the fluorescent imaging of cancer cells to potentially monitor the drug-releasing behavior from the carrier.

Chen et al. encapsulated C_6H_6 into the mesopores of MSNs for intelligent ^{19}F MR imaging. Typically, the mesopores were sealed by Au nanoparticles *via* a pH-responsive hydrazone linker (Chen et al. 2014a). In neutral condition, the encapsulated C_6H_6 could not act as the ^{19}F CAs for MRI imaging because it was firmly sealed by Au nanoparticles. Under the mild acidic microenvironment, the hydrazone bond was broken, which opened the mesopores to release C_6H_6. The released C_6H_6 acted as the CAs for ^{19}F MR imaging. The performance of these pH-responsive ^{19}F MRI CAs was demonstrated preliminarily *in vitro*. Significant contrast-enhanced ^{19}F MR imaging could be obtained either in human lung cancer cells (A549) or PBS solution at the pH of 6.0. Comparatively, the ^{19}F CAs were not activated at neutral PBS solution (pH = 7.4) as demonstrated by no obvious signal changes in MR image. The mild acidic microenvironment of cancer cells broke the hydrazone bond to release C_6H_6 and activate the CAs for ^{19}F MR imaging. As compared to ^{1}H MRI, the progress of ^{19}F MRI based on mesoporous materials is at the preliminary stage. The potential of mesoporous materials for ^{19}F MR imaging still requires more research to demonstrate their advantages for diagnosis as compared to other nanosystems and diagnostic imaging modalities.

Mesoporous Silica-Based Composite Materials for MRI-Based Multi-Modality Imaging

In addition to endowing MSNs with MRI performance, their well-defined mesoporous structure and abundant surface chemistry can be further used for integrating more functionalities. Such a unique multi-modality diagnostic imaging can overcome the drawbacks of each single-imaging modality (Ali et al. 2011; Choi et al. 2008; McCarthy and Weissleder 2008). For example, the simultaneous integration of fluorescent quantum dots and magnetic nanoparticles into MSNs could bring with the dual-modality diagnostic imaging performance (Kim et al. 2006). The rational design of the mesoporous silica-based hybrid materials can endow them with contrast-enhanced capability for concurrent ultrasound imaging and MRI. Chen et al. dispersed ultrasmall MnO_x nanoparticles into the mesopores of hollow MSNs (designated as MnO_x-HMSNs) (Chen et al. 2012d). The integrated MnO_x nanoparticles acted as the T_1-weighted MRI CAs. Simultaneously, the unique hollow nanostructure could be used as the CAs for ultrasound imaging, which was demonstrated both *in vitro* and *in vivo*. The DOX-loaded MnO_x-HMSNs efficiently circumvented the multidrug resistance of MCF-7/ADR breast cancer cells.

Furthermore, Chen et al. integrated triple functional moieties into MSNs for near-infrared optical, MR and positron emission tomography (PET) imaging by embedding near-infrared dye ZW800, Gd^{3+} chelates and radionuclide ^{64}Cu into the MSNs matrix (Huang et al. 2012). The nanocomposites showed high stability and long intracellular retention time. More importantly, all three diagnostic imaging modalities exhibited the contrast enhancement of sentinel lymph nodes in 4T1 tumor metastatic mice xenografts up to three weeks where the differences in uptake rate, particle accumulation and contrast enhancement between metastatic and contra-lateral sentinel lymph modes were clearly observed. Lin et al. functionalized MSNs with luminescent $GdVO_4$: Eu^{3+} for concurrent dual fluorescent imaging and T_1-wegited MRI (Huang et al. 2013). We have recently developed a layer-by-layer self-assembly strategy to coat negatively charged fluorescent quantum dots onto the surface of magnetic $Fe_3O_4@SiO_2@mSiO_2$ ellipsoidal nanoparticles for dual fluorescent cell labeling and T_2-weighted MR imaging (Chen et al. 2011b). Fluorescent $NaYF_4$: Yb^{3+}, Er^{3+}/Tm^{3+} layer was also coated onto the surface of core/shell structured $Fe_3O_4@SiO_2@mSiO_2$ for dual-mode diagnostic imaging (Gai et al. 2010). The unique bright up-conversion fluorescent emission was excited by 980 nm laser. In addition, the mesoporous silica shell of core/shell $Fe_3O_4@$ $mSiO_2$ could be labeled with fluorescent dye molecules (e.g., fluorescein isothiocyanate) for simultaneous T_2-weighted MRI and cell labeling (Liong et al. 2008).

The advances in nano-synthetic chemistry promote the generation of diverse multifunctional mesoporous silica-based composite nanosystems for multi-modality imaging. Other imaging modules, such as CT, photoacoustic imaging, etc., are also expected to be integrated into MSNs for some specific diagnostic-imaging purposes. However, most of the multi-modality imaging protocols are still in the conceptual stage, which means that their practical applications are sustainably limited by the lack of adequate integrated apparatus. Their high imaging performance actually provides great potential for future clinical applications for precise and sensitive diagnosis of cancer.

Conclusions and Outlook

The construction of multifunctional theranostic nanosystems is of high importance for efficient cancer treatment. Mesoporous silica-based nanoparticles provide promising platforms for the integration of MRI module to generate composite nanosystems for efficient MR imaging and drug delivery. This book chapter summarizes the developments of synthetic strategies to fabricate diverse mesoporous silica-based composite nanosystems for MR imaging and their extensive biomedical applications in T_1/T_2-weighted and ^{19}F-based MR imaging. The elaborate design of these composite materials for multi-modality imaging has also been included. Although these mesoporous composite nanosystems have showed excellent performance in cancer theranostics, their extensive clinical translations suffer from the following several critical issues to be resolved in the near future.

(1) The biocompatibility of mesoporous silica-based multifunctional nanosystems should be further clearly clarified in detail. Most of the reports concentrate on the imaging performance and therapeutic efficiency of achieved theranostic nanosystems. The preliminary biocompatibility was also simply included. For example, Mou et al. observed no obvious abnormalities in mice after the administration of magnetic composites at a dose of 2 mg Fe/kg body weight (Wu et al. 2008). Chen et al. also showed that rattle-type magnetic $Fe_3O_4@mSiO_2$ nanocapsules were non-toxic against MCF-7 cells even at the high concentration of 400 μg/mL (Chen et al. 2010c). The urinal excretion of Gd-labeled mesoporous silica microparticles was observed (Steinbacher et al. 2010). However, these biocompatibility data are far from the final conclusion that these multifunctional mesoporous nanoparticles are indeed biocompatible and non-toxic. The systematic biosafety evaluations, including biodistribution, excretion, biodegradation, histo/hemo-compatibility, etc., should also be conducted to promote their further clinical translations. It should be noted that the integration of the multiple functions of MSNs may bring with them unique diagnostic or therapeutic performances of the nanosystems. However, the biocompatibility of these composite nanosystems can be reduced because of the complicated

composition. Therefore, the practical design and fabrication of mesoporous silica-based composite nanosystems should be firmly correlated to the specific application requirements in clinic.

(2) Most of the researchers in this field are chemists, material scientists and biologists. Very few researchers from clinical biomedicine and pharmaceutical companies are involved. This phenomenon directly brings with it the results that most of the related researches remain in the research stage instead of progressing into the clinical step. Therefore, the participation of clinical researchers in the promotion of the above-mentioned mesoporous composite nanosystems is highly desirable, as it can provide guidance about which aspects of these nanosystems should be further improved to promote their clinical translation.

(3) There is still a lack of adequate synthetic strategies to fabricate MSNs-based theranostic nanosystems with highly efficient, simple, economic and large-scale features. Multiple steps are generally required for the synthesis, making further industrial translation almost impossible. Based on the advances in nano-synthetic chemistry, it is highly expected that more simple synthetic methodologies will be developed to fabricate mesoporous composite nanosystems with desirable nanostructure, composition and function. In addition, the targeting efficiency of synthesized mesoporous composite nanoparticles should be further enhanced to guarantee high accumulation within tumor tissues. The biodegradation issue of these inorganic nanosystems should also be considered because the biodegradation of inorganic nanoparticles are still in debate for further clinical translation. The biodegradation mechanism, biodegradation-rate control, final biodegradation products, biodegradation-product excretion and biodegradation-product toxicity should be systematically investigated and revealed before clinical experiment.

MSNs are now one of the most promising inorganic nanosystems for biomedical applications. The integration of other multifunctional modules with MSNs can exert some unique theranostic functions. The development of synthetic chemistry will also promote the generation of diverse mesoporous composite nanosystems for MRI-based theranostic applications. Although these are still some critical issues to be resolved in the near future, it is believed that these mesoporous composite nanoparticles will finally enter the clinical stage in order to benefit human health based on their high performance in theranostic nanomedicine.

Acknowledgements

We acknowledge financial support from the National Natural Science Foundation of China (51302293), Natural Science Foundation of Shanghai

(13ZR1463500), Shanghai Rising-Star Program (14QA1404100), Foundation for Youth Scholar of State Key Laboratory of High Performance Ceramics and Superfine Microstructures (Grant No. SKL201203) and Biomedical ECR Grant (BM-2014) of the University of Queensland.

References

Ali, Z., Abbasi, A.Z., Zhang, F., Arosio, P., Lascialfari, A., Casula, M.F., et al. 2011. Multifunctional nanoparticles for dual imaging. Analytical Chemistry 83: 2877-2882.

Allen, T.M. and Cullis, P.R. 2004. Drug delivery systems: entering the mainstream. Science 303: 1818-1822.

Ambrogio, M.W., Thomas, C.R., Zhao, Y.L., Zink, J.I. and Stoddartt, J.F. 2011. Mechanized silica nanoparticles: a new frontier in theranostic nanomedicine. Acc Chem Res 44: 903-913.

Arvizo, R.R., Bhattacharyya, S., Kudgus, R.A., Giri, K., Bhattacharya, R. and Mukherjee, P. 2012. Intrinsic therapeutic applications of noble metal nanoparticles: past, present and future. Chem Soc Rev 41: 2943-2970.

Boehm-Sturm, P., Mengler, L., Wecker, S., Hoehn, M. and Kallur, T. 2011. In vivo tracking of human neural stem cells with F-19 magnetic resonance imaging. Plos One. DOI: 10.1371/journal.pone.0029040.

Boisselier, E. and Astruc, D. 2009. Gold nanoparticles in nanomedicine: preparations, imaging, diagnostics, therapies and toxicity. Chem Soc Rev 38: 1759-1782.

Chen, F., Hong, H., Zhang, Y., Valdovinos, H.F., Shi, S., Kwon, G.S., et al. 2013a. In vivo tumor targeting and image-guided drug delivery with antibody-conjugated, radio labeled mesoporous silica nanoparticles. ACS Nano 7: 9027-9039.

Chen, J.J., Lanza, G.M. and Wickline, S.A. 2010a. Quantitative magnetic resonance fluorine imaging: today and tomorrow. Wiley Interdiscip Rev Nanomed Nanobiotechnol 2: 431-440.

Chen, S., Yang, Y., Li, H., Zhou, X. and Liu, M. 2014a. pH-Triggered Au-fluorescent mesoporous silica nanoparticles for F-19 MR/fluorescent multimodal cancer cellular imaging. Chem Commun 50: 283-285.

Chen, Y., Chen, H.R., Guo, L.M., He, Q.J., Chen, F., Zhou, J., et al. 2010b. Hollow/rattle-type mesoporous nanostructures by a structural Difference-based selective etching strategy. ACS Nano 4: 529-539.

Chen, Y., Chen, H.R., Zeng, D.P., Tian, Y.B., Chen, F., Feng, J.W., et al. 2010c. Core/shell structured hollow mesoporous nanocapsules: a potential platform for simultaneous cell imaging and anticancer drug delivery. ACS Nano 4: 6001-6013.

Chen, Y., Chen, H.R., Sun, Y., Zheng, Y.Y., Zeng, D.P., Li, F.Q., et al. 2011a. Multifunctional mesoporous composite nanocapsules for highly efficient MRI-guided high-intensity focused ultrasound cancer surgery. Angew Chem Int Ed 50: 12505-12509.

Chen, Y., Chen, H.R., Zhang, S.J., Chen, F., Zhang, L.X., Zhang, J.M., et al. 2011b. Multifunctional mesoporous nanoellipsoids for biological bimodal imaging and magnetically targeted delivery of anticancer drugs. Adv Func Mater 21: 270-278.

Chen, Y., Chu, C., Zhou, Y.C., Ru, Y.F., Chen, H.R., Chen, F., et al. 2011c. Reversible pore-structure evolution in hollow silica nanocapsules: large pores for siRNA delivery and nanoparticle collecting. Small 7: 2935-2944.

Chen, Y., Chen, H., Zhang, S., Chen, F., Sun, S., He, Q., et al. 2012a. Structure-property relationships in manganese oxide – mesoporous silica nanoparticles used for T1-weighted MRI and simultaneous anti-cancer drug delivery. Biomaterials 33: 2388-2398.

Chen, Y., Chen, H.R., Zhang, S.J., Chen, F., Sun, S.K., He, Q.J., et al. 2012b. Structure-property relationships in manganese oxide – mesoporous silica nanoparticles used for T-1-weighted MRI and simultaneous anti-cancer drug delivery. Biomaterials 33: 2388-2398.

Chen, Y., Gao, Y., Chen, H.R., Zeng, D.P., Li, Y.P., Zheng, Y.Y., et al. 2012c. Engineering inorganic nanoemulsions/nanoliposomes by fluoride-silica chemistry for efficient delivery/co-delivery of hydrophobic agents. Adv Funct Mater 22: 1586-1597.

Chen, Y., Yin, Q., Ji, X.F., Zhang, S.J., Chen, H.R., Zheng, Y.Y., et al. 2012d. Manganese oxide-based multifunctionalized mesoporous silica nanoparticles for pH-responsive MRI, ultrasonography and circumvention of MDR in cancer cells. Biomaterials 33: 7126-7137.

Chen, Y., Chen, H.R. and Shi, J.L. 2013b. In vivo bio-safety evaluations and diagnostic/therapeutic applications of chemically designed mesoporous silica nanoparticles. Adv Mater 25: 3144-3176.

Chen, Y., Xu, P.F., Chen, H.R., Li, Y.S., Bu, W.B., Shu, Z., et al. 2013c. Colloidal HPMO nanoparticles: silica-etching chemistry tailoring, topological transformation and nano-biomedical applications. Adv Mater 25: 3100-3105.

Chen, Y., Chen, H. and Shi, J. 2014a. Inorganic nanoparticle-based drug codelivery nanosystems to overcome the multidrug resistance of cancer cells. Mol Pharm11: 2495-2510.

Chen, Y., Chen, H.R. and Shi, J.L. 2014b. Construction of homogenous/heterogeneous hollow mesoporous silica nanostructures by silica-etching chemistry: principles, synthesis and applications. Acc Chem Res 47: 125-137.

Chen, Y., Jiang, L., Wang, R., Lu, M., Zhang, Q., Zhou, Y., et al. 2014c. Injectable smart phase-transformation implants for highly efficient in vivo magnetic-hyperthermia regression of tumors. Adv Mater 26: 7468-7473.

Chen, Y., Meng, Q., Wu, M., Wang, S., Xu, P., Chen, H., et al. 2014d. Hollow mesoporous organosilica nanoparticles: a generic intelligent framework-hybridization approach for biomedicine. J Am Chem Soc 136: 16326-16334.

Chen, Y., Xu, P., Shu, Z., Wu, M., Wang, L., Zhang, S., et al. 2014e. Multifunctional graphene oxide-based triple stimuli-responsive nanotheranostics. Adv Funct Mater 24: 4386-4396.

Chen, Y., Xu, P., Wu, M., Meng, Q., Chen, H., Shu, Z., et al. 2014f. Colloidal RBC-shaped, hydrophilic and hollow mesoporous carbon nanocapsules for highly efficient biomedical engineering. Adv Mater 26: 4294-4301.

Chen, Y., Ye, D., Wu, M., Chen, H., Zhang, L., Shi, J., et al. 2014g. Break-up of two-dimensional MnO2 nanosheets promotes ultrasensitive pH-triggered theranostics of cancer. Adv Mater 26: 7019-7026.

Chen, Y., Tan, C., Zhang, H. and Wang, L. 2015. Two-dimensional graphene analogues for biomedical applications. Chem Soc Rev 44: 2681-2701.

Choi, J.S., Park, J.C., Nah, H., Woo, S., Oh, J., Kim, K.M., et al. 2008. A hybrid nanoparticle probe for dual-modality positron emission tomography and magnetic resonance imaging. Angew Chem Int Ed 47: 6259-6262.

Chung, C., Kim, Y.-K., Shin, D., Ryoo, S.-R., Hong, B.H. and Min, D.-H. 2013. Biomedical applications of graphene and graphene oxide. Acc Chem Res 46: 2211-2224.

Daniel, M.C. and Astruc, D. 2004. Gold nanoparticles: assembly, supramolecular chemistry, quantum-size-related properties and applications toward biology, catalysis and nanotechnology. Chem Rev 104: 293-346.

Duncan, A.K., Klemm, P.J., Raymond, K.N. and Landry, C.C. 2012. Silica microparticles as a solid support for gadolinium phosphonate magnetic resonance imaging contrast agents. J Am Chem Soc 134: 8046-8049.

Ebner, B., Behm, P., Jacoby, C., Burghoff, S., French, B.A., Schrader, J., et al. 2010. Early assessment of pulmonary inflammation by F-19 MRI in vivo. Circ Cardiovasc Imaging 3: 202-U109.

Farokhzad, O.C. and Langer, R. 2009. Impact of nanotechnology on drug delivery. ACS Nano 3: 16-20.

Gai, S., Yang, P., Li, C., Wang, W., Dai, Y., Niu, N., et al. 2010. Synthesis of magnetic, up-conversion luminescent and mesoporous core-shell-structured nanocomposites as drug carriers. Adv Funct Mater 20: 1166-1172.

Gao, Y., Chen, Y., Ji, X.F., He, X.Y., Yin, Q., Zhang, Z.W., et al. 2011. Controlled intracellular release of doxorubicin in multidrug-resistant cancer cells by tuning the shell-pore sizes of mesoporous silica nanoparticles. ACS Nano 5: 9788-9798.

Hartono, S.B., Gu, W.Y., Kleitz, F., Liu, J., He, L.Z., Middelberg, A.P.J., et al. 2012. Poly-L-lysine functionalized large pore cubic mesostructured silica nanoparticles as biocompatible carriers for gene delivery. ACS Nano 6: 2104-2117.

Hsiao, J.K., Tsai, C.P., Chung, T.H., Hung, Y., Yao, M., Liu, H.M., et al. 2008. Mesoporous silica nanoparticles as a delivery system of gadolinium for effective human stem cell tracking. Small 4: 1445-1452.

Hu, H., Zhou, H., Du, J., Wang, Z.Q., An, L., Yang, H., et al. 2011. Biocompatiable hollow silica microspheres as novel ultrasound contrast agents for in vivo imaging. J Mater Chem 21: 6576-6583.

Huang, C.-C., Khu, N.-H. and Yeh, C.-S. 2010. The characteristics of sub 10 nm manganese oxide T-1 contrast agents of different nanostructured morphologies. Biomaterials 31: 4073-4078.

Huang, C.C., Tsai, C.Y., Sheu, H.S., Chuang, K.Y., Su, C.H., Jeng, U.S., et al. 2011. Enhancing transversal relaxation for magnetite nanoparticles in MR imaging using Gd3+-chelated mesoporous silica shells. ACS Nano 5: 3905-3916.

Huang, S.S., Cheng, Z.Y., Ma, P.A., Kang, X.J., Dai, Y.L. and Lin, J. 2013. Luminescent GdVO4:Eu3+ functionalized mesoporous silica nanoparticles for magnetic resonance imaging and drug delivery. Dalton Trans 42: 6523-6530.

Huang, X.L., Zhang, F., Lee, S., Swierczewska, M., Kiesewetter, D.O., Lang, L.X., et al. 2012. Long-term multimodal imaging of tumor draining sentinel lymph nodes using mesoporous silica-based nanoprobes. Biomaterials 33: 4370-4378.

Jokerst, J.V., Thangaraj, M., Kempen, P.J., Sinclair, R. and Gambhir, S.S. 2012. Photoacoustic imaging of mesenchymal stem cells in living mice via silica-coated gold nanorods. ACS Nano 6: 5920-5930.

Kim, B.H., Lee, N., Kim, H., An, K., Park, Y.I., Choi, Y., et al. 2011a. Large-scale synthesis of uniform and extremely small-sized iron oxide nanoparticles for high-resolution T-1 magnetic resonance imaging contrast agents. J Am Chem Soc133: 12624-12631.

Kim, J., Lee, J.E., Lee, J., Yu, J.H., Kim, B.C., An, K., et al. 2006. Magnetic fluorescent delivery vehicle using uniform mesoporous silica spheres embedded with monodisperse magnetic and semiconductor nanocrystals. J Am Chem Soc 128: 688-689.

Kim, J., Kim, H.S., Lee, N., Kim, T., Kim, H., Yu, T., et al. 2008. Multifunctional uniform nanoparticles composed of a magnetite nanocrystal core and a mesoporous silica shell for magnetic resonance and fluorescence imaging and for drug delivery. Angew Chem Int Ed 47: 8438-8441.

Kim, T., Momin, E., Choi, J., Yuan, K., Zaidi, H., Kim, J., et al. 2011b. Mesoporous silica-coated hollow manganese oxide nanoparticles as positive T(1) contrast agents for labeling and MRI tracking of adipose-derived mesenchyrnal stem cells. J Am Chem Soc 133: 2955-2961.

Kostarelos, K., Bianco, A. and Prato, M. 2009. Promises, facts and challenges for carbon nanotubes in imaging and therapeutics. Nat Nanotech 4: 627-633.

Lal, S., Clare, S.E. and Halas, N.J. 2008. Nanoshell-enabled photothermal cancer therapy: impending clinical impact. Acc Chem Res 41: 1842-1851.

Lalwani, G., Henslee, A.M., Farshid, B., Lin, L.J., Kasper, F.K., Qin, Y.X., et al. 2013. Two-dimensional nanostructure-reinforced biodegradable polymeric nanocomposites for bone tissue engineering. Biomacromolecules 14: 900-909.

Lammers, T., Aime, S., Hennink, W.E., Storm, G. and Kiessling, F. 2011. Theranostic nanomedicine. Acc Chem Res 44: 1029-1038.

Laurent, S., Forge, D., Port, M., Roch, A., Robic, C., Elst, L.V., et al. 2008. Magnetic iron oxide nanoparticles: synthesis, stabilization, vectorization, physicochemical characterizations and biological applications. Chem Rev 108: 2064-2110.

LaVan, D.A., McGuire, T. and Langer, R. 2003. Small-scale systems for in vivo drug delivery. Nat Biotech 21: 1184-1191.

Lee, C.H., Cheng, S.H., Wang, Y.J., Chen, Y.C., Chen, N.T., Souris, J., et al. 2009. Near-infrared mesoporous silica nanoparticles for optical imaging: characterization and in vivo biodistribution. Adv Funct Mater 19: 215-222.

Lee, J.E., Lee, N., Kim, H., Kim, J., Choi, S.H., Kim, J.H., et al. 2010. Uniform mesoporous dye-doped silica nanoparticles decorated with multiple magnetite nanocrystals for simultaneous enhanced magnetic resonance imaging, fluorescence imaging and drug delivery. J Am Chem Soc 132: 552-557.

Lee, J.E., Lee, N., Kim, T., Kim, J. and Hyeon, T. 2011. Multifunctional mesoporous silica nanocomposite nanoparticles for theranostic applications. Acc Chem Res 44: 893-902.

Li, S.A., Liu, H.A., Li, L., Luo, N.Q., Cao, R.H., Chen, D.H., et al. 2011. Mesoporous silica nanoparticles encapsulating Gd2O3 as a highly efficient magnetic resonance imaging contrast agent. Appl Phys Lett 98: 093704.

Li, X., Shi, J.L., Dong, X.P., Zhang, L.X. and Zeng, H.Y. 2008. A mesoporous bioactive glass/polycaprolactone composite scaffold and its bioactivity behavior. J Biomed Mater Res A 84A: 84-91.

Li, Z., Yi, P.W., Sun, Q., Lei, H., Zhao, H.L., Zhu, Z.H., et al. 2012a. Ultrasmall water-soluble and biocompatible magnetic iron oxide nanoparticles as positive and negative dual contrast agents. Adv Func Mater 22: 2387-2393.

Li, Z.X., Barnes, J.C., Bosoy, A., Stoddart, J.F. and Zink, J.I. 2012b. Mesoporous silica nanoparticles in biomedical applications. Chem Soc Rev 41: 2590-2605.

Liong, M., Lu, J., Kovochich, M., Xia, T., Ruehm, S.G., Nel, A.E., et al. 2008. Multifunctional inorganic nanoparticles for imaging, targeting and drug delivery. ACS Nano 2: 889-896.

Luo, T., Huang, P., Gao, G., Shen, G., Fu, S., Cui, D., et al. 2011. Mesoporous silica-coated gold nanorods with embedded indocyanine green for dual mode X-ray CT and NIR fluorescence imaging. Opt Express 19: 17030-17039.

Mamaeva, V., Sahlgren, C. and Linden, M. 2013. Mesoporous silica nanoparticles in medicine-recent advances. Adv Drug Deliv Rev 65: 689-702.

Martinez, H.P., Kono, Y., Blair, S.L., Sandoval, S., Wang-Rodriguez, J., Mattrey, R.F., et al. 2010. Hard shell gas-filled contrast enhancement particles for colour Doppler ultrasound imaging of tumors. Medchemcomm 1: 266-270.

McCarthy, J.R. and Weissleder, R. 2008. Multifunctional magnetic nanoparticles for targeted imaging and therapy. Adv Drug Deliv Rev 60: 1241-1251.

Na, H.B., Song, I.C. and Hyeon, T. 2009. Inorganic Nanoparticles for MRI Contrast Agents. Adv Mater 21: 2133-2148.

Nakamura, T., Sugihara, F., Matsushita, H., Yoshioka, Y., Mizukami, S. and Kikuchi, K. 2015. Mesoporous silica nanoparticles for 19F magnetic resonance imaging, fluorescence imaging and drug delivery. Chem Sci 6: 1986-1990.

Ninjbadgar, T. and Brougham, D.F. 2011. Epoxy ring opening phase transfer as a general route to water dispersible superparamagnetic Fe3O4 nanoparticles and their application as positive MRI contrast agents. Adv Funct Mater 21: 4769-4775.

Pan, L.M., He, Q.J., Liu, J.N., Chen, Y., Ma, M., Zhang, L.L., et al. 2012. Nuclear-targeted drug delivery of TAT peptide-conjugated monodisperse mesoporous silica nanoparticles. J Am Chem Soc 134: 5722-5725.

Park, K., Lee, S., Kang, E., Kim, K., Choi, K. and Kwon, I.C. 2009. New generation of multifunctional nanoparticles for cancer imaging and therapy. Adv Funct Mater 19: 1553-1566.

Patra, C.R., Bhattacharya, R., Mukhopadhyay, D. and Mukherjee, P. 2010. Fabrication of gold nanoparticles for targeted therapy in pancreatic cancer. Adv Drug Deliv Rev 62: 346-361.

Penfield, J.G. and Reilly, R.F. 2007. What nephrologists need to know about gadolinium. Nat Clin Pract Nephrol 3: 654-668.

Peng, Y.-K., Lai, C.-W., Liu, C.-L., Chen, H.-C., Hsiao, Y.-H., Liu, W.-L., et al. 2011. A new and facile method to prepare uniform hollow MnO/functionalized mSiO(2) core/shell nanocomposites. ACS Nano 5: 4177-4187.

Perez-Rodriguez, J., Lai, S., Ehst, B.D., Fine, D.M. and Bluemke, D.A. 2009. Nephrogenic systemic fibrosis: incidence, associations and effect of risk factor assessment-report of 33 cases. Radiology 250: 371-377.

Piao, Y., Kim, J., Bin Na, H., Kim, D., Baek, J.S., Ko, M.K., et al. 2008. Wrap-bake-peel process for nanostructural transformation from beta-FeOOH nanorods to biocompatible iron oxide nanocapsules. Nat Mater 7: 242-247.

Riehemann, K., Schneider, S.W., Luger, T.A., Godin, B., Ferrari, M. and Fuchs, H. 2009. Nanomedicine-challenge and perspectives. Angew Chem Int Ed 48: 872-897.

Schwert, D.D., Davies, J.A. and Richardson, N. 2002. Non-gadolinium-based MRI contrast agents. Contrast Agents I 221: 165-199.

Shao, Y.Z., Tian, X.M., Hu, W.Y., Zhang, Y.Y., Liu, H., He, H.Q., et al. 2012. The properties of Gd2O3-assembled silica nanocomposite targeted nanoprobes and their application in MRI. Biomaterials 33: 6438-6446.

Shi, J.L., Chen, Y. and Chen, H.R. 2013. Progress on the multifunctional mesoporous silica-based nanotheranostics. J Inorg Mater 28: 1-11.

Shin, J.M., Anisur, R.M., Ko, M.K., Im, G.H., Lee, J.H. and Lee, I.S. 2009. Hollow manganese oxide nanoparticles as multifunctional agents for magnetic resonance imaging and drug delivery. Angew Chem Int Ed 48: 321-324.

Skrabalak, S.E., Chen, J.Y., Sun, Y.G., Lu, X.M., Au, L., Cobley, C.M., et al. 2008. Gold nanocages: synthesis, properties and applications. Acc Chem Res 41: 1587-1595.

Slowing, II, Vivero-Escoto, J.L., Wu, C.W. and Lin, V.S.Y. 2008. Mesoporous silica nanoparticles as controlled release drug delivery and gene transfection carriers. Adv Drug Deliv Rev 60: 1278-1288.

Srinivas, M., Boehm-Sturm, P., Aswendt, M., Pracht, E.D., Figdor, C.G., de Vries, I.J., et al. 2013. In vivo 19F MRI for cell tracking. J Vis Exp (81): e50802. doi: 10.3791/50802.

Steinbacher, J.L., Lathrop, S.A., Cheng, K., Hillegass, J.M., Butnor, K., Kauppinen, R.A., et al. 2010. Gd-labeled microparticles in MRI: in vivo imaging of microparticles after intraperitoneal injection. Small 6: 2678-2682.

Sun, C., Lee, J.S.H. and Zhang, M.Q. 2008. Magnetic nanoparticles in MR imaging and drug delivery. Adv Drug Deliv Rev 60: 1252-1265.

Taboada, E., Rodriguez, E., Roig, A., Oro, J., Roch, A. and Muller, R.N. 2007. Relaxometric and magnetic characterization of ultrasmall iron oxide nanoparticles with high magnetization. Evaluation as potential T-1 magnetic resonance imaging contrast agents for molecular imaging. Langmuir 23: 4583-4588.

Tarn, D., Ashley, C.E., Xue, M., Carnes, E.C., Zink, J.I. and Brinker, C.J. 2013. Mesoporous silica nanoparticle nanocarriers: biofunctionality and biocompatibility. Acc Chem Res 46: 792-801.

Taylor, K.M.L., Kim, J.S., Rieter, W.J., An, H., Lin, W.L. and Lin, W.B. 2008. Mesoporous silica nanospheres as highly efficient MRI contrast agents. J Am Chem Soc 130: 2154-2155.

Terreno, E., Castelli, D.D., Viale, A. and Aime, S. 2010. Challenges for molecular magnetic resonance imaging. Chem Rev 110: 3019-3042.

Tian, B., Wang, C., Zhang, S., Feng, L.Z. and Liu, Z. 2011a. Photothermally enhanced photodynamic therapy delivered by nano-graphene oxide. ACS Nano 5: 7000-7009.

Tian, Q., Tang, M., Sun, Y., Zou, R., Chen, Z., Zhu, M., et al. 2011b. Hydrophilic flower-like CuS superstructures as an efficient 980 nm laser-driven photothermal agent for ablation of cancer cells. Adv Mater 23: 3542-3547.

Torchilin, V.P. 2005. Recent advances with liposomes as pharmaceutical carriers. Nat Rev Drug Discov 4: 145-160.

Tromsdorf, U.I., Bruns, O.T., Salmen, S.C., Beisiegel, U. and Weller, H. 2009. A highly effective, nontoxic T-1 MR contrast agent based on ultrasmall PEGylated iron oxide nanoparticles. Nano Lett 9: 4434-4440.

Vallet-Regi, M. 2006. Ordered mesoporous materials in the context of drug delivery systems and bone tissue engineering. Chem Eur J 12: 5934-5943.

Viswanathan, S., Kovacs, Z., Green, K.N., Ratnakar, S.J. and Sherry, A.D. 2010. Alternatives to gadolinium-based metal chelates for magnetic resonance imaging. Chem Rev 110: 2960-3018.

Vivero-Escoto, J.L., Taylor-Pashow, K.M.L., Huxford, R.C., Della Rocca, J., Okoruwa, C., An, H.Y., et al. 2011. Multifunctional mesoporous silica nanospheres with cleavable Gd(III) chelates as MRI contrast agents: synthesis, characterization, target-specificity and renal clearance. Small 7: 3519-3528.

Wagner, V., Dullaart, A., Bock, A.K. and Zweck, A. 2006. The emerging nanomedicine landscape. Nat Biotech 24: 1211-1217.

Wang, X., Chen, H.R., Chen, Y., Ma, M., Zhang, K., Li, F.Q., et al. 2012. Perfluorohexane-encapsulated mesoporous silica nanocapsules as enhancement agents for highly efficient high intensity focused ultrasound (HIFU). Adv Mater 24: 785-791.

Wang, X., Chen, H., Zhang, K., Ma, M., Li, F., Zeng, D., et al. 2013. An intelligent nanotheranostic agent for targeting, redox-responsive ultrasound imaging and imaging-guided high-intensity focused ultrasound synergistic therapy. Small 10: 1403-1411.

Wang, Y. and Gu, H. 2015. Core–shell-type magnetic mesoporous silica nanocomposites for bioimaging and therapeutic agent delivery. Adv Mater 27: 576-585.

Wu, C., Ramaswamy, Y., Zhu, Y.F., Zheng, R., Appleyard, R., Howard, A., et al. 2009. The effect of mesoporous bioactive glass on the physiochemical, biological and drug-release properties of poly(DL-lactide-co-glycolide) films. Biomaterials 30: 2199-2208.

Wu, M., Meng, Q., Chen, Y., Xu, P., Zhang, S., Li, Y., et al. 2014. Ultrasmall confined iron oxide nanoparticle MSNs as a pH-responsive theranostic platform. Adv Funct Mater 24: 4273-4283.

Wu, M., Meng, Q., Chen, Y., Du, Y., Zhang, L., Li, Y., et al. 2015. Large-pore ultrasmall mesoporous organosilica nanoparticles: micelle/precursor co-templating assembly and nuclear-targeted gene delivery. Adv Mater 27: 215-222.

Wu, S.H., Lin, Y.S., Hung, Y., Chou, Y.H., Hsu, Y.H., Chang, C., et al. 2008. Multifunctional mesoporous silica nanoparticles for intracellular labeling and animal magnetic resonance imaging studies. Chembiochem 9: 53-57.

Xia, T.A., Kovochich, M., Liong, M., Meng, H., Kabehie, S., George, S., et al. 2009. Polyethyleneimine coating enhances the cellular uptake of mesoporous silica nanoparticles and allows safe delivery of siRNA and DNA constructs. ACS Nano 3: 3273-3286.

Yang, P.P., Gai, S.L. and Lin, J. 2012. Functionalized mesoporous silica materials for controlled drug delivery. Chem Soc Rev 41: 3679-3698.

Zeng, L., Ren, W., Zheng, J., Cui, P. and Wu, A. 2012. Ultrasmall water-soluble metal-iron oxide nanoparticles as T-1-weighted contrast agents for magnetic resonance imaging. Phys Chem Chem Phys 14: 2631-2636.

Zhao, W.R., Gu, J.L., Zhang, L.X., Chen, H.R. and Shi, J.L. 2005. Fabrication of uniform magnetic nanocomposite spheres with a magnetic core/mesoporous silica shell structure. J Am Chem Soc 127: 8916-8917.

Zhou, M., Zhang, R., Huang, M.A., Lu, W., Song, S.L., Melancon, M.P., et al. 2010. A chelator-free multifunctional Cu-64 CuS nanoparticle platform for simultaneous micro-PET/CT imaging and photothermal ablation therapy. J Am Chem Soc 132: 15351-15358.

Zhu, M., Wang, H.X., Liu, J.Y., He, H.L., Hua, X.G., He, Q.J., et al. 2011. A mesoporous silica nanoparticulate/beta-TCP/BG composite drug delivery system for osteo-articular tuberculosis therapy. Biomaterials 32: 1986-1995.

Zhu, Y., Ikoma, T., Hanagata, N. and Kaskel, S. 2010. Rattle-type Fe3O4@SiO2 hollow mesoporous spheres as carriers for drug delivery. Small 6: 471-478.

Zhu, Y.F., Shi, J.L., Shen, W.H., Dong, X.P., Feng, J.W., Ruan, M.L., et al. 2005. Stimuli-responsive controlled drug release from a hollow mesoporous silica sphere/polyelectrolyte multilayer core-shell structure. Angew Chem Int Ed 44: 5083-5087.

Zhu, Y.F., Kockrick, E., Ikoma, T., Hanagata, N. and Kaskel, S. 2009. An efficient route to rattle-type Fe3O4@SiO2 hollow mesoporous spheres using colloidal carbon spheres templates. Chem Mater 21: 2547-2553.

16

Recent Advances in The Engineering of Silica-Based Core@Shell Structured Hybrid Nanoparticles

Feng Chen[1,*,#] and *Weibo Cai*[1,2,3,4,*,#]

INTRODUCTION

Coating an existing functional nanoparticle with well-selected shell(s) is an easy and effective way for building hybrid nanomaterials (Schartl 2010). Depending on the final functionality of as-designed nanomaterials, the cores can be magnetic nanocrystals (such as superparamagnetic iron oxide nanoparticle [SPION]), photothermal sensitive nanoparticles (such as gold nanorods [AuNR], copper sulfide [CuS] nanoparticles), optical nanoparticles (such as quantum dots [QDs], upconversion nanoparticle [UCNP]), to name a few (Guerrero-Martinez et al. 2010; Huang and Davis 2011; Li et al. 2009; Liu et al. 2015a; Piao et al. 2008; Reiss et al. 2009). To date, silica-shell-coating strategy remains one of the most used, economical and practical techniques for the design and synthesis of hybrid nanomaterials (Piao et al. 2008). Since the last decade, the coating technique has evolved greatly with the shell being no longer limited to dense silica ($dSiO_2$). Well-established methods have also been reported for coating inorganic nanocrystals with thickness controllable mesoporous silica and hollow mesoporous silica shells (Chen et al. 2013b; Liu et al. 2015a; Tang et al. 2012; Wu et al. 2013; Yang et al. 2012). According to the literature (Liu et al. 2015a; Piao et al. 2008; Shi et al. 2013; Yang et al. 2012), the advantages of silica coating include: 1) silica is generally

[1] Department of Radiology, University of Wisconsin – Madison, WI, USA.
[2] Department of Medical Physics, University of Wisconsin – Madison, WI, USA.
[3] Materials Science Program, University of Wisconsin – Madison, WI, USA.
[4] University of Wisconsin Carbone Cancer Center, Madison, WI, USA.
[#] Address for Correspondence: Departments of Radiology and Medical Physics, University of Wisconsin – Madison, Room 7137, 1111 Highland Avenue, Madison, WI 53705-2275, USA.
[*] Corresponding authors: chenf@mskcc.org and wcai@uwhealth.org

recognized as safe by the U.S. Food and Drug Administration (FDA) and is highly biocompatible; 2) the coating process can be easily regulated, 3) it provides controllable porosity and optical transparency, 4) it protects the cores from harsh conditions, 5) it has well-established surface modification chemistry, 6) it is relatively easy for large-scale production and others.

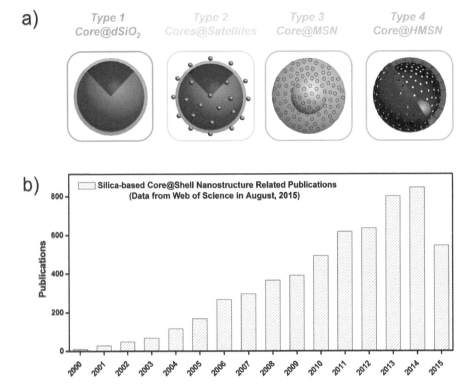

Fig. 16.1 (a) Four types of silica-based core@shell structured hybrid nanomaterials. (b) Annual publication of silica-based core@shell structured nanomaterials (Data from Web of Science in August, 2015). Search keywords were silica, core@shell, nanoparticles.

A quick search in the Web of Science by using the keywords; "silica", "core@shell" and "nanoparticles" reveals an exponential growth in publications starting from the year 2000 (Fig. 16.1). Generally, there are four types of silica-based core@shell structured hybrid nanomaterials (Fig. 16.1a). They are:

- **Type 1: Core@dSiO$_2$**, where the core can be dye-doped silica matrix or functional nanocrystals (such as QDs, Au, UCNP, SPION, etc.) and the shell is dense silica with no pores;

- **Type 2: Cores@Satellites**, where the cores can be pure dense silica or Core@dSiO$_2$ and the satellites can be ultrasmall functional nanoparticles such as Au, CuS, or others;

- **Type 3: Core@MSN**, where MSN stands for mesoporous silica nanoshell. The core can be functional nanocrystals, while the shell is thickness controllable MSN with high surface area and larger pore volume for anti-cancer drug loading and releasing;

- **Type 4: Core@HMSN**, where HMSN stands for hollow mesoporous silica nanoshell. Different from Type 3 hybrid nanomaterials, the core is a big cavity with any type of functional nanoparticles and the shell is only a thin layer of MSN with pores which function as the diffusion channels for anti-cancer drugs.

In this chapter, we intend to present the most recent advances in the engineering of silica-based core@shell structured hybrid nanomaterials for biomedical imaging and therapy.

Translational Research of Dye-Doped Silica Hybrid Nanoparticles (or C Dots)(Type 1)

Among the numerous silica-based core@shell structured nanoparticles, ultrasmall fluorescent dye-doped silica hybrid nanoparticles (also known as C dots) are perhaps one of the most promising types of nanoparticles which have already shown great potential in future tumor-targeted cancer detection and image-guided surgery. C dots are so far the only renal clearable inorganic nanoparticles that have received FDA Investigational New Drug (IND) approval for first-in-human clinical trials (Benezra et al. 2011; Phillips et al. 2014).

Developed in 2005 by Wiesner and co-workers in Cornell University, the first generation of these dye-doped silica nanoparticles had a physical size in the range of 20-30 nm (Fig. 16.2b) and were not perfected for renal clearance (Ow et al. 2005). Although encapsulating organic dyes into the silica matrix through covalent binding was reported back in the year 1992 (Van Blaaderen and Vrij 1992), previously reported dye-doped silica nanoparticles were usually in the hundreds of nanometers size range. The engineering of ultrasmall sized dye-doped silica nanoparticles was still one of the major challenges during that time.

In Wiesner's study, the dye encapsulation process was achieved based on a modified Stöber method, where organic dyes were first covalently attached to a silica precursor to form the adduct of the core materials, followed by co-condensation with the sol-gel precursor to form a protective shell (Figs. 16.2a, b) (Ow et al. 2005). By using tetramethylrhodamine isothiocyanate (TRITC) as the encapsulated organic dye, the authors demonstrated a 30-fold brightness enhancement in core@shell structured nanoparticle (when compared with free TRITC). Although further investigation is needed to explain such enhancement, the protection of dyes from the solvent after shell coating was believed to be one of the possible contributions. One of the other advantages of fluorescent dye-doped silica hybrid nanoparticles is that the synthetic protocol can be extended to incorporate organic dyes with different spectral characteristics, covering the entire UV-Vis absorption and emission wavelengths, as shown in Fig. 16.2c, making them highly useful tools for multiplex imaging studies.

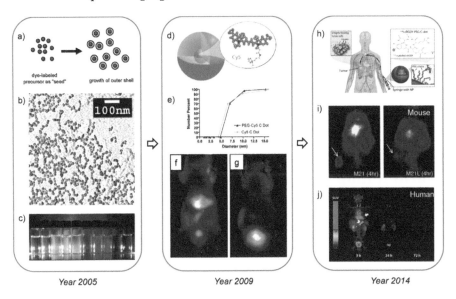

Year 2005   Year 2009   Year 2014

Fig. 16.2 (a) A schematic illustration showing the synthesis of fluorescent core@shell silica nanoparticles. (b) A TEM image of fluorescent core@shell silica nanoparticles. (c) A digital photo showing the fluorescent core@shell silica nanoparticles which were incorporated with different organic dyes, covering the entire UV-Vis absorption and emission wavelengths. (d) A schematic illustration of Cy5 incorporated into the core of silica nanoparticle. (e) DLS plot of particle size for bare Cy5-C dots (gray) and PEG-coated Cy5-C dots (black). (f) *In vivo* imaging of bare C dots (45 min post-injection), showing particle accumulation in the liver and bladder. (g) *In vivo* imaging of PEGylated C dots (45 min post-injection), showing particle accumulation in the bladder. (h) A schematic showing tumor active targeting of the hybrid (positron emission tomography [PET]-optical) imaging nanoparticle, i.e., [124]I-cRGDY–PEG–C dots. (i) *In vivo* PET imaging of [124]I-cRGDY–PEG–C dots in M21 (left, $\alpha_v\beta_3$ positive) and M21L (right, $\alpha_v\beta_3$ negative) mouse models. (j) Maximum intensity projection PET images at 3, 24 and 72 hours after intravenous injection of [124]I-cRGDY–PEG–C dots in a human patient, revealing probe activity in bladder, heart and bowel. Reproduced with permissions from (Benezra et al. 2011; Burns et al. 2009; Ow et al. 2005; Phillips et al. 2014).

To demonstrate their potential for clinical translation, great efforts have been devoted towards the engineering of fluorescent silica nanoparticles with efficient urinary excretion in a follow-up study (Burns et al. 2009). Two major upgrades were achieved in the second generation of C dots. Firstly, the hydrodynamic (HD) size was tunable down to ~3 nm (which was smaller than the renal clearance cutoff: 5.5 nm (Soo Choi et al. 2007) (Figs. 16.2d, e).

Secondly, nanoparticles were covalently coated with methoxy-terminated poly (ethylene glycol) chains (PEG, ~0.5 kDa) to prevent opsonization and to further enhance particle clearance (while maintaining ultrasmall HD size). As shown in Fig. 16.2f, fast and high liver uptake was observed at 45 min post-injection of non-PEGylated C dots, indicating the opsonization of C dots by serum proteins (which effectively increased the HD size and prevented renal excretion). Significantly, decreased liver uptake and dominant bladder uptake were observed after the injection of PEGylated C dots (Fig. 16.2g).

By using Cy5 as the organic dyes, the authors further compared the whole body biodistribution and clearance rate of PEGylated C dots with two different HD sizes (i.e., 3.3 and 6.0 nm). As expected, the smaller sized C dots (HD = 3.3 nm) exhibited a shorter blood circulation time ($t_{1/2}$ = 190 min) when compared with the larger ones (HD = 6.0 nm, $t_{1/2}$ = 350 min). The estimated total excreted fractions after 48 h post-injection were found to be 73% ID and 64% ID for 3.3 and 6.0 nm sized C dots, respectively (Burns et al. 2009). This was the first and the only type of silica-based nanoparticle which showed renal clearable capability, paving the way for its coming clinical translation studies.

An important benchmark for any substance undergoing the FDA IND approval is that the particles must be safe for human use and must leave no trace after renal excretion. Since silica (or silicon dioxide) is "generally recognized as safe" by the FDA (ID Code: 14808-60-7) (http://www.fda.gov/Food/Ingredients Packaging Labeling/GRAS/SCOGS/ucm261095.html) and since PEGylated C dots have demonstrated their efficient renal clearance capability in small animals (Burns et al. 2009), in 2010 FDA approved the first-in-human clinical trials of C dots, which was considered to be a major milestone in their bench-to-bedside journey.

Before injecting C dots into human patients, a systematic pre-clinical study (in both small and large animals) was carried out in order to better understand the toxicity, pharmacokinetics, tumor active targeting and clearance of functionalized C dots (Benezra et al. 2011). Nanoparticles were first conjugated with cyclic arginine–glycine–aspartic acid-tyrosine (cRGDY) peptide ligands and labeled with iodine-124 (^{124}I, $t_{1/2}$ = 4.2 d) for targeting integrin $\alpha_v\beta_3$ and *in vivo* PET imaging. The final HD size was well-controlled to be around 7 nm with about 6-7 cRGDY ligands in each nanoparticle. Systematic *in vitro* receptor-binding studies showed high and specific binding affinity of ^{124}I-cRGDY-PEG-C dots (when compared with native controls of ^{124}I-PEG-C dots and ^{124}I-cRADY-PEG-C dots). Further investigation showed a multivalent interaction between cRGDY-PEG-C dots and integrin $\alpha_v\beta_3$ receptors.

In vivo tumor targeting, biodistribution and clearance studies were then performed in M21 tumor xenograft models. When compared with small peptides, which usually have less than 10 min blood circulation half-life, ^{124}I-cRGDY-PEG-C dots showed a nearly 6 h half-life *in vivo*. Efficient renal excretion was found in both targeted and non-targeted groups, with nearly 50% ID excreted within 24 h and over 70% ID by 96 h. Whole body PET imaging revealed a 3-fold higher nanoparticle uptake in M21 tumors in the targeted group than the non-targeted group (in M21L tumor mice) (Fig. 16.2i). The highest tumor uptake was found to be ~2% ID/g. Blocking study further confirmed the specific targeting of ^{124}I-cRGDY-PEG-C dots in M21 tumors. *In vivo* optical imaging was also performed to show their potential in local and regional lymph node mapping.

Locally injected ^{124}I-cRGDY-PEG-C dot tracer in larger-animal spontaneous melanoma miniswine model was later conducted to assess the feasibility of performing real-time, intraoperative image-guided metastatic disease detection and staging. Results showed that ^{124}I-cRGDY-PEG-C dot tracer was able to confirm the ^{18}F-FDG imaging findings and even managed to identify 2 additional hypermetabolic nodes using dynamic PET scanning, clearly indicating the advantages of using ^{124}I-cRGDY-PEG-C dot for a more accurate early metastatic disease detection in the future (Benezra et al. 2011).

In a follow-up study, *in vivo* safety, pharmacokinetics, clearance properties and radiation dosimetry of ^{124}I-cRGDY–PEG–C dots were further assessed by using serial PET and CT imaging after intravenous administration of a single-injection particle tracer dose in a total of five human patients with metastatic melanoma (Fig. 16.2h) (Phillips et al. 2014). Systematic investigation showed relatively low tissue activities (in a range of zero to < 0.05% ID/g) in most of the patients. A significantly higher activity accumulation (~0.02 to 0.04% ID/g) was measured within the thyroid gland in one of the patients. Whole body tracer clearance half-life was estimated to range from 13 to 21 h with no notable accumulation observed in the reticuloendothelial system (RES) (Fig. 16.2j). Radio-TLC analyses of plasma samples matched with the intact radiolabeled nanoparticle before injection, indicating that no substantial loss of radiolabel occurred during the study. Radiation dosimetry study further showed an average effective dose of 0.183 ± 0.065 mSv/MBq, which was comparable to the estimated values from their previous preclinical data (~0.157 mSv/MBq) (Benezra et al. 2011).

Although the tracer was not yet optimized for active tumor targeting, lesion uptake and localization were seen in several patients. Specific tracer localization in liver metastasis was observed in patient #1. Multimodality (PET/MRI) imaging showed the accumulation of ^{124}I-cRGDY–PEG–C dots in a pituitary lesion in patient #2. The same tracer also showed high and retained uptake in both renal cortices over several days in patient #4, who had a history of impaired renal function and mildly elevated creatinine levels due to the chemotherapy (cisplatin)–related nephrotoxicity. Renal function assessment study showed no substantial change in average blood urea nitrogen and creatinine concentrations over the 2-week study interval, indicating that renal function was unaffected by the particle tracer. No changes in liver function were found either. Although no *ex vivo* histology data could be provided due to the biopsy risk, these first-in-human results clearly suggest the safe use and great potential of [124]I-cRGDY–PEG–C dots in human cancer diagnostics.

Besides optimizing the clearance profiles and the *in vivo* targeting efficacy, massive production of C dots with a satisfied reproducibility is another vital aspect during the clinical translation process. Both the first and second generation of C dots were synthesized by using a modified Stöber method (Benezra et al. 2011; Burns et al. 2009; Ow et al. 2005; Phillips et al. 2014), where alcohol was used as the solvent. A greatly simplified synthesis and cleaning protocols are needed for reducing the volatile waste and rendering particle production substantially faster and more cost-effective. To address these challenges, a water-based synthetic approach was developed very recently by the same group and showed not only improved control over particle size and compositions, but also enhanced performance characteristics (Ma et al. 2015). In order to distinguish the materials synthesized in water from the past particles synthesized via a modified Stöber process in alcohol, the new third generation water-based fluorescent core@shell silica nanoparticles were described as Cornell prime dots, or simply C' dots.

Detailed water-based particle synthesis pathways are shown in Fig. 16.3a, together with the chemical and physical structures of the products (Ma et al. 2015). Tetramethyl orthosilicate (TMOS) was selected as the silica source and ammonium hydroxide as the base catalyst. Nanoparticles were modified *in situ* with PEG for better *in vivo* stability by adding PEG-silane (e.g., Methoxy-terminated polyethylene glycol, molecular weight: ~500). A well-established dialysis process was then introduced to remove the unreacted reagents. Particle size (ranging from 2.4 to 7.3 nm) could be well-controlled by varying reaction parameters including the concentration of TMOS, the concentration of ammonium hydroxide, the length of particle growth period and the reaction temperature. A representative TEM image of 4.2 nm sized C' dots is also shown in Fig. 16.3b.

Results also showed successful control over encapsulating different organic dyes with varied optical characteristics (tuned from the visible into the NIR part of the optical spectrum) by co-condensing different silane-conjugated fluorophores into the silica matrix (Figs. 16.3c, d). To encapsulate NIR dyes with larger negative charge into the silica matrix, aluminum alkoxide together with TMOS were used as the co-precursors and the aqueous growth solution pH was adjusted to ~1.5 (slightly below the isoelectric point of silica, pH ~2.7). Additional silica shells can also be added to improve the optical property while keeping the overall HD below 10 nm. A modified protocol was also used to synthesize functional ligands (such as cRGDyC peptide) conjugated C' dots. Although no *in vivo* renal clearance study was provided in the study, as proposed new synthetic protocols are believed to have a great impact on the future C' dots GMP process.

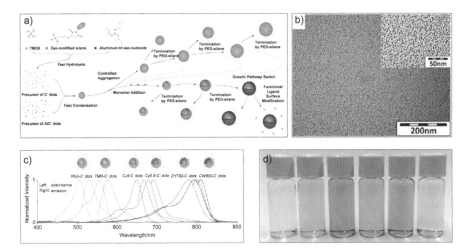

Fig. 16.3 (a) An illustration of water-based synthesis of C′ dots. (b) A representative TEM image of 4.2 nm sized C′ dots. (c) Normalized absorbance and emission spectra of C′ dots with different dyes/colors. (d) Photograph of C′ dots encapsulated with different organic dyes (from left to right: RhG, TMR, Cy5, Cy5.5, DY782 and CW800). Reproduced with permission from (Ma et al. 2015).

Silica Meets Gold Nano-Shell (-Sphere and Nanostar)(Type 1)

Dense silica coating has been well accepted as one of the most widely adapted techniques for the surface modification of inorganic functional nanocrystals, such as UCNP (Chen et al. 2013a; Chen et al. 2011; Chen et al. 2012; Chen et al. 2010a), SPION (Chen et al. 2009b), Au nanospheres (Kircher et al. 2012) and QDs (Yi et al. 2005), to name a few. Initially invented by Naomi Halas and Jennifer West from Rice University in the mid-1990s, PEGylated silica-cored Au nanoshells are the first photothermal nanoparticles that have advanced into clinical trials, appearing as AuroShell® Particles in 2008 (Loo et al. 2004). In this section, we are going to focus on the recent advances in the engineering of silica core Au nanoshell (i.e., SiO_2@Au) or Au nanosphere core silica shell (i.e., Au@SiO_2) hybrid nanomaterials for photothermal therapy and multimodality imaging.

Drug loaded PEGylated gold nanoshell coated silica nanorattle (denoted as pGSNs) has been developed by researchers for investigating the photothermal and chemo-therapeutic synergistic effects (Liu et al. 2011). The shell thickness of gold nanoshell can be easily tuned by altering the precursors and reaction time. One of the advantages of this hybrid nanomaterial is its high potential for anti-cancer drug loading due to the presence of a big cavity inside each pGSNs (Li et al. 2011). To further ensure an enhanced therapeutic effect, the same group further reported a transferrin (Tf) targeting ligands conjugated pGSNs (Fig. 16.4) (denoted as pGSNs-Tf) in a follow-up study (Liu et al. 2012a). Although enhanced tumor uptake was expected after the injection of pGSNs-Tf, inductively coupled optical emission spectrometry (ICP-OES) of Au failed to provide any significant enhancement in the targeted group, indicating that cell-based active targeting strategy might not work as expected due to limited nanoparticle extravasation *in vivo*. The observed enhanced therapeutic effect was suggested to be due to an enhanced intracellular localization within cancer cells of pGSNs-Tf, although more solid data is required to support the hypothesis. Moreover, systematic *in vivo* non-invasive imaging (such as serial PET imaging) and quantitative tumor uptake quantification are needed in the future to better determine the best therapeutic time window for this interesting nanoparticle.

Silica nanorattle@Au-Tf

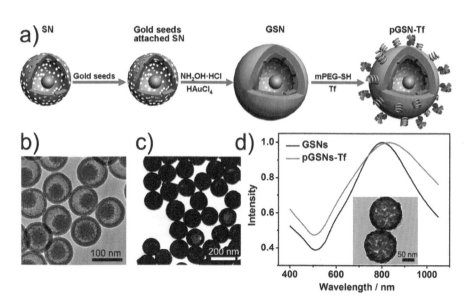

Fig. 16.4 (a) A schematic illustration of GSNs synthesis and bioconjugation. (b) TEM image of SNs. (c) TEM image of GSNs. (d) Absorption spectra of pGSNs and pGSNs-Tf, the inset is the TEM image of pGSNs-Tf. Reproduced with permission from (Liu et al. 2012a).

Besides using gold nanoshell for photothermal therapeutic application, due to the presence of localized surface plasmon resonance effect (Yang et al. 2015), functionalized gold nanosphere can also be used for multimodality imaging. To address the limitations (e.g., inadequate sensitivity, specificity and spatial resolution) of traditional imaging techniques, a unique triple-modality magnetic resonance imaging–photoacoustic imaging–Raman imaging nanoparticle (termed here MPR nanoparticle) based on silica-coated Au nanosphere has been developed to accurately delineate the margins of brain tumors in living mice both preoperatively and intraoperatively (Fig. 16.5a) (Kircher et al. 2012).

Au@dSiO$_2$-DOTA-Gd

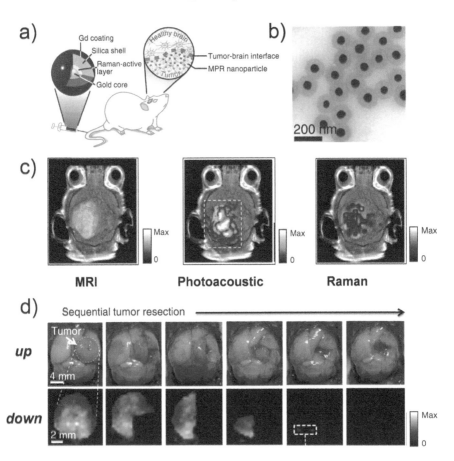

Fig. 16.5 (a) A schematic illustration showing the concept of using MPR nanoparticle for triple-modality tumor imaging. A Raman-active outer layer was engineered in between Au and dense silica. MPRs are injected intravenously into a mouse bearing an orthotopic brain tumor. As the nanoparticles circulate in the bloodstream, they diffuse through the disrupted blood-brain barrier and are then sequestered and retained by the tumor. (b) A TEM image of Au@dSiO$_2$-DTPA-Gd. (c) Two-dimensional axial MRI, photoacoustic and Raman images. (d) Raman-guided intraoperative surgery using MPRs. Living tumor-bearing mice underwent craniotomy under general anesthesia. Quarters of the tumor were then sequentially removed (up) and intraoperative Raman imaging was performed after each resection step (down) until the entire tumor had been removed. Reproduced with permission from (Kircher et al. 2012).

To synthesize MPR nanoparticle, a 60 nm sized Au nanosphere was first covered with a thin Raman-active outer layer, followed by the coating of a 30 nm-thick silica protective layer. Chelators (i.e., DOTA, or 1, 4, 7, 10-tetraazacyclododecane-1, 4, 7, 10-tetraacetic acid) were then conjugated to the outer silica surface through maleimide linkage. The final MPR nanoparticle (average size: ~120 nm) can be achieved by labeling with Gd^{3+} ions for MR imaging (Fig. 16.5b). High detection sensitivity was demonstrated both *in vitro* and *in vivo*. *In vivo* triple-modality study was then performed in orthotopic brain tumor-bearing mice. Intravenous injection of MPRs could lead to their accumulation and retention by the brain tumors, with no MPR accumulation in the surrounding healthy tissue, allowing for a noninvasive tumor delineation using all three modalities through the intact skull (Fig. 16.5c). Successful accumulation of nanoparticles in the brain was further validated by using *ex vivo* histology study. As shown in Fig. 16.5d, the authors further demonstrated intraoperative tumor resection under the guidance of Raman imaging. This was the first example of combining MRI–photoacoustic imaging–Raman imaging together into one single Au-silica-based hybrid nanoparticle. The authors expected such a probe to help future radiologists and neurosurgeons to better 'see' the tumors before and during surgery, thus allowing for more accurate brain tumor resection.

Although high clinical translation potential has been suggested (considering the relatively low toxicity of Au nanosphere core and the high biocompatibility of silica shell), due to the large particle size (over 100 nm), the majority of MPR nanoparticles are still expected to be accumulated in the liver and spleen, provoking potential long-term toxicity concerns. No whole body biodistribution data could be found in this study. Another concern was the stability of MPR nanoparticles in the blood. Without suitable PEGylation, any silica-based nanoparticle will be aggregated in the blood (due to highly negatively charged surface), resulting in rapid and high liver uptake after intravenous injection.

The inability to visualize the true extent of cancers represents a significant challenge in many areas of oncology. In a follow-up study, the same group showed the precise visualization of tumor margins, microscopic tumor invasion and multifocal locoregional tumor spread by using a new generation of surface-enhanced resonance Raman scattering nanoparticles (termed as SERRS nanostars). The SERRS nanostars were formed by coating silica onto a star-shaped gold core, which was covered by a layer of Raman reporters. Based on the enhanced permeability and retention (EPR) effect, as-developed SERRS nanostars enabled the accurate detection of macroscopic malignant lesions in genetically engineered mouse models of pancreatic cancer, breast cancer, prostate cancer and sarcoma and in one human sarcoma xenograft model with an extremely high sensitivity (~1.5 fM limit of detection, ~400-fold lower than that of previous generations of Raman nanoparticles (Kircher et al. 2012)). Although PEGylation was introduced in this study, the final

PEGylated SERRS nanostars still showed highly negatively charged surfaces (–45.4 mV) when compared with samples before PEGylation (–48.7 mV). This also explains the fast and high RES uptake in RES organs (such as liver and spleen) at 16 h post-injection. Although evidence of hepatic clearance has been shown to alleviate the long-term toxicity concerns, successful translation of these SERRS nanostars might be hindered due to their relatively large particle size (> 70 nm). Ultrasmall PEGylated Au@dSiO$_2$ recently developed by Wiesner and co-workers might provide a great solution to these limitations (Sun et al. 2014).

Silica-Based Core-Multi-Shell Hybrid Nanostructures (Type 2)

Core-multi-shell structured nanomaterials have attracted growing research interests for developing hybrid nanomedicine with both diagnosis and therapeutic functionalities (Liu et al. 2006; Xiao et al. 2012; Xing et al. 2012). For example, by using NaYF$_4$: Yb/Er/Tm@NaGdF$_4$ nanocrystal as the inner core, Shi and co-workers demonstrated an efficient way to add CT imaging modality by firstly coating it with dense silica and then decorating with a fluorescence-transparent TaO$_x$ shell (Xiao et al. 2012). As-design water-soluble litchi-shaped trimodal imaging probe showed extraordinarily high r_1 (11.45 mM^{-1}s^{-1}) and r_2 (147.3 mM^{-1}s^{-1}) values thanks to the ultra-thin NaGdF$_4$ shell. An obvious CT contrast enhancement could be obtained by the combined effect of radiopaque TaO$_x$ shell and UCNP itself (Xiao et al. 2012).

Besides decorating gold and TaO$_x$ nanoparticles for multimodality imaging, negatively charged ultrasmall CuS nanoparticles with attractive photothermal property have also been decorated to UCNP@SiO$_2$ for a novel multifunctional core/satellite nanotheranostic (CSNT, Figs. 16.6a, b). As reported by Shi and co-workers, these CSNTs could not only convert near-infrared light (wavelength: 980 nm) into heat for effective thermal ablation, but could also induce a highly localized radiation dose boost to trigger substantially enhanced radiation damage both *in vitro* and *in vivo*. Due to the synergistic interaction between photothermal ablation (PTA) and the enhanced radiotherapy (RT), *in vivo* therapeutic study showed that tumors could be eradicated without visible recurrence in 120 days (Figs. 16.6c, d). Negligible toxicity to the mice within a month at a dose of 10.5 mg/kg was observed after the hematological analysis. Moreover, the novel CSNTs could facilitate upconversion luminescence/magnetic resonance/computer tomography trimodal imaging. Although as-prepared hybrid nanoparticles were modified with PEG to increase their blood circulation time *in vivo*, only treatment study with intratumoral injection was presented. The decoration of CuS nanoparticles based on electrostatic attraction might also provoke concerns about their stability when circulating in the blood. Systematic investigation of these issues as well as the active or passive tumor targeting efficacy after intravenous injection are needed in the future studies.

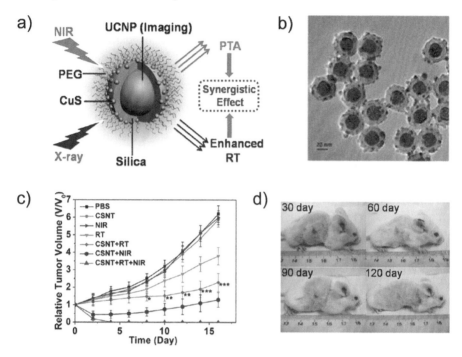

Fig. 16.6 (a) A schematic showing using CSNT for an enhanced PTA/RT synergistic therapy. The UCNP core was introduced for upconversion luminescence imaging and CuS satellites were decorated over silica shielded UCNP for converting the 980 nm laser into heat for PTA. The hybrid nanoparticles were finally modified with PEG for an improved stability *in vivo*. (b) A TEM image of CSNTs. (c) Time-dependent tumor growth curves of different groups of mice with various treatments. In all experiments, the power density of 980 nm NIR-laser and the radiation dose used for RT were kept constant at 1.5 W/cm² and 6 Gy, respectively. (d) The photographs of mice in 30, 60, 90 and 120 days of treatment (group CSNT+RT+PAT) showing the complete eradication of the tumor and no visible recurrences of the tumors in at least 120 days. Reproduced with permissions from (Xiao et al. 2013).

Coating Nanocrystals with Mesoporous Silica Nanoshell (Type 3)

Mesoporous silica nanoparticles exhibit controllable pore size, high specific surface area, large pore volume and well-established surface modification chemistry (Shen et al. 2014) and thus have been accepted as highly attractive drug carriers since 2001 (Vallet-Regi et al. 2000). Coating functional nanoparticles with a layer of MSN could form a new type of hybrid nanomaterial with not only their intrinsic magnetic or optical properties, but also an additional drug delivery capability. Pioneered by Gorelikov and Matsuura, a generalized, single-step synthesis procedure to coat individual cetyltrimethyl ammonium bromide- (CTAB) capped nanoparticles with a thin layer of mesoporous silica was developed in the year 2008 (Gorelikov and Matsuura 2008). A similar technique has been used for the successful synthesis of AuNR@MSN (Zhang et al. 2012) SPION@MSN (Kim et al. 2008) MSN coated hollow manganese oxide nanoparticles (HMnO@MSN) (Kim et al. 2011), UCNP@MSN (Liu et al. 2013; Liu et al. 2012b), to name a few. More complex hybrid structures, such as dye-doped MSN decorated with SPION (Lee et al. 2010) and Au decorated magnetic core and mesoporous silica shell hybrid nanoparticles (Ma et al. 2012), have also been developed for simultaneous imaging and therapy. Although great progress in the engineering of MSN coated hybrid nanoparticles has been made during the last 5 years, most of the previous works have been focused on nanomaterial synthesis. Engineering of silica-based Type 3 hybrid nanomaterials for *in vivo* tumor-targeted imaging and therapy is still one of major challenges in this field.

In one study, researchers demonstrated that enhanced photodynamic therapy (PDT) effect could be achieved by encapsulating not one but two different types of photosensitizers (PSs) (i.e., ZnPC and MC540) into mesoporous silica coated $NaYF_4$: Er/Yb nanoparticles (i.e., UCNP@MSN) (Idris et al. 2012). By using UCNP as the internal light source for the PSs, *in vitro* experiments demonstrated the higher generation rate of singlet oxygen and lower cell viability when shedding 980 nm light on co-PS-loaded nanoparticles. After the injection of B10-F0 melanoma cell that was pre-labeled with co-PS-loaded UCNP under the skin of C57BL/6 mice, they succeeded in inhibiting the tumor growth rate significantly by irradiating the injected site with a 980 nm laser. However, in the case of injecting the drug either intratumorally or intravenously, only slightly tumor growth inhibition could be achieved. Since the half-life of singlet oxygen is very short, on the order of microseconds and can therefore act only on target structures that are in close proximity to it, in order to guarantee an effective PDT therapeutic outcome, sufficient amount of PDT agents need to be accumulated in the tumor site and somehow further internalized by the cells.

To demonstrate the potential of using MSN coated multifunctional nanomaterials for both imaging and photothermal therapy, in another study, the design, synthesis, surface engineering and *in vivo* active vasculature targeting of a new category of theranostic nanoparticles for future cancer management were reported (Chen et al. 2015). Water-soluble photothermally sensitive CuS nanoparticles (capped with cetyltrimethylammonium chloride or CTAC) were encapsulated in biocompatible mesoporous silica shells, followed by multi-step surface engineering to form the final theranostic nanoparticle, i.e., ^{64}Cu-NOTA-CuS@MSN-PEG-TRC105 (Fig. 16.7a). TRC105 is a human/murine chimeric IgG1 monoclonal antibody, which binds to both human and murine CD105 on tumor neovasculature (Rosen et al. 2012). Radioisotope ^{64}Cu is a positron emitter with a 12.7 h half-life, for PET imaging and biodistribution studies.

The initial attempt to encapsulate CuS-citrate nanoparticles into MSN was not successful possibly due to its negatively charged surface (-21.3 ± 4.0 mV). To facilitate the coating of MSN over CuS, positively charged CTAC stabilized CuS (i.e., CuS-CTAC) was used instead. The surfactant CTAC functions as not only the capping agent for stabilizing the CuS nanoparticles, but also as the template for the successful growth of the mesoporous silica shell over CuS. Fig. 16.7b shows the TEM image of core@shell structured CuS@MSN with an average particle size of 65 nm. The successful synthesis of CuS-CTAC and CuS@MSN was further confirmed by their characteristic green color (Fig. 16.7b) and strong absorption peak at around 980 nm from the UV-Vis absorbance spectra (Fig. 16.7c). As-synthesized CuS@MSN possesses significantly higher specific surface areas (495 m^2/g) and larger pore volume (0.68 cm^3/g) with average pore sizes further found to be ~2.2 nm. An anti-cancer drug loading study showed about 465.1 mg/g drug loading capacity by using small molecule anti-cancer drug doxorubicin (DOX) as the model drug. Efficient energy conversion from light to heat was also confirmed by their photothermal images (Fig. 16.7c, inset) with the control water sample showing only slightly increased temperature after the same 980 nm laser irradiation.

To date, the engineering of biocompatible theranostic nanoparticles with highly specific *in vivo* tumor active targeting capabilities is still in its very early stages with only a few examples of *in vivo* tumor actively targeted theranostic nanoparticles being reported (Chen et al. 2014a). To demonstrate the *in vivo* tumor targeting efficacy of as-designed ^{64}Cu-NOTA-CuS@MSN-PEG-TRC105 nanoconjugates, *in vivo* tumor vasculature targeted PET imaging was then carried out in 4T1 murine breast tumor-bearing mice (which express a high level of CD105 on the tumor neovasculature (Seon et al. 2011)). The accumulation of ^{64}Cu-NOTA-CuS@MSN-PEG-TRC105 in the 4T1 tumor was found to be $4.9 \pm 0.7\%$ ID/g at 4 h post-injection and peaked at $6.0 \pm 0.4\%$ ID/g at 24 h post-injection as shown in Fig. 16.7d (targeted). In contrast, without the conjugation of TRC105 (i.e., EPR effect alone, non-targeted group in Fig. 16.7d), the 4T1 tumor uptake of ^{64}Cu-NOTA-CuS@MSN-PEG was found to be around 1% ID/g at all of the time points examined, indicating that TRC105 conjugation is likely the controlling factor for enhanced tumor accumulation of ^{64}Cu-CuS@MSN-TRC105. A blocking study was further performed to study the CD105 targeting specificity of ^{64}Cu-NOTA-CuS@MSN-PEG-TRC105 *in vivo*. A significantly reduced tumor uptake (3.0 ± 0.1 and $2.3 \pm 0.3\%$ ID/g at 4 and 24 h post-injection, respectively) was observed in the blocking group as shown in Fig. 16.7d, clearly demonstrating the CD105 specificity of ^{64}Cu-CuS@MSN-TRC105 *in vivo*.

As a proof of concept study, we also demonstrated an effective *in vivo* photothermal ablation of tumors after a single intratumoral injection of

CuS@MSN in 4T1 tumor-bearing mice (n = 5, dose: 33 mg/kg, Fig. 16.7e). Decreases in the tumor size in these two groups were also observed after laser treatment, a more prominent shrinking rate was found in the *CuS@MSN+980 nm laser* group during the first four days (Fig. 16.7e). Tumors from *CuS@MSN+980 nm laser* group continued to shrink and completely vanished on Day 14 with no re-growth in the next 67 days. Also, *in vivo* long-term toxicity studies in healthy female BALB/c mice showed no noticeable toxicity effects within 60 days after intravenous injection of PEGylated CuS@MSN at a dose of 90 mg/kg. To the best of our knowledge, as-developed ⁶⁴Cu-NOTA-CuS@MSN-PEG-TRC105 is one of the very few theranostic nanoparticles that possess the capability for *in vivo* active tumor targeting. We are currently investigating the potential of these nanoparticles for *in vivo* tumor-targeted photothermally enhanced drug delivery and thermo-chemotherapy.

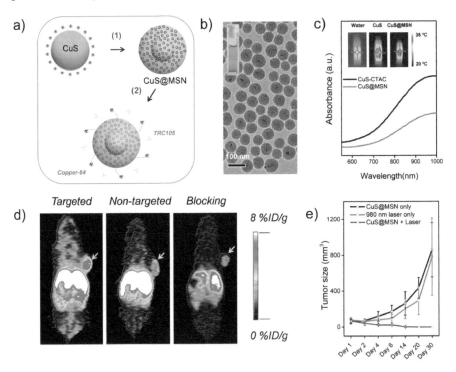

Fig. 16.7   (a) A schematic illustration showing two major steps for the synthesis of ⁶⁴Cu-NOTA-CuS@MSN-PEG-TRC105 theranostic nanoparticle. Step 1: encapsulating CuS-CTAC inside mesoporous silica shell, resulting in CuS@MSN nanoparticle; Step 2: surface engineering of CuS@MSN to form ⁶⁴Cu-NOTA-CuS@MSN-PEG-TRC105 nanoconjugates. (b) A TEM image of CuS@MSN. Inset shows the digital photo of CuS@MSN in water. (c) UV-Vis absorption spectra of CuS-CTAC and CuS@MSN. Inset shows the photothermal images of water, CuS-CTAC and CuS@MSN (from left to right). (d) *In vivo* PET imaging of different nano-conjugates in 4T1 tumor bearing mice. (e) Changes of tumor size of mice from 3 different groups after laser treatment (n = 5). Laser dose: 4 W/cm², 15 min. CuS@MSN dose: 33 mg/kg. Reproduced with permission from (Chen et al. 2015).

Yolk-Shell Structured Hybrid Nanomaterials (Type 4)

Hollow mesoporous silica nanoparticles with a large cavity inside each original mesoporous silica nanoparticle have recently been developed to greatly enhance the drug loading capacity (Chen et al. 2014b; Shi et al. 2013; Tang et al. 2012). To further improve the drug loading capacity, a new type of hybrid nanomaterial named yolk-shell structured nanoparticles have also been developed. Each nanoparticle possesses an inorganic functional core (which can be silica (Chen et al. 2009a; Liu et al. 2010), UCNP (Fan et al. 2015a; Fan et al. 2014; Fan et al. 2013; Fan et al. 2015b; Liu et al. 2014a; Liu et al. 2014b; Liu et al. 2015b; Liu et al. 2015c), iron oxide nanoparticle (Chen et al. 2010b), Au nanoparticle (Zhang et al. 2008), to name a few), a big cavity for drug loading and a thin mesoporous silica shell with tunable pore size. In this section, we will discuss the very recent advances in the engineering of UCNP@HMSN based yolk-shell structured nanoparticles for both imaging and therapy.

Upconversion luminescence (UCL) is a unique process where low energy continuous-wave (CW) of near-infrared (NIR) light is converted into higher energy light (emission bands ranging from ultraviolet [UV] to NIR region) through the sequential absorption of multiple photons or energy transfer (Auzel 2004). UCNPs have aroused a lot of interest owing to their unique UCL features (such as sharp emission lines (Haase and Schafer 2011), long lifetimes (~ms) (Ju et al. 2012), large anti-Stokes shift (Haase and Schafer 2011), superior photostability (Wu et al. 2009), high detection sensitivity (Cheng et al. 2012), non-blinking and non-bleaching (Park et al. 2009; Wu et al. 2009), high tissue penetration depth (Chatterjee et al. 2008), minimal photo-damage (Nam et al. 2011) and extremely low auto-fluorescence (Xiong et al. 2009)) that are highly suitable for multimodal imaging in living subjects (Zhou et al. 2012). All of these have made UCNPs one of the most promising nanoparticles for future cancer imaging and therapy.

To demonstrate the synergetic chemo-/radiotherapy effect in cancer treatment, Shi and co-workers developed an interesting new type of UCNP-based yolk-shell structured hybrid nanomaterial, which combines the MRI/UCL dual-modality imaging, chemo- and radio-therapy together in one sub-100 nm sized nanoparticle (Fig. 16.8a) (Fan et al. 2013). Gd^{3+} doped core@shell structured UCNPs (i.e., $NaYF_4$: $Yb/Er@NaGdF_4$) were selected as the cores for providing the MRI/UCL imaging functionality. Two dense silica shells were then coated on the pre-prepared UCNPs by a modified water-in-oil reverse microemulsion method and a water-phase silica regrowth method. To create a cavity for drug loading, polyvinylpyrrolidone (PVP, Mw = 40000) was introduced as a protecting agent, which protects the outer shell of $dSiO_2$ better during the "surface-protected etching" process (Zhang et al. 2008). A representative TEM image of UCNP-based yolk-shell structured hybrid nanomaterials is shown in Fig. 16.8b. Cisplatin, a well-known anti-cancer drug and radiosensitizer, was selected as the model drug. The highest loading capacity of cisplatin was estimated to be about 10 wt%. Systematic *in vitro* and *in vivo* studies demonstrated the enhanced therapeutic effect when combing cisplatin-loaded nanoparticles with radiation (Fig. 16.8c). The regrowth of tumors from Day 7 after treatment was observed even in the most effective group. Although most of the *in vivo* studies in current work were based on intratumoral injection to quickly prove the concept, the authors managed to provide a pilot treatment study by using folic acid conjugated nanoparticles via intravenous injection. No suitable PEGylation was used in the research and detailed investigations on tumor active targeting efficacy and *in vivo* biodistribution of nanoparticles are needed. A similar UCNP-based yolk-shell hybrid nanoparticle was later developed for synergetic chemo-/radio-/photodynamic therapy (Fan et al. 2014), for real-time *in vivo* quantitative monitoring of drug release (Liu et al. 2014a) and for selective *in vivo* hypoxia imaging (Liu et al. 2014b).

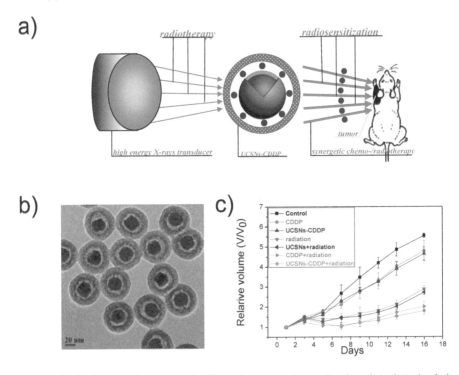

Fig. 16.8 (a) A schematic illustration of radiosensitization using as-developed cisplatin-loaded yolk-shell structured nanoparticles. (b) A TEM image of yolk-shell structured nanoparticles. (c) Tumor growth curves of HeLa tumor xenografts after different treatments. Reproduced with permission from (Fan et al. 2013).

Conclusion

In this chapter, we have discussed the latest research advances in the engineering of silica-based core@shell hybrid nanomaterials for cancer imaging and therapy. Although most of the current studies are still focusing on pre-clinical small animal research, some nanoparticles have already shown great potential for clinical translation. C dots are currently the only type of nanoparticles that have received the FDA IND approval for first-in-human clinical trials (Benezra et al. 2011; Phillips et al. 2014). Despite these promising results, there are still a number of substantial challenges in the development of silica-based hybrid nanomaterials for biomedical applications.

Firstly, although silica is generally recognized as safe by the FDA, the toxicity of encapsulated inorganic nanocrystals still provokes great concerns. Systematic *in vivo* long-term toxicity studies are still needed for silica coated inorganic functional nanoparticles, such as Au@SiO$_2$, UCNP@MSN, UCNP@HMSN, etc. Secondly, the engineering of silica-based hybrid nanoparticles with high *in vivo* stability, suitable long *in vivo* blood circulation half-life and active tumor targeting efficacy are still major challenges in this field. Thirdly, the radiolabeling and PET imaging of hybrid nanomaterials will soon be accepted as a highly useful quantitative tool for studying the fate of hybrid nanoparticles *in vivo*. Fourthly, in order to integrate different functionalities into one multifunctional nanosystem, most of the current silica-based hybrid nanomaterials have a HD size larger than the renal cutoff (~5.5 nm), causing rapid and high liver uptake after intravenous injection. We believe that the engineering of ultrasmall and renal clearable silica-based hybrid nanoparticles should hold a greater chance for future clinical translation.

Acknowledgements

This work is supported, in part, by the University of Wisconsin – Madison, the National Institutes of Health (NIBIB/NCI 1R01CA169365 & P30CA014520) and the American Cancer Society (125246-RSG-13-099-01-CCE).

References

Auzel, F. 2004. Upconversion and anti-Stokes processes with f and d ions in solids. Chem Rev 104: 139-173.

Benezra, M., Penate-Medina, O., Zanzonico, P.B., Schaer, D., Ow, H., Burns, A., et al. 2011. Multimodal silica nanoparticles are effective cancer-targeted probes in a model of human melanoma. J Clin Invest 121: 2768-2780.

Burns, A.A., Vider, J., Ow, H., Herz, E., Penate-Medina, O., Baumgart, M., et al. 2009. Fluorescent silica nanoparticles with efficient urinary excretion for nanomedicine. Nano Lett 9: 442-448.

Chatterjee, D.K., Rufaihah, A.J. and Zhang, Y. 2008. Upconversion fluorescence imaging of cells and small animals using lanthanide doped nanocrystals. Biomaterials 29: 937-943.

Chen, D., Li, L.L., Tang, F.Q. and Qi, S.O. 2009a. Facile and scalable synthesis of tailored silica "nanorattle" structures. Adv Mater 21: 3804-3807.

Chen, F., Bu, W., Chen, Y., Fan, Y., He, Q., Zhu, M., et al. 2009b. A sub-50-nm mono-sized superparamagnetic Fe3O4@SiO2 T2-weighted MRI contrast agent: highly reproducible synthesis of uniform single-loaded core-shell nanostructures. Chem Asian J 4: 1809-1816.

Chen, F., Zhang, S., Bu, W., Liu, X., Chen, Y., He, Q., et al. 2010a. A "neck-formation" strategy for an antiquenching magnetic/upconversion fluorescent bimodal cancer probe. Chemistry 16: 11254-11260.

Chen, F., Bu, W., Zhang, S., Liu, X., Liu, J., Xing, H., et al. 2011. Positive and negative lattice shielding effects co-existing in Gd (III) ion doped bifunctional upconversion nanoprobes. Adv Funct Mater 21: 4285-4294.

Chen, F., Zhang, S., Bu, W., Chen, Y., Xiao, Q., Liu, J., et al. 2012. A uniform sub-50 nm-sized magnetic/upconversion fluorescent bimodal imaging agent capable of generating singlet oxygen by using a 980 nm laser. Chemistry 18: 7082-7090.

Chen, F., Bu, W., Zhang, S., Liu, J., Fan, W., Zhou, L., et al. 2013a. Gd3+ -ion-doped upconversion nanoprobes: relaxivity mechanism probing and sensitivity optimization. Adv Funct Mater 23: 298-307.

Chen, F., Ehlerding, E.B. and Cai, W. 2014a. Theranostic nanoparticles. J Nucl Med 55: 1919-1922.

Chen, F., Hong, H., Shi, S., Goel, S., Valdovinos, H.F., Hernandez, R., et al. 2014b. Engineering of hollow mesoporous silica nanoparticles for remarkably enhanced tumor active targeting efficacy. Sci Rep 4: 5080.

Chen, F., Hong, H., Goel, S., Graves, S.A., Orbay, H., Ehlerding, E.B., et al. 2015. In vivo tumor vasculature targeting of CuS@MSN based theranostic nanomedicine. ACS Nano 9: 3926-3934.

Chen, Y., Chen, H., Guo, L., He, Q., Chen, F., Zhou, J., et al. 2010b. Hollow/rattle-type mesoporous nanostructures by a structural difference-based selective etching strategy. ACS Nano 4: 529-539.

Chen, Y., Chen, H. and Shi, J. 2013b. In vivo bio-safety evaluations and diagnostic/therapeutic applications of chemically designed mesoporous silica nanoparticles. Adv Mater 25: 3144-3176.

Cheng, L., Wang, C. and Liu, Z. 2012. Upconversion nanoparticles and their composite nanostructures for biomedical imaging and cancer therapy. Nanoscale 5: 23-37.

Fan, W., Shen, B., Bu, W., Chen, F., Zhao, K., Zhang, S., et al. 2013. Rattle-structured multifunctional nanotheranostics for synergetic chemo-/radiotherapy and simultaneous magnetic/luminescent dual-mode imaging. J Am Chem Soc 135: 6494-6503.

Fan, W., Shen, B., Bu, W., Chen, F., He, Q., Zhao, K., et al. 2014. A smart upconversion-based mesoporous silica nanotheranostic system for synergetic chemo-/radio-/photodynamic therapy and simultaneous MR/UCL imaging. Biomaterials 35: 8992-9002.

Fan, W., Bu, W., Zhang, Z., Shen, B., Zhang, H., He, Q., et al. 2015a. X-ray Radiation-Controlled NO-Release for On-Demand Depth-Independent Hypoxic Radiosensitization. Angew Chem Int Ed Engl 54: 14026-14030.

Fan, W., Shen, B., Bu, W., Zheng, X., He, Q., Cui, Z., et al. 2015b. Intranuclear biophotonics by smart design of nuclear-targeting photo-/radio-sensitizers co-loaded upconversion nanoparticles. Biomaterials 69: 89-98.

Gorelikov, I. and Matsuura, N. 2008. Single-step coating of mesoporous silica on cetyltrimethyl ammonium bromide-capped nanoparticles. Nano Lett 8: 369-373.

Guerrero-Martinez, A., Perez-Juste, J. and Liz-Marzan, L.M. 2010. Recent progress on silica coating of nanoparticles and related nanomaterials. Adv Mater 22: 1182-1195.

Haase, M. and Schafer, H. 2011. Upconverting nanoparticles. Angew Chem Int Ed 50: 5808-5829.

http://www.fda.gov/Food/IngredientsPackagingLabeling/GRAS/SCOGS/ucm261095.htm.

Huang, W.Y. and Davis, J.J. 2011. Multimodality and nanoparticles in medical imaging. Dalton Trans 40: 6087-6103.

Idris, N.M., Gnanasammandhan, M.K., Zhang, J., Ho, P.C., Mahendran, R. and Zhang, Y. 2012. In vivo photodynamic therapy using upconversion nanoparticles as remote-controlled nanotransducers. Nat Med 18: 1580-1585.

Ju, Q., Tu, D., Liu, Y., Li, R., Zhu, H., Chen, J., et al. 2012. Amine-functionalized lanthanide-doped KGdF4 nanocrystals as potential optical/magnetic multimodal bioprobes. J Am Chem Soc 134: 1323-1330.

Kim, J., Kim, H.S., Lee, N., Kim, T., Kim, H., Yu, T., et al. 2008. Multifunctional uniform nanoparticles composed of a magnetite nanocrystal core and a mesoporous silica shell for magnetic resonance and fluorescence imaging and for drug delivery. Angew Chem Int Ed Engl 47: 8438-8441.

Kim, T., Momin, E., Choi, J., Yuan, K., Zaidi, H., Kim, J., et al. 2011. Mesoporous silica-coated hollow manganese oxide nanoparticles as positive T1 contrast agents for labeling and MRI tracking of adipose-derived mesenchymal stem cells. J Am Chem Soc 133: 2955-2961.

Kircher, M.F., de la Zerda, A., Jokerst, J.V., Zavaleta, C.L., Kempen, P.J., Mittra, E., et al. 2012. A brain tumor molecular imaging strategy using a new triple-modality MRI-photoacoustic-Raman nanoparticle. Nat Med 18: 829-834.

Lee, J.E., Lee, N., Kim, H., Kim, J., Choi, S.H., Kim, J.H., et al. 2010. Uniform mesoporous dye-doped silica nanoparticles decorated with multiple magnetite nanocrystals for simultaneous enhanced magnetic resonance imaging, fluorescence imaging and drug delivery. J Am Chem Soc 132: 552-557.

Li, D., He, Q. and Li, J. 2009. Smart core/shell nanocomposites: intelligent polymers modified gold nanoparticles. Adv Colloid Interface Sci 149: 28-38.

Li, L., Guan, Y., Liu, H., Hao, N., Liu, T., Meng, X., et al. 2011. Silica nanorattle-doxorubicin-anchored mesenchymal stem cells for tumor-tropic therapy. ACS Nano 5: 7462-7470.

Liu, H., Chen, D., Li, L., Liu, T., Tan, L., Wu, X., et al. 2011. Multifunctional gold nanoshells on silica nanorattles: a platform for the combination of photothermal therapy and chemotherapy with low systemic toxicity. Angew Chem Int Ed 50: 891-895.

Liu, H., Liu, T., Wu, X., Li, L., Tan, L., Chen, D., et al. 2012a. Targeting gold nanoshells on silica nanorattles: a drug cocktail to fight breast tumors via a single irradiation with near-infrared laser light. Adv Mater 24: 755-761.

Liu, J., Qiao, S.Z., Budi Hartono, S. and Lu, G.Q. 2010. Monodisperse yolk-shell nanoparticles with a hierarchical porous structure for delivery vehicles and nanoreactors. Angew Chem Int Ed Engl 49: 4981-4985.

Liu, J., Bu, W., Zhang, S., Chen, F., Xing, H., Pan, L., et al. 2012b. Controlled synthesis of uniform and monodisperse upconversion core/mesoporous silica shell nanocomposites for bimodal imaging. Chemistry 18: 2335-2341.

Liu, J., Bu, W., Pan, L. and Shi, J. 2013. NIR-triggered anticancer drug delivery by upconverting nanoparticles with integrated azobenzene-modified mesoporous silica. Angew Chem Int Ed Engl 52: 4375-4379.

Liu, J., Bu, J., Bu, W., Zhang, S., Pan, L., Fan, W., et al. 2014a. Real-time in vivo quantitative monitoring of drug release by dual-mode magnetic resonance and upconverted luminescence imaging. Angew Chem Int Ed Engl 53: 4551-4555.

Liu, J., Liu, Y., Bu, W., Bu, J., Sun, Y., Du, J., et al. 2014b. Ultrasensitive nanosensors based on upconversion nanoparticles for selective hypoxia imaging in vivo upon near-infrared excitation. J Am Chem Soc 136: 9701-9709.

Liu, J.N., Bu, W.B. and Shi, J.L. 2015a. Silica coated upconversion nanoparticles: a versatile platform for the development of efficient theranostics. Acc Chem Res 48: 1797-1805.

Liu, N., Prall, B.S. and Klimov, V.I. 2006. Hybrid gold/silica/nanocrystal-quantum-dot superstructures: synthesis and analysis of semiconductor-metal interactions. J Am Chem Soc 128: 15362-15363.

Liu, Y., Liu, Y., Bu, W., Cheng, C., Zuo, C., Xiao, Q., et al. 2015b. hypoxia induced by upconversion-based photodynamic therapy: towards highly effective synergistic bioreductive therapy in tumors. Angew Chem Int Ed Engl 54: 8105-8109.

Liu, Y., Liu, Y., Bu, W., Xiao, Q., Sun, Y., Zhao, K., et al. 2015c. Radiation-/hypoxia-induced solid tumor metastasis and regrowth inhibited by hypoxia-specific upconversion nanoradiosensitizer. Biomaterials 49: 1-8.

Loo, C., Lin, A., Hirsch, L., Lee, M.H., Barton, J., Halas, N., et al. 2004. Nanoshell-enabled photonics-based imaging and therapy of cancer. Technol Cancer Res Treat 3: 33-40.

Ma, K., Mendoza, C., Hanson, M., Werner-Zwanziger, U., Zwanziger, J. and Wiesner, U. 2015. Control of ultrasmall sub-10 nm ligand-functionalized fluorescent core–shell silica nanoparticle growth in water. Chemistry of Materials 27: 4119-4133.

Ma, M., Chen, H., Chen, Y., Wang, X., Chen, F., Cui, X., et al. 2012. Au capped magnetic core/mesoporous silica shell nanoparticles for combined photothermo-/chemo-therapy and multimodal imaging. Biomaterials 33: 989-998.

Nam, S.H., Bae, Y.M., Park, Y.I., Kim, J.H., Kim, H.M., Choi, J.S., et al. 2011. Long-term real-time tracking of lanthanide ion doped upconverting nanoparticles in living cells. Angew Chem Int Ed 50: 6093-6097.

Ow, H., Larson, D.R., Srivastava, M., Baird, B.A., Webb, W.W. and Wiesner, U. 2005. Bright and stable core-shell fluorescent silica nanoparticles. Nano Lett 5: 113-117.

Park, Y.I., Kim, J.H., Lee, K.T., Jeon, K.S., Bin Na, H., Yu, J.H., et al. 2009. Nonblinking and nonbleaching upconverting nanoparticles as an optical imaging nanoprobe and T1 magnetic resonance imaging contrast agent. Adv Mater 21: 4467-4471.

Phillips, E., Penate-Medina, O., Zanzonico, P.B., Carvajal, R.D., Mohan, P., Ye, Y., et al. 2014. Clinical translation of an ultrasmall inorganic optical-PET imaging nanoparticle probe. Sci Transl Med 6: 260ra149.

Piao, Y., Burns, A., Kim, J., Wiesner, U. and Hyeon, T. 2008. Designed fabrication of silica-based nanostructured particle systems for nanomedicine applications. Adv Funct Mater18: 3745-3758.

Reiss, P., Protiere, M. and Li, L. 2009. Core/shell semiconductor nanocrystals. Small 5: 154-168.

Rosen, L.S., Hurwitz, H.I., Wong, M.K., Goldman, J., Mendelson, D.S., Figg, W.D., et al. 2012. A phase I first-in-human study of TRC105 (Anti-Endoglin Antibody) in patients with advanced cancer. Clin Cancer Res 18: 4820-4829.

Schartl, W. 2010. Current directions in core-shell nanoparticle design. Nanoscale 2: 829-843.

Seon, B.K., Haba, A., Matsuno, F., Takahashi, N., Tsujie, M., She, X., et al. 2011. Endoglin-targeted cancer therapy. Curr Drug Deliv 8: 135-143.

Shen, D., Yang, J., Li, X., Zhou, L., Zhang, R., Li, W., et al. 2014. Biphase stratification approach to three-dimensional dendritic biodegradable mesoporous silica nanospheres. Nano Lett 14: 923-932.

Shi, S., Chen, F. and Cai, W. 2013. Biomedical applications of functionalized hollow mesoporous silica nanoparticles: focusing on molecular imaging. Nanomedicine (Lond) 8: 2027-2039.

Soo Choi, H., Liu, W., Misra, P., Tanaka, E., Zimmer, J.P., Itty Ipe, B., et al. 2007. Renal clearance of quantum dots. Nat Biotech 25: 1165-1170.

Sun, Y., Sai, H., von Stein, F., Riccio, M. and Wiesner, U. 2014. Water-based synthesis of ultrasmall PEGylated gold–silica core–shell nanoparticles with long-term stability. Chemistry of Materials 26: 5201-5207.

Tang, F., Li, L. and Chen, D. 2012. Mesoporous silica nanoparticles: synthesis, biocompatibility and drug delivery. Adv Mater 24: 1504-1534.

Vallet-Regi, M., Rámila, A., del Real, R.P. and Pérez-Pariente, J. 2000. A new property of MCM-41: drug delivery system. Chem Mater 13: 308-311.

Van Blaaderen, A. and Vrij, A. 1992. Synthesis and characterization of colloidal dispersions of fluorescent, monodisperse silica spheres. Langmuir 8: 2921-2931.

Wu, S., Han, G., Milliron, D.J., Aloni, S., Altoe, V., Talapin, D.V., et al. 2009. Nonblinking and photostable upconverted luminescence from single lanthanide-doped nanocrystals. Proc Natl Acad Sci USA 106: 10917-10921.

Wu, S.H., Mou, C.Y. and Lin, H.P. 2013. Synthesis of mesoporous silica nanoparticles. Chem Soc Rev 42: 3862-3875.

Xiao, Q., Bu, W., Ren, Q., Zhang, S., Xing, H., Chen, F., et al. 2012. Radiopaque fluorescence-transparent TaOx decorated upconversion nanophosphors for in vivo CT/MR/UCL trimodal imaging. Biomaterials 33: 7530-7539.

Xiao, Q., Zheng, X., Bu, W., Ge, W., Zhang, S., Chen, F., et al. 2013. A core/satellite multifunctional nanotheranostic for in vivo imaging and tumor eradication by radiation/photothermal synergistic therapy. J Am Chem Soc 135: 13041-13048.

Xing, H., Bu, W., Zhang, S., Zheng, X., Li, M., Chen, F., et al. 2012. Multifunctional nanoprobes for upconversion fluorescence, MR and CT trimodal imaging. Biomaterials 33: 1079-1089.

Xiong, L., Chen, Z., Tian, Q., Cao, T., Xu, C. and Li, F. 2009. High contrast upconversion luminescence targeted imaging in vivo using peptide-labeled nanophosphors. Anal Chem 81: 8687-8694.

Yang, P., Gai, S. and Lin, J. 2012. Functionalized mesoporous silica materials for controlled drug delivery. Chem Soc Rev 41: 3679-3698.

Yang, X., Yang, M., Pang, B., Vara, M. and Xia, Y. 2015. Gold nanomaterials at work in biomedicine. Chem Rev 115: 10410-10488.

Yi, D.K., Selvan, S.T., Lee, S.S., Papaefthymiou, G.C., Kundaliya, D. and Ying, J.Y. 2005. Silica-coated nanocomposites of magnetic nanoparticles and quantum dots. J Am Chem Soc 127: 4990-4991.

Zhang, Q., Zhang, T., Ge, J. and Yin, Y. 2008. Permeable silica shell through surface-protected etching. Nano Lett 8: 2867-2871.

Zhang, Z., Wang, L., Wang, J., Jiang, X., Li, X., Hu, Z., et al. 2012. Mesoporous silica-coated gold nanorods as a light-mediated multifunctional theranostic platform for cancer treatment. Adv Mater 24: 1418-1423.

Zhou, J., Liu, Z. and Li, F. 2012. Upconversion nanophosphors for small-animal imaging. Chem Soc Rev 41: 1323-1349.

Index

About the Editors

Feng Chen, Ph.D.

Feng Chen is a Research Associate in the Department of Radiology at the University of Wisconsin – Madison under the supervision of Dr. Weibo Cai. He received his Ph.D. degree in Materials Physics and Chemistry from the Shanghai Institute of Ceramics, Chinese Academy of Sciences in 2012. Dr. Chen has published more than 60 peer-reviewed papers in top-ranked scholarly journals, 2 book chapters and more than 20 conference abstracts. In addition, he has also co-edited a special issue entitled *"Image-Guided drug Delivery"* in the journal *Current Drug Targets*. He has performed peer review for more than 30 journals since 2012. Dr. Chen's primary research focus is on the engineering of hybrid nanomaterials for tumor targeted imaging and therapy.

Weibo Cai, Ph.D.

Weibo Cai is an Associate Professor of Radiology, Medical Physics and Biomedical Engineering at the University of Wisconsin – Madison. He received a BS degree in Chemistry from Nanjing University, China (1995) and a Ph.D. degree in Chemistry from the University of California, San Diego (2004). Between 2005 and 2008, Dr. Cai did his post-doctoral research in the laboratory of Prof. Xiaoyuan (Shawn) Chen at the Molecular Imaging Program at Stanford (MIPS). In February 2008, Dr. Cai joined the University of Wisconsin – Madison as a Biomedical Engineering Cluster Hire and was promoted to Associate Professor with Tenure in 2014. Dr. Cai's research at UW-Madison (http://mi.wisc.edu/) is primarily focused on molecular imaging and nanotechnology.

Dr. Cai has authored more than 200 peer-reviewed articles, 20 book chapters and 200 conference abstracts. His publications have been cited more than 10,000 times with an H-index of greater than 50. He has edited 3 books, guest-edited ~10 special topic issues for various scientific journals and given more than 150 talks. Dr. Cai has received many awards, including the Society of Nuclear Medicine and Molecular Imaging (SNMMI) Young Professionals

Committee Best Basic Science Award (2007), the European Association of Nuclear Medicine (EANM) Springer Prize (2011 & 2013), American Cancer Society Research Scholar (2013-2017), among many others. Dr. Cai has served on the Editorial Board of more than 20 scientific journals, performed peer review for over 130 journals and participated in many grant review panels (NIH, Cancer Prevention and Research Institute of Texas [CPRIT], Susan G. Komen, Prostate Cancer Canada, many European grants, etc.). He is currently the Editor-in-Chief of the American Journal of Nuclear Medicine and Molecular Imaging (http://www.ajnmmi.us), a journal that was launched in 2011 and fully indexed in PubMed.

Dr. Cai is an active member of several scientific societies and he has served on various committees in these societies such as SNMMI Committee on Awards, SNMMI Radiopharmaceutical Sciences Council Board of Directors, etc. Dr. Cai's trainees at UW – Madison have received more than 50 awards to date, such as the 2012 Berson-Yalow Award from SNMMI, multiple Young Investigator Awards from SNMMI and several post-doctoral fellowships.

Printed and bound by CPI Group (UK) Ltd, Croydon, CR0 4YY

01/11/2024

01782624-0017